T0137094

Emergence, Complexity and Computation

Volume 32

The Emergence, Complexity and Computation (ECC) series publishes new developments, advancements and selected topics in the fields of complexity, computation and emergence. The series focuses on all aspects of reality-based computation approaches from an interdisciplinary point of view especially from applied sciences, biology, physics, or chemistry. It presents new ideas and interdisciplinary insight on the mutual intersection of subareas of computation, complexity and emergence and its impact and limits to any computing based on physical limits (thermodynamic and quantum limits, Bremermann's limit, Seth Lloyd limits…) as well as algorithmic limits (Gödel's proof and its impact on calculation, algorithmic complexity, the Chaitin's Omega number and Kolmogorov complexity, non-traditional calculations like Turing machine process and its consequences,…) and limitations arising in artificial intelligence field. The topics are (but not limited to) membrane computing, DNA computing, immune computing, quantum computing, swarm computing, analogic computing, chaos computing and computing on the edge of chaos, computational aspects of dynamics of complex systems (systems with self-organization, multiagent systems, cellular automata, artificial life,…), emergence of complex systems and its computational aspects, and agent based computation. The main aim of this series it to discuss the above mentioned topics from an interdisciplinary point of view and present new ideas coming from mutual intersection of classical as well as modern methods of computation. Within the scope of the series are monographs, lecture notes, selected contributions from specialized conferences and workshops, special contribution from international experts.

More information about this series at http://www.springer.com/series/10624

Andrew Adamatzky

Editor

Shortest Path Solvers. From Software to Wetware

 Springer

Editor
Andrew Adamatzky
Unconventional Computing Centre
University of the West of England
Bristol
UK

ISSN 2194-7287   ISSN 2194-7295   (electronic)
Emergence, Complexity and Computation
ISBN 978-3-030-08473-8   ISBN 978-3-319-77510-4   (eBook)
https://doi.org/10.1007/978-3-319-77510-4

Printed on acid-free paper

This Springer imprint is published by the registered company Springer International Publishing AG part of Springer Nature
The registered company address is: Gewerbestrasse 11, 6330 Cham, Switzerland

Preface

Shortest path problem is a fundamental and classical problem of graph theory and computer science and is applied in transport and logistics, telecommunication networks, virtual reality and gaming, geometry, social networks analysis, and robotics. The book offers advanced parallel and distributed algorithms and experimental laboratory prototypes of chemical, physical, and living unconventional shortest path and maze solvers.

Chapter by Matic—"A Parallel Algorithm for the Constrained Shortest Path Problem on Lattice Graphs"—opens the book. The chapter shows how to solve a constrained shortest path problem in parallel using a paradigm of water percolating through the graph. These ideas are somewhat echoed in several other chapters devoted to physical and chemical shortest path solvers. Pascoal presents breadth-first search tree algorithms for computation of the maximal and the minimal sets of non-dominating paths in "The MinSum-MinHop and the MaxMin-MinHop Bicriteria Path Problems".

Optimal algorithm for moving a swarm of robots into a common place via shortest paths—"Gathering a Swarm of Robots Through Shortest Paths"—is analysed by Cicerone, Di Stefano, Navarra. These results are augmented by constructs of mobile lattice automata forming a checkerboard pattern, where each agent moves along the shortest path in Hoffmann's chapter "Checkerboard Pattern Formed by Cellular Automata Agents". D'Angelo, D'Emidio, and Frigioni review distributed shortest path algorithms for large-scale Ethernet networks, where each node of the network stores distance to every other node in "Distance-Vector Algorithms for Distributed Shortest Paths Computation in Dynamic Networks". An algorithm for multi-objective problem of vehicle routing inspired by virus propagation is proposed and analyzed by Psychas, Delimpasi, Marinaki and Marinakis in "Influenza Virus Algorithm for Multiobjective Energy Reduction Open Vehicle Routing Problem". Overview of pragmatic algorithms of shortest path computation is given in "Practical Algorithms for the All-Pairs Shortest Path Problem" by Brodnik and Marko Grgurovič.

Computing of shortest path with uniform arrays of processing units is discussed by Akl in "Computing Shortest Paths with Cellular Automata" and by Tsompanas, Dourvas, Ioannidis, Sirakoulis, Hoffmann, Adamatzky in "Cellular Automata Applications in Shortest Path Problem". von Thienen and Czaczkes compare the shortest pathfinding mechanisms in real ant colonies and artificial ant colony optimisation techniques in "Do Ants Use Ant Colony Optimization?".

Next two chapters make a nice transition from software to experimental laboratory prototypes of distributed path solvers. Both chapters relate to slime *Physarum polycephalum*, a large single cell which spans sources of nutrients with an optimal network of protoplasmic tubes. Zhang and Yan—"Physarum-Inspired Solutions to Network Optimization Problems"—present traffic flow assignment and supply chain network design solutions using a mathematical model of the slime mould. Multi-agent model of *Physarum polycephalum* and volumetric topological material optimisation are employed in computation of shortest paths by Jones and Safonov in "Slime Mould Inspired Models for Path Planning: Collective and Structural Approaches".

Irimia—"Maze-Solving Cells"—shows experimental laboratory results on how a single human cell solves a maze crawling in a self-sustainable loop of externally and internally generated gradients. Insight into the inefficiency and irreproducibility of shortest path solving by living creatures—slime mould *Physarum polycephalum* and ciliate *Paramecium caudatum*—in experimental laboratory conditions is provided by Mayne in his chapter "When the Path is Never Shortest: A Reality Check on Shortest Path Biocomputation". Čejková, Tóth, Braun, Branicki, Ueyama, and Lagzi—"Shortest Path Finding in Mazes by Active and Passive Particles"—present their chemistry-based concepts for maze solving which rely on surface tension-driven phenomena at the air–liquid interface. Chapter "The Electron in the Maze" by Ayrinhac addresses a computation of the shortest path in maze of conductive channels with electrical current. The book closes with a discussion on common physical/chemical principles of experimental laboratory prototypes of maze solves, including fluid mappers and slime mould—"Maze Solvers Demystified and Some Other Thoughts" by Adamatzky.

The book is a pleasure to explore for readers from all walks of life, from undergraduate students to university professors, from mathematicians, computers scientists, and engineers to chemists and biologists.

Bristol, UK   Andrew Adamatzky
February 2018

Contents

A Parallel Algorithm for the Constrained Shortest Path Problem on Lattice Graphs 1
Ivan Matic

Gathering a Swarm of Robots Through Shortest Paths 27
Serafino Cicerone, Gabriele Di Stefano and Alfredo Navarra

The MinSum-MinHop and the MaxMin-MinHop Bicriteria Path Problems 73
Marta Pascoal

Distance-Vector Algorithms for Distributed Shortest Paths Computation in Dynamic Networks 99
Gianlorenzo D'Angelo, Mattia D'Emidio and Daniele Frigioni

Influenza Virus Algorithm for Multiobjective Energy Reduction Open Vehicle Routing Problem 145
Iraklis-Dimitrios Psychas, Eleni Delimpasi, Magdalene Marinaki and Yannis Marinakis

Practical Algorithms for the All-Pairs Shortest Path Problem 163
Andrej Brodnik and Marko Grgurovič

Computing Shortest Paths with Cellular Automata 181
Selim G. Akl

Cellular Automata Applications in Shortest Path Problem 199
Michail-Antisthenis I. Tsompanas, Nikolaos I. Dourvas, Konstantinos Ioannidis, Georgios Ch. Sirakoulis, Rolf Hoffmann and Andrew Adamatzky

Checkerboard Pattern Formed by Cellular Automata Agents 239
Rolf Hoffmann

Do Ants Use Ant Colony Optimization? 265
Wolfhard von Thienen and Tomer J. Czaczkes

**Slime Mould Inspired Models for Path Planning: Collective
and Structural Approaches** 293
Jeff Jones and Alexander Safonov

Physarum-Inspired Solutions to Network Optimization Problems 329
Xiaoge Zhang and Chao Yan

Maze-Solving Cells 365
Daniel Irimia

**When the Path Is Never Shortest: A Reality Check on Shortest
Path Biocomputation** 379
Richard Mayne

Shortest Path Finding in Mazes by Active and Passive Particles 401
Jitka Čejková, Rita Tóth, Artur Braun, Michal Branicki, Daishin Ueyama
and István Lagzi

The Electron in the Maze 409
Simon Ayrinhac

Maze Solvers Demystified and Some Other Thoughts 421
Andrew Adamatzky

Index 439

A Parallel Algorithm for the Constrained Shortest Path Problem on Lattice Graphs

Ivan Matic

Abstract The edges of a graph are assigned weights and passage times which are assumed to be positive integers. We present a parallel algorithm for finding the shortest path whose total weight is smaller than a pre-determined value. In each step the processing elements are not analyzing the entire graph. Instead they are focusing on a subset of vertices called *active vertices*. The set of active vertices at time t is related to the boundary of the ball B_t of radius t in the first passage percolation metric. Although it is believed that the number of active vertices is an order of magnitude smaller than the size of the graph, we prove that this need not be the case with an example of a graph for which the active vertices form a large fractal. We analyze an OpenCL implementation of the algorithm on GPU for cubes in $\mathbb{Z}^d$.

1 Definition of the Problem

The graph $G(V, E)$ is undirected and the function $f : E \to \mathbb{Z}_+^2$ is defined on the set of its edges. The first component $f_1(e)$ of the ordered pair $f(e) = (f_1(e), f_2(e))$ for a given edge $e \in E$ represents the time for traveling over the edge e. The second component $f_2(e)$ represents the weight of e.

A path in the graph G is a sequence of vertices $(v_1, v_2, \ldots, v_k)$ such that for each $i \in \{1, 2, \ldots, k-1\}$ there is an edge between v_i and v_{i+1}, i.e. $(v_i, v_{i+1}) \in E$. For each path $\pi = (v_1, \ldots, v_k)$ we define $F_1(\pi)$ as the total time it takes to travel over π and $F_2(\pi)$ as the sum of the weights of all edges in π. Formally,

$$F_1(\pi) = \sum_{i=1}^{k-1} f_1(v_i, v_{i+1}) \quad \text{and} \quad F_2(\pi) = \sum_{i=1}^{k-1} f_2(v_i, v_{i+1}).$$

I. Matic (✉)
Department of Mathematics, Baruch College, CUNY, One Bernard Baruch Way,
New York, NY 10010, USA
e-mail: ivan.matic@baruch.cuny.edu

© Springer International Publishing AG, part of Springer Nature 2018
A. Adamatzky (ed.), *Shortest Path Solvers. From Software to Wetware*,
Emergence, Complexity and Computation 32,
https://doi.org/10.1007/978-3-319-77510-4_1

1

Let $A, B \subseteq V$ be two fixed disjoint subsets of V and let $M \in \mathbb{R}_+$ be a fixed positive real number. Among all paths that connect sets A and B let us denote by $\hat{\pi}$ the one (or one of) for which $F_1(\pi)$ is minimal under the constraint $F_2(\pi) < M$. We will describe an algorithm whose output will be $F_1(\hat{\pi})$ for a given graph G.

The algorithm belongs to a class of label correcting algorithms [11, 20]. The construction of labels will aim to minimize the memory consumption on SIMD devices such as graphic cards. Consequently, the output will not be sufficient to determine the exact minimizing path. The reconstruction of the minimizing path is possible with subsequent applications of the method, because the output can include the vertex $X \in B$ that is the endpoint of $\hat{\pi}$, the last edge x on the path $\hat{\pi}$, and the value $F_2(\hat{\pi})$. Once X and x are found, the entire process can be repeated for the graph $G'(V', E')$ with

$$V' = V \setminus B, \quad A' = A, \quad B' = \{X\}, \quad \text{and} \quad M' = F_2(\hat{\pi}) - f_2(x).$$

The result will be second to last vertex on the minimizing path $\hat{\pi}$. All other vertices on $\hat{\pi}$ can be found in the same way.

Although the algorithm works for general graphs and integer-valued functions f, its implementation on SIMD hardware requires the vertices to have bounded degree. This requirement is satisfied by subgraphs of $\mathbb{Z}^d$.

Finding the length of the shortest path in graph is equivalent to finding the shortest passage time in first passage percolation. Each of the vertices in A can be thought of as a source of water. The value $f_1(e)$ of each edge e is the time it takes the water to travel over e. Each drop of water has its *quality* and each drop that travels through edge e looses $f_2(e)$ of its quality. Each vertex P of the graph has a label $Label(P)$ that corresponds to the quality of water that is at the vertex P. Initially all vertices in A have label M while all other vertices have label 0. The drops that get their quality reduced to 0 cannot travel any further. The time at which a vertex from B receives its first drop of water is exactly the minimal $F_1(\pi)$ under the constraint $F_2(\pi) < M$.

Some vertices and edges in the graph are considered *active*. Initially, the vertices in A are *active*. All edges adjacent to them are also called *active*. Each cycle in algorithm corresponds to one unit of time. During one cycle the water flows through active edges and decrease their time components by 1. Once an edge gets its time component reduced to 0, the edge becomes *used* and we look at the source S and the destination D of this water flow through the edge e. The destination D becomes *triggered*, and its label will be *corrected*. The label correction is straight-forward if the edge D was inactive. We simply check whether $Label(S) - f_2(e) > Label(D)$, and if this is true then the vertex D gets its label updated to $Label(S) - f_2(e)$ and its status changed to *active*. If the vertex D was active, the situation is more complicated, since the water has already started flowing from the vertex D. The existing water flows correspond to water of quality worse than the new water that has just arrived to D. We resolve this issue by introducing phantom edges to the graph that are parallel to the existing edges. The phantom edges will carry this new high quality water, while old edges will continue carrying their old water flows. A vertex stops being active if all of its edges become used, but it may get activated again in the future.

2 Related Problems in the Literature

The assignment of phantom edges to the vertices of the graph and their removal is considered a label correcting approach in solving the problem. Our particular choice of label correction is designed for large graphs in which the vertices have bounded degree. Several existing serial computation algorithms can find the shortest path by maintaining labels for all vertices. The labels are used to store the information on the shortest path from the source to the vertex and additional preprocessing of vertices is used to achieve faster implementations [5, 9]. The ideas of first passage percolation and label correction have naturally appeared in the design of *pulse algorithms* for constrained shortest paths [17]. All of the mentioned algorithms can also be parallelized but this task would require a different approach in designing a memory management that would handle the label sets in programming environments where dynamical data structures need to be avoided.

The method of aggressive edge elimination [22] can be parallelized to solve the Lagrange dual problems. In the case of road and railroad networks a substantial speedup can be achieved by using a preprocessing of the network data and applying a generalized versions of Dijkstra's algorithm [12].

The parallel algorithm that is most similar in nature to the one discussed in this paper is developed for wireless networks [16]. There are two features of wireless networks that are not available to our model. The first feature is that the communication time between the vertices can be assumed to be constant. The other feature is that wireless networks have a processing element available to each vertex. Namely, routers are usually equipped with processors. Our algorithm is build for the situations where the number of processing cores is large but not at the same scale as the number of vertices. On the other hand our algorithm may not be effective for the wireless networks since the underlying graph structure does not imply that the vertices are of bounded degree. The increase of efficiency of wireless networks can be achieved by solving other related optimization problems. One such solution is based on constrained node placement [21].

The execution time of the algorithm is influenced by the sizes of the sets of active vertices, active edges, and phantom edges. The sizes of these sets are order of magnitude smaller than the size of the graph. Although this cannot be proved at the moment, we will provide a justification on how existing conjectures and theorems from the percolation theory provide some estimates on the sizes of these sets. The set of active vertices is related to the limit shape in the model of first passage percolation introduced by Hammersley and Welsh [10]. The first passage percolation corresponds to the case $M = \infty$, i.e. the case when there are no constraints. If we assume that $A = \{0\}$, for each time t we can define the ball of radius t in the first passage percolation metric as:

$$B_t = \{x : \tau(0, x) \le t\},$$

where $\tau(0, x)$ is the *first passage time*, i.e. the first time at which the vertex x is reached.

The active vertices at time t are located near the boundary of the ball B_t. It is known that for large t the set $\frac{1}{t}B_t$ will be approximately convex. More precisely, it is known [7] that there is a convex set B such that

$$\mathbb{P}\left((1 - \varepsilon)B \subseteq \frac{1}{t}B_t \subseteq (1 + \varepsilon)B \text{ for large } t\right) = 1.$$

However, the previous theorem does not guarantee that the boundary of B_t has to be of zero volume. In fact the boundary can be non-polygonal as was previously shown [8].

The set of active vertices does not coincide with the boundary of B_t, but it is expected that if ∂B_t is of small volume then the number of active vertices is small in most typical configurations of random graphs. We provide an example for which the set of active vertices is a large fractal, but simulations suggest that this does not happen in average scenario.

The fluctuations of the shape of B_t are expected to be of order $t^{2/3}$ in the case of $\mathbb{Z}^2$ and the first passage time $\tau(0, n)$ is proven to have fluctuations of order at least $\log n$ [23]. The fluctuations are of order at most $n/\log n$ [3, 4] and are conjectured to be of order $t^{2/3}$. They can be larger and of order n for modifications of $\mathbb{Z}^2$ known as thin cylinders [6].

The scaling of $t^{2/3}$ for the variance is conjectured for many additional interface growth models and is related to the Kardar-Parisi-Zhang equation [1, 15, 25].

The constrained first passage percolation problem is a discrete analog to Hamilton-Jacobi equation. The large time behaviors of its solutions are extensively studied and homogenization results are obtained for a class of Hamiltonians [2, 13, 14, 26]. Fluctuations in dimension one are of order t [24] while in higher dimensions they are of lower order although only the logarithmic improvement to the bound has been achieved so far [19].

3 Example

Before providing a more formal description of the algorithm we will illustrate the main ideas on one concrete example of a graph. Consider the graph shown in Fig. 1 that has 12 vertices labeled as $1, 2, \ldots, 12$. The set A contains the vertices 1, 2, and 3, and the set B contains only the vertex 12. The goal is to find the length of the shortest path from A to B whose total weight is smaller than 19.

The vertices are drawn with circles around them. The circles corresponding to the vertices in A are painted in blue and have the labels 19. The picture contains the time and weight values for each of the edges. The time parameter of each edge is

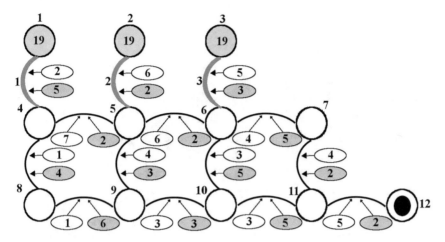

Fig. 1 The initial state of the graph

written in the empty oval, while the weight parameter is in the shaded oval. Since the number of edges in this graph is relatively small it is not difficult to identify the minimizing path (3, 6, 10, 11, 12). The time required to travel over this path is 16 and the total weight is 15.

Initially, the vertices in set A are called *active*. Active vertices are of blue color and edges adjacent to them are painted in blue. These edges are considered *active*. Numbers written near their centers represent the sources of water. For example, the vertex 2 is the source of the flow that goes through the edge (2, 5).

Notice that the smallest time component of all active edges is 2. The first cycle of the algorithm begins by decreasing the time component of each active edge by 2. The edge (1, 4) becomes *just used* because its time component is decreased to 0. The water now flows from the vertex 1 to the vertex 4 and its quality decreases by 5, since the weight of the edge (1, 4) is equal to 5. The vertex 4 becomes active and its label is set to

$$Label(4) = Label(1) - f_2(1, 4) = 19 - 5 = 14.$$

The edge (1, 4) becomes used, and the vertex 1 turns into inactive since there are no active edges originating from it. Hence, after 2 s the graph turns into the one shown in Fig. 2.

The same procedure is repeated until the end of the 5th second and the obtained graph is the top one in Fig. 3. In the 6th second the edge (2, 5) gets its time parameter decreased to 0 and the vertex 5 gets activated. Its label becomes

$$Label(5) = Label(2) - f_2(2, 5) = 19 - 2 = 17.$$

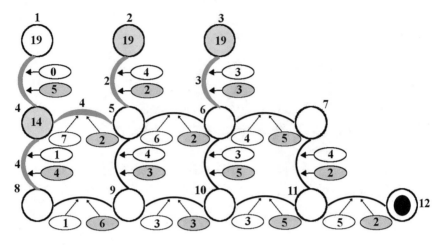

Fig. 2 The configuration after the second 2

However, the edges $(4, 5)$, $(5, 9)$, and $(5, 6)$ were already active and the water was flowing through them towards the vertex 5.

The old flow of water through the edge $(4, 5)$ will complete in additional 5 s. However, when it completes the quality of the water that will reach the vertex 5 will be

$$Label(4) - f_2(4, 5) = 14 - 2 = 12 < Label(5)$$

because the label of the vertex 5 is 17. Thus there is no point in keeping track of this water flow. On the other hand, the water flow that starts from 5 and goes towards 4 will have quality

$$Label(5) - f_2(4, 5) = 17 - 2 = 15$$

which is higher than the label of the vertex 4. Thus the edge $(4, 5)$ will change its source from 4 to 5 and the time parameter has to be restored to the old value 7. At this point the vertex 4 becomes inactive as there is no more flow originating from it.

The same reversal of the direction of the flow happens with the edge $(5, 9)$. On the other hand, something different happens to the edge $(5, 6)$: it stops being active. The reason is that the old flow of water from 6 to 5 will not be able to increase the label of the vertex 5. Also, the new flow of water from 5 to 6 would not be able to change the label of vertex 6.

A special care has to be taken when a water flow reaches a vertex that is already active. In the case of the graph G such situation happens after the 11th second. The configuration is shown in the top picture of Fig. 4. The edge $(7, 11)$ has the smallest time parameter 2. The time will progress immediately to 13 and all active edges get their time parameters decreased by 2. In the 13th second the water from the edge $(7, 11)$ reaches the vertex 11. The label of vertex 7 is $Label(7) = 11$, while

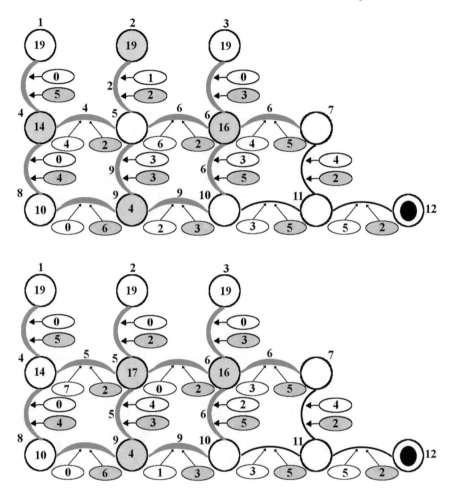

Fig. 3 The configurations after the seconds 5 and 6

$Label(11) = 6$. The weight of the flow over the edge between these two vertices is 2, hence this new water is of higher quality than the one present at the vertex 11.

In this situation we consider every active edge originating from 11 and create a phantom edge through which this new water will flow. We will create a new vertex $11'$ with label

$$Label(11') = Label(7) - f_2(7, 11) = 9$$

and connect it with each of the neighbors of 11 that can get their label increased with the new flow. The only one such neighbor is 12 and we obtain the graph as shown in the lower part of Fig. 4.

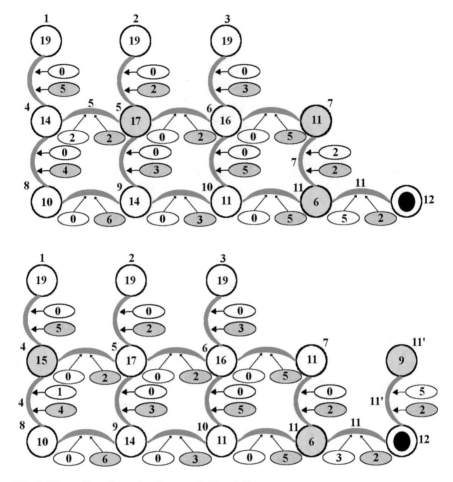

Fig. 4 The configurations after the seconds 11 and 13

It can be now easily verified that after additional 3 s, i.e. in the end of the second 16 the vertex 12 becomes active with the label 4. Thus we conclude that it takes water to travel 16 s over the shortest path. The total weight of the shortest path is $19 - 4 = 15$. The minimizing path is $(3, 6, 10, 11, 12)$.

4　Pseudo-code of the Algorithm

We will organize the algorithm by dividing it into smaller components. The first component is the initialization, and the others are performed in the main loop that consists of 9 major steps. The parallelization will happen only in these individual steps of the main loop. The pseudo-code for the main function is presented in Algorithm 1. Each step will be described in full details and accompanied by a pseudo-code that outlines the main ideas. For the sake of brevity, some data structures in pseudo-code will be modeled with sets. However, the usage of sets is avoided as they cannot support insertion and deletion of elements in parallel. The sets are replaced by indicator sequences for which appropriate operations are easier to parallelize. For the full source code the reader is referred to [18].

Algorithm 1 Main function

Input: Graph $G = (V, E)$; $A, B \subset G$ such that $A \cap B = \emptyset$, $M \in \mathbb{R}$, two functions $f_1, f_2 : E \to \mathbb{R}$.

$\qquad$ $f_1(e)$ is the time to travel over the edge e and $f_2(e)$ is the weight of the edge e.

Output: The shortest time to travel from A to B over a path whose weight is less than M.

1: **function** MAIN
2: $\quad$ Initialization
3: $\quad$ $L =$ShortestTravelTimeAndTerminalConditionCheck
4: $\quad$ **while** $L = 0$ **do**
5: $\quad\quad$ TriggerVertices
6: $\quad\quad$ AnalyzeTriggeredVertices
7: $\quad\quad$ GetInputFromPhantoms
8: $\quad\quad$ TriggerEdges
9: $\quad\quad$ TreatTriggeredEdges
10: $\quad\quad$ $L =$ShortestTravelTimeAndTerminalConditionCheck
11: $\quad\quad$ FinalTreatmentOfPhantoms
12: $\quad\quad$ FinalTreatmentOfVertices
13: $\quad\quad$ FinalTreatmentOfActiveEdges
14: $\quad$ **return** L

---

5 Memory Management and Initialization

5.1 Labels for Vertices and Edges

In this section we will describe the memory management of variables necessary for the implementation of the algorithm. Before providing the precise set of variables let us describe the information that has to be carried throughout the execution process. As we have seen before, the vertices will have labels assigned to them. Initially we label each vertex of G with 0 except for vertices in A which are labeled by M.

To each vertex and edge in G we assign a *State*. The vertices have states in the set {*active, inactive*}. Initially all vertices in A are *active*, while the other vertices are *inactive*. The states of the edges belong to the set {*active, passive, used, just used*}. Initially the edges adjacent to the vertices in A are set to active while all other are passive.

To each edge we associate a pointer to one of its endpoints and call it *Source*. This variable is used at times when the water is flowing through the edge and it records the source of the current water flow. Initially, to each edge that originates from a vertex in A we set the source to be the pointer to the vertex in A. All other edges have their source initially set to 0.

There is additional variable *Time* that represents the time and is initially set to 0.

5.2 Termination

The algorithm terminates if one of the following two conditions is satisfied:

1° A vertex from B becomes *active*. The variable *Time* contains the time it takes to reach this vertex along the shortest path $\hat{\pi}$, i.e.

$$Time = F_1\left(\hat{\pi}\right).$$

The label of the last vertex $\hat{B}$ on the path allows us to determine the value $F_2\left(\hat{\pi}\right)$. Namely,

$$F_2\left(\hat{\pi}\right) = M - \text{Label}(\hat{B}).$$

We will not go into details on how to recover the exact shortest path. Instead we will just outline how this can be done. We need to identify the *used* edge f (or one of the used edges, if there are more than one) that is adjacent to $\hat{B}$. This edge can help us in finding the second to last point of the path $\hat{\pi}$. Let us denote by F the other endpoint of f. It could happen that F is a phantom vertex (i.e. a copy of another vertex), and we first check whether $F \in PhantomVertices$. If this is not the case, then F is the second to last element of the path $\hat{\pi}$. If $F \in PhantomVertices$ then the vertex F is a copy of some other vertex in the graph and the phantom vertex F has the pointer to the original based on which it is created. This original vertex is the second to last point on the path $\hat{\pi}$.

2° There is no *active* edge in the graph. In this case there is no path that satisfies the constraint $F_2 \leq M$.

5.3 Sequences Accessible to All Processing Elements

It is convenient to store the vertices and edges in sequences accessible to all processing elements. We will assume here that the degree of each vertex is bounded above by d.

Algorithm 2 Function that checks whether the algorithm has finished and returns the time for travel over the shortest path

---

```
 1: function SHORTESTPATHLENGTHANDTERMINALCONDITIONCHECK
 2:     // Returns 0 if the path is not found yet.
 3:     // Returns −1 if there is no path with weight smaller than M.
 4:     // Returns the weight of the shortest path if it is found.
 5:     // A non-zero return value is the indication that the algorithm is over.
 6:     #Performed in parallel
 7:     if ∃B₀ ∈ B such that B₀ = active then
 8:         result ← Time
 9:     else
10:         if there are no active vertices then
11:             result ← −1
12:         else
13:             result ← 0
14:     #barrier
15:     return result
```

5.3.1 Vertices

Each vertex takes 5 integers in the sequence of vertices. The first four are name, label, status, and the location of the first edge in the sequence of edges. The fifth element is be used to store a temporary replacement label. Initially, and between algorithm steps, this label is set to -1.

When a first drop of water reaches an inactive vertex V, we say that the vertex is *triggered*, and that state exists only temporarily during an algorithm cycle. In the end of the algorithm cycle some triggered vertices become active. However it could happen that a triggered vertex does not get a water flow of higher quality than the one already present at the vertex. The particular triggered vertex with this property does not get activated.

5.3.2 Edges

Each edge e takes 8 integers in the sequence of edges. Although the graph is undirected, each edge is stored twice in the memory. The 8 integers are the start point, the end point, remaining time for water to travel over the edge (if the edge is active), the weight of the travel $f_2(e)$, the initial passage time $f_1(e)$, the label of the vertex that is the source of the current flow through the edge (if there is a flow), status, and the location of the same edge in the opposite direction.

5.3.3 Active Vertices

The sequence contains the locations of the vertices that are active. This sequence removes the need of going over all vertices in every algorithm step. The locations are

sorted in decreasing order. In the end of the sequence we will add triggered vertices
that will be joined to the active vertices in the end of the cycle.

5.3.4 Active Edges

The role of the sequence is similar to the one of active vertices. The sequence maintains the location of the active edges. Each edge is represented twice in this sequence.
The second appearance is the one in which the endpoints are reversed. The locations are sorted in decreasing order. During the algorithm cycle we will append the
sequence with triggered edges. In the end of each cycle the triggered edges will be
merged to the main sequence of active edges.

5.3.5 Sequence of Phantom Edges

The phantom edges appear when an active vertex is triggered with a new drop of
water. Since the vertex is active we cannot relabel the vertex. Instead each of the edges
going from this active triggered vertex need to be doubled with the new source of
water flowing through these new edges that are called phantoms. They will disappear
once the water finishes flowing through them.

5.3.6 Sequence of Elements in B

Elements in B have to be easily accessible for quick check whether the algorithm
has finished. For this reason the sequence should be in global memory.

Listing 3 summarizes the initializing procedures.

6   Graph Update

The algorithm updates the graph in a loop until one vertex from B becomes active.
Each cycle consists of the following nine steps.

6.1   Step 1: Initial Triggering of Vertices

In this step we go over all active edges and decrease their time parameters by m,
where m is the smallest remaining time of all active edges. If for any edge the time
parameter becomes 0, the edge becomes *just used* and its destination triggered.

To avoid the danger of two processing elements writing in the same location of
the sequence of active vertices, we have to make sure that each processing element

---

Algorithm 3 Initialization procedure

1: **procedure** INITIALIZATION
2:     **for** $e \in E$ **do**
3:         $State(e) \leftarrow passive$
4:         $Source(e) \leftarrow 0$
5:         $TimeRemaining(e) \leftarrow 0$
6:     **for** $v \in V \setminus A$ **do**
7:         $State(v) \leftarrow inactive$
8:         $Label(v) \leftarrow 0$
9:     **for** $v \in A$ **do**
10:         $State(v) \leftarrow active$
11:         $Label(v) \leftarrow M$
12:         **for** $e \in Edges(v)$ **do**
13:             $State(e) \leftarrow active$
14:             $Source(e) \leftarrow v$
15:             $TimeRemaining(e) \leftarrow f_1(e)$
16:     $TriggeredVertices \leftarrow \emptyset$
17:     $PhantomVertices \leftarrow \emptyset$
18:     $PhantomEdges \leftarrow \emptyset$
19:     $Time \leftarrow 0$

---

that runs concurrently has pre-specified location to write. This is accomplished by first specifying the number of threads in the separate variable *nThreads*. Whenever kernels are executed in parallel we are using only *nThreads* processing elements. Each processing element has its id number which is used to determine the memory location to which it is allowed to write. The sequence of triggered vertices has to be cleaned after each parallel execution and at that point we take an additional step to ensure we don't list any of the vertices as triggered twice.

Algorithm 4 Procedure TriggerVertices

1: **procedure** TRIGGERVERTICES
2:     $TriggeredEdges \leftarrow \emptyset$
3:     **#Performed in parallel**
4:     $m \leftarrow \min \{TimeRemaining(e) : e \in ActiveEdges\}$
5:     **#barrier**
6:     $Time \leftarrow Time + m$
7:     **#Performed in parallel**
8:     **for** $e \in ActiveEdges$ **do**
9:         $TimeRemaining(e) \leftarrow TimeRemaining(e) - m$
10:         **if** $TimeRemaining(e) = 0$ **then**
11:             $State(e) = just\ used$
12:             $S_e \leftarrow Source(e)$
13:             $D_e \leftarrow TheTwoEndpoints(e) \setminus \{S_e\}$
14:             $TriggeredVertices \leftarrow TriggeredVertices \cup \{D_e\}$
15:     **#barrier**

6.2   Step 2: Analyzing Triggered Vertices

For each triggered vertex Q we look at all of its edges that are just used. We identify the largest possible label that can result from one of just used edges that starts from Q. That label will be stored in the sequence of vertices at the position reserved for temporary replacement label. The vertex is labeled as just triggered. If the vertex Q is not active, this label will replace the current label of the vertex in one of the later steps. If the vertex Q is active, then this temporary label will be used later to construct an appropriate phantom edge.

We are sure that different processing elements are not accessing the same vertex at the same time, because before this step we achieved the state in which there are no repetitions in the sequence of triggered vertices.

Algorithm 5 Analysis of triggered vertices
___
1: **procedure** ANALYZETRIGGEREDVERTICES
2:     *TempLabel* $\leftarrow \emptyset$
3:     **#Performed in parallel**
4:     **for** $Q \in$ *TriggeredVertices* **do**
5:         $TempLabel(Q) \leftarrow \max \{Label(P) - f_2(P, Q) : State(P, Q) = just\ used\}$
6:     **#barrier**

6.3   Step 3: Gathering Input from Phantoms

The need to have this step separated from the previous ones is the current architecture of graphic cards that creates difficulties with dynamic memory locations. It is more efficient to keep phantom edges separate from the regular edges. The task is to look for all phantom edges and decrease their time parameters. If a phantom edge gets its time parameter equal to 0, its destination is studied to see whether it should be added to the sequence of triggered vertices. We calculate the new label that the vertex would receive through this phantom. We check whether this new label is higher than the currently known label and the temporary label from possibly previous triggering of the vertex. The phantoms will not result in the concurrent writing to memory locations because each possible destination of a phantom could have only one edge that has time component equal to 0.

6.4   Step 4: Triggering Edges

In this step we will analyze the triggered vertices and see whether each of their neighboring edges needs to change the state. Triggered vertices are analyzed using

---

Algorithm 6 Input from phantoms

1: **procedure** GETINPUTFROMPHANTOMS
2:     **#Performed in parallel**
3:     Decrease time parameters of fantom edges (as in Listing 4)
4:     Trigger the destinations of phantom edges (as in Listing 4)
5:     **#barrier**
6:     **#Performed in parallel**
7:     Analyze newly triggered vertices, in a way similar to Listing 5
8:     **#barrier**

---

separate processing elements. A processing element analyzes the vertex Q in the following way.

Each edge j of Q will be considered triggered if it can cause the other endpoint to get better label in future through Q. The edge j is placed in the end of the sequence of active edges.

Algorithm 7 Procedure that triggers the edges

1: **procedure** TRIGGEREDGES
2:     **#Performed in parallel**
3:     **for** $Q \in TriggeredVertices$ **do**
4:         **for** $P \in Neighbors(Q)$ **do**
5:             **if** $TempLabel(Q) - f_2(P, Q) > Label(P)$ **then**
6:                 $State(P, Q) \leftarrow active$
7:                 $TriggeredEdges \leftarrow TriggeredEdges \cup \{(P, Q)\}$
8:     **#barrier**

6.5   Step 5: Treatment of Triggered Edges

Consider a triggered edge j. We first identify its two endpoints. For the purposes of this step we will identify the endpoint with the larger label, call it the source, and denote by S. The other will be called the destination and denoted by D. In the end of the cycle, this vertex S will become the source of the flow through j.

Notice that at least one of the endpoints is triggered. If only one endpoint is triggered, then we are sure that this triggered endpoint is the one that we designated as the source S.

We then look whether the source S was active or inactive before it was triggered.

6.5.1 Case in Which the Source S Was Inactive Before Triggering

There are several cases based on the prior status of j. If j was passive, then it should become active and no further analysis is necessary. If it was used or just used, then it should become active and the time component should be restored to the original one. Assume now that the edge j was active. Based on the knowledge that S was inactive vertex we can conclude that the source of j was D. However we know that the source of j should be S and hence the time component of j should be restored to the backup value.

Consequently, in the case that S was inactive, regardless of what the status of j was, we are sure its new status must be active and its time component can be restored to the original value. This restoration is not necessary in the case that j was passive, although there is no harm in doing it.

If the edge j was not active before, then the edge j should be added to the list of active edges. If the edge j was active before, then it should be removed from the list of triggered edges because all triggered edges will be merged into active edges. The edge j already appears in the list of active edges and need not be added again.

6.5.2 Case in Which the Source S Was Active Before Triggering

In this case we create phantom edges. Each such triggered edge generates four entries in the phantom sequence. The first one is the source, the second is the destination, the third is the label of the source (or the label stored in the temporary label slot, if higher), and the fourth is the original passage time through the edge j.

Algorithm 8 Treatment of triggered edges

1: **procedure** TREATTRIGGEREDEDGES
2:     **#Performed in parallel**
3:     **for** $j \in TriggeredEdges$ **do**
4:         $S \leftarrow$ The endpoint of j with larger label
5:         $D \leftarrow$ The endpoint of j with smaller label
6:         $OldStateOfS \leftarrow State(S)$
7:         $OldStateOfJ \leftarrow State(j)$
8:         **if** $OldStateOfS = inactive$ **then**
9:             $State(j) \leftarrow active$
10:            $Source(j) \leftarrow S$
11:            $TimeRemaining(j) \leftarrow f_1(j)$
12:        **if** $OldStateOfS = active$ **then**
13:            Create a phantom vertex S' and connect it to D
14:            $TimeRemaining(S', D) \leftarrow f_1(S, D)$
15:    **#barrier**

6.6 Step 6: Checking Terminal Conditions

In this step we take a look whether a vertex from B became active or if there are no active edges. These would be the indications of the completion of the algorithm. The function that checks the terminal conditions is presented earlier in Listing 2.

6.7 Step 7: Final Treatment of Phantoms

In this step we go once again over the sequence of phantoms and remove each one that has its time parameter equal to 0.

Algorithm 9 Final treatment of phantoms

1: **procedure** FINALTREATMENTOFPHANTOMS
2:    **#Performed in parallel**
3:    **for** $j \in PhantomEdges$ **do**
4:       **if** $TimeRemaining(j) = 0$ **then**
5:          Remove j and its source from the sequence of phantoms
6:    **#barrier**

6.8 Step 8: Final Treatment of Vertices

In this step of the program the sequence of active vertices is updated so it contains new active vertices and looses the vertices that may cease to be active.

6.8.1 Preparation of Triggered Vertices

For each triggered vertex Q we first check whether it was inactive before. If it was inactive then its label becomes equal to the label stored at the temporary storing location in the sequence of vertices. If it was active, its label remains unchanged. The phantoms were created and their labels are keeping track of the improved water quality that has reached the vertex Q.

We may now clean the temporary storing location in the sequence of vertices so it now contains the symbol for emptiness (some pre-define negative number).

6.8.2 Merging Triggered with Active Vertices

Triggered vertices are now merged to the sequence of active vertices.

6.8.3 Check Active Vertices for Potential Loss of Activity

For each active vertex Q look at all edges from Q. If there is no active edge whose source is Q, then Q should not be active any longer.

6.8.4 Condensing the Sequence of Active Vertices

After previous few steps some vertices may stop being active in which case they should be removed from the sequence.

Algorithm 10 Final treatment of vertices

1: **procedure** FINALTREATMENTOFVERTICES
2:     **#Performed in parallel**
3:     **for** $Q \in$ *TriggeredVertices* **do**
4:         **if** *State*$(Q) =$ *inactive* **then**
5:             *State*$(Q) \leftarrow$ *active*
6:             *Label*$(Q) \leftarrow$ *TempLabel*(Q)
7:     **#barrier**
8:     *TempLabel* $\leftarrow \emptyset$
9:     **#Performed in parallel**
10:     Merge triggered vertices to active vertices
11:     **#barrier**
12:     **#Performed in parallel**
13:     **for** $Q \in$ *TriggeredVertices* **do**
14:         **if** there are no active edges starting from Q **then**
15:             *State*$(Q) \leftarrow$ *inactive*
16:     **#barrier**

6.9   Step 9: Final Treatment of Active Edges

We first need to merge the triggered edges with active edges. Then all just used edges have to become used and their source has to be re-set so it is not equal to any of the endpoints. Those used edges should be removed from the sequence of active edges.

The remaining final step is to condense the obtained sequence so there are no used edges in the sequence of active edges.

Algorithm 11 Final treatment of active edges

1: **procedure** FINALTREATMENTOFACTIVEEDGES
2:    #**Performed in parallel**
3:       Merge triggered edges to active edges
4:    #**barrier**
5:    #**Performed in parallel**
6:       Transform all *just used* into *used* and erase their *Source* components
7:    #**barrier**
---

7 Large Sets of Active Vertices

In this section we will prove that it is possible for the set of active vertices in dimension 2 to contain more than $O(n)$ elements. We will construct examples in the case when the time to travel over each vertex is from the set $\{1, 2\}$ and when $M = +\infty$.

We will consider the subgraph $V_n = [-n, n] \times [0, n]$ of $\mathbb{Z}^2$. At time 0 the water is located in all vertices of the x axis. For sufficiently large n we will provide an example of configuration ω of passage times for the edges of the graph V_n such that the number of active vertices at time n is of order $n \log n$. This would establish a lower bound on the probability that the number of active vertices at time t is large.

Let us assume that each edge of the graph has the time component assigned from the set $\{1, 2\}$ independently from each other. Assume that the probability that 1 is assigned to each edge is equal to p, where $0 < p < 1$.

Theorem 1 *There exists $t_0 \geq 0$, $\mu > 0$, and $\alpha > 0$ such that for each $t > t_0$ there exists n such that the number A_t of active vertices at time t in the graph V_n satisfies*

$$\mathbb{P}\left(A_t \geq \alpha t \log t\right) \geq e^{-\mu t^2}.$$

To prepare for the proof of the theorem we first study the evolution of the set of active edges in a special case of a graph. Then we will construct a more complicated graph where the set of active edges will form a fractal of length $t \log t$.

Lemma 1 *If all edges on the y-axis have time parameter equal to 1 and all other edges have their time parameter equal to 2, then at time T the set of active vertices (Fig. 5) is given by*

$$A_T = \{(0, T)\} \cup \{(0, T - 1)\} \cup \bigcup_{k=1}^{\lfloor \frac{T+1}{4} \rfloor} \{(-k, T - 2k), (k, T - 2k)\}$$

$$\cup \bigcup_{z \in \mathbb{Z} \setminus \{-\lfloor \frac{T+1}{4} \rfloor, \dots, \lfloor \frac{T+1}{4} \rfloor\}} \left\{\left(z, \left\lfloor \frac{T}{2} \right\rfloor\right)\right\}.$$

Fig. 5 The active edges at time T

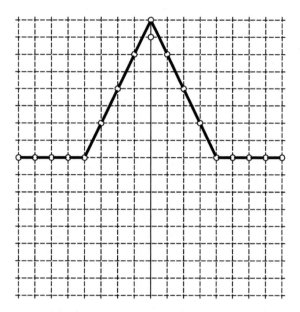

Proof After $T - 2k$ units of time the water can travel over the path γ_k that consists of vertices $(0, 0), (0, 1), \ldots, (0, T - 2k)$. In additional $2k$ units of time the water travels over the path γ'_k that consists of vertices $(0, T - 2k), (1, T - 2k), \ldots, (k, T - 2k)$.

Consider any other path that goes from x axis to the point $(k, T - 2k)$ for some fixed $k \leq \lfloor \frac{T+1}{4} \rfloor$. If the path takes some steps over edges that belong to y axis then it would have to go over at least k horizontal edges to reach y axis, which would take $2k$ units of time. The path would have to take at least $T - 2k$ vertical edges, which would take at least $T - 2k$ units of time. Thus the travel would be longer than or equal to T.

However, if the path does not take steps over the edges along y axis then it would have to take at least $T - 2k$ steps over edges that have passage time equal to 2. This would take $2(T - 2k) = 2T - 4k$ units of time. If $T + 1$ is not divisible by 4, then $k < \frac{T+1}{4}$ and

$$2T - 4k > 2T - T - 1 = T - 1,$$

which would mean that the travel time is at least T. If $T + 1$ is divisible by 4 and $k = \lfloor \frac{T+1}{4} \rfloor$ then the vertical path would reach $(k, T - 2k)$ at time $T - 1$. However, the vertex $(k, T - 2k)$ would still be active because the water would not reach $(k + 1, T - 2k)$ which is a neighbor of $(k, T - 2k)$.

Fig. 6 The set of active
edges in configuration ω_2

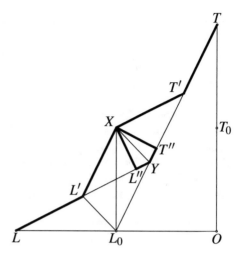

Let us denote by N_t the number of active vertices at time t whose x coordinate is between $-t$ and t,

$$N_t = \left\{ (x, y) \in \{-t, -t+1, \ldots, t-1, t\} \times \mathbb{Z}_0^+ : (x, y) \text{ is active at time } t \right\}.$$

Theorem 2 *There exist real numbers α and $t \geq 0$ and an environment ω for which*

$$N_t(\omega) \geq \alpha t \log t.$$

Proof Assume that $t = 2^k$ for some $k \in \mathbb{N}$. Let us define the following points with their coordinates $T = (0, t)$, $L = \left(-\frac{t}{2}, 0\right)$, and $O = \left(0, \frac{t}{2}\right)$ (Fig. 6). We will recursively construct the sequence of pairs $(\omega_1, \mathscr{I}_1)$, $(\omega_2, \mathscr{I}_2)$, $\ldots$, $(\omega_k, \mathscr{I}_k)$ where ω_j is an assignment of passage times to the edges and $\mathscr{I}_j$ is a subgraph of $\mathbb{Z}^2$. This subgraph will be modified recursively. All edges in $\mathscr{I}_j$ have passage times equal to 2 in the assignment ω_j. Having defined the pair $(\omega_j, \mathscr{I}_j)$ we will improve passage times over some edges in the set $\mathscr{I}_j$ by changing them from 2 to 1. This way we will obtain a new environment ω_{j+1} and we will define a new set $\mathscr{I}_{j+1}$ to be a subset of $\mathscr{I}_j$. The new environment ω_{j+1} will satisfy

$$N_t(\omega_{j+1}) \geq N_t(\omega_j) + \beta t,$$

for some $\beta > 0$.

Let us first construct the pair $(\omega_1, \mathscr{I}_1)$. We will only construct the configuration to the left of the y axis and then reflect it across the y axis to obtain the remaining configuration.

All edges on the y axis have the passage times equal to 1, and all edges on the segment LO have the passage times equal to 1. All other edges have the passage times equal to 2. Define $\mathscr{I}_1 = \triangle LOT$. Then the polygonal line LYT contains the active vertices whose x coordinate is between $-t$ and 0.

The environment ω_2 is constructed in the following way. Let us denote by L_0 and T_0 the midpoints of LO and TO. Let X be the midpoint of LT. We change all vertices on L_0X and T_0X to have the passage time equal to 1. We define $\mathscr{I}_2 = \triangle LL_0X \cup \triangle XT_0T$.

Let L_1 and L_2 be the midpoints of LL_0 and L_0O and let L' and L'' be the intersections of XL_1 and XL_2 with LY. The points T' and T'' are defined in an analogous way: first T_1 and T_2 are defined to be the midpoints of TT_0 and OT_0 and T' and T'' are the intersections of XT_1 and XT_2 with TY.

The polygonal line $LL'XL''YT''XT'T$ is the set of active edges that are inside the triangle LOT. The following lemma will allow us to calculate $N_t(\omega_2) - N_t(\omega_1)$.

Lemma 2 *Let Λ and λ denote the lengths of the polygonal lines $LL'XL''YT''XT'T$ and LYT respectively. If t is the length of OT then*

$$\Lambda = \lambda + \frac{4}{3\sqrt{5}}t.$$

Proof It suffices to prove that $LL' + L'X + XL'' + L''Y = LY + \frac{2}{3\sqrt{5}}t$. From the similarities $\triangle LL_0X \sim \triangle LOT$ and $\triangle LL_0L' \sim LOX$ we have that $L_0L'\|OX$. Therefore L' is the midpoint of LY and $LY = LL' + L'Y = LL' + L'X$. It remains to prove that $XL'' + L''Y = \frac{2}{3\sqrt{5}}t$. From

$$\angle L_0XL'' = \angle L'XL_0 = \angle L_0LL''$$

we conclude that the quadrilateral $LL_0L''X$ is inscribed in a circle. The segment LX is a diameter of the circle hence $\angle LL''X = \angle LL_0X = 90°$. We also have $\angle L''XY = \angle L_0XY - \angle L_0XL'' = 45° - \angle OLT_0 = 45° - \arctan\frac{1}{2}$.

The point Y is the centroid of $\triangle LOT$ hence $XY = \frac{1}{3}XO = \frac{1}{3\sqrt{2}}t$. Therefore

$$
\begin{aligned}
XL'' + L''Y &= XY \cos\left(45° - \arctan\frac{1}{2}\right) + XY \sin\left(45° - \arctan\frac{1}{2}\right) \\
&= \frac{\cos\left(45° - \arctan\frac{1}{2}\right) + \sin\left(45° - \arctan\frac{1}{2}\right)}{3\sqrt{2}}t \\
&= \frac{\cos\left(45° - \arctan\frac{1}{2}\right)\cos 45° + \sin\left(45° - \arctan\frac{1}{2}\right)\sin 45°}{3}t \\
&= \frac{\cos\left(45° - \arctan\frac{1}{2} - 45°\right)}{3}t = \frac{\cos\left(\arctan\frac{1}{2}\right)}{3}t \\
&= \frac{2}{3\sqrt{5}}t.
\end{aligned}
$$

Fig. 7 The set of active edges in configuration ω_3

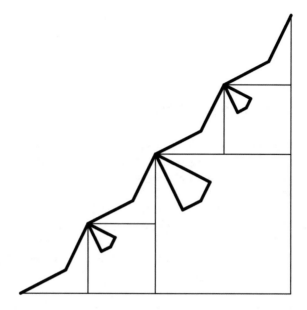

The number of edges on each of the segments of the polygonal lines we obtained is equal to $\frac{u}{\sqrt{5}}$, where u is the length of the segment. Using this fact with the previous lemma applied to both $\triangle LOT$ and its reflection along OT gives us

$$N_t\,(\omega_2) - N_t\,(\omega_1) = \frac{4}{3\sqrt{5}}t \cdot \frac{1}{\sqrt{5}} = \frac{4}{15}t.$$

We now continue in the same way and in each of the triangles LL_0X and XT_0T we perform the same operation to obtain ω_3 and $\mathscr{I}_3$. Since the side length of LL_0X is $\frac{t}{2}$, the increase in the number of elements in the new set of active vertices is $\frac{2}{15} \cdot \frac{t}{2}$. However, this number has to be now multiplied by 4 because there are 4 triangles to which the lemma is applied: $\triangle LL_0X$, $\triangle XT_0T$, and the reflections of these two triangles with respect to OT. Therefore the increase in the number of active vertices is $N_t\,(\omega_3) - N_t\,(\omega_2) = 4 \cdot \frac{2}{15} \cdot \frac{t}{2} = \frac{4}{15}t$ (Fig. 7).

This operation can be repeated k times and we finally get that

$$N_t\,(\omega_k) = N_t\,(\omega_1) + (k-1) \cdot \frac{4}{15}t \ge k \cdot \frac{4}{15}t.$$

Thus the theorem holds if we set $\alpha = \frac{4}{15\log 2}$.

Proof (Proof of Theorem 1) Recall that p is the probability that the time 1 is assigned to each edge. Let $\rho = \min\{p, 1-p\}$. The configuration provided in the proof of Theorem 2 has its probability greater than or equal to ρ^{t^2}. Therefore

$$P\left(A_t \geq \alpha t \log t\right) \geq \rho^{t^2} = e^{t^2 \ln \rho}.$$

Therefore we may take $\mu = -\ln \rho$.

8 Performance Analysis

The algorithm was implemented in C++ and OpenCL. The hardware used has a quad core Intel i5 processor with clock speed of 3.5 GHz and AMD Radeon R9 M290X graphic card with 2 gigabytes of memory. The graphic card has 2816 processing elements.

The table provides a comparison of the performance of the algorithm on 4 samples of three dimensional cubes with edges of lengths 50, 75, 100, and 125. The initial configuration for each of the graphs assumes that there is water on the boundary of the cube, while the set B is defined to be the center of the cube. The same program was executed on graphic card and on CPU.

Graph	GPU time (s)	CPU time (s)
$50 \times 50 \times 50$	3	10
$75 \times 75 \times 75$	8	61
$100 \times 100 \times 100$	21	275
$125 \times 125 \times 125$	117	1540

The graph that corresponds to the cube $100 \times 100 \times 100$ has 1000000 vertices and 2970000 edges, while the graph corresponding to the cube $125 \times 125 \times 125$ has 1953125 vertices and 5812500 edges.

9 Conclusion

The algorithm described in this chapter solves the constrained shortest path problem using parallel computing. It is suitable to implement on graphic cards and CPUs that have large number of processing elements. The algorithm is implemented in C++ and OpenCL and the parallelization improves the speed tenfold.

The main idea is to follow the percolation of water through the graph and assign different qualities to drops that travel over different edges. Each step of the algorithm corresponds to a unit of time. It suffices to analyze only those vertices and edges through which the water flows. We call them active vertices and active edges. Therefore, the performance of the algorithm is tied to the sizes of these active sets.

Theorem 1 proves that it is possible to have at time t an active set of size $O(t \log t)$. The proof of the theorem relied on constructing one such set. It is an open problem to find the average size of the active set at time t.

Problem 1 If the weights and travel times of the edges are chosen independently at random, what is the average size of the active set at time t?

At some stages of the execution, the program needs additional memory to store phantom edges in the graph. It would be interesting to know how many phantom edges are allocated during a typical execution. This can be formally phrased as an open problem.

Problem 2 If the weights and travel times of the edges are chosen independently at random, what is the average number of phantoms that need to be created during the execution of the algorithm?

Acknowledgements The author was supported by PSC-CUNY grants #68387 − 0046, #69723 − 0047 and Eugene M. Lang Foundation.

References

1. G. Amir, I. Corwin, J. Quastel, Probability distribution of the free energy of the continuum directed random polymer in $1 + 1$ dimensions. Comm. Pure Appl. Math. **64**, 466–537 (2011)
2. S. Armstrong, H. Tran, Y. Yu, Stochastic homogenization of a nonconvex Hamilton–Jacobi equation. Calc. Var. Partial Differential Equations, **54**, 1507–1524 (2015). (Submitted) arXiv:1311.2029
3. M. Benaim, R. Rossignol, Exponential concentration for first passage percolation through modified poincaré inequalities. Ann. Inst. Henri Poincaré Probab. Stat. **44**(3), 544–573 (2008)
4. I. Benjamini, G. Kalai, O. Schramm, First passage percolation has sublinear distance variance. Ann. Probab. **31**(4), 1970–1978 (2003)
5. N. Boland, J. Dethridge, I. Dumitrescu, Accelerated label setting algorithms for the elementary resource constrained shortest path problem. Oper. Res. Lett. **34**, 58–68 (2006)
6. S. Chatterjee, P.S. Dey, Central limit theorem for first-passage percolation time across thin cylinders. Probab. Theory Relat Fields **156**(3), 613–663 (2013)
7. J.T. Cox, R. Durrett, Some limit theorems for percolation processes with necessary and sufficient conditions. Ann. Probab. **9**(4), 583–603 (1981)
8. M. Damron, M. Hochman, Examples of nonpolygonal limit shapes in i.i.d. first-passage percolation and infinite coexistence in spatial growth models. Ann. Appl. Probab. **23**(3), 1074–1085 (2013)
9. M. Desrochers, J. Desrosiers, M. Solomon, A new optimization algorithm for the vehicle routing problem with time windows. Oper. Res. **40**(2), 342–354 (1992)
10. J. Hammersley, D. Welsh, First-passage percolation, subadditive processes, stochastic networks, and generalized renewal theory. in *Bernoulli-Bayes-Laplace Anniversary Volume* 1965
11. S. Irnich, G. Desaulniers, Shortest path problems with resource constraints. in *Column Generation*, ed. by G. Desaulniers, J. Desrosiers, M. M. Solomon. GERAD 25th Anniversary Series (Springer, 2005), pp. 33–65
12. E. Köhler, R.H. Möhring, H. Schilling, Acceleration of shortest path and constrained shortest path computation. Lect. Notes Comput. Sci. **3503**, 126–138 (2005)
13. E. Kosygina, F. Rezakhanlou, S.R.S. Varadhan, Stochastic homogenization of Hamilton-Jacobi-Bellman equations. Comm. Pure Appl. Math. **59**(10), 1489–1521 (2006)

14. E. Kosygina, F. Yilmaz, O. Zeitouni, Nonconvex homogenization of a class of one-dimensional stochastic viscous Hamilton-Jacobi equations, in *preparation* (2017)
15. J. Krug H. Spohn, Kinetic roughening of growing surfaces. Solids Far Equilib. 412–525 (1991)
16. X.-Y. Li, P.-J. Wan, Y. Wang, O. Frieder, Constrained shortest paths in wireless networks, in *IEEE MilCom* (2001), pp. 884–893
17. L. Lozano, A.L. Medaglia, On an exact method for the constrained shortest path problem. Comput. Oper. Res. **40**(1), 378–384 (2013)
18. I. Matic, Parallel algorithm for constrained shortest path problem in C++/OpenCL. https://github.com/maticivan/parallel_constrained_shortest_path
19. I. Matic, J. Nolen, A sublinear variance bound for solutions of a random Hamilton-Jacobi equation. J. Stat. Phys. **149**, 342–361 (2012)
20. K. Mehlhorn, M. Ziegelmann, Resource constrained shortest paths. Lect. Notes Comput. Sci. **1879**, 326–337 (2000)
21. S. Misra, N.E. Majd, H. Huang, Approximation algorithms for constrained relay node placement in energy harvesting wireless sensor networks. IEEE Trans. Comput. **63**(12), 2933–2947 (2014)
22. R. Muhandiramge, N. Boland, Simultaneous solution of lagrangean dual problems interleaved with preprocessing for the weight constrained shortest path problem. Networks **53**, 358–381 (2009)
23. C.M. Newman, M.S.T. Piza, Divergence of shape fluctuations in two dimensions. Ann. Probab. **23**(3), 977–1005 (1995)
24. F. Rezakhanlou, Central limit theorem for stochastic Hamilton-Jacobi equations. Commun. Math. Phys. **211**, 413–438 (2000)
25. T. Sasamoto, H. Spohn, One-dimensional kardar-parisi-zhang equation: An exact solution and its universality. Phys. Rev. Lett. **104** (2010)
26. P.E. Souganidis, Stochastic homogenization of Hamilton-Jacobi equations and some applications. Asymptot. Anal. **20**(1), 1–11 (1999)

Gathering a Swarm of Robots Through Shortest Paths

Serafino Cicerone, Gabriele Di Stefano and Alfredo Navarra

Abstract The gathering problem has been largely studied in the last years with respect to different environments. The requirement is to move a team of robots initially placed at different locations toward a common point. Robots move based on the so called Look-Compute-Move model. Each time a robot wakes up, it perceives the current configuration in terms of robots' positions (Look), it decides whether and where to move (Compute), and makes the computed move (Move) in the case that the decision was affirmative. All the phases are performed asynchronously for each robot. Robots are oblivious, anonymous, silent, and execute the same distributed and deterministic algorithm. So far, the goal has been mainly to detect the minimal assumptions that allow to accomplish the gathering task, without taking care of any cost measure of the provided solutions. We provide an overview of recent results that first extend the classic notion of optimization problem to the context of robot-based computing systems, and then show that the gathering problem can be optimally solved. As cost measure, the overall traveled distance performed by all robots is considered. This implies that the provided optimal algorithms must be able to solve the gathering by moving robots through shortest paths. The presented optimal algorithms refer to robots moving on either the plane or graphs. In the latter case, different topologies are considered, like trees, rings, and infinite grids.

S. Cicerone (✉) · G. Di Stefano
Dipartimento di Ingegneria e Scienze dell'Informazione e Matematica,
Università degli Studi dell'Aquila, Via Vetoio, 67100 Coppito, L'Aquila, Italy
e-mail: serafino.cicerone@univaq.it

G. Di Stefano
e-mail: gabriele.distefano@univaq.it

A. Navarra
Dipartimento di Matematica e Informatica, Università degli Studi di Perugia,
Via Vanvitelli 1, 06123 Perugia, Italy
e-mail: alfredo.navarra@unipg.it

© Springer International Publishing AG, part of Springer Nature 2018
A. Adamatzky (ed.), *Shortest Path Solvers. From Software to Wetware*,
Emergence, Complexity and Computation 32,
https://doi.org/10.1007/978-3-319-77510-4_2

27

1   Introduction

The gathering task is a basic primitive in robot-based computing systems. It has been extensively studied in the literature under different assumptions. The problem asks to design a distributed algorithm that allows a swarm of robots to meet at some common place (not determined in advance). Depending on the environment where they move as well as on the capabilities of the robots, very different and challenging aspects must be faced (see, e.g. [5, 11–14, 17, 18, 23, 25], and references therein). Concerning the environment, in this work we consider robots placed on either the *Euclidean plane* or the *vertices of graphs*. Concerning the robots' capability, we follow the research trend of admitting the minimal setting required to accomplish the gathering task (as it is considered one of the main issues in robot-based computing systems). Then, we consider systems in which initially each robot occupies a different location and robots are endowed with very few capabilities. They are assumed to be:

- *Dimensionless*: modeled as geometric points in the plane or considered as placed on vertices of a graph;
- *Anonymous*: no unique identifiers for robots, no labels for vertices and edges;
- *Autonomous*: no centralized control;
- *Oblivious*: no memory of past events;
- *Homogeneous*: they all execute the same deterministic algorithm;
- *Asynchronous*: there is no global clock that synchronizes their actions;
- *Silent*: no direct way of communicating, no possibilities of leaving any mark at visited vertices;
- *Unoriented*: no common coordinate system, no common compass, no common left-right orientation (i.e., no chirality/handedness).

Robots are equipped with sensors and motion actuators, and operate in *Look-Compute-Move* cycles (see, e.g. [18]). In the Look phase a robot takes a snapshot of the current global configuration (when a robot moves on the plane, it performs this operation in terms of the relative robots' positions, according to its own coordinate system). Successively, in the Compute phase it decides whether to move toward a specific direction or not, and in the affirmative case it moves (Move).

A Look-Compute-Move cycle is called a *computational cycle* of a robot.

Computational cycles are performed asynchronously, i.e., the time between Look, Compute, and Move phases is finite but unbounded, and it is decided by an *adversary* for each robot. Moreover, during the Look phase, a robot does not perceive whether other robots are moving or not. Hence, robots may move based on outdated perceptions.[1]

The concept of the adversary is rather common in distributed computing where the evolving of the system might be subject to different unconstrained events.

---

[1] In fact, asynchrony implies that, based on the configuration perceived during the Look phase at some time t, a robot r computes a destination at some time $t' > t$, starts to move at an even later time $t'' > t'$, eventually stopping at time $t''' \geq t''$; thus it might be possible that at time t'' some robots are in different positions from those previously perceived by r at time t, because in the meantime they performed their Move operations (possibly several times).

The environment in which robots move may affect the last part of the computational cycle: in the plane a robot can move toward any point and the distance traveled within a move is neither infinite nor infinitesimally small[2]; in graphs, a robot can move only to one of its adjacent vertices and the move is instantaneous (notice this implies that during the Look phase robots are always detected on vertices and not on edges).

During the Look phase, robots are assumed to perceive *multiplicities* [11, 18, 20]. The multiplicity detection capability has been exploited in various forms. In any case, a robot perceives whether a location is occupied by robots or not, but in the *global-strong* version, a robot is able to perceive the exact number of robots that occupy the same location. In the *global-weak* version, a robot perceives only whether a location is occupied by one robot or if a multiplicity occurs, i.e., the location is occupied by an undefined number of robots greater than one. In the *local-strong* version, a robot can perceive only whether a location is occupied or not, but it is able to perceive the exact number of robots occupying the location where it resides. Finally, in the *local-weak* version, a robot can perceive the multiplicity only on the location where it resides but not the exact number of robots composing it.

The scheduler (decided by the adversary) determining the computational cycles timing is assumed to be fair, that is, each robot performs its cycle within finite time and infinitely often. In the literature, this kind of scheduler is called Asynchronous (ASYNC). Different options for the scheduler are:

- Fully-synchronous (FSYNC): all robots are awake and run their computational cycle concurrently. Each phase of the cycle has exactly the same duration for all robots. This is equivalent to a fully synchronized system in which robots are activated simultaneously and all operations happen instantaneously.
- Semi-synchronous (SSYNC): It coincides with the FSYNC model with the only difference that not all robots are necessarily activated during a cycle, but those who are activated are fully synchronized.

Contribution. The classical gathering problem, where robots are free to gather anywhere in the Euclidean plane, has been solved in [8] for any number of robots $n > 2$, even assuming the few capabilities recalled above. In the same paper, authors posed the following interesting problem: "*In all existing investigations on the Gathering Problem, the focus has been on computability (i.e., feasibility), while the complexity of the solutions has never been an issue; indeed, there is a general absence of cost measures. An interesting fundamental research question is the definition of cost measures and their use in the analysis of the complexity of the solution protocols.*"

In this chapter, we survey recent results (cf. [7, 15, 16]) that try to answer to such a research question not only on the plane but also on discrete structures like graphs. In

---

[2]More precisely, the adversary has also the power to stop a moving robot before it reaches its destination, but there exists an (unknown arbitrarily small) constant $\delta > 0$ such that if the destination point is closer than δ, the robot will reach it, otherwise the robot will be closer to it by at least δ. Note that, without this assumption, an adversary would make it impossible for any robot to ever reach its destination.

particular, when the environment is the plane, the problem is modified by assuming that robots must meet at some predetermined points, herein called *meeting-points*.[3] During the Look phase, robots detect the meeting-points in the snapshot of the system as sort of landmarks. Contrarily to the classic gathering problem, the introduction of meeting-points implies that some configurations turn out to be ungatherable.

As a contribution, we show that it is possible to use a generalization of the classic definitions of optimization problem and approximation algorithm to the context of robot-based computing systems. This generalization has been introduced in [7] and allow us to formally define the concept of *optimal algorithm* for solving both the gathering problem on meeting-points and the gathering problem on (specific topologies of) graphs. In both the cases, the cost measure is the *minimum total traveled distance of all robots*. This measure implies that any optimal gathering algorithm must move robots through shortest paths until the final gathering point.

The defined cost measure is strictly related to the concept of *Weber point* [3, 9, 24]. The Weber point is in fact defined as the point minimizing the sum of distances between itself and a given set of points in the plane (the robots in our scenario). A well-known characterization says that if the provided points are not collinear, the Weber point is unique, but its computation is usually unfeasible, even for just five points [9].

For the gathering problem on meeting-points, the concept of Weber point has to be restricted to the set of meeting-points, and this leads to define the concept of *discrete Weber point*. Of course, it is always possible to compute the discrete Weber point where to finalize the gathering since it must be chosen among the finite set of meeting-points. Differently from the classical notion of Weber point, the discrete Weber points are not necessarily unique even when the robots are not collinear, and hence in general constitute a subset of the meeting-points. In order to calculate and manipulate such a subset it is possible to exploit results about so-called k-ellipses [26–28]. For the gathering problem on graphs, the same observations hold: it is possible that more than one discrete Weber point exists and all the discrete Weber points can be easily computed. Notice that both on meeting-points and on graphs, discrete Weber points can be computed as long as a robot is empowered with the global-strong multiplicity detection.

This works shows that, for each optimization problem addressed and for each environment considered, the following holds: (1) some input configurations cannot be optimally gathered on discrete Weber points even though they are potentially gatherable, and (2) for all configurations admitting optimal gathering there exists a distributed optimal algorithm that gathers on a discrete Weber point by letting robots move along the shortest paths.

Outline. This chapter is organized as follows. Section 2 formally defines the two addressed versions of the gathering problem, namely that on meeting-point on the plane and that on graphs. Section 3 formalizes a general notion of optimization prob-

---

[3]Meeting-points for gathering purposes are interesting not only from a theoretical point of view, but also for practical reasons when not all places can be candidate to serve as gathering points.

lems in robot-based computing systems and then provides the optimization version of the gathering problems defined in the previous section.

Section 4 is devoted to present an optimal algorithm for the problem of gathering on meeting-points. The section starts by providing some details on the possible information that robots can deduce from the Look phase. In particular, we recall the concepts of configuration view and configuration symmetry, which in turn are used to provide general impossibility results. Then, we discuss the concept of discrete Weber point (i.e., a Weber point chosen only among the finite set of meeting-points) and show useful properties that are successively exploited by the gathering algorithms. Finally, we present a distributed optimal gathering algorithm that solves the problem for all configurations where optimal gathering can be achieved.

Section 5 is devoted to an optimal algorithm for the problem of gathering on different graph topologies. The section starts by reconsidering the notions of configuration automorphisms and symmetries to be applied to general graphs. Accordingly, we provide general impossibility results. We also reconsider the notion of Weber points for graphs and show useful properties that are successively exploited by the gathering algorithms. Finally, the optimal gathering algorithms when the graph topologies correspond to trees, rings and infinite grids are provided.

Section 6 concludes the chapter and outlines some possible research directions.

2   Gathering in Different Environments

In this work we consider two different environments where robots move: the Euclidean plane and undirected graphs. According to such environments, we address two different versions of the classical problem, namely the *Gathering on Meeting Points on the plane* denoted as GMP, and the *Gathering on Graphs* denoted as GG.

Definition of GMP. The system is composed of n mobile *robots* freely moving on the plane. At any time, the multiset $R = \{r_1, r_2, \ldots, r_n\}$, with $r_i \in \mathbb{R}^2$, contains the *positions* of all the robots. The set $U(R) = \{x \mid x \in R\}$ contains the *unique* robots' positions. M is a finite set of fixed *meeting-points* in the plane representing the only locations in which robots can be gathered (notice that the concept of meeting-points as been already considered in [6, 19], but for addressing the pattern formation problem). The center of gravity of points in M, that is the point whose coordinates are the mean values of the coordinates of the points of the set, is denoted by $cg(M)$. The pair $C = (R, M)$ represents a *configuration*.

Similarly to [19], a robot is said to be stationary in a configuration C if it is (1) inactive, or (2) active, and it has not taken the snapshot yet, or it has taken snapshot C, or it has taken snapshot $C' \neq C$ (with C' being produced by the algorithm before C) which leads to a null movement. A configuration C is said to be *stationary* if all robots are stationary at C. A configuration C is *initial* at time t if it is stationary and at time t all robots have distinct positions (i.e., $|U(R)| = n$). Unlike the initial configurations, in general, not all robots are stationary at a non-initial configuration

C, but at least one robot that takes the snapshot C is stationary by definition. A configuration C is *final* at time t if at that time there exists a point $m \in M$ such that $r_i = m$ for each $r_i \in R$, and at each time $t' \geq t$ each robot performing a move makes a null movement; in this case we say that the robots have gathered on point m at time t.

Problem GMP can be formally defined as the problem of transforming an initial configuration into a final one. A *gathering algorithm* for the GMP problem is a deterministic distributed algorithm that brings the robots in the system to a final configuration in a finite number of cycles from any given initial configuration, regardless of the adversary. We say that an initial configuration $C = (R, M)$ is *ungatherable* if there are no gathering algorithms for GMP with respect to C. Moreover, a gathering algorithm $\mathscr{A}$ for GMP with respect to C *ensures the gathering on* $T \subseteq M$ if it finalizes the gathering on some point of T. Note that if $\mathscr{A}$ ensures the gathering on T, then an execution of $\mathscr{A}$ will provide a solution in T, regardless of the adversary.

Definition of GG. A simple undirected graph $G = (V, E)$, with vertex set V and edge set E, will represent the topology where robots are placed on. A function $\mu : V \longrightarrow \mathbb{N}$, represents the number of robots on each vertex of G, and we call (G, μ) a *configuration* whenever $\sum_{v \in V} \mu(v)$ is bounded and greater than zero.

A configuration is *initial* if each robot lies on a different vertex (i.e., $\mu(v) \leq 1 \; \forall v \in V$). A configuration is *final* if all the robots are on a single vertex u (i.e., $\mu(u) > 0$ and $\mu(v) = 0, \; \forall v \in V \setminus \{u\}$).

Similarly to GMP, the GG problem can be formally defined as the problem of transforming an initial configuration into a final one. A *gathering algorithm* for the GG problem is a deterministic distributed algorithm that brings the robots in the system to a final configuration in a finite number of cycles from any given initial configuration, regardless of the adversary. We say that an initial configuration $C = (G, \mu)$ is *ungatherable* if there are no gathering algorithms for GG with respect to C.

3 Optimization Problems for Robot-Based Computing Systems

In this section, we recall from [7] how the classical notion of optimization problems (cf. [2]) is extended to optimization problems solvable in the context of robot-based computing systems. According to the new framework, we describe how both GMP and GG can be reformulated as optimization problems.

Let Π be an optimization problem for robot-based computing systems. For the sake of clarity, we address the minimization case only. The maximization case can be derived analogously. Problem Π consists of a triple $(\mathscr{I}, sol, mis)$, where:

- $\mathscr{I}$ is the set of instances (i.e., all possible initial configurations);
- sol is a function that maps each initial configuration $C \in \mathscr{I}$ to the set $sol(C)$ of feasible solutions of C;

- given an instance $C \in \mathscr{I}$ and a solution $s \in sol(C)$, then $mis(C, s)$ denotes the real positive measure of s, and the function mis is called *objective function* for Π.

The goal of Π with respect to an instance C is to find an optimal solution, that is, a feasible solution $s \in sol(C)$ such that

$$mis(C, s) = \min\{mis(C, s') \ : \ s' \in sol(C)\}.$$

In the following, *opt* will denote the function mapping an instance C to the measure of an optimum solution in $sol(C)$. Given an instance C and a solution $s \in sol(C)$, we define the *performance ratio* of s with respect to C as

$$R(C, s) = \frac{mis(C, s)}{opt(C)}.$$

The performance ratio is always a number greater than or equal to 1 and is as close to 1 as the solution s is close to the optimal solution.

Now, let $\mathscr{A}$ be an algorithm for Π and C be an initial configuration. Even though we are dealing with deterministic algorithms, different executions of $\mathscr{A}$ starting from the same initial configuration C can lead to different solutions. In fact, in the described asynchronous setting, an execution depends on the time required by the scheduled activities, and this is implemented by the behavior of the adversary. Then, there exists a set $sol_{\mathscr{A}}(C) \subseteq sol(C)$ of solutions, each corresponding to a possible execution of $\mathscr{A}$ starting from C. If $\mathscr{A}(C)$ is a solution $s \in sol_{\mathscr{A}}(C)$ which maximizes $mis(C, s)$, that is

$$\mathscr{A}(C) = \arg \max_{s \in sol_{\mathscr{A}}(C)} mis(C, s),$$

then:

- we say that $\mathscr{A}$ is an *optimal* algorithm for Π if $R(C, \mathscr{A}(C)) = 1$ for each instance $C \in \mathscr{I}$;
- given a function $f : \mathbb{N} \to (1, \infty)$, we say that $\mathscr{A}$ is a $f(n)$-*approximation* for Π if $R(C, \mathscr{A}(C)) \le f(|C|)$ for each instance $C \in \mathscr{I}$. Here $|C|$ denotes the size of a configuration C.

GMP as an optimization problem. We can now formalize GMP as an optimization problem $\text{GMP}^+ = (\mathscr{I}, sol, mis)$. Of course, $\mathscr{I}$ is the set of all the possible initial configurations to be gathered. To formalize the set $sol(C)$ for any given initial configuration $C = (R, M)$ for GMP, we need to remark what an execution of a gathering algorithm $\mathscr{A}$ produces. In particular, an execution of $\mathscr{A}$ for C can be seen as a set $\mathscr{P}$ containing $|R|$ polycurves in the plane. Each polycurve in $\mathscr{P}$ models all the movements performed by a specific robot in R starting from its initial position and ending on the final gathering point $m \in M$. A movement performed by a robot during a computational cycle corresponds to a part of the polycurve, and the endpoints of each curve are the positions of the robot at the beginning and at the end, respectively, of a computational cycle. Let us call *gathering solution* a set of polycurves each one starting from the initial position of a different robot in R and ending on the same final gathering point $m \in M$. The set of all the gathering solutions for a configuration C defines $sol(C)$. Concerning, $mis(C, \mathscr{P})$ we use the following measure:

- $mis(C, \mathscr{P})$ corresponds to $\sum_{P \in \mathscr{P}} length(P)$, where $length(P)$ is the length of the polycurve P.

GG as an optimization problem. Also GG can be expressed as an optimization problem $GG^+ = (\mathscr{I}, sol, mis)$, where $\mathscr{I}$ is the set of all the possible initial configurations for GG. Given an initial configuration $C = (G, \mu)$, an execution of a gathering algorithm $\mathscr{A}$ for C can be seen as a set $\mathscr{P}$ containing paths in the graph G. Each path in $\mathscr{P}$ models all the movements performed by a specific robot starting from its initial position and ending on the final gathering point. A movement performed by a robot during a computational cycle corresponds to a subpath made of just two adjacent vertices. So, a gathering solution is a set of paths, each one starting from a distinct vertex and ending on the same final gathering vertex. The set of all the gathering solutions for a configuration C defines $sol(C)$. Concerning, $mis(C, \mathscr{P})$ we use the following measure:

- $mis(C, \mathscr{P})$ corresponds to $\sum_{P \in \mathscr{P}} length(P)$, where $length(P)$ is the length of the path P expressed as the number of edges in the path.

4 Optimal Gathering for GMP$^+$

In this section an optimal gathering algorithm for GMP$^+$ is described (cf. [7]). Before presenting the algorithm, we give some detail on the possible information that robots can deduce from the Look phase. In particular, we recall the concepts of *configuration view* and *configuration symmetry*, which in turn are used to provide general impossibility results. Then, we discuss the concept of discrete Weber-point (i.e., a Weber-point chosen only among the finite set of meeting-points, and not among the infinite set of points in the plane) and show useful properties that are successively exploited by our gathering algorithms.

4.1 Configuration View

Given two distinct points u and v in the plane, denote by $d(u, v)$ their distance, $line(u, v)$ the straight line passing through them, and (u, v) (resp. $[u, v]$) the open (resp. closed) segment containing all points of this line that lie between u and v. The half-line starting at point u (but excluding the point u) and passing through v is denoted by $hline(u, v)$. We denote by $\sphericalangle(u, c, v)$ the angle centered in c and with sides $hline(c, u)$ and $hline(c, v)$. The angle $\sphericalangle(u, c, v)$ is measured from u to v in clockwise or counter-clockwise direction, and the measure is always positive and ranges from $0°$ to less than $360°$ and the direction in which it is taken will be clear from the context.

Given a set P of n points in the plane and an additional point $c \notin P$, let $\bigcup_{p \in P} hline(c, p)$ be the set of all half-lines starting from c and passing through each point in P. The successor of $p \in P$ with respect to c, denoted by $succ(p, c)$, is defined as the point $q \in P$ such that

- either q is the closest point to p on $hline(c, p)$, with $d(c, q) > d(c, p)$;
- or $hline(c, q)$ is the half-line following $hline(c, p)$ in the order implied by the clockwise direction, and q is the closest point to c on $hline(c, q)$.

Let us now assume that, during its last Look phase, a robot $r \in R$ has just taken a snapshot of the current configuration $C = (R, M)$. Since in our model robots do not have a common understanding of the handedness (chirality), we assume that r arbitrarily chooses one direction as "clockwise" and accordingly uses the function $succ()$ to define the "view" of the configuration C (that is, of the multiset $R \cup M$) from a given observation point p. In particular, if $p \in (R \cup M) \setminus cg(M)$ and $P = (R \cup M) \setminus \{p\}$, then the function $succ()$ allows r to compute the sequence $V_r^+(p) = (p_0 = p, p_1, \ldots, p_{f-1})$,[4] where $f = |U(R)| + |M|$, $p_i = succ(p_{i-1}, p_0)$, $i \geq 1$, and the first half-line needed by $succ()$ to identify p_1 is $hline(p_0, cg(M))$.[5] In other words, if r "simulates" its position as it were in the point p, then $V_r^+(p)$ represents the order in which r views all the points in C starting from p itself and turning clockwise $hline(p, cg(M))$ according to $succ()$. Similarly, $V_r^-(p)$ can be computed by r by turning in counter-clockwise direction[6] (see Fig. 1).

Now, from each sequence $V_r^+(p)$, r can directly get the sequence $\mathscr{V}_r^+(p)$, that is the *clockwise view* of p as observed by r, as follows:

- for each $i \geq 1$, point p_i in $V_r^+(p)$ is replaced by the triple α_i, d_i, x_i in $\mathscr{V}_r^+(p)$, where $\alpha_i = \sphericalangle(cg(M), p, p_i)$, $d_i = d(p, p_i)$, and $x_i \in \{r, m, x\}$ according whether p_i is a robot position not in a multiplicity, a meeting-point, or a robot position where a multiplicity occurs, respectively.
- for $i = 0$, $\alpha_0 = 0$, $d_0 = d(p_0, cg(M))$, and x_0 is an element in $\{r, m, x\}$ defined as above.

Note that in $\mathscr{V}_r^+(p)$ there are no elements of type x if robots do not have the multiplicity detection capability. Similarly, the robot r can compute the *counter-clockwise view* of p, denoted by $\mathscr{V}_r^-(p)$, by considering the sequence $V_r^-(p)$.

By defining $r < m < x$ for the third component in the triples used to define $\mathscr{V}_r^+(p)$ from $V_r^+(p)$, any set of strings encoding the views of a robot can be ordered lexico-

---

[4]The subscript in the symbol $V_r^+(p)$ is used to remark who is computing the view (in this case r), while the argument indicates the point from which the view is computed.

[5]If two points $r' \in U(R)$ and $m \in M$, different from p, are coincident, then points r', m will appear in this order in $V_r^+(p)$.

[6]Remember that the terms clockwise and counter-clockwise always refer to the local coordinate system of the robot that computes the view. During a computational cycle, r maintains the same local orientation to compute the view of each point $p \in R \cup M$, but the orientation could change between two different computational cycles.

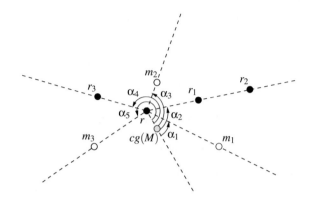

Fig. 1 Robot r computing its view: $V_r^-(r) = (r, m_1, r_1, r_2, m_2, r_3, m_3)$ and hence $\mathcal{V}_r^-(r) = (0°, d(r, cg(M)), \mathtt{r}, \alpha_1, d(r, m_1), \mathtt{m}, \alpha_2, d(r, r_1), \mathtt{r}, \alpha_2, d(r, r_2), \mathtt{r}, \alpha_3, d(r, m_2), \mathtt{m}, \alpha_4, (r, r_3), \mathtt{r}, \alpha_5, d(r, m_3), \mathtt{m})$. Black circles represent robots, white circles represent meeting-points

graphically, and hence the minimum (i.e., the *minimum view*) can be computed. In particular the *view of p* computed by r is defined as $\mathcal{V}_r(p) = \min\{\mathcal{V}_r^+(p), \mathcal{V}_r^-(p)\}$, the *view of the configuration C* computed by r is defined as $\mathcal{V}_r(C) = \bigcup_{p \in R \cup M}\{\mathcal{V}_r(p)\}$, and the minimum of $\mathcal{V}_r(C)$ can be computed.

Until now, $\mathcal{V}_r(p)$ has been defined for points p not coincident with $cg(M)$. If p coincides with $cg(M)$ then the first half-line to be used to build $V_r^+(p)$ and $V_r^-(p)$ is $hline(cg(M), q)$, being q the point in $P = (R \cup M) \setminus \{p\}$ with minimum view. If there is more than one point with minimum view in P, we can choose one of them since the outcome will be the same (this will be confirmed by Lemma 1). In this case, $\mathcal{V}_r(p)$ is minimum by definition in the set of views associated to points in $R \cup M$, as $\alpha_0 = 0$ and $d_0 = 0$, whereas for each other point q not in $cg(M)$, $\mathcal{V}_r(q)$ is such that $\alpha_0 = 0$ and $d_0 > 0$. If two points in $R \cup M$ lie on $cg(M)$ the minimum is determined by the value of x_0.

Notice that, even if robots do not have a common understanding of the handedness (chirality), by computing their view of C they all get the same information. This is better expressed by the following statement.

Property 1 If two distinct robots r_1 and r_2 take a snapshot of the same configuration C, then they compute the same view of C, that is $\mathcal{V}_{r_1}(C) = \mathcal{V}_{r_2}(C)$.

Two additional concepts about views will be used in the following:

- if $p \in U(R) \cup M$ and $S \subseteq U(R) \cup M$, then *min_view*(p, S) says whether p is a point with minimum view in S or not;
- if $m \in M$, then *start*(m) represents the point(s) in R closest to m but not on it, and having the minimum view in case of ties.

Note that, in symmetric configurations, *start*(m), $m \in M$, may define more than one robot.

4.2 Configuration Automorphisms and Symmetries

Let $\varphi : \mathbb{R}^2 \to \mathbb{R}^2$ be a map from points to points in the plane. It is called an *isometry* or distance preserving if for any $a, b \in \mathbb{R}^2$ one has $d(\varphi(a), \varphi(b)) = d(a, b)$. Examples of isometries in the plane are *translations, rotations* and *reflections*. An isometry φ is a translation if there exists no point x such that $\varphi(x) = x$; it is a rotation if there exists a unique point x such that $\varphi(x) = x$ (and x is called *center of rotation*); it is a reflection if there exists a line ℓ such that $\varphi(x) = x$ for each point $x \in \ell$ (and ℓ is called *axis of symmetry*).

An *automorphism* of a configuration $C = (R, M)$ is an isometry from $\mathbb{R}^2$ to itself, that maps multiplicities to multiplicities, single robots to single robots (i.e., points of R into R), and meeting-points to meeting-points (i.e., points of M into M). Note that, by considering the global-weak multiplicity detection, the mapping of multiplicities to multiplicities is done regardless of the number of robots. The set of all automorphisms of C forms a group with respect to the composition called *automorphism group* of C and it is denoted by $\mathrm{Aut}(C)$.

The isometries in $\mathrm{Aut}(C)$ are the identity, rotations, reflections and their compositions (translations are not possible as the sets R and M are finite). Note that the existence of two reflections implies the existence of a rotation.

If $|\mathrm{Aut}(C)| = 1$, that is C admits only the identity automorphism, then C is said to be *asymmetric*, otherwise it is said to be *symmetric* (i.e., C admits rotations or reflections). If C is symmetric due to an automorphism φ, a robot cannot distinguish its position at $r \in R$ from $r' = \varphi(r)$. As a consequence, no algorithm can distinguish r from r', and then it cannot avoid that the two robots start the computational cycle simultaneously. In such a case, there might be a so called *pending move*, that is one of the two robots performs its entire computational cycle while the other has not started or not yet finished its Move phase, i.e. its move is pending. Clearly, any other robot is not aware whether there is a pending move, that is it cannot deduce such an information from its view. This fact greatly increases the difficulty to devise a gathering algorithm for symmetric configurations.

Given an isometry $\varphi \in \mathrm{Aut}(C)$, the *cyclic subgroup* of order k generated by φ is given by $\{\varphi^0, \varphi^1 = \varphi, \varphi^2 = \varphi \circ \varphi, \ldots, \varphi^{k-1}\}$ where φ^0 is the identity. A reflection ρ generates a cyclic subgroup $H = \{\rho^0, \rho\}$ of order two. The cyclic subgroup generated by a rotation ρ can have any order $k > 1$, with k fixed and depending on the size of C. If H is a cyclic subgroup of $\mathrm{Aut}(C)$, the *orbit* of a point $p \in R \cup M$ is $Hp = \{\gamma(p) \mid \gamma \in H\}$. Note that the orbits Hr, for each $r \in R$ form a partition of R. The associated equivalence relation is defined by saying that r and r' are *equivalent* if and only if their orbits are the same, that is $Hr = Hr'$. Equivalent robots are indistinguishable by any algorithm.

Next lemmata provide relationships between isometries and configuration views.

Lemma 1 *Let $C = (R, M)$, $|M| > 1$, be a configuration without multiplicities and let $r \in R$ be a robot that has taken a snapshot of C during its last Look phase. Then:*

- *C admits a reflection if and only if there exist two points $p, q \in R \cup M$, not necessarily distinct, such that $\mathcal{V}_r^+(p) = \mathcal{V}_r^-(q)$;*

- *C admits a rotation if and only if there exist two distinct points $p, q \in R \cup M$, such that $\mathcal{V}_r^+(p) = \mathcal{V}_r^+(q)$.*

Lemma 2 *Let $C = (R, M)$ be a configuration without multiplicities and ℓ be a line passing through $cg(M)$. If C is asymmetric or ℓ is the only axis of reflection for C, then all robots that take a snapshot of C agree on the same North-South orientation of ℓ.*

According to this lemma, in the remainder we can assume that, under certain conditions, all robots can agree about the North of a line or axis ℓ passing through $cg(M)$, and in case, about the "northernmost robot or meeting-point" on ℓ.

4.3 Ungatherability Results

In this section we state a theorem providing a sufficient condition for a configuration to be ungatherable: if this condition applies then GMP is not solvable. Actually, results are even stronger, as they hold also for the case of the synchronous environments FSYNC. We first need the following definition:

Definition 1 *Let $C = (R, M)$ be a configuration. An isometry $\varphi \in Aut(C)$ is called partitive on $\mathbb{R}^2 \setminus P$ if the cyclic subgroup H generated by φ has order $k > 1$, and $|Hp| = k$ for each $p \in \mathbb{R}^2 \setminus P$.*

Notice that the identity is not partitive. A reflection ρ with axis of symmetry ℓ generates a cyclic group $H = \{\rho^0, \rho\}$ of order two and is partitive on $\mathbb{R}^2 \setminus \ell$. A rotation ρ is partitive on $\mathbb{R}^2 \setminus \{c\}$, where c is the center of rotation, and the cyclic subgroup generated by ρ can have any order $k > 1$, with k fixed and depending on the size of C. In the following, we say that an isometry φ *fixes* a point p when $\varphi(p) = p$. The following theorem provides us a sufficient condition for establishing when a configuration is ungatherable.

Theorem 1 *In the FSYNC setting, consider an initial configuration $C = (R, M)$ and a subset of points $P \subset \mathbb{R}^2$ with $P \cap R = \emptyset$. If there exists an isometry $\varphi \in Aut(C)$ that is partitive on $\mathbb{R}^2 \setminus P$ and fixes the points of P, then there is no GMP gathering algorithm for C that ensures the gathering on $M \setminus P$.*

Theorem 1 is given in a form that can be easily generalized to spaces of more than two dimensions. For the Euclidean plane, the following corollary characterizes the initial configurations of the GMP that are ungatherable.

Corollary 1 *In the FSYNC setting, consider an initial configuration $C = (R, M)$ admitting an isometry $\varphi \in Aut(C)$. C is ungatherable if one of the following holds:*

- *φ is a rotation with center c and $c \notin R \cup M$;*
- *φ is a reflection with axis ℓ and $\ell \cap (R \cup M) = \emptyset$.*

Another consequence of Theorem 1 for initial symmetric configurations is that, if a gathering algorithm exists in any setting, and in particular in the ASYNC one, then it must gather the robots on a point in the set P defined in the statement of the theorem. In particular, when the configuration admits a rotation whose center c is not occupied by a robot, then the set P contains only c, and hence the following corollary can be stated.

Corollary 2 *Consider an initial configuration $C = (R, M)$ admitting a rotation with center c and $c \notin R$. If there exists a gathering algorithm for GMP with respect to C, then there exists a meeting-point m on the center c where the gathering is finalized.*

Similarly, when the configuration admits a reflection then the set P of Theorem 1 contains only the points of the axis of reflection, and hence the following corollary can be stated.

Corollary 3 *Consider an initial configuration $C = (R, M)$ admitting a reflection with axis ℓ and $\ell \cap R = \emptyset$. If there exists a gathering algorithm for GMP with respect to C, then there exists a meeting-point m on the axis ℓ where the gathering is finalized.*

4.4 Weber Points for GMP

A typical approach used to solve the classic gathering problem is to choose as destination a point which is invariant with respect to the robots movements toward it. The only known point with such a property is the unique point in the plane that minimizes the sum of the distances between itself and all positions of the robots (actually, such a point may be not unique when the robots are collinear). In fact, this point, known as the Weber point, does not change when moving any of the robots straight toward it [27, 28]. Unfortunately, it has been proven in [3] that the Weber point is not expressible as an algebraic expression involving radicals since its computation requires finding zeros of high-order polynomials even for the case $n = 5$ (see also [9]).

It is worth to note that when the Weber point of a set of robots is restricted to belong to a finite set of discrete points (i.e., the set M), then it fulfills the following properties: (1) it can be easily computed, and (2) it may become not unique even though the robots are not collinear. Such a restriction leads to our concept of *discrete Weber points*. In particular, if $C = (R, M)$ is a configuration without multiplicities, then we define the *Weber-distance of C* as the value

$$wd(C) = \min_{m \in M} \sum_{r \in R} d(r, m).$$

The *Weber-distance* of a point $m \in M$ in C is denoted by $wd(C, m)$ and is defined as $wd(C, m) = \sum_{r \in R} d(r, m)$. Hence, a point $m \in M$ is called *discrete Weber point* of C if $wd(C, m)$ is minimum, that is $wd(C, m) = wd(C)$. We denote by $WP(C)$ the set containing all the discrete Weber points of C.

In the following, we provide some useful properties about the discrete Weber points in $WP(C)$. We use the simple sentence "robot r moves toward a meeting-point m" to mean that r *performs a straight move toward m and the final position of r lies on the interval $(r, m]$*. We start by observing that it is easy to verify (see also [8]) the following result.

Lemma 3 *Let $C = (R, M)$ be a configuration, $m \in WP(C)$, and $r \in R$. If $C' = (R', M)$ represents the configuration obtained after r moved toward m, then m is in $WP(C')$.*

A consequence, the above lemma implies that after the movement of r toward a discrete Weber point m, the set of discrete Weber points is restricted to the meeting-points lying on the half-line $hline(r, m)$.

Lemma 4 *Let $C = (R, M)$ be a configuration, $m \in WP(C)$, and $r \in R$. If $C' = (R', M)$ represents the configuration obtained after r moved toward m, then all the discrete Weber points in $WP(C')$ lie on $hline(r, m)$.*

We are now interested in estimating how many points in $WP(C)$ are still discrete Weber points after the move of r toward m. To this end, we pose the following general question: "How many discrete Weber points in $WP(C)$ lie on a given line in the plane?"

It is well known that the ellipse is the plane curve consisting of all points p whose sum of distances from two given points p_1 and p_2 (i.e., the *foci*) is a fixed number d. Generalizing, a k-ellipse is the plane curve consisting of all points p whose sum of distances from k given points $p_1, p_2, \ldots, p_k$ is a fixed number (cf. Fig. 2). In [26], it is shown that a k-ellipse is a strictly-convex curve, provided the foci p_i are not collinear. This implies that a line intersects a k-ellipse in at most two points. Now, if we apply the notion of k-ellipse to the GMP problem, we easily get that

$$\left\{ p : \sum_{r \in R} d(p, r) = d \right\} \tag{1}$$

Fig. 2 A 3-ellipse and two 4-ellipse each with its foci. In the first 4-ellipse has a focus on it

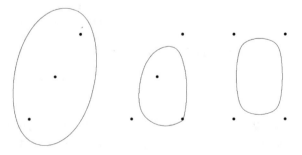

is a $|R|$-ellipse consisting of all points p whose sum of distances from all robots is a fixed number d. If we set $d = wd(C)$, then Eq. (1) represents the $|R|$-ellipse containing all the discrete Weber points in $WP(C)$. In the following, such a curve will be denoted by $\mathscr{E}(C)$. The next results characterize $\mathscr{E}(C)$ and, in turn, the set of all discrete Weber points after a robot moved toward one of such points.

If $C = (R, M)$ is a configuration in which points in R are collinear, then the *median segment* of R, denoted by $med(R)$, is:

- the segment $[r_1, r_2]$, when $|R|$ is even and r_1 and r_2 are the median points of R;
- the single point r, when $|R|$ is odd and r is the median point of R.

Lemma 5 *Let $C = (R, M)$ be a configuration.*

- *If points in R are not collinear, then $\mathscr{E}(C)$ is either a single point or a strictly-convex curve with non-empty interior;*
- *If points in R are collinear, then $\mathscr{E}(C)$ is either $med(R)$ or a strictly-convex curve with non-empty interior.*

By using Lemmata 4 and 5, we get the next result that characterizes the discrete Weber points remaining in a configuration after a robot r moved toward a discrete Weber point m. In particular, in case of non-collinear robots, it states that after the move at most two discrete Weber points remain: one is m and the other, if any, lies on $hline(r, m)$; moreover, in any case, the remaining discrete Weber points form a subset of the initial set of discrete Weber points.

Lemma 6 *Let $C = (R, M)$ be a configuration. Assume that a robot $r \in R$ moves toward a point $m \in WP(C)$ and this move creates a configuration $C' = (R', M)$. Then:*

(a) *if $\mathscr{E}(C)$ is a strictly-convex curve with non-empty interior, then $WP(C')$ contains one or two discrete Weber points only. In one case $WP(C') = \{m\}$, while in the other $WP(C') = \{m, m'\}$, with $m' \in WP(C)$ and m' lying on $hline(r, m)$;*

(b) *if $\mathscr{E}(C)$ is a single point m (i.e, $WP(C) = \{m\}$), then $WP(C')$ contains m only;*

(c) *if $\mathscr{E}(C)$ is $med(R)$ (i.e., $WP(C) = med(R) \cap M$), then $WP(C') = med(R') \cap M$, with $med(R') \subseteq med(R)$.*

The next lemma characterizes the discrete Weber points in case of a particular rotation.

Lemma 7 *Let $C = (R, M)$ be an initial configuration that admits a rotation with center $c \in M$. Then $c \in WP(C)$, and if all robots are not collinear then $WP(C) = \{c\}$.*

For the sake of readability, in the rest of the paper, each time we use the term Weber points, we will always refer to discrete Weber points, that is points in the set $WP(C)$, with $C = (R, M)$ being the current configuration.

4.5   The Algorithm

Let $C = (R, M)$ be an initial configuration for the problem GMP$^+$. According to the measure $mis(C, \mathscr{P}) = \sum_{P \in \mathscr{P}} length(P)$ used in the definitions of GMP$^+$, we observe that the *minsum-distance* of C, defined as the value

$$\Delta^+(C) = \min_{m \in M} \sum_{r \in R} d(r, m),$$

represents a lower bound for $mis(C, \mathscr{P})$ for any polycurve $\mathscr{P}$ representing a solution for C. Since the minsum-distance of C is nothing else that the Weber-distance, then it follows that any optimal gathering algorithm for GMP$^+$ has to select a Weber point as gathering point and make all robots move toward it along shortest paths. As a consequence, Corollaries 2 and 3 imply that in some cases there are no optimal gathering algorithms for GMP$^+$, as stated in the following result.

Corollary 4 *Let $C = (R, M)$ be an initial configuration for* GMP$^+$. *There exists no optimal gathering algorithm for* GMP$^+$ *with respect to C if one of the following holds:*

- *C admits a rotation with center c and $c \notin R \cup WP(C)$;*
- *C admits a reflection with axis ℓ and $\ell \cap (R \cup WP(C)) = \emptyset$.*

To the aim of defining an optimal gathering algorithm for the GMP$^+$ problem, we start by providing a partition of the set $\mathscr{I}$. According to Corollary 1 there are configurations in $\mathscr{I}$ that are ungatherable. The class of such configurations is denoted by $\mathscr{U}$ and contains any C fulfilling one of the following conditions:

- C admits a rotation with center c, and there are neither robots nor meeting-points on c;
- C admits a reflection on axis ℓ, and there are neither robots nor meeting-points on ℓ.

According to Corollary 4, there are configurations in $\mathscr{I} \setminus \mathscr{U}$ that cannot be gathered by any optimal gathering algorithms for GMP$^+$. The class of such configurations is denoted by $\mathscr{N}^+$ and contains any C such that:

- C admits a reflection on axis ℓ, and there are meeting-points on ℓ but neither robots nor Weber points on ℓ.

Notice that Corollary 4 also concerns configurations admitting rotations. Actually, Lemma 7 ensures that such configurations are in $\mathscr{U}$. In particular, the lemma states that any configuration admitting a rotation with center $c \in M$ is such that $c \in WP(C)$.

In the rest of this section we provide an optimal gathering algorithm for the GMP$^+$ problem when the input is restricted to $\mathscr{I} \setminus (\mathscr{U} \cup \mathscr{N}^+)$. We assume $|R| > 1$ and $|M| > 1$: in fact, if $|R| = 1$ it is sufficient that the only robot reaches a Weber point and if $|M| = 1$ all the robots can move toward the only meeting-point.

All the initial configurations processed by the algorithm, along with configurations created during the execution, are partitioned in the following classes:

- $\mathcal{S}_1^+$: any configuration C with one multiplicity;
- $\mathcal{S}_2^+$: any $C = (R, M)$ with $|WP(C)| = 1$, and $C \notin \mathcal{S}_1^+$;
- $\mathcal{S}_3^+$: any $C = (R, M)$ with $cg(M) \in WP(C)$, and $C \notin \bigcup_{i=1}^2 \mathcal{S}_i^+$;
- $\mathcal{S}_4^+$: any C admitting a rotation, and $C \notin \bigcup_{i=1}^3 \mathcal{S}_i^+$;
- $\mathcal{S}_5^+$: any $C = (R, M)$ with all points in R and all points in $WP(C)$ lying on a line ℓ, and $C \notin \bigcup_{i=1}^4 \mathcal{S}_i^+$;
- $\mathcal{S}_6^+$: any C admitting a reflection with at least one robot and one Weber point on the axis, and $C \notin \bigcup_{i=1}^5 \mathcal{S}_i^+$;
- $\mathcal{S}_7^+$: any C admitting a reflection with at least one robot on the axis, and $C \notin \bigcup_{i=1}^6 \mathcal{S}_i^+$;
- $\mathcal{S}_8^+$: any C admitting a reflection with at least one Weber point on the axis, and $C \notin \bigcup_{i=1}^7 \mathcal{S}_i^+$;
- $\mathcal{S}_9^+$: any asymmetric configuration C, and $C \notin \bigcup_{i=1}^5 \mathcal{S}_i^+$;
- $\mathcal{S}_0^+$: $\mathcal{S}_1^+ \cup \mathcal{S}_2^+$ (class defined for the sake of convenience only).

Note that $\mathcal{S}_1^+$ is the only class not containing initial configurations. Moreover, according to the definition of the above classes, it easily follows that the set $\{\mathcal{U}, \mathcal{N}^+, \mathcal{S}^+, \mathcal{S}^+, \ldots, \mathcal{S}^+\}$ is a partition of the configurations without multiplicities, and then it induces a partition of the set $\mathcal{I}$.

The main strategy of the algorithm is to select and move robots straight toward a Weber point m so that, after a certain number of moves, m remains the only Weber point (hence reaching a configuration in class $\mathcal{S}_2^+$). Once only one Weber point exists, all robots move toward it. According to the multiplicity detection, once a multiplicity is created, robots are no longer able to compute the Weber points accurately. Hence, our strategy ensures to create the first multiplicity over m, and once this happens all robots move toward it without creating other multiplicities. Note that, in the initial configuration, it is possible that there is already a robot on m. Hence, it is possible to create a configuration in class $\mathcal{S}_1^+$ without creating a configuration in class $\mathcal{S}_2^+$.

The general algorithm is shown in Fig. 3. It is divided into various sub-procedures, that will be defined later in this section, each of them designed to process configurations belonging to a given class $\mathcal{S}_i^+, i \geq 1$. Priorities among procedures are implicitly defined by the subscripts in the name of the classes. Moves are always computed without creating undesired multiplicities, as described by the following remark.

Remark 1 In Procedure COMPUTE$^+$, and in any procedure called there-in, robots move without creating undesired multiplicities. In fact, robots move straight toward a Weber point, then two robots meet only at the final destination point, unless they move along the same direction. In such a case, robots move without overtaking each other. In particular, if a robot r is moving toward a point p and there is another robot r' in the open segment (r, p), then r moves toward a point p' on (r, p) such that $d(r, p') = \frac{d(r,r')}{2}$.

In this way, undesired multiplicities are never created. Once a multiplicity is created on a meeting-point m, it is then easy to move all other robots toward it, by exploiting the multiplicity detection. Hence the gathering is easily finalized.

Given a configuration $C = (R, M)$, all procedures are invoked after having computed the class $\mathscr{S}_i^+$ which C belongs to. For this task any robot can exploit the multiplicity detection capability (for class $\mathscr{S}_1^+$), the computation of $WP(C)$ and $cg(M)$ (for classes $\mathscr{S}_2^+$ and $\mathscr{S}_3^+$), whether its view contains all robots and Weber points associated with a same angle (for class $\mathscr{S}_5^+$), and Lemma 1 (for all the remaining classes).

Classes $\mathscr{S}_1^+$, $\mathscr{S}_2^+$, and $\mathscr{S}_3^+$. In classes $\mathscr{S}_1^+$, $\mathscr{S}_2^+$, and $\mathscr{S}_3^+$, robots can move concurrently toward the unique multiplicity, the only Weber point, or $cg(M)$, respectively. Class $\mathscr{S}_1^+$ is the only one not containing initial configurations (they are created by our strategy). According to the multiplicity detection, once a multiplicity is created, robots are no longer able to compute the Weber points accurately. Hence, the strategy ensures to create the first multiplicity over a Weber point, and once this happens all robots move toward it without creating other multiplicities.

Class $\mathscr{S}_4^+$. In class $\mathscr{S}_4^+$ the algorithm handles configurations that admit a rotation (these include also configurations where all robots are collinear). In such a cases, Procedure COMPUTE$^+$ leads to a configuration C' in class $\mathscr{S}_0^+$ with a single Weber point and with at most one robot moving toward it.

In particular, the algorithm makes the robot r on the center to move toward an arbitrary point $m \in WP(C)$ (see Line 5 in Fig. 3 of Procedure COMPUTE$^+$). By Lemma 5, r is inside $\mathscr{E}(C)$ and hence $hline(r, m)$ intersects $\mathscr{E}(C)$ at m only. Then, by Lemma 6(a), once the robot has moved, only one Weber point remains. It follows that a configuration $C' \in \mathscr{S}_0^+$ is created.

Class $\mathscr{S}_5^+$. In class $\mathscr{S}_5^+$ the algorithm considers any configuration C where all robots and all Weber points lie on a line. After the Look phase, a robot can detect whether the current configuration admits such a property by checking if there exists a robot whose view contains all robots and all Weber points associated with a same angle. In such a cases, Procedure COMPUTE$^+$ calls the subroutine LINE (Fig. 4), which leads to

---

Procedure: COMPUTE$^+$
Input: Configuration $C = (R, M)$

1 Compute $cg(M)$, C, $WP(C)$;
2 **if** $C \in \mathscr{S}_1^+$ **then** move toward the unique multiplicity ;
3 **if** $C \in \mathscr{S}_2^+$ **then** move toward $WP(C)$;
4 **if** $C \in \mathscr{S}_3^+$ **then** move toward $cg(M)$;
5 **if** $C \in \mathscr{S}_4^+ \wedge r$ *lies in the center of the rotation* **then** move toward any $m \in WP(C)$;
6 **if** $C \in \mathscr{S}_5^+$ **then** LINE(C) ;
7 **if** $C \in \mathscr{S}_6^+$ **then** REFLECTION$_{RW}$(C);
8 **if** $C \in \mathscr{S}_7^+$ **then** REFLECTION$_R$(C);
9 **if** $C \in \mathscr{S}_8^+$ **then** REFLECTION$_W$(C);
10 **if** $C \in \mathscr{S}_9^+$ **then** ASYMMETRIC(C);

Fig. 3 Procedure COMPUTE$^+$ executed by any robot r during the Compute phase. In each procedure, moves are performed accordingly to Remark 1

Procedure: LINE
Input: Configuration $C = (R,M) \in \mathscr{S}_5^+$ with $R \bigcup WP(C)$ lying on a line ℓ

1 **if** *M admits a reflection with axis ℓ' perpendicular to ℓ and the intersection is $w \in WP(C)$*
   then
2 $\quad$ move toward w
3 **if** *C admits a reflection with the axis perpendicular to ℓ passing through a robot r', or C*
   admits a rotation with the center lying on ℓ occupied by a robot r' **then**
4 $\quad$ **if** $r = r'$ **then**
5 $\quad\quad$ move toward a Weber point in any direction on ℓ

6 **else**
7 $\quad$ **if** *there are robots in between $WP(C)$ on ℓ* **then**
8 $\quad\quad$ **if** *r has minimum view among robots in between $WP(C)$* **then**
9 $\quad\quad\quad$ move toward a Weber point in any direction on ℓ

10 $\quad$ **else**
11 $\quad\quad$ Let r_1 and r_2 be the robots such that $[r_1, r_2]$ is the smallest segment containing all
   the points in $WP(C)$;
12 $\quad\quad$ Let $m_1, m_2 \in WP(C)$ be the closest meeting-points to r_1 and r_2, respectively;
13 $\quad\quad$ Let $k_1 = |(M \setminus (WP(C))) \cap (\ell \setminus hline(r_1, r_2))|$;
14 $\quad\quad$ Let $k_2 = |(M \setminus (WP(C))) \cap (\ell \setminus hline(r_2, r_1))|$;
15 $\quad\quad$ **if** $r = r_1 \wedge ((k_1 > k_2) \vee (k_1 = k_2 \wedge d(r, m_1) < d(r_2, m_2)) \vee$
   $(k_1 = k_2 \wedge d(r, m_1) = d(r_2, m_2) \wedge min\_view(r, \{r, r_2\})))$ **then**
16 $\quad\quad\quad$ move toward m_2

Fig. 4 Procedure LINE for moving robot r in case of configurations in class $\mathscr{S}_5^+$

a configuration C' in class $\mathscr{S}_0^+$, with robots moving toward the Weber point or the multiplicity, eventually.

Class $\mathscr{S}_6^+$. In class $\mathscr{S}_6^+$ the algorithm considers any configuration C admitting reflections with robots and Weber points on the axis. In such a cases, Procedure COMPUTE$^+$ calls the subroutine REFLECTION$_{RW}$ (Fig. 5), which leads to a configuration C' in class $\mathscr{S}_0^+$, possibly with one robot moving toward the Weber point, eventually.

Class $\mathscr{S}_7^+$. In class $\mathscr{S}_7^+$ the algorithm considers any configuration C admitting reflections with robots but no Weber points on the axis. In such a cases, Procedure COMPUTE$^+$ calls the subroutine REFLECTION$_R$ (Fig. 6), which leads to a configuration C' either in class $\mathscr{S}_0^+$ or in class $\mathscr{S}_9^+$, possibly with one robot moving toward a Weber point.

Class $\mathscr{S}_8^+$. In class $\mathscr{S}_8^+$ the algorithm considers any configuration C admitting reflections with Weber points but no robots on the axis. In such a cases, Procedure COMPUTE$^+$ calls the subroutine REFLECTION$_W$ (Fig. 7), which leads to a configuration C' in class $\mathscr{S}_0^+$, possibly with robots moving toward the Weber point, or in class $\mathscr{S}_9^+$, possibly with one moving robot and one pending robot both with the same target Weber point.

Procedure: REFLECTION$_{RW}$
Input: Configuration $C = (R, M)$ in class $\mathscr{S}_6^+$ with reflection axis ℓ

1 **if** $WP(C) \cap \ell = \{m\}$ **then**
2   **if** $min\_view(r, R \cap \ell)$ **then**
3   move toward m

4 **else**
5   Let $m_1, m_2 \in M$ such that $WP(C) \cap \ell = \{m_1, m_2\}$;
6   **if** $r \in (m_1, m_2) \wedge min\_view(r, R \cap (m_1, m_2))$ **then**
7   move toward m_1 ;   /* or m_2, equivalently */
8   **else**
9   **if** $R \cap (m_1, m_2) = \emptyset$ **then**
10              $d_1 = \min\{d(m_1, r), d(m_2, r)\}$;
11              $d_2 = \min_{r_j \in R \setminus \{r\} \cap \ell} \{d(m_i, r_j) : m_i \in \{m_1, m_2\}\}$;
12              $r' = \operatorname{argmin}_{r_j \in R \setminus \{r\} \cap \ell} \{d(m_i, r_j) : m_i \in \{m_1, m_2\}\}$;
13   **if** $d_1 < d_2 \vee (d_1 = d_2 \wedge \mathscr{V}_r(r) < \mathscr{V}_r(r'))$ **then**
14   move toward the farthest Weber point among m_1 and m_2;

Fig. 5 Procedure REFLECTION$_{RW}$ for moving robot r in case of configurations in class $\mathscr{S}_6^+$

Procedure: REFLECTION$_R$
Input: Configuration $C = (R, M)$ in class $\mathscr{S}_7^+$ with reflection axis ℓ

1 $R' = \{p \in R \cap \ell \mid \exists\, m \in WP(C) : hline(p, m) \cap WP(C) = \{m\}\}$;
2 **if** $|R'| > 0$ **then**
3   **if** $min\_view(r, R')$ **then**
4   move toward any $m \in WP(C)$ s.t. $hline(r, m) \cap WP(C) = \{m\}$

5 **else**
6   Let $p \in R \cap \ell$;
7   let $m_p, m_p' \in WP(C)$ s.t. p, m_p and m_p' lie (in order) on a line ℓ' and the angle between ℓ and ℓ' is minimum;
8   let N_p be the number of robots between p and m_p on $hline(p, m_p)$;
9   **if** $\exists p'$ *being the* $(N_p + 1)$-*th robot, outside* $\mathscr{E}(C)$, *on* $hline(m_p, m_p') \wedge$ $d(p, m_p) > d(p', m_p') \wedge (R \setminus \{p, p'\}, M)$ *admits a reflection* **then**
10   **if** $r = p$ **then** move toward $x \in hline(r, m_p)$ s.t. $d(r, x) = \frac{d(r, m_p) - d(p', m_p')}{2}$;
11   **else**
12   **if** $r = p$ **then** move toward m_p ;

Fig. 6 Procedure REFLECTION$_R$ for moving robot r in case of configurations in class $\mathscr{S}_7^+$

Class $\mathscr{S}_9^+$ In class $\mathscr{S}_9^+$. the algorithm considers any asymmetric configuration C. In such a cases, Procedure COMPUTE$^+$ calls the subroutine ASYMMETRIC (Fig. 8). If C is generated from a configuration in $\mathscr{S}_7^+$ or $\mathscr{S}_8^+$ with possibly a pending robot with the same target point, then ASYMMETRIC leads to a configuration in class $\mathscr{S}_0^+$, possibly with robots moving toward the Weber point, eventually.

Procedure: REFLECTION$_W$
Input: Configuration $C = (R, M)$ in class $\mathscr{S}_8^+$ with reflection axis ℓ

1 Let $m \in WP(C)$ s.t. $min\_view(m, WP(C) \cap \ell)$ is true;
2 $p = start(m)$;
3 **if** $\exists q \in R, \exists m' \in WP(C)$: $m' \in hline(p, m) \wedge m' \notin [p, m] \wedge q \in hline(m, m') \wedge q \notin [m, m'] \wedge$
   $d(q, m') < d(p, m)$ **then**
4 $\quad \lfloor$ **if** $r = q$ **then** move toward m ;

5 **else**
6 $\quad \lfloor$ **if** $r = p$ **then** move toward m ;

Fig. 7 Procedure REFLECTION$_W$ for moving robot r in case of configurations in class $\mathscr{S}_8^+$

Correctness. Figure 9 shows all transitions among classes defined by the algorithm, and, in particular, it shows that from each class $\mathscr{S}_i^+$, $i \geq 3$, a configuration in class $\mathscr{S}_1^+$ or $\mathscr{S}_2^+$ (i.e., in class $\mathscr{S}_0^+$) is reached. The next theorem states the correctness of the algorithm.

Theorem 2 *(Optimal gathering on fixed points on the plane) Procedure* COMPUTE$^+$ *is an optimal gathering algorithm that solves the* GMP$^+$ *problem for an initial configuration C if and only if $C \in \mathscr{I} \setminus (\mathscr{U} \cup \mathscr{N}^+)$.*

5 Optimal Gathering for GG$^+$

In this section an optimal gathering algorithm for GG$^+$ is described (cf. [15, 16]). Before presenting the algorithm, we reconsider the notions of configuration automorphisms and symmetries to be applied to general graphs, and accordingly we provide general impossibility results. We also reconsider the notion of Weber points for graphs and show useful properties that are successively exploited by our gathering algorithms. In particular, we provide optimal gathering algorithm for GG$^+$ when the graph topologies correspond to trees, rings and infinite grids.

5.1 Configuration Automorphisms, Symmetries and Ungatherabilty Results

Two graphs $G = (V_G, E_G)$ and $H = (V_H, E_H)$ are *isomorphic* if there is a bijection φ from V_G to V_H such that $uv \in E_G$ if and only if $\varphi(u)\varphi(v) \in E_H$. An *automorphism* on a graph G is an isomorphism from G to itself, that is a permutation of the vertices of G that maps edges to edges and non-edges to non-edges. The set of all automorphisms of G forms a group called *automorphism group* of G and denoted by Aut(G).

The concept of isomorphism can be extended to configurations in a natural way: two configurations (G, μ) and (G', μ') are isomorphic if G and G' are isomorphic via a bijection φ and for each vertex v in G, $\mu(v) = \mu'(\varphi(v))$. An *automorphism*

Procedure: ASYMMETRIC
Input: Configuration $C = (R, M)$ in class $\mathscr{S}_9^+$

1 FROM_ASYM = true;
2 **if** $WP(C) = \{m_1, m_2\} \wedge |R|$ *is even* **then**
3     $\quad$ $\ell = line(m_1, m_2)$;
4     $\quad$ $L = \ell \setminus (m_1, m_2)$;
5     $\quad$ Let $p \in R \cap L$ and $m_1 \in WP(C)$, such that $d(p, m_1)$ is minimum;
6     $\quad$ **if** *p exists and it is unique* **then**
7        $\quad\quad$ Let $m_2 \neq m_1$ be the second point in $WP(C)$;
8        $\quad\quad$ $\ell_1 = line(cg(M), m_1)$;
9        $\quad\quad$ $\ell_2 = line(cg(M), m_2)$;
10       $\quad\quad$ $\ell' = $ line specular to ℓ wrt ℓ_1;
11       $\quad\quad$ $\ell'' = $ line specular to ℓ wrt ℓ_2;
12       $\quad\quad$ Let $p'' \in R \cap \ell''$ on the half-line specular to $hline(m_2, p)$ such that $d(p'', m_2) > d(p, m_2)$ is minimum;
13       $\quad\quad$ **if** *p'' exists and $(R \setminus \{p, p''\}, M)$ is a reflection* **then**
14          $\quad\quad\quad$ FROM_ASYM = false;
15          $\quad\quad\quad$ **if** $r = p''$ **then**
16             $\quad\quad\quad\quad$ move toward m_1

17       $\quad\quad$ **else**
18          $\quad\quad\quad$ Let $p' \in R \cap \ell'$ on the half-line specular to $hline(m_1, p)$ such that $d(p', m_1)$ is minimum;
19          $\quad\quad\quad$ **if** *p' exists and $(R \setminus \{p, p'\}, M)$ is a reflection* **then**
20             $\quad\quad\quad\quad$ FROM_ASYM = false;
21             $\quad\quad\quad\quad$ **if** $r = p'$ **then**
22                $\quad\quad\quad\quad\quad$ move toward m_1

```
/* basic strategy - case (a)                                              */
```
23 **if** FROM_ASYM $\wedge$ $WP(C) = \{m_1, m_2\}$, *with* $\mathscr{V}_r(m_1) < \mathscr{V}_r(m_2)$ **then**
24    $\quad$ **if** *r is the first robot occurring in $\mathscr{V}_r(m_1)$ not collinear with both m_1 and m_2* **then**
25       $\quad\quad$ move toward m_1

```
/* basic strategy - case (b)                                              */
```
26 **if** FROM_ASYM $\wedge$ $|WP(C)| > 2$ **then**
27    $\quad$ Let $m \in WP(C)$ s.t. $\forall m' \in WP(C) \setminus \{m\}$, $d(start(m), m) < d(start(m'), m') \vee (d(start(m), m) = d(start(m'), m') \wedge \mathscr{V}_r(m) < \mathscr{V}_r(m'))$;
28    $\quad$ **if** $r = start(m)$ **then**
29       $\quad\quad$ Let p be the point on the segment (r, m) at distance $\frac{d(r,m)}{2}$ from r;
30       $\quad\quad$ move toward p

Fig. 8 Procedure ASYMMETRIC for moving robot r in case of configurations in class $\mathscr{S}_9^+$

on a configuration (G, μ) is an isomorphism from (G, μ) to itself and the set of all automorphisms of (G, μ) forms a group that we call *automorphism group* of (G, μ), denoted by $\text{Aut}((G, \mu))$.

Given an isomorphism $\varphi \in \text{Aut}((G, \mu))$, the *cyclic subgroup* of order p generated by φ is given by $\{\varphi^0, \varphi^1 = \varphi, \varphi^2 = \varphi \circ \varphi, \ldots, \varphi^{p-1}\}$ where φ^0 is the identity. If H is a subgroup of $\text{Aut}((G, \mu))$, the *orbit* of a vertex v of G is $Hv = \{\gamma(v) \mid \gamma \in H\}$.

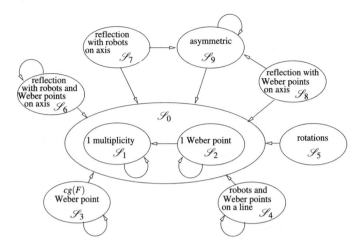

Fig. 9 Schematization of the optimal gathering algorithm for GMP$^+$ along with priorities

If $|\text{Aut}(G)| = 1$, that is, G admits only the identity automorphism, then G is said *asymmetric*, otherwise it is said *symmetric*. Analogously, if $|\text{Aut}((G, \mu))| = 1$, we say that (G, μ) is *asymmetric*, otherwise it is *symmetric*.

The next theorem provides a sufficient condition for a configuration to be not gatherable, but we first need the following definition:

Definition 2 Let $C = ((V, E), \mu)$ be a configuration. An isomorphism $\varphi \in \text{Aut}(C)$ is called *partitive* on $V' \subseteq V$ if the cyclic subgroup $H = \{\varphi^0, \varphi^1 = \varphi, \varphi^2 = \varphi \circ \varphi, \ldots, \varphi^{p-1}\}$ generated by φ has order $p > 1$ and is such that $|Hu| = p$ for each $u \in V'$.

Note that, in the above definition, the orbits Hu, for each $u \in V'$ form a partition of V'. The associated equivalence relation is defined by saying that x and y are equivalent if and only if there exists a $\gamma \in H$ with $\gamma(x) = y$. The orbits are then the equivalence classes under this relation; two elements x and y are equivalent if and only if their orbits are the same; i.e., $Hx = Hy$. Moreover, note that $\mu(u) = \mu(v)$ whenever u and v are equivalent.

Theorem 3 *Let $C = ((V, E), \mu)$ be a non-final configuration. If there exists $\varphi \in \text{Aut}(C)$ partitive on V then C is not gatherable.*

In Fig. 10a, it is shown a partitive configuration where each vertex belongs to an orbit of size three. By the above theorem we deduce that the gathering cannot be assured, since each move allowed by an algorithm can be executed synchronously by all the three robots due to an adversary. This would always produce a new partitive configuration.

Figure 10b, shows a configuration admitting an isomorphism which is not partitive. In this case the gathering is possible even though not the optimal one. In fact, each

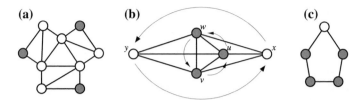

Fig. 10 A gray vertex indicates the presence of one robot. **a** Configuration admitting a partitive isomorphism: the sets of the partition are the three central vertices, the vertices with robots, and the three remaining vertices. **b** Configuration admitting a non-partitive isomorphism that maps v in u, u in w, w in v, x in y and y in x. **c** Configuration admitting a non-partitive isomorphism with two sets of the partition of size two, and one of size one

of the three occupied vertices are Weber points, but moving from one to another may produce the same configuration if the three robots move concurrently in the same direction. Hence, a gathering algorithm can move the three robots towards the two empty vertices. Once all the three robots have moved, a multiplicity is created. The multiplicity either contains all the robots or just two. In the first case the gathering has been accomplished. In the second case, the gathering is finalized by letting the single robot move towards the multiplicity.

Finally, Fig. 10c shows a configuration admitting a non-partitive isomorphism but the gathering cannot be assured as shown in [21]. It follows that, there exist configurations not admitting partitive isomorphisms but still not gatherable.

It is worth noting how most of the configurations proved to be not gatherable for the rings [22], trees [11], and grids [10] fall into the hypothesis of Theorem 3. Considering the ring case [22], for instance, *periodic* configurations (i.e., invariant with respect to not full rotations), or configurations admitting an *edge-edge symmetry* (i.e., invariant to reflection on an even ring) are not gatherable.

The next theorem suggests the gathering point in some circumstances.

Theorem 4 *Given a configuration $C = ((V, E), \mu)$, and a subset of nodes $V' \subset V$, if there exists an automorphism $\varphi \in Aut(C)$ that is partitive on $V \setminus \{V'\}$, with $l(v) = 0$ for any $v \in V'$, then, any gathering algorithm can not assure the gathering on a vertex in $V \setminus V'$.*

The above theorem implies that some configurations can be gathered only at some predetermined vertices, regardless of whether they are Weber points or not. Hence, in such cases the optimality of the provided solutions cannot be measured with respect to the minimum distances of the robots towards Weber points.

5.2 Weber Points for GG

Definition 3 Given a configuration (G, μ), with $G = (V, E)$, the *centrality* of each $v \in V$, is $c_{G,\mu}(v) = \sum_{u \in V} d(u, v) \cdot \mu(u)$. A vertex $v \in V$ is a *Weber point* if it has the minimal centrality, that is, $c_{G,\mu}(v) = \min\{c_{G,\mu}(u) \mid u \in V\}$.

Whenever clear by the context, we refer to the centrality of a vertex v by $c_G(v)$, $c_\mu(v)$, or simply $c(v)$. By definition, a Weber point is a vertex that has the overall minimal distance from all the robots in the configuration. Then, an algorithm that gathers all the robots on a Weber point via shortest paths is optimum with respect to the total number of moves. More formally, a gathering algorithm must define the sequence of moves for each robot, leading to a final configuration. A move is the change of the position of a single robot from a vertex u to an adjacent vertex v. This equals to change the configuration from, say (G, μ) to (G, μ'), where $\mu'(w) = \mu'(w) \; \forall w \in V \backslash \{u, v\}$, $\mu'(u) = \mu(u) - 1$ and $\mu'(v) = \mu(v) + 1$. A robot perceives its position on the graph G if (G, μ) is asymmetric. Whereas, if (G, μ) admits a non-identity isomorphism φ, a robot cannot distinguish its position at u from $\varphi(u)$. As a consequence, two robots (e.g., one on u and one on $\varphi(u)$) can decide to move simultaneously, as any algorithm is unable to distinguish between them. This fact greatly increases the difficulty to devise a gathering algorithm for symmetric configurations.

We say that an algorithm *assures* the gathering if it achieves the gathering regardless any possible sequence of the moves it allows, and possible simultaneous moves.

In the remainder of this section, we provide general results that allow to define optimal gathering algorithms. We start by observing that the GG problem can be characterized as follows:

Proposition 1 *Gathering is achieved on a configuration $((V, E), \mu)$ if and only if there exists a vertex $v \in V$ such that $c(v) = 0$.*

Along the text, we say that a robot on a vertex u *moves towards a vertex v* if it moves to a vertex adjacent to u along a shortest path between u and v.

Theorem 5 *Given a configuration $((V, E), \mu)$ with Weber points in $X \subseteq V$, a move of a robot towards a Weber point x gives rise to a configuration $((V, E), \mu')$ with Weber points in $X' \subseteq V$ such that:*

1. *$c_{\mu'}(v) = c_\mu(v) - 1$ for each $v \in X'$;*
2. *$x \in X'$;*
3. *$X' \subseteq X$.*

When the configuration admits a unique Weber point (or a Weber point can be uniquely determined), the above theorem suggests an optimal gathering algorithm that also exploits concurrency among robots. In fact, regardless other robots, each one can move towards the only Weber point via the shortest path, until finalizing the gathering.

Corollary 5 *Let $C = ((V, E), \mu)$ be a configuration. Then:*

- *if C admits only one Weber point then the gathering can be achieved by an optimal algorithm;*
- *if there exists a real function $f : V \longrightarrow \mathbb{R}^+$ such that f admits only one minimum on the set of Weber points, then gathering can be achieved by an optimal algorithm.*

5.3   Optimal Gathering on Trees

In this section, we characterize the gathering on tree topologies. We provide a general algorithm that always achieves the optimal gathering starting from configurations not falling into the hypothesis of Theorem 3. To this aim, we exploit interesting properties resulting from the tree topology.

Let (T, μ) be a configuration for a tree T, and a and b two of its vertices. We denote by P_{ab} the path between a and b of length $|P_{ab}|$. Tree T can be decomposed into three subtrees, see Fig. 11. The one containing a when removing from T the edge incident to a in P_{ab}, and denoted by T_a; The one containing b when removing the edge incident to b in P_{ab}, and denoted by T_b; And the third one obtained from T by removing both T_a and T_b, and denoted by T_{ab}. Let $L_a = \sum_{v \in T_a} \mu(v)$ and $L_b = \sum_{v \in T_b} \mu(v)$, that is the number of robots in T_a and T_b, respectively.

Theorem 6 *Let (T, μ) be a configuration for a tree T. Then, the following properties hold:*

- *given two distinct Weber points a and b, T_{ab} does not contain any robots;*
- *given two distinct Weber points a and b, $L_a = L_b$;*
- *the Weber points form a path;*
- *if the number of robots is odd, then there exists only one Weber point.*

The above properties, imply the existence of a simple optimal gathering algorithm when the number of robots is odd. A complete characterization about the existence of optimal gathering algorithms on trees is given by the next theorem. It shows that an optimal algorithm exists unless there is an automorphism that maps each vertex to a different one. This is a lighter condition with respect to the case given in Theorem 3.

Theorem 7 (Optimal gathering on trees) *Let $C = (T, \mu)$ be a configuration for a tree $T = (V, E)$. There exists an optimal gathering algorithm for C if and only if for each $\varphi \in Aut(C)$ there exists $v \in V$ such that $\varphi(v) = v$.*

The algorithm provided in the proof of Theorem 7 works as follows.
From Theorem 6, let P_{ab} be the path of Weber points.

Fig. 11 Partitioning of a tree into three subtrees

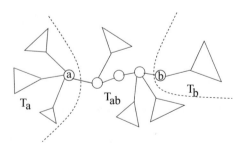

If T has one center x, there must be a vertex v in P_{ab} nearest to x (possibly, x coincides with v). By Corollary 5, it follows that all the robots can move towards v via the shortest paths, and eventually finalizing the optimal gathering.

If T has two centers joined by edge xy, then three cases arise: (1) xy is in T_a (T_b, respectively); (2) xy is in T_{ab} but either x or y is not in P_{ab}; (3) xy is an edge of P_{ab}.

In case (1), by Corollary 5, all the robots can move towards a (b, resp.), the gathering point.

In case (2), there must exist a vertex w in P_{ab} nearest to xy (possibly, w coincides either with x or with y). Again by Corollary 5, all the robots can move towards w and gather there.

In case (3), by the hypothesis of Theorem 7, for each $\varphi \in \text{Aut}(C)$ there exists $v \in V$ such that $\varphi(v) = v$. Then, the two subtrees T_x and T_y obtained by removing xy cannot be isomorphic. In this case, it is always possible to determine which tree between T_x and T_y is less than the other with respect to a natural ordering on labeled trees (see, e.g. [1, 4]), where the label of a vertex is given by function μ, and each tree is represented by the string obtained by reading the labels from the root downwards to the leaves. Without loss of generality, assuming T_x greater than T_y, all the robots in T_x can move towards x. In this way, after each move, T_x remains always greater than T_y. Once all the robots in T_x are at x, they move to y. As soon as one robot moves from x to y, the path of Weber points will be P_{yb}, xy is not in P_{yb} and we can proceed as before: all the robots can move towards y and gather there.

The gathering algorithm provided above exploits similar properties of that presented in [11]. However, this new version accomplishes the optimal gathering while the old one always gather robots on a center of the underlying tree. Considering Fig. 12a, it is easy to provide configurations where this algorithm performs the gathering in two moves, while the old algorithm requires n moves.

Before concluding this section, it is interesting to characterize the disposal of Weber points on the degenerate case of paths. This will be of practical interest in the next section for characterizing Weber points on rings.

Lemma 8 *Given a configuration (P, μ) where P is a path graph, the set of Weber points is constituted either by one occupied vertex, or by one subpath whose extremes are occupied.*

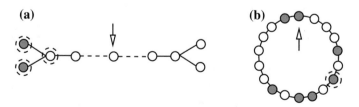

(a)   **(b)**

Fig. 12 A gray vertex indicates the presence of one robot; Dashed circled vertices are Weber points. Dashed line stands for an undefined sequence of empty vertices. Vertices pointed by an arrow represent the gathering vertices with respect to algorithms in [11] for (**a**), and in [22] for (**b**)

5.4   Optimal Gathering on Rings

In this section, we fully characterize optimal gathering on ring topologies. As we are going to show, optimal gathering is achievable when the input configuration is asymmetric, or it admits an axis of symmetry passing through a robot or a Weber point.

From Theorem 3, gathering is unfeasible on periodic configurations or those admitting an edge-edge symmetry. Another impossibility result concerning gathering on rings comes from [22]:

Theorem 8 *If $C = (R, \mu)$ is a configuration on ring R with only two robots, then C is ungatherable.*

Moreover, from Theorem 4, we have that some configurations can be gathered only at some predetermined vertices, regardless of whether they are Weber points or not. Hence, in such cases optimal gathering can be accomplished only if the predetermined vertex is a Weber point. On rings, Theorem 4 applies on configurations admitting a vertex-edge or a vertex-vertex symmetry and any vertex lying on the axis is empty. It follows that from configurations satisfying the hypothesis of Theorem 4, optimal gathering can be accomplished only if there is at least a Weber point on the axis of symmetry.

Before providing a general algorithm for achieving optimal gathering whenever possible, we provide some useful properties concerning the disposal of Weber points on rings.

Theorem 9 *Given a configuration (R, μ) on a ring R, if an empty vertex u is a Weber point then also its neighbors are Weber points.*

By the above theorem, as for the path case, if there exists a sequence of vertices that are Weber points, then the extremes of such a sequence are Weber points occupied by robots. It is worth noting that on rings there might occur more than one of such sequences.

As first result on rings, the next theorem provides an algorithm to assure optimal gathering from asymmetric configurations. Actually, the algorithm defined by the next theorem will be used later as part of the general algorithm for solving optimal gathering from any configuration on rings where it is achievable.

Let $C = (\mu(v_0), \mu(v_1), \ldots, \mu(v_{n-1}))$ be one of the possible views computed by a robot occupying vertex v_0 during its Look phase according its clockwise direction. We denote by $\overline{C} = (\mu(v_0), \mu(v_{n-1}), \mu(v_{n-2}), \ldots, \mu(v_1))$, and by C_i the configuration obtained by reading C starting from v_i, that is

$$C_i = (\mu(v_i), \mu(v_{(i+1) \bmod n}), \ldots, \mu(v_{(i+j) \bmod n})).$$

By referring to Fig. 12b, the robot pointed by the arrow has view $C = (1, 0, 0, 0, 1, 0, 1, 0, 1, 1, 0, 1, 0, 0, 0, 0, 0, 1)$ if it reads in the clockwise direction. Then, $\overline{C} = (1, 1, 0, 0, 0, 0, 0, 1, 0, 1, 1, 0, 1, 0, 1, 0, 0, 0)$. Its lexicographical maximum view is $\overline{C}$, while the absolute maximum view of the configuration is $\overline{C_9} = (1, 1, 0, 1, 0, 1, 0, 0, 0, 1, 1, 0, 0, 0, 0, 0, 1, 0)$.

Theorem 10 *Given an asymmetric configuration (R, μ) without multiplicities on a ring R of n vertices, it is always possible to assure optimal gathering.*

The algorithm provided in the proof of Theorem 10 works as follows.

From Theorem 9, there exists in R at least one Weber point occupied by a robot. Consider two different cases: either there are only isolated Weber points (i.e., no two Weber points are adjacent) or not.

In the first case, as the configuration is asymmetric, among all Weber points which are occupied by robots, there must be one whose view represents the lexicographical maximum among all the views. Let r be the robot occupying such a Weber point, and without loss of generality, let $(\mu(v_0), \mu(v_1), \ldots, \mu(v_{n-1}))$ be its maximum view.

Let r' be a robot on v_i, where $i > 0$ is the smallest index such that $\mu(v_i) = 1$. The gathering algorithm makes r' move towards r. By Theorem 5, v_0 remains a Weber point. Moreover the view of r remains maximum as it has been increased. The obtained configuration is still asymmetric unless only one Weber point is left. In fact, since a possible axis should pass through the unique Weber point with the lexicographical maximum view, it is enough to observe that other Weber points can reside only at one side of such an axis since all those on the side where r' resides have disappeared, if any. This is repeated until a multiplicity is created on v_0.

In the second case, the algorithm considers the robots on Weber points with views such that v_1 or v_{n-1} is a Weber point. If both v_1 and v_{n-1} are Weber points, the robot chooses the maximum view, otherwise it chooses the view in the direction of the adjacent Weber point. Let r be the robot with the maximal view chosen in the described way, and r' the first robot seen by r according to its view. By Theorem 9, r and r' determine a path P of Weber points.

Consider the two views of r and r', respectively, in the opposite direction with respect to P. The algorithm makes move r (r', resp.) towards r' (r resp.) if its view is lexicographically bigger than that of r' (r, resp.). In doing so, the path P is shortened and will be again selected in the subsequent steps as r has increased its maximum view. The obtained configuration is still asymmetric unless a multiplicity is created. In fact, since a possible axis should pass through the unique Weber point detected by the proposed strategy where r resides, it is enough to observe that only one neighbor of r is a Weber point. As above, this is repeated until a multiplicity is created.

In both cases, once a multiplicity M is created, either it coincides with the only Weber point left, or the algorithm makes move towards M the robot closer to M on the side closest to other Weber points. In this way, after this movement, only one Weber point remains and optimal gathering can be finalized by exploiting Corollary 5.

The main difference of the described algorithm with that in [22] is in the choice of the vertex where a multiplicity is created. Once this is done, the two algorithms finalize the gathering on the created multiplicity by moving robots along the shortest paths towards it.

The algorithm proposed in [22] considers the longest interval I of empty vertices. Among the two intervals of empty vertices neighboring to I, the shortest one was reduced by moving the robot delimiting I. Ties were broken by the asymmetry of the configuration. The described move was repeated until creating a multiplicity.

It is also worth noting that once a multiplicity is created, our algorithm requires one further move to obtain a single Weber point, and from there all robots can move concurrently towards the gathering node, while in [22] at most two robots at the two sides of the multiplicity can move concurrently.

Figure 12b shows a configuration where our algorithm requires 25 moves while the algorithm in [22] takes 35 moves. It is easy to provide worsen instances where I is far apart from Weber points, hence resulting in a much larger difference with our algorithm in terms of computed moves.

It is worth noting that the gathering vertices selected by those algorithms turn out to be the right choice according to Theorem 4. However, such algorithms are not optimal as the performed moves are not always along the shortest paths towards the gathering vertex. For a correct comparison, we remark that those papers deal with the global weak multiplicity detection while here we are assuming the global strong version.

In what follows, we provide a gathering algorithm for all symmetric configurations that admit optimal gathering. This makes use of the algorithm designed for asymmetric configurations defined in the proof of Theorem 10, hence providing a full characterization of optimal gathering on rings.

Some further definitions and useful properties about Weber points on rings are still required.

Definition 4 Given a configuration (R, μ) on a ring R of n vertices, two vertices are said *antipodal* if their distance is $\lfloor \frac{n}{2} \rfloor$. Two robots are said *antipodal* if they lie on two antipodal vertices.

Lemma 9 *Given a configuration $C = (R, \mu)$ on a ring R with an even number of vertices, if vertex u is a Weber point then it is a Weber point also in the configuration C' obtained from C by removing all the antipodal robots.*

Corollary 6 *Given a configuration $C = (R, \mu)$ on a ring R with an even number of vertices, if it contains only pairs of antipodal robots then C is periodic.*

It is worth noting that when a configuration satisfies the hypothesis of Corollary 6 then by Lemma 9 all its vertices are Weber points since all the vertices of an empty ring have centrality nil.

Theorem 11 *Given an aperiodic configuration $C = (R, \mu)$ on a ring R with an even number of vertices, if vertex v is a Weber point in C then its antipodal vertex is not a Weber point.*

Theorem 12 *Given a configuration C on a ring R with an odd number of vertices, if two adjacent vertices u and v are two Weber points, and at most one of them is occupied, then vertex w whose antipodal vertices are u and v is not a Weber point.*

We are now ready to define another part of the main algorithm to deal with symmetric configurations with an odd number of robots.

Theorem 13 *Given a symmetric configuration $C = (R, \mu)$ on a ring R with an odd number of robots, then optimal gathering can be accomplished.*

The algorithm provided in the proof of Theorem 13 works as follows.

If C is symmetric, there must be exactly one robot lying on the axis of symmetry since the number of robots is odd. If there is only one Weber point, then optimal gathering is achieved by exploiting Corollary 5. If there are at least two Weber points, the algorithm then moves the robot on the axis towards one of the two possible directions, indiscriminately. By Theorem 5, all the Weber points contained in the semi-ring where the robot moved are maintained, while all the Weber points in the other semi-ring (that was originally symmetric) disappear. The only exception might be the antipodal vertex (if any) with respect to the original location of the moved robot when the ring is composed of an even number of vertices. Again, if after the move there is only one Weber point, then optimal gathering is accomplished by exploiting Corollary 5. If there are at least two Weber points, the obtained configuration can be symmetric or asymmetric (possibly containing a multiplicity). In the former case, the algorithm moves again the new robot on the axis, and by [22] we are assured that this can happen a finite number of times until reaching an asymmetric configuration. From asymmetric configurations, Theorem 9 can be exploited to finalize the gathering.

For the case of even number of robots on symmetric configurations, we need two more definitions.

Definition 5 Given a symmetric configuration on a ring R of n vertices, a vertex v is called *north* if it lies on the axis and it is a Weber point. The edge whose endpoints are the antipodal vertices (in case of vertex-edge symmetry) of v or its antipodal vertex (in case of vertex-vertex symmetry) is called *south*.

It is worth nothing that we use the above definition only for symmetric configurations with single axis of type vertex-vertex or vertex-edge, with a Weber point on the axis. In particular, in case of vertex-vertex symmetry the definition is not ambiguous. In fact, from Theorem 11 the two vertices on the axis cannot be both Weber points. Contrary, if both are not Weber points and are empty, by Theorem 4 optimal gathering cannot be accomplished, hence we do not need such definitions. As we are going to see, if both are not Weber points but are occupied, optimal gathering can be assured but the strategy does not require to define north and south. Whereas, another definition required together with north and south is the following.

Definition 6 Given a symmetric configuration on a ring R, the line orthogonal to the axis of symmetry, cutting R on two edges into two subrings whose size differ of at most one vertex in favor of the southern side is called the *horizon*.

We are now ready to provide an optimal gathering algorithm for all configurations where this is possible. Let $\mathscr{U}$ be the set of initial configurations where optimal gathering cannot be assured, that is, configurations with two robots, or periodic, or admitting an axis of symmetry not passing through a Weber point nor a robot. The algorithm is described from the perspective of a generic robot after having performed its Look phase. The output of the algorithm is then the decision of the robot to move somewhere or to stay idle.

Theorem 14 (Optimal gathering on rings) *There exist an optimal gathering algorithm for a configuration $C = (R, \mu)$ on a ring R of n vertices if and only if $C \notin \mathcal{U}$.*

The algorithm provided in the proof of Theorem 14 works as follows.

Configurations with an odd number of robots have been solved by Theorem 13, and robots can always recognize they are in such a case by computing $k = \sum_{i=0}^{n-1} \mu(v_i)$.

The asymmetric case has been already solved by Theorem 10. The proposed technique must be slightly modified in order to integrate it with symmetric cases, hence obtaining a unique optimal gathering algorithm characterizing all possible configurations.

If configuration C admits a single axis of symmetry passing through two robots, by Theorem 4, the vertices where such robots lie cannot be both Weber points. If one is a Weber point and it is the only one among all vertices, then by Corollary 5 optimal gathering can be accomplished. If there is more than one Weber point in C then the algorithm makes move one of the robots on the axis (towards any direction) as follows.

If there is an odd number of Weber points (the north is a Weber point, necessarily), then the robot occupying the south is moved. In doing so, the number of Weber points remains odd since those initially residing at one side of the axis of symmetry have disappeared but not the one on the north. The obtained configuration can be still symmetric (of type robot-robot or node-node) with a Weber point on the axis and less Weber points than the original one. The case of node-node-symmetry will be discussed later, while for the case of robot-robot the same arguments can be applied again.

If there is an even number of Weber points (the north is not a Weber point), then the robot occupying the north is moved unless it creates a new axis of symmetry. In such a case, the one on the south is moved and we are sure that the configuration becomes asymmetric.

If C is symmetric without robots on the axis, the algorithm allows only moves towards north where the gathering will be eventually accomplished. The north, which is a Weber point, must exist as otherwise, by Theorem 4, optimal gathering is not possible.

Consider an even number of robots greater than four. The case of four robots will be handled later. From Theorem 9 the set of Weber points in R is given by a set of paths, and by hypothesis there is at least one path of Weber points (possibly made of just one vertex) containing the north. Moreover, due to symmetry and by Theorems 11 and 12, the number of such paths is odd and the two adjacent vertices at one side of R divided by the horizon cannot be both Weber points unless they are both occupied. If they are not occupied, the algorithm then moves the robots lying on the two Weber points closest to the horizon—and in case of ties the farthest from the axis of symmetry—towards north. Due to asynchrony, either one or two robots move. In either cases, by Theorem 5 all Weber points below the horizon disappear while all those above are preserved. Hence the number of paths of Weber points is still odd, and the obtained configuration C' maintains the original axis of symmetry if both robots have moved, or it is asymmetric at one step from the original symmetry if only one robot has moved. When only one robot moves, we are sure that the new configuration cannot be symmetric. Then, robots can detect whether the current asymmetric configuration may have been obtained from a symmetric one, and hence they can recognize the unique robot that can (re)-establish the original symmetry. Note that, the algorithm leads to a symmetric configuration even though the initial configuration is asymmetric but obtainable from a symmetric one. This is the only modification required to the algorithm provided in the proof of Theorem 10 for asymmetric configurations. This technique is applied until creating two symmetric multiplicities. Since we are considering at least six robots, if there are two robots above the multiplicities, the algorithm makes the two northern robots move towards north where they will create a third multiplicity, eventually. If there are no robots norther than the multiplicities, then the multiplicities are moved towards north until they join. In both cases, once a multiplicity is created on the north, by Theorem 5 this is the only Weber point, and hence the gathering can be easily finalized by applying Corollary 5.

In the case where in C the two pairs of adjacent vertices divided by the horizon are all occupied, the two southern ones move towards north. If both move, two multiplicities are created and similar arguments as above hold. If only one moves, only one multiplicity is created, but again similar arguments as above permit to create the second symmetric multiplicity, hence the gathering can be finalized on north.

It remains to consider the case of exactly four robots. Since the number of paths of Weber points must be odd there can be only one path of Weber points containing the north. The algorithm moves the two robots closest to the north towards it until creating a multiplicity. This can be realized since the other two robots are symmetric with respect to the original axis of symmetry, and cannot be as such with respect to another axis. Again, once a multiplicity is created on the north, this is the only Weber point, and hence the gathering can be finalized by applying Corollary 5.

5.5   Optimal Gathering on Infinite Grids

In this section, we fully characterize optimal gathering on *infinite grids*. Let an infinite path be the graph $P = (\mathbb{Z}, E)$ with $E = \{\{i, i + 1\} \ : \ i \in \mathbb{Z}\}$. An infinite grid is defined as the Cartesian product $G = P \times P$. A vertex of the grid is then an ordered pair of integers called *coordinates*. If G is an infinite grid then $C = (G, \mu)$ is a configuration on G.

Notice that on infinite grids the topology does not help in detecting a gathering vertex. Nonetheless, the interest in infinite grids also arises from the fact that they represent a natural discretization of the plane. We detect all the specific configurations where gathering cannot be performed. For all other configurations, we devise a distributed algorithm that exploit the global-strong multiplicity detection and, assures gathering on a Weber point by letting robots move along the shortest paths toward such a vertex, i.e., our algorithm is optimal in terms of moves.

Let $C = (G, \mu)$ be a configuration and S_C be the minimal (finite) sub-grid containing all the occupied vertices of the infinite grid G, and (S_C, μ) be the corresponding configuration. It is worth mentioning that S_C may change while robots move. As a consequence, even though S_C is a finite grid, the approach of [10] cannot be applied (it is strongly dependent on the dimensions of the grid where robots reside).

During the Look phase, a robot perceives (S_C, μ) and it is able to recognize its position on S_C if (G, μ) is asymmetric. Whereas, if (G, μ) admits an isometry φ different from the identity, a robot cannot distinguish its position at u from $\varphi(u)$, unless $u = \varphi(u)$. As a consequence, two robots (e.g., one on u and one on $\varphi(u)$) can decide to move simultaneously, as any algorithm is unable to distinguish between them. This fact greatly increases the difficulty to devise a gathering algorithm for symmetric configurations.

In an infinite grid, the center of a rotation can be a vertex, or the center of an edge, or the center of the area surrounded by four vertices, whereas the angle of rotation can be of 90° or 180°. Reflections axis can be horizontal (vertical), passing through vertices or through the middle of edges, or diagonal (45°), passing through vertices. If we assume the infinite grid embedded in a Cartesian plane, it is not difficult to see that other than rotations and reflections it admits also *translations*, that is a shifting of the vertices by applying the same displacement to each vertex. Regarding translations, even if they are possible for infinite grids, they do not belong to any automorphism group of configurations as these are defined for a finite (not null) number of robots. Note that, an infinite number of robots (or no robots at all) is required also when the configuration admits two parallel axis of symmetry, one axis and one center of rotation not lying on the axis, or two distinct centers of rotation. Moreover the automorphism group of a configuration with a finite number of robots is finite.

In order to check whether the current configuration could have been obtained from a symmetric one, we introduce the concept of *previous position* for a robot. Sometimes, an algorithm simulates itself by considering a configuration C' which is identical to the current configuration C but for the position of one robot r. If an execution of the algorithm can lead from C' to C then the simulated position of r in C' is called a *previous position* for r. This method will be used to detect possible pending moves when C' is symmetric.

Definition 7 Given a configuration $C = (G, \mu)$, $G_{\mathrm{WP}}(C)$ is the subgraph induced by its Weber points.

According to Corollary 5, in a configuration that admits only one Weber point the gathering can be achieved by an optimal algorithm. In what follows, configurations with one single Weber point are called of type S.

Impossibility results. We have already observed that, in general, a partitive configuration is ungatherable. In infinite grids, this result implies that all initial configurations with an axis of symmetry not passing through vertices or admitting a rotation with a center not coinciding with a vertex, are ungatherable. In fact, all such configurations are partitive with orbits of size at least two, and only those admitting rotations of 90° have orbits of size four.

In what follows we say that a symmetry is *allowed* if it is not partitive.

Theorem 15 *Let $C = (G, \mu)$ be a configuration, Then,*

- *If C contains only two robots (or equivalently, two multiplicities of the same size), then it is ungatherable.*
- *If C contains only four robots (or equivalently, four multiplicities of the same size) disposed on the corners of S_C, then C is ungatherable.*

In this section we denote by $\mathscr{U}$ the set of all initial configurations proved to be ungatherable. The set $\mathscr{U}$ includes all partitive configurations, those with only two robots, and those with four robots disposed as the corners of a rectangle. We will show a gathering algorithm for all the remaining initial configurations.

It is worth noting (see [10]) that gathering on finite grids was possible without any multiplicity detection due to the existence of special vertices like corners. The following results states that in our context the multiplicity detection is mandatory.

Theorem 16 *If robots are not empowered by any multiplicity detection capability, then the gathering problem is unsolvable on infinite grids.*

Here, we are assuming robots empowered by global-strong multiplicity detection. As shown by Fig. 13, relaxing such an assumption to the global-weak or any local multiplicity detection, makes ungatherable in the optimal way some configurations. Figure 13a shows a symmetric configuration which may lead to two different configurations. Only two moves in fact can be defined in order to gather all robots via shortest paths at one of the nine Weber points available. It is easy to check that $G_{\text{WP}}(C)$ is constituted by the central subgrid of dimension 3×3. Any optimal algorithm can allow each robot to move toward either the farthest or the closest robot that shares one coordinate with it, any other move would lead robots away from Weber points.

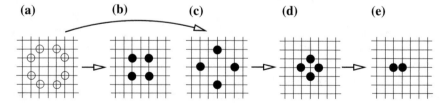

(a)   **(b)**   **(c)**   **(d)**   **(e)**

Fig. 13 Empty circles represent single robots; filled circles represent multiplicities. **a** Symmetric configuration with nine Weber points in the center. **b** Symmetric configuration obtainable from (**a**) with nine Weber points. **c** Symmetric configuration obtainable from (**a**) with one Weber point in the center. **d** Symmetric configuration obtainable from (**c**) with one Weber point recognizable only if robots are empowered by global-strong multiplicity detection. **e** Symmetric configuration obtainable from (**d**) with one Weber point recognizable only if robots are empowered by global-strong multiplicity detection

Note that, the adversary can make all robots synchronously move since for each pair of robots there exists a symmetry with respect to which their occupied vertices are equivalent. If all robots move synchronously, in the first case (in the second case, resp.) configuration of Fig. 13b (Fig. 13c, resp.) is reached. By Theorem 15, configuration in Fig. 13b is ungatherable. From configuration in Fig. 13c, only the move toward the center (the unique Weber point left) can be allowed. If the adversary makes all the robots move for one step, configuration in Fig. 13d is obtained. From there, if all the robots but those belonging to only one multiplicity make another step, then configuration in Fig. 13e is reached. From this configuration, without global-strong multiplicity detection, the two multiplicities are indistinguishable and by Theorem 15 the reached configuration in ungatherable.

5.5.1   One-Dimensional Grids

We first consider infinite paths as grids with one row and infinitely many columns.

Lemma 10   *If the number of robots k is odd, then there exists only one Weber point. If k is even, then all vertices of the subpath delimited by the two central robots (including the vertices where such robots lie) are Weber points.*

Theorem 17   *Optimal gathering on one-dimensional grids is always achievable except for configurations with only two robots or admitting partitive automorphisms.*

In the previous theorem, the ungatherable cases simply follow from Theorem 3 and 15. When the number of robots is odd, from Lemma 10 there exists only one Weber point and optimal gathering can be achieved by Corollary 5.

Let us observe how the optimal gathering is achievable when the number of robots is even. In such a case, if the configuration is symmetric, then the subpath of Weber points must be odd as otherwise the configuration is partitive. The idea is then to move the robots delimiting the Weber points toward the central vertex. If both move synchronously, the configuration remains symmetric but the interval of Weber points is reduced until only the Weber point at the central vertex remains. If only one moves, it is possible to recognize the robot that has to move to (re)-establish the symmetry. In fact, considering the two intervals of free vertices neighboring the robots delimiting the Weber points, the algorithm allows to move the robot delimiting the shortest interval.

When the number of robots is even, but the configuration is asymmetric, then either it is at one move from a possible symmetry which is allowed, or one of the two robots delimiting the Weber points can be chosen to move toward the other one without creating a symmetry until only one Weber point remains.

Finally, when there is only one Weber point, from Corollary 5, all robots can move safely toward it. It is worth to notice that such an algorithm also works when the input configuration admits multiplicities.

5.5.2 Two-Dimensional Grids

We now describe a general optimal algorithm to solve the gathering problem for each configuration $C = (G, \mu)$ such that $C \notin \mathcal{U}$. From Corollary 5, if the configuration C admits only one Weber point (that is, $C \in S$), then optimal gathering can be accomplished. Another characterization is provided by considering S_C, and in particular the projections of the robots to the two generating paths P_1 and P_2 of G. Given a robot on a generic vertex (i, j) of G, its projections on P_1 and P_2 are a robot on vertex i and a robot on vertex j, respectively. This gives rise to two configurations (P_1, μ_1) and (P_2, μ_2) such that $\mu_1(v) = \sum_j \mu((v, j))$ and $\mu_2(v) = \sum_i \mu((i, v))$. As the movements on a grid are either vertical or horizontal, solving the gathering with respect to the two dimensions separately, solves the general problem.

Theorem 18 *Given a configuration $C = (G, \mu)$ with $G = P_1 \times P_2$, if (P_1, μ_1) and (P_2, μ_2) are optimally gatherable, then also C is optimally gatherable.*

The optimal gathering considered in the previous theorem is obtained by simply considering (P_1, μ_1) and (P_2, μ_2) separately. Each time a robot wakes-up, it can move with respect to any of the two instances indiscriminately, as they are independent to each other. Theorem 17 guarantees optimal gathering on both the instances even though they might contain multiplicities.

Note that there are gatherable configurations that do not satisfy the assumptions of Theorem 18. Hence, a more general strategy must be designed in order to cope with all the gatherable configurations. The next theorem provides a useful characterization about the arrangement of Weber points in a configuration.

Theorem 19 *Given a configuration $C = (G, \mu)$ with $G = P_1 \times P_2$, $G_{WP}(C)$ is a finite grid defined by the Cartesian product of the subpaths induced by the Weber points belonging to (P_1, μ_1) and (P_2, μ_2).*

By referring to Fig. 14, it is worth noting that $G_{WP}(C)$, for some configuration C, is in general a finite grid where robots can occupy only the corners. Moreover,

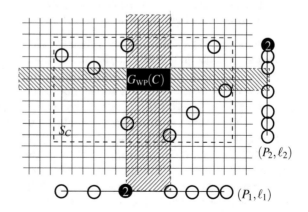

Fig. 14 A sample configuration C which induces S_C, $G_{WP}(C)$, and its projections to the sides of S_C

all the vertices belonging to the strips from $G_{\text{WP}}(C)$ to the borders of S_C cannot be occupied, but for the ones sharing coordinates with the border of $G_{\text{WP}}(C)$. These robots will be said to *determine* $G_{\text{WP}}(C)$. Note that, given a configuration C with k robots, evaluating the set of Weber points has time complexity $O(|S_C| \times k)$.

Grids with an odd number of robots. By Lemma 10, an odd number of robots implies a single Weber point for each instance on the two paths generating G. By Theorem 19, the Cartesian product of those two Weber points constitutes the only Weber point of the configuration, hence by Corollary 5 optimal gathering can be assured. Then, the following results follows.

Corollary 7 *If the number of robots in a grid G is odd, then optimal gathering can be accomplished.*

Grids with an even number of robots for which S_C has all sides of odd length and a center that is a Weber point. Here we describe a general strategy that solves the optimal gathering problem for all the configurations with an even number of robots not in $\mathscr{U}$.

First of all, if S_C has both sides odd and the center is a Weber point then we can gather all the robots in the center. The idea at the basis of the strategy is to move all the robots not lying on the border of S_C toward the center that becomes the only Weber point of the current configuration. From there on, all the other robots can join the unique Weber point. This can be easily realized if the number of robots is "sufficiently" large, while for few robots specific strategies are required. The next lemma holds when the number of robots is at least 6.

Lemma 11 *Let $C = (G, \mu)$ be an initial configuration inducing S_C with both sides odd and the center being a Weber point, then if the number of robots is at least 6, there exists an optimal gathering algorithm.*

Let us now describe the optimal gathering algorithm whose existence is guaranteed by the previous theorem.

For each side of S_C, consider the robot (or the two robots) farthest from the central vertex of the chosen side. The algorithm first makes all other robots move toward the center of S_C. At this point, we consider different cases according to the number of robots occupying the corners of S_C:

- *No corners occupied.* Each robot lying on a side of S_C is moved to the center of the side. The obtained configuration is then composed of one or two robots at the center of each side and zero or more robots at the center of S_C.
- *One corner occupied.* Since one corner is occupied, the robots (if any) on the sides defining the occupied corner have been moved by the algorithm toward the center of S_C. For each of the other two sides, let r be the closest robot to the occupied corner, the algorithm makes the other robot (if any) move toward the center of S_C. Then, r is moved to the center of its side. The obtained configuration is then composed by the robot on the corner, no other robots on the two sides sharing one coordinate with the corner occupied, one robot for each of the other two sides at their centers, and at least three robots in the center of S_C.

- *Two corners occupied.* If the two corners occupied do not share any coordinate, then the algorithm makes all the robots move but those in the corners to the center of S_C, where there will be at least four robots.

  If the two corners occupied share one coordinate, then consider the only side s of S_C whose endpoints are not occupied. All the robots except the two on the occupied corners and the one(s) on s farthest from the center of S_C, are moved to the center of S_C. After completion of these moves, the algorithm makes the remaining robot(s) on s move to its center. The obtained configuration is then composed by the two robots on the corners, no other robots on the three sides sharing one coordinate with the corners occupied, one robot or two robots in the center of s, and at least two robots in the center of S_C.
- *Three corners occupied.* In this case, there are two robots in two different corners not sharing any coordinate, hence the algorithm makes all the other robots move toward the center of S_C, where there will be at least four robots.
- *Four corners occupied.* The algorithm makes all robots move but those in the corners toward the center of S_C, where there will be at least two robots.

  In all the above cases, the obtained configuration admits only one Weber point at the center of S_C. This can be easily checked considering (G, μ) as the Cartesian product of two configurations on two paths (P_1, μ_1) and (P_2, μ_2) where $\mu_1(x) = \sum_y \mu((x, y))$ and $\mu_2(y) = \sum_x \mu((x, y))$, and by applying Lemma 10 and Theorem 19.

  By Corollary 5, at this point all the robots can move toward the only Weber point, hence achieving the gathering. Moreover, as all the computed moves have been performed toward the final Weber point, then the proposed algorithm is optimal.

For the case of just 4 robots, the next lemma holds.

Lemma 12 *Let $C = (G, \mu)$ be an initial configuration inducing S_C with both sides odd and the center being a Weber point, then if there are only 4 robots, there exists an optimal gathering algorithm unless each robot occupies a different corner of S_C.*

Remark 2 Configurations admitting rotations but not in $\mathcal{U}$ are solved by the above two lemmata since the center of rotation is a vertex in such cases (i.e., S_C has both sides odd) and it is a Weber point.

Grid with an even number of robots for which S_C has a side with an even length or a center that is not a Weber point. Before proceeding with the characterization of the algorithm, we need to better specify the view of the robots during their Look phase when no multiplicities occur.

Let us consider the eight sequences of distances (number of empty vertices) between occupied vertices obtained by traversing S_C starting from its four corners and proceeding toward the two possible directions. If proceeding vertically, all columns will be considered sequentially. Similarly for the rows if proceeding horizontally. Note that the two sequences associated to a corner occupied by a robot start with 0. For instance, by referring to Fig. 14, the top-left corner is associated with $(1, 27, 24, 6, 36, 15, 11, 12, 4)$ by reading the configuration vertically, and with (6, 7, 1, 18, 43, 28, 9, 19, 5) by reading the configuration horizontally. To better understand how such sequences are computed, consider the one obtained by the vertical reading. Starting from the top-left corner of S_C and moving down the first column it takes 1 empty vertex to reach the first robot. Thus, 1 is the first term of the sequence. Going further down we see 7 empty vertices in the first column. Then we start from the top of the second column and walk down. We pass additional 9 empty vertices. We then walk down the third column and pass another 9 empty vertices. Finally, starting from the top of the fourth column we meet 2 empty vertices before arriving at the next robot. Thus the total number of empty vertices between the first and the second robot is $7 + 9 + 9 + 2 = 27$, and this is the second term of the sequence, and so forth.

We associate for each corner the lexicographically largest sequence between the two readings from such corner. Note that, in square grids such two sequences are always different, but for the two corners through which a possible axis of symmetry passes. In rectangular grids, if the two sequences are equal, we assume as larger the one when read in the direction of the largest side.

We define the maximal sequence as the largest one among the four sequences associated to the four corners. We refer to the corner(s) defining the maximal sequence as *preferred corner(s)*, and to the direction(s) that implies the maximal sequence as *preferred direction(s)*.

In Fig. 14, the preferred corner of S_C is the bottom-left one, the preferred direction is horizontal, and the maximal sequence is (10, 11, 21, 34, 19, 12, 21, 7, 1).

We are now ready to describe the gathering algorithm for each configuration $C \notin \mathcal{U}$ with more than one Weber point, where S_C has at least one side even or its center is not a Weber point.

In general, if a configuration is symmetric, the algorithm may allow the movement of two symmetric robots. If both move, the configuration remains symmetric. If only one moves, the algorithm always forces to move the one that can (re)-establish the symmetry. In fact, from asymmetric configurations at one step from an allowed symmetry it is always possible to detect one unique robot that has to move in order to (re)-establish the symmetry.

Let C be a configuration. According to the number of corners of $G_{\mathrm{WP}}(C)$ occupied by robots, different strategies are applied (see Fig. 15 for a visualization of the configurations that will be considered).

- *Type F: No corners occupied.* Among these configurations, $F1$ are the asymmetric ones, $F2$ the symmetric configurations with a horizontal/vertical axis, and $F3$ the symmetric configurations with a diagonal axis.
  First we consider the cases when $G_{\mathrm{WP}}(C)$ is not a path. If C is in $F2$, then among the robots determining $G_{\mathrm{WP}}(C)$, consider those closest to $G_{\mathrm{WP}}(C)$. Among such robots consider those closest to $G_{\mathrm{WP}}(C)$ with respect to the preferred direction, if any. Only two robots are then selected by considering those closest to the preferred corners. Such robots move toward $G_{\mathrm{WP}}(C)$. In this case, either two symmetric robots move synchronously, or only one moves, and the other one is possibly pending. We will show that our algorithm always force the possible pending robot to move.

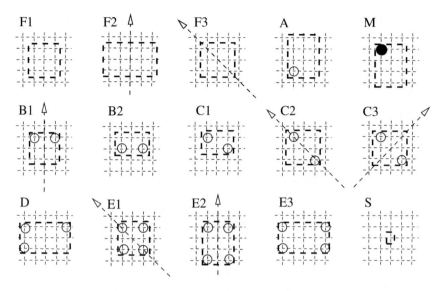

Fig. 15 Types of configurations according to the number of corners of $G_{WP}(C)$ occupied by robots/multiplicities. Dashed lines delimit $G_{WP}(C)$, empty circles represent single robots, and filled circles represent multiplicities

Eventually, this process leads to symmetric configurations with two corners of $G_{WP}(C)$ occupied. The axis of reflection is the bisector of the line determined by $G_{WP}(C)$.

If C is in $F3$, then among the robots determining $G_{WP}(C)$, consider those closest to $G_{WP}(C)$. Among such robots consider those closest to $G_{WP}(C)$ with respect to the preferred direction, if any. Only two robots are then selected by considering those closest to the preferred corner(s). Such robots move toward $G_{WP}(C)$. As above, our algorithm forces the two selected robots to move symmetrically. Again, we will show that our algorithm always forces the possible pending robot to move. Eventually, this process leads to symmetric configurations with one corner occupied by a multiplicity.

If C is in $F1$ at more than one move from an allowed symmetry, the closest robot to $G_{WP}(C)$ moves toward it. Ties are solved by considering the preferred direction and the preferred corner.

When $G_{WP}(C)$ consists of just a path, the strategy is the same as above but limited to the robot(s) lying on the extension of $G_{WP}(C)$, and not those determining its endpoints.

- *Type A*: *One corner occupied by a single robot*. If the configuration is asymmetric, first robots check whether an allowed symmetry can be (re)-established. We will show that this is possible by finding the previous positions of the unique robot on the corner of $G_{WP}(C)$. If a symmetry can be (re)-established, then the possible pending robot is forced to perform its move. In any other case, the single robot on the corner moves in one of the two directions that reduce $G_{WP}(C)$, until obtaining only one Weber point.

- *Type M*: *One corner occupied by a multiplicity and possibly other corners occupied.* First, all robots sharing one coordinate with the multiplicity move in turn (i.e., without creating another multiplicity) until joining it, toward the shared coordinate. After that, if there is more than one Weber point, among the robots determining $G_{WP}(C)$, consider those closest to it. Such robots—at most four—move (even concurrently) along the direction that reduces $G_{WP}(C)$, until they share one coordinate with the multiplicity.

- *Types B and C*: *Two corners occupied.* Among these configurations we denote by B the set of configurations where the two corners occupied share one coordinate except for symmetric configurations with the axis passing through the two occupied corners. These last configurations and the remaining ones with two corners occupied are denoted by C. $B1 \subset B$ represents symmetric configurations, $B2 \subset B$ the asymmetric ones. $C1 \subset C$ represents asymmetric configurations, $C2 \subset C$ the symmetric ones with the axis passing through the occupied corners, and $C3 \subset C$ the remaining symmetric configurations.

  From $B1$, the two robots on $G_{WP}(C)$ move toward each other, still maintaining the symmetry. Note that, the two robots are separated by an odd path, as otherwise the configuration is ungatherable by Theorem 3.

  From $B2$, the algorithm (re)-establishes an allowed symmetry, if any. Otherwise, the robot r on the corner of $G_{WP}(C)$, closest to the preferred corner, moves toward the other robot r'. If such a move would generate an ungatherable configuration, then r' moves toward r.

  From $C1$, the algorithm (re)-establishes an allowed symmetry, if any. Otherwise, if $G_{WP}(C)$ is composed of more than four vertices, the first robot r met among the two on the corners of $G_{WP}(C)$ from the preferred corner along the preferred direction moves toward the other robot r' unless it makes $G_{WP}(C)$ as a path or the move brings to an ungatherable configuration. In which case, r' moves. When $G_{WP}(C)$ is a square grid of four vertices, then we ensures that the selected robot can safely move toward the other one.

  From $C2$, if $G_{WP}(C)$ is not a path, the robot that moves is the one at the corner of $G_{WP}(C)$ closest to the corner of S_C associated with the maximal sequence among the two corners on the axis. If $G_{WP}(C)$ is a path, the robot that moves is the one closest to the corners of S_C associated with the maximal sequence.

  From $C3$, the two robots on the corners of $G_{WP}(C)$ move toward the corner of $G_{WP}(C)$ on the axis, closest to the corner of S_C associated with the maximal sequence.

  In all the cases, a configuration in S with a multiplicity occupying the only Weber point will be reached, eventually.

- *Type D*: *Three corners occupied.* From D, the robot on the middle corner moves toward one of the two other occupied corners. This leads to a configuration with two corners occupied or with one corner occupied by a multiplicity. We will show that if the configuration is at one move from an allowed symmetry, then the defined move involves exactly the robot that can (re)-establish the symmetry.

Procedure: COMPUTE
Input: $C = ((V,E),\mu)$

1 Compute S_C, $G_{\text{WP}}(C)$, and $k = \sum_{v \in V} \mu(v)$;
2 **if** $C \notin \mathcal{U}$ **then**
3   **if** $C \in S$ **then**
4   any robot moves towards the unique Weber point;
5   **else**
6   **if** S_C *has both sizes odd and its center is a Weber point* **then**
7   **if** $k = 4$ **then** Apply the strategy defined by Lemma 12;
8   **else** Apply the strategy defined by Lemma 11;
9   **else**
10   **case** $C \in \mathcal{X}$, with $\mathcal{X} \in \{A,B,C,D,E,F,M\}$
11   Apply the strategy designed for type $\mathcal{X}$;

Fig. 16 Procedure COMPUTE

- *Type E*: *Four corners occupied.* Among configurations E, we denote by $E1$ the symmetric ones with a diagonal axis, by $E2$ the remaining symmetric ones, and by $E3$ the asymmetric ones.
  From $E1$, the robot on $G_{\text{WP}}(C)$ closest to the corner of S_C on the axis associated with the maximal sequence, moves reducing the Weber points.
  From $E2$, the two robots on $G_{\text{WP}}(C)$ closest to the preferred corners move toward each other. Note that, if both move synchronously, then a symmetric configuration with two corners occupied or with a multiplicity is obtained. If only one moves, then a configuration of type D is obtained. According to that case, the robot that will be allowed to move is the one with a possible pending move, hence it does not create further pending moves.
  From $E3$, the robot on $G_{\text{WP}}(C)$ closest to the preferred corner moves reducing the Weber points.

Summarizing, the algorithm applied by the robots during the compute phase is shown in Fig. 16.

Correctness. When $C \in S$, the correctness is guaranteed by Corollary 5. By Theorem 19, S includes all configurations with an odd number of robots. When there is more than one Weber point, if C contains an even number of robots with S_C admitting both sides odd and its center being a Weber point, the correctness is guaranteed by Lemmata 11 and 12.

When S_C has at least one side even or its center is not a Weber point, we are going to prove that the defined strategy always leads to configurations in S, and from there the gathering is finalized by applying Corollary 5. Moreover, along this transition, the target vertex where gathering is finalized never changes, hence robots always move along their shortest paths toward the gathering vertex.

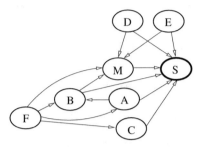

Fig. 17 Transitions among types of configurations allowed by the optimal gathering algorithm when S_C has at least one side even or its center is not a Weber point

The general scheme of the transitions obtainable by our algorithm when S_C has at least one side even or its center is not a Weber point is shown in Fig. 17. For the ease of visualization, self-loops are not shown in the figure since from each class except S, an exit transition is always taken, eventually. From S, instead, gathering is always finalized. We have to prove that the only possible transitions generated by our algorithm are those depicted in the figure, hence S being the only sink node that will be reached, eventually.

The following theorem summarizes the described results about infinite grids.

Theorem 20 (Optimal gathering on infinite grids) *Given an initial configuration $C = (G, \mu)$ on an infinite grid G, optimal gathering can be assured unless $C \in \mathcal{U}$.*

6 Conclusion

We have reviewed recent results about the gathering problem of anonymous, asynchronous, and oblivious robots placed on different environments, like the Euclidean plane and some graph topologies (trees, rings, and infinite grids). Robots operate in the Look-Compute-Move model possibly empowered with the multiplicity detection, and they are required to gather at one of some predetermined meeting-points (when the environment is the plane) or at a vertex of the graph.

These results provide a possible answer to a recent research question that asks for gathering algorithms designed by taking into account not only the feasibility but also some kind of cost measure for the solution. To this aim, we have shown that it is possible to use a generalization of the classic definitions of optimization problem and approximation algorithm to the context of robot-based computing systems. The considered cost measure is the minimum total traveled distance of all robots, which implies that any optimal gathering algorithm must move robots through shortest paths.

We have shown that, for each considered environment in which robots move, there exists an optimal distributed gathering algorithm that solves the problem for all configurations but those proved to be (optimally) ungatherable.

These results suggest to investigate a new branch of research that concerns the definition of new optimization problems. In fact, previous strategies/settings can be now reconsidered with respect to the new optimization tasks, or other objective functions. Actually, the introduced concepts of meeting-points or optimality requirements can be also explored with respect to other problems concerning robot-based computing systems. For instance, a challenging investigation may concern basic coordination problems like pattern formation, scattering, leader election or dynamic tasks like flocking and sequential pattern formation.

Acknowledgements The work has been supported in part by the European project "Geospatial based Environment for Optimisation Systems Addressing Fire Emergencies" (GEO-SAFE), contract no. H2020-691161 and by the Italian project "RISE: un nuovo framework distribuito per data collection, monitoraggio e comunicazioni in contesti di emergency response", Fondazione Cassa Risparmio Perugia, code 2016.0104.021.

References

1. A. Aho, J. Hopcroft, J. Ullman, *Data Structures and Algorithms*. (Addison Wesley, 1983)
2. G. Ausiello, P. Crescenzi, G. Gambosi, V. Kann, A. Marchetti-Spaccamela, M. Protasi, *Complexity and Approximation: Combinatorial Optimization Problems and Their Approximability Properties* (Springer, Berlin Heidelberg, 1999)
3. C. Bajaj, The algebraic degree of geometric optimization problems. Discret. Comput. Geom. **3**(1), 177–191 (1988)
4. S. Buss, Alogtime algorithms for tree isomorphism, comparison, and canonization, in Kurt Gödel *Colloquium, LNCS*, vol. 1289 (Springer, 1997), pp. 18–33
5. J. Chalopin, Y. Dieudonné, A. Labourel, A. Pelc, Rendezvous in networks in spite of delay faults. Distrib. Comput. **29**(3), 187–205 (2016)
6. S. Cicerone, G. Di Stefano, A. Navarra, Asynchronous embedded pattern formation without orientation, in: *Proceeding of the 30th International Symposium on Distributed Computing (DISC), LNCS*, vol. 9888 (Springer 2016) pp. 85–98
7. S. Cicerone, G. Di Stefano, A. Navarra, Gathering of robots on meeting-points: Feasibility and optimal resolution algorithms. Distrib. Comput. **31**(1), 1–50 (2018)
8. M. Cieliebak, P. Flocchini, G. Prencipe, N. Santoro, Distributed computing by mobile robots: Gathering. SIAM J. Comput. **41**(4), 829–879 (2012)
9. E.J. Cockayne, Z.A. Melzak, Euclidean constructibility in graph-minimization problems. Math. Mag. **42**(4), 206–208 (1969)
10. G. D'Angelo, G. Di Stefano, R. Klasing, A. Navarra, Gathering of robots on anonymous grids and trees without multiplicity detection. Theor. Comput. Sci. **610**, 158–168 (2016)
11. G. D'Angelo, G. Di Stefano, Navarra, A.: Gathering asynchronous and oblivious robots on basic graph topologies under the look-compute-move model. in: *Search Theory: A Game Theoretic Perspective*, (Springer, 2013), pp. 197–222
12. G. D'Angelo, G. Di Stefano, A. Navarra, N. Nisse, K. Suchan, Computing on rings by oblivious robots: A unified approach for different tasks. Algorithmica **72**(4), 1055–1096 (2015)
13. G. D'Angelo, A. Navarra, N. Nisse, A unified approach for gathering and exclusive searching on rings under weak assumptions. Distrib. Comput. **30**(1), 17–48 (2017)
14. B. Degener, B. Kempkes, T. Langner, F. Meyer auf der Heide, P. Pietrzyk, R. Wattenhofer, A tight runtime bound for synchronous gathering of autonomous robots with limited visibility, *in Proceedings of the 23rd annual ACM symposium on Parallelism in algorithms and architectures (SPAA)* (2011) pp. 139–148

15. G. Di Stefano, A. Navarra, Gathering of oblivious robots on infinite grids with minimum traveled distance. Inf. Comput. **254**, 377–391 (2017)
16. G. Di Stefano, A. Navarra, Optimal gathering of oblivious robots in anonymous graphs and its application on trees and rings. Distrib. Comput. **30**(2), 75–86 (2017)
17. A. Farrugia, L. Gasieniec, L. Kuszner, E. Pacheco, Deterministic rendezvous in restricted graphs. in *Proceeding of the 41st International Conference on Current Trends in Theory and Practice of Informatics (SOFSEM), LNCS*, vol. 8939, (Springer, 2015), pp. 189–200
18. P. Flocchini, G. Prencipe, N. Santoro, Distributed Computing by Oblivious Mobile Robots. Synth. Lect. Distrib. Comput. Theory (2012)
19. N. Fujinaga, Y. Yamauchi, H. Ono, S. Kijima, M. Yamashita, Pattern formation by oblivious asynchronous mobile robots. SIAM J. Comput. **44**(3), 740–785 (2015)
20. T. Izumi, T. Izumi, S. Kamei, F. Ooshita, Feasibility of polynomial-time randomized gathering for oblivious mobile robots. IEEE Trans. Parallel Distrib. Syst. **24**(4), 716–723 (2013)
21. R. Klasing, A. Kosowski, A. Navarra, Taking advantage of symmetries: Gathering of many asynchronous oblivious robots on a ring. Theor. Comput. Sci. **411**, 3235–3246 (2010)
22. R. Klasing, E. Markou, A. Pelc, Gathering asynchronous oblivious mobile robots in a ring. Theor. Comput. Sci. **390**, 27–39 (2008)
23. E. Kranakis, D Krizanc, E. Markou, *The Mobile Agent Rendezvous Problem in the Ring*. Morgan and Claypool (2010)
24. Y. Kupitz, H. Martini, Geometric aspects of the generalized Fermat-Torricelli problem in Intuitive Geometry. Bolyai Soc. Math Stud. (6), (1997)
25. A. Pelc, Deterministic rendezvous in networks: A comprehensive survey. Networks **59**(3), 331–347 (2012)
26. J. Sekino, n-ellipses and the minimum distance sum problem. Amer. Math. Mon. **106**(3), 193–202 (1999)
27. E. Weiszfeld, Sur le point pour lequel la somme des distances de n points donnés est minimum. Tohoku Math. **43**, 355–386 (1936)
28. E. Weiszfeld, F. Plastria, On the point for which the sum of the distances to n given points is minimum. Ann. Oper. **167**(1), 7–41 (2009)

The MinSum-MinHop and the MaxMin-MinHop Bicriteria Path Problems

Marta Pascoal

Abstract The number of hops (or arcs) of a path is a frequent objective function with applications to problems where network resources utilization is to be minimized. In this chapter we solve bicriteria path problems involving this objective function and two other common metrics, the path cost and the path capacity. Labeling algorithms are introduced, which use a breadth-first search tree in order to compute the maximal and the minimal sets of non-dominated paths. Dominance rules are derived for the two bicriteria problems and the properties of this data structure are explored to better suit the number of hops objective function and thus simplify the labeling process. Computational experiments comparing the new methods with standard approaches on randomly generated test instances and on instances that simulate video traffic are presented and discussed. Results show a significant speed-up over generic standard methods.

1 Introduction

Optimal path problems are classical network optimization problems which arise in several contexts and with different types of objective functions [1]. However, very often a single objective function cannot completely characterize a problem. Two of the most common objective functions used in these problems are the path cost, given by an additive function, in general to be minimized, and the path capacity, which is a bottleneck function, in general to be maximized. These two problems will be designated by MinSum and MaxMin, respectively. In 1980 Hansen [11] presented a list of several bicriteria path problems, studied their complexity, and adapted labeling algorithms to solve them. These problems include the MinSum-MinSum path prob-

M. Pascoal (✉)
CMUC, Department of Mathematics, University of Coimbra,
3001-501 Coimbra, Portugal
e-mail: marta@mat.uc.pt

M. Pascoal
Institute for Systems and Computers Engineering, Coimbra, Portugal

lem (which minimizes two cost functions) and the MinSum-MaxMin path problem (which minimizes the cost and maximizes the capacity). Later Martins [13] generalized labeling algorithms for the MinSum path problem with more than two objective functions, and in [14] the same author studied the MinSum-MaxMin case and developed both an algorithm for finding the maximal set of non-dominated paths and another algorithm for finding simply the minimum set of non-dominated paths, thus computing the non-dominated objective values. Recently the algorithm presented by Martins [13] was extended by Gandibleux et al. [9] to cope with more than a single cost function and one bottleneck function. Other labeling algorithms [4, 18] have also been presented for the MinSum-MinSum case, as well as for path problems with a higher number of objective functions to optimize [3, 12, 15].

The number of arcs in a path, or the number of hops borrowing from the telecommunications terminology, is another useful in practice and common objective function with particular interest to telecommunications. In general this criterion should be minimized, thus we designate the corresponding problem by MinHop. In such problems it is frequent to look for a route using the maximum available bandwidth, or having the minimum cost (sometimes also related with the arcs bandwidth). The hop-constrained shortest path problem, where the goal is to seek the shortest path with at most H arcs, for a given integer H, is another version of the problem, which arises often as a subproblem in telecommunication network design problems, where a given specified level of service with respect to certain measures (such as delay or reliability) must be met by each commodity [2]. This particular problem has been addressed, for instance, by Dahl and Gouveia [8]and by Riedl in [17], where the problem's polytope is characterized for particular cases of H.

When combining the objective function number of hops with the cost or with the bandwidth, the MinHop-MinSum or the MinHop-MaxMin path problems are obtained. These are particular cases of the MinSum-MinSum or the MinSum-MaxMin path problems, since the number of hops can be seen as an additive function where all arcs have cost equal to 1. Some problems related to these two variants can be found in the works by Cheng and Ansari, Guerin and Orda, and Randriamasy, Fourni and Hong [5, 10, 16]. Still, despite their large number of potential applications, to our knowledge no specific method have been developed to deal with them.

This manuscript focuses on labeling algorithms for the MinHop-MinSum and the MinHop-MaxMin path problems, aiming to determine the minimal and the maximal sets of non-dominated paths. It is known that breadth-search allows to scan the nodes of a tree by order of the level they belong to, and therefore the proposed methods use a queue to manage the labels associated with each generated path. This results in a simplification of the labeling procedure, namely regarding the dominance test and in the decrease of the CPU time.

The remainder of the text is organized as follows. In Sect. 2 notation, preliminary concepts and the problems are introduced. Section 3 is dedicated to the presentation of labeling algorithms for the minimum hop-shortest path problem from two points of view, the determination of all non-dominated paths or simply of those with distinct objective values, while Sect. 4 is devoted to the minimum hop-maximum capacity path problem. In Sect. 5 the results of computational experiments are reported and discussed. Conclusions follow in Sect. 6.

2 Preliminaries

Let (N,A) be a directed network consisting of a set $N = \{1, \ldots, n\}$ of nodes and of a set $A = \{1, \ldots, m\}$ of arcs. Let s, the initial node, and t, the terminal node, be two distinct nodes in (N,A). A path from s to t in (N,A) is a sequence of the form $p = \langle v_1, \ldots, v_\ell \rangle$ where $v_1 = s$, $v_\ell = t$, $v_i \in N$, $i = 1, \ldots, \ell$, and $(v_i, v_{i+1}) \in A$, $i = 1, \ldots, \ell - 1$. For simplicity we write $(i,j) \in p$ if i and j are consecutive nodes and i precedes j in p. Let P denote the set of all paths from s to t in (N,A). With each arc $(i,j) \in A$ are associated a cost $c_{ij} \in \mathbb{R}$ and a capacity $u_{ij} \in \mathbb{R}^+$. Then the cost of path p is defined by

$$c(p) = \sum_{(i,j)\in p} c_{ij},$$

its capacity by

$$u(p) = \min_{(i,j)\in p} \{u_{ij}\},$$

and its number of arcs by

$$h(p) = \ell - 1.$$

Usually the functions c and h are minimized, whereas u is maximized. The single criterion problems thus obtained will be called MinSum, MinHop and MaxMin, respectively. These functions are combined in bicriteria path problems. The following sections focus on two of these problems, the MinHop-MinSum path problem, where h and c are minimized, and the MinHop-MaxMin problem, where h is minimized and u is maximized.

In general, the two objective functions of a bicriteria problem are not correlated and there is no solution optimizing them simultaneously. Instead, the set of non-dominated paths, for which there is no other solution that improves one of the objectives without worsening the other, is computed. In the following definitions we borrow some terminology used by Gandibleux et al. [9] for multicriteria path problems. As we focus on the optimization of different objective functions we present a general definition of dominance.

Definition 1 Let p_1, p_2 be two paths between the same pair of nodes in (N,A) and f_1, f_2 two functions defined for any path.

1. Path p_1 dominates path p_2 (denoted $p_1 D p_2$) if and only if $f_i(p_1)$ is better than or equal to $f_i(p_2)$, $i = 1, 2$, and it is strictly better for at least one of the objective functions. In that case it can also be said that $(f_1(p_1), f_2(p_1))$ dominates $(f_1(p_2), f_2(p_2))$ (denoted $(f_1(p_1), f_2(p_1))_D (f_1(p_2), f_2(p_2))$).

2. Path p_1 strictly dominates path p_2 if and only if it is better than p_2 both for f_1 and f_2. Paths p_1, p_2 are equivalent when $f_i(p_1) = f_i(p_2)$, $i = 1, 2$.

Definition 2 A path $p \in P$ is non-dominated, or efficient, if and only if it is not dominated by any other. If there is no path that strictly dominates p, then p is said to be weakly non-dominated, or weakly efficient.

Definition 3 Let P_D be the subset of dominated paths in P. Then $P_N = P - P_D$ denotes the maximal complete set of non-dominated paths in P, while $\bar{P}_N$ denotes the minimal complete set of non-dominated paths, the largest subset of P_N that contains no equivalent solutions.

3 MinHop-MinSum Path Problem

In labeling algorithms for bicriteria path problems several labels can be assigned to a network node. Each label corresponds to a path in the network starting from s, and thus a tree of paths rooted at s can be constructed. Some branches of this tree can be pruned by testing the dominance of new nodes and of those that are already in the tree, corresponding to paths. Like for the single criterion case, label setting and label correcting algorithms for these problems differ on the strategy they use to pick the next label to scan. If the lexicographically smallest label is taken, in the first case, then it is permanent, that is, non-dominated, after being scanned, while with label correcting algorithms non-dominated paths can only be known when all labels have been scanned. In the following the labels used for each problem, as well as the dominance rules to compare them, are defined.

For the MinHop-MinSum path problem a label associated with a node x in the search tree of paths from s to other nodes has the form $l_x = [\pi_x^h, \pi_x^c, \xi_x, \beta_x]$, where

- π_x^h denotes the number of arcs in the path from the root node to x,
- π_x^c denotes its cost,
- ξ_x is the node preceding x in that tree, and
- β_x is the network node that corresponds to x.

Given $\beta_x = i \in N$ and $(i,j) \in A$, a new node y can be considered in the tree, with the label

$$l_y = [\pi_x^h + 1, \pi_x^c + c_{ij}, x, j].$$

Let X denote the set of labels that are eligible to be scanned.

By definition of dominance between two labels l_x, l_y corresponding to paths ending at the same network node can be compared by saying that l_x is dominated by l_y if and only if

$$(\pi_x^h > \pi_y^h \text{ and } \pi_x^c \geq \pi_y^c) \text{ or } (\pi_x^h \geq \pi_y^h \text{ and } \pi_x^c > \pi_y^c). \tag{1}$$

Therefore, a new label l_y can be discarded if there is another node x already in the tree such that $\beta_x = \beta_y$ and

$$(\pi_x^h < \pi_y^h \text{ and } \pi_x^c \leq \pi_y^c) \text{ or } (\pi_x^h \leq \pi_y^h \text{ and } \pi_x^c < \pi_y^c). \tag{2}$$

On the other hand, whenever (1) holds for some label l_x in the search tree, then l_x can be replaced by l_y.

Now, considering the computation of the maximal set of non-dominated paths and assuming X is manipulated as a First In First Out list (FIFO), that is, a queue, the number of arcs of the analyzed paths, i.e. the π_x^h values, forms a non-decreasing sequence [7]. Thus for a new label l_y condition $\pi_x^h \leq \pi_y^h$ always holds, so that the first part of (1) is never satisfied and (1) can be replaced by

$$\pi_x^h = \pi_y^h \text{ and } \pi_x^c > \pi_y^c. \tag{3}$$

Furthermore the labels associated with a certain node have particular properties, as shown in Lemma 1.

Lemma 1 *If X is a FIFO, then for each level of the tree of paths rooted at s a tree node is associated with several labels, if and only if all have the same objective function values.*

Proof Let l_x and l_y be two labels at the same level in X, that correspond to the same node, i.e. $\beta_x = \beta_y$ and $\pi_x^h = \pi_y^h$. By contradiction, assuming that $\pi_x^c \neq \pi_y^c$, then:

1. either $\pi_x^c < \pi_y^c$, which implies $l_x {}_D l_y$, so y should not belong to X,
2. or else $\pi_x^c > \pi_y^c$, therefore $l_y {}_D l_x$, so x should not belong to X.

Moreover, if a network node has several non-dominated labels with the same objective values, then they share the same value of h and belong to the same paths tree level.   □

This result yields a simplification of the acceptance condition of a new label, l_y (2), which can be restated as

$$(\pi_x^h < \pi_y^h \text{ and } \pi_x^c > \pi_y^c) \text{ or } (\pi_x^h = \pi_y^h \text{ and } \pi_x^c \geq \pi_y^c), \tag{4}$$

for any $x \in X$ such that $\beta_x = \beta_y$, while if (3) holds for some node $x \in X$, this node can be removed from the tree since the corresponding label is dominated. Using the FIFO data structure to represent X, and conditions (3) and (4) as dominance rules, the following result holds.

Corollary 1 *If X is a FIFO, then the sequence of labels extracted from X and associated with a network node is in lexicographic order.*

Another consequence of Lemma 1 and Corollary 1 is that each label picked in X is non-dominated.

Algorithm 1: MinHop-MinSum maximal set of non-dominated paths determination

---

1 **for** $i \in N$ **do** $Last_i \leftarrow 0$
2 $nX \leftarrow 1; l_{nX} \leftarrow [0, 0, -, s]; X \leftarrow \{1\}$
3 $Last_s \leftarrow 1$
4 $P_N \leftarrow \emptyset$
5 **while** $X \neq \emptyset$ **do**
6 $x \leftarrow$ first node in $X; X \longleftarrow X - \{x\}; i \leftarrow \beta_x$
7   **if** $i = t$ **then** $P_N \leftarrow P_N \cup \{x\}$
8   **for** $j \in N$ such that $(i, j) \in A$ **do**
9 $y \leftarrow Last_j$
10   **if** $y = 0$ or ($y \neq 0$ and $\pi_x^h + 1 > \pi_y^h$ and $\pi_x^c + c_{ij} < \pi_y^c$) **then**
11 $nX \leftarrow nX + 1; l_{nX} \leftarrow [\pi_x^h + 1, \pi_x^c + c_{ij}, x, j]$; Insert nX at the end of X
             $Last_j \leftarrow nX$
12   **else**
13   **if** $\pi_x^h + 1 = \pi_y^h$ and $\pi_x^c + c_{ij} = \pi_y^c$ **then**
14 $nX \leftarrow nX + 1; l_{nX} \leftarrow [\pi_y^h, \pi_y^c, x, j]$; Insert nX at the end of X
15   **else**
16   **if** $\pi_x^h + 1 = \pi_y^h$ and $\pi_x^c + c_{ij} < \pi_y^c$ **then**
17 $l_y \leftarrow [\pi_y^h, \pi_x^c + c_{ij}, x, j]$
18   Remove from X other labels associated with j with the objective values of x

---

Corollary 2 *If X is a FIFO, then a label chosen in X is permanent.*

Two conclusions can be drawn from this latter result. First, non-dominated paths from s to t can be known from the moment labels that correspond to t are chosen in X. Second, in order to check the dominance of a new label it is sufficient to compare it with the latest label inserted in X associated with the same node. In the pseudo-code presented below this information is maintained in the n elements array $Last$.

Although X is manipulated as a FIFO, in the sense that new elements are inserted at the end of the queue while the first one is scanned, an adaptation might have to be done when a new label l_y dominates a previous one, l_x. If l_x is the only label at level π_x^h, it is sufficient to replace l_x by l_y. However, if there are multiple labels associated with β_x at level π_x^h, then they are all dominated and should be deleted. An alternative is to include an additional comparison between labels that are picked in X and the correspondent $Last$, in order to check their dominance.

The pseudo-code of the method just described is presented in Algorithm 1. Lines 1 to 4 outline the initialization of the used variables. Besides the aforementioned variables, P_N is used to store the non-dominated labels of paths from node s to node t. One such path is identified whenever a node x corresponding to node t is selected in

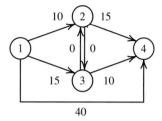

Fig. 1 Network G_1

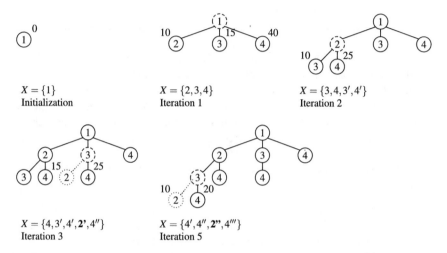

Fig. 2 MinHop-MinSum maximal set of non-dominated paths from node 1 to node 4 in network G_1

X. Lines 8 to 19 describe the analysis of the arcs emerging from the selected node, i. In particular, lines 10 and 14 correspond to the first and the second parts of condition (4), the case where the new label is inserted in set X. The latter case is implemented by line 17, combined with line 13, and means that the new label dominates the current as well as their equivalents. In that case the label l_y replaces the label l_x and the dominated labels are removed from set X.

Figure 2 illustrates the application of Algorithm 1 to the network G_1 in Fig. 1 when the origin and the destination nodes are $s = 1$ and $t = 4$, respectively. In particular, each plot in Fig. 2 represents the tree of paths that have been found in each iteration of the algorithm. The dashed circles highlight the nodes that are scanned at each iteration and the number attached to each circle is the cost of the path it corresponds to.

At the initialization step of Algorithm 1, naturally the first, the queue that stores the nodes yet to scan is set only with to the initial node, $X = \{1\}$. Afterwards:

- In the next iteration node 1 is picked in X and, because $(1, 2)$, $(1, 3)$ and $(1, 4)$ are arcs in G_1, the nodes 2, 3, and 4, are then stored in X, to be scanned in a forthcoming iteration.
- Assuming the new nodes are inserted in X in that order, and because X is managed as a FIFO list, the node to scan at iteration 2 is node 2. In this case new labels are created for nodes that already have a label. These nodes, $3'$ and $4'$, are marked with an additional prime, to distinguish them from the previous labels corresponding to other paths. It can be remarked that $l_3 = [1, 15, 1, 3]$ and $l_4 = [1, 40, 1, 4]$, whereas $l_{3'} = [2, 10, 2, 3]$ and $l_{4'} = [2, 25, 2, 4]$. The latter labels are not dominated by the first, nor the latter dominate the first.
- Node 3 is scanned at iteration 3. The arcs $(3, 2)$ and $(3, 4)$ emerge from node 3, however the label of the new path from node 1 to node 2 would be $[2, 15, 3, 2']$, which is dominated by $l_2 = [1, 10, 1, 2]$. Thus, this new label, marked by a dotted circle in Fig. 2 and in bold in the queue X, is discarded.
- The paths formed by the algorithm do not change in the next iteration because no arcs emerge from node 4, the next one to be scanned. This step is omitted in Fig. 2. The node selected in X at iteration 5 is node 3'. Once again the node $2''$ is not included in the queue X, given that its label is dominated. Still, a new path from node 1 to node 4 is found, represented in X as node $4'''$.
- All the nodes that are still stored in the queue X correspond to the network node 4, therefore no additional labels are created in the remaining iterations of the method.

At the end of Algorithm 1 the maximal set of non-dominated paths from node 1 to node 4 is $\{\langle 1, 4 \rangle, \langle 1, 2, 4 \rangle, \langle 1, 3, 4 \rangle, \langle 1, 2, 3, 4 \rangle\}$. Two of them, $\langle 1, 2, 4 \rangle$ and $\langle 1, 3, 4 \rangle$, both have 2 arcs each and the same cost, 25, thus they are equivalent.

If the goal is to find the non-dominated objective values, i.e. the minimal set of non-dominated paths, then it is sufficient to compute a single path for each objective values pair, which allows to make certain simplifications to the method above. The main difference now is that in such a case at most one label is used for each node and each number of arcs, so Lemma 1 is now replaced by the following result.

Lemma 2 *If X is a FIFO, then at most one label per level is associated with a network node.*

Moreover, Corollaries 1 and 2 are still valid. Thus, concerning the minimal set of non-dominated paths computation a new label l_y should be inserted in X if and only if

$$(\pi_x^h < \pi_y^h \text{ and } \pi_x^c > \pi_y^c) \text{ or } (\pi_x^h = \pi_y^h \text{ and } \pi_x^c > \pi_y^c), \tag{5}$$

for any $x \in X$ such that $\beta_x = \beta_y$, and in the second case, i.e., if

$$\pi_x^h = \pi_y^h \text{ and } \pi_x^c > \pi_y^c, \tag{6}$$

l_x can be discarded. The fact that there is a single label associated with each pair of objective values also allows to replace labels, with no need for deletions, if they become dominated by others.

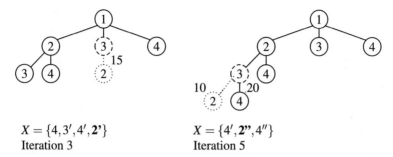

$X = \{4, 3', 4', 2'\}$
Iteration 3

$X = \{4', 2'', 4''\}$
Iteration 5

Fig. 3 MinHop-MinSum minimal set of non-dominated paths from node 1 to node 4 in network G_1

The pseudo-code of Algorithm 2 is a simplified version of Algorithm 1 for finding the minimal set of non-dominated paths. The variables used in the pseudo-code have similar meaning, except P_N which is replaced by $\bar{P}_N$. Line 10 of the pseudo-code corresponds to the case of l_y being added as a possible non-dominated label, whereas Line 14 is executed if l_y replaces the current label of node i in set X.

Algorithm 2: MinHop-MinSum minimal set of non-dominated paths determination

1 **for** $i \in N$ **do** $Last_i \leftarrow 0$
2 $nX \leftarrow 1; l_{nX} \leftarrow [0, 0, -, s]; X \leftarrow \{1\}$
3 $Last_s \leftarrow 1$
4 $\bar{P}_N \leftarrow \emptyset$
5 **while** $X \neq \emptyset$ **do**
6      $x \leftarrow$ first node in X; $X \longleftarrow X - \{x\}$; $i \leftarrow \beta_x$
7   **if** $i = t$ **then** $\bar{P}_N \leftarrow \bar{P}_N \cup \{x\}$
8   **for** $j \in N$ such that $(i, j) \in A$ **do**
9          $y \leftarrow Last_j$
10   **if** $y = 0$ or $(y \neq 0$ and $\pi_x^h + 1 > \pi_y^h$ and $\pi_x^c + c_{ij} < \pi_y^c)$ **then**
11             $\lfloor nX \leftarrow nX + 1; l_{nX} \leftarrow [\pi_x^h + 1, \pi_x^c + c_{ij}, x, j]$; Insert nX at the end of X
12         $Last_j \leftarrow nX$
13   **else**
14             $\lfloor$ **if** $\pi_x^h + 1 = \pi_y^h$ and $\pi_x^c + c_{ij} = \pi_y^c$ **then** $l_y \leftarrow [\pi_y^h, \pi_x^c + c_{ij}, x, j]$

The initialization and the first two iterations of Algorithm 2 when applied to the previous example, the network G_1 in Fig. 1, are similar to the determination of the maximal set of non-dominated paths shown in Fig. 2. Then, when scanning node 3 it is possible to reach node 4 with the label $l_{4''} = [2, 25, 3, 4]$. As seen before, this label is equivalent to $l_{4'} = [2, 2, 4, 25]$, therefore it is discarded, as shown in Fig. 3. The minimal set of non-dominated paths output by this algorithm is $\{\langle 1, 4 \rangle, \langle 1, 2, 4 \rangle, \langle 1, 2, 3, 4 \rangle\}$.

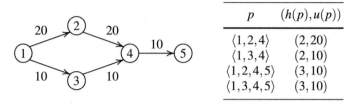

p	$(h(p), u(p))$
$\langle 1,2,4\rangle$	$(2,20)$
$\langle 1,3,4\rangle$	$(2,10)$
$\langle 1,2,4,5\rangle$	$(3,10)$
$\langle 1,3,4,5\rangle$	$(3,10)$

Fig. 4 Efficient solutions formed by weakly efficient subpaths

Regarding the number of operations performed by the algorithms above, we first note that the MinHop-MinSum path problem has up to $n(n-2)$ non-dominated pairs of objective values, if there is a path with every number of arcs between 1 and $n-2$ from s to any other node i. Therefore the tree of paths obtained by Algorithm 2 can have at most $n-2$ levels. In the worst case, every of the m network arcs needs to be scanned once for each of those levels. Analyzing an arc implies the insertion and deletion of an element in X, which can be done in constant time. Therefore the worst-case time complexity of Algorithm 2 is $O(nm)$.

As for Algorithm 1, its theoretical complexity also depends on the number of levels the paths tree can have and on the total number of non-dominated labels. Even though the tree can have up to $n-2$ levels, because there may be multiple labels with the same objective values, the number of labels cannot be polynomially bounded.

4 MinHop-MaxMin Path Problem

The MinHop-MaxMin path problem can be viewed as a special case of the problems solved by Martins and by Gandibleux et al. [9, 13] considering a single cost function and $c_{ij} = 1$ for any $(i,j) \in A$. Like in the previous section, labeling algorithms can also be designed for the MinSum-MaxMin path problem, by including the number of hops and the capacity values in each label and modifying the label dominance test. However, non-dominated paths may contain weakly non-dominated subpaths [9]. Therefore weakly non-dominated labels may be necessary to determine the maximal set of non-dominated paths. As an example, the paths $\langle 1,2,4,5\rangle$ and $\langle 1,3,4,5\rangle$ between nodes 1 and 5 in the network depicted in Fig. 4 are both non-dominated with respect to the MinHop-MaxMin path problem, although the same does not hold for their subpaths given that $\langle 1,2,4\rangle_D\langle 1,3,4\rangle$.

Maintaining the previous notation a label associated with a node x in the search tree of paths from s to other nodes has now the form $l_x = [\pi_x^h, \pi_x^u, \xi_x, \beta_x]$, where π_x^u denotes the path capacity. Furthermore, if $\beta_x = i \in N$ and $(i,j) \in A$, a new label associated with node y can be created, such that

$$l_y = [\pi_x^h + 1, \min\{\pi_x^u, u_{ij}\}, x, j].$$

Now consider that instead of minimizing a cost function, we maximize a capacity function. Given l_x, l_y two labels corresponding to a path starting at s and ending at the same network node, we say l_x is dominated by l_y if and only if

$$(\pi_x^h > \pi_y^h \text{ and } \pi_x^u \leq \pi_y^u) \text{ or } (\pi_x^h \geq \pi_y^h \text{ and } \pi_x^u < \pi_y^u). \qquad (7)$$

Taking into account that in this problem weakly non-dominated labels can be used to obtain non-dominated solutions, namely solutions with the same number of hops but different capacities, a new label l_y should only be discarded if there is another node x in X such that $\beta_x = \beta_y$ and

$$\pi_x^h < \pi_y^h \text{ and } \pi_x^u \geq \pi_y^u. \qquad (8)$$

On the other hand, label l_x can be replaced by label l_y whenever

$$\pi_x^h > \pi_y^h \text{ and } \pi_x^u \leq \pi_y^u. \qquad (9)$$

Back to the determination of the maximal set of non-dominated paths using a FIFO list, because the number of arcs of the scanned paths is non-decreasing condition (9) never holds. This means that no label should be deleted and the dominance test whenever a new label is formed can be replaced simply by (8).

The acceptance of weakly non-dominated labels that may be dominated implies that Corollaries 1 and 2 no longer hold true. For this reason the set P_N can only be known after the labeling process is over. Moreover, it is not enough to compare a new label with the last one observed for that node, and two cases should be distinguished:

1. $\pi_x^h = \pi_y^h$ for some x in X, then node β_x already has a label at level π_y^h of the search tree and neither l_x strictly dominates l_y nor l_y strictly dominates l_x, therefore node y is inserted in set X;
2. $\pi_x^h < \pi_y^h$ for every x in X, then label l_y belongs to a different level than label l_x and it should be inserted if and only if $\pi_x^u < \pi_y^u$, that is, if and only if

$$\max\{\pi_x^u : x \in X \text{ and } \pi_x^h < \pi_y^h\} < \pi_y^u. \qquad (10)$$

In short, a new candidate label will only be accepted if its capacity improves the capacity of the labels with fewer arcs. An auxiliary array storing the best capacity value found for each network node until the latest scanned level, *Best*, is used in Algorithm 3, where the pseudo-code for finding the maximal set of MinHop-MaxMin paths is summarized.

The application of Algorithm 3 to the network G_2 in Fig. 5 is illustrated in Fig. 6. At the initialization phase, naturally the first, the queue that stores the nodes yet to scan is set only to the initial node, $X = \{1\}$. Afterwards:

- In the next iteration node 1 is picked in X and the nodes 2, 3, and 4 are then stored in X, so that they can be scanned in a future iteration.

Algorithm 3: MinHop-MaxMin maximal set of non-dominated paths determination

---

1 **for** $i \in N$ **do**
2 $\quad$ $Best_i \leftarrow 0; \pi_i^u \leftarrow 0$
3 $\quad$ $Last_i \leftarrow 0$
4 $nX \leftarrow 1; l_{nX} \leftarrow [0, +\infty, -, s]; X \leftarrow \{1\}$
5 $Last_s \leftarrow 1$
6 **while** $X \neq \emptyset$ **do**
7 $\quad$ $x \leftarrow$ first node in $X; X \longleftarrow X - \{x\}; i \leftarrow \beta_x$ **for** $j \in N$ such that $(i,j) \in A$ **do**
8 $\quad\quad$ **if** $\min\{\pi_x^u, u_{ij}\} > Best_j$ **then**
9 $\quad\quad\quad$ $y \leftarrow Last_j$
10 $\quad\quad\quad$ **if** $y = 0$ or $(y \neq 0$ and $\pi_x^h + 1 > \pi_y^h)$ **then**
11 $\quad\quad\quad\quad$ $Best_j \leftarrow \pi_{Last_j}^u$
12 $\quad\quad\quad\quad$ **if** $\min\{\pi_x^u, u_{ij}\} > Best_j$ **then**
13 $\quad\quad\quad\quad\quad$ $nX \leftarrow nX + 1; l_{nX} \leftarrow [\pi_x^h + 1, \min\{\pi_x^u, u_{ij}\}, x, j];$ Insert nX
14 $\quad\quad\quad\quad\quad$ at the end of X
15 $\quad\quad\quad\quad\quad$ $Last_j \leftarrow nX$

16 $\quad\quad$ **else**
17 $\quad\quad\quad$ **if** $\pi_x^h + 1 = \pi_y^h$ **then**
18 $\quad\quad\quad\quad$ $nX \leftarrow nX + 1; l_{nX} \leftarrow [\pi_y^h, \min\{\pi_x^u, u_{ij}\}, x, j];$ Insert nX at the end of X
19 $\quad\quad\quad\quad$ **if** $\pi_x^h + 1 = \pi_y^h$ **then** $Last_j \leftarrow nX$

20 $P_N \leftarrow \{$non-dominated paths from s to $x \in X$ where $\beta_x = t\}$

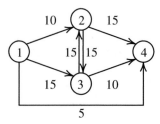

Fig. 5 Network G_2

- Assuming the new nodes are inserted in X in that order, the node 2 is scanned at iteration 2. Then new labels are created for nodes 3 and 4, denoted by $3'$ and $4'$, having $l_{3'} = [2, 10, 2, 3]$ and $l_{4'} = [2, 10, 2, 4]$ as labels. Note that $l_{3'}$ is dominated by the former label of node 3, $l_3 = [1, 15, 1, 3]$.
- Node 3 is scanned at iteration 3. Two new labels are created $l_{2'} = [2, 15, 3, 2]$ and $l_{4''} = [2, 10, 3, 4]$. Additionally, $2'$ and $4''$ are inserted in queue X.
- No new paths are formed in the next two iterations, which correspond to scanning nodes 4 and $4'$. Node $2'$ is the selected node at iteration 5. Once again the node $3''$

is not included in X, given that its label is dominated, but a new path from node 1 to node 4 is found, represented in X as node $4'''$.

- No further paths are created when scanning the nodes remaining in X, $4''$ and $4'''$.

At the end of Algorithm 3 the maximal set of non-dominated paths from node 1 to node 4 is $\{\langle 1,4 \rangle, \langle 1,2,4 \rangle, \langle 1,3,4 \rangle, \langle 1,3,2,4 \rangle\}$, and both $\langle 1,2,4 \rangle$ and $\langle 1,3,4 \rangle$ are equivalent.

Even though weakly non-dominated subpaths have to be generated, node t can be treated differently from the others because only non-dominated t labels are necessary (otherwise the label corresponds to a solution that contains a loop and is dominated). A different dominance test can then be applied, because a new label l_y, $\beta_y = t$, should be discarded if and only if there is another one l_x, $\beta_x = t$, such that

$$(\pi_x^h < \pi_y^h \text{ and } \pi_x^u \geq \pi_y^u) \text{ or } (\pi_x^h = \pi_y^h \text{ and } \pi_x^u > \pi_y^u), \tag{11}$$

and it should replace l_x if

$$\pi_x^h = \pi_y^h \text{ and } \pi_x^u < \pi_y^u. \tag{12}$$

This variation of Algorithm 3 ensures that t labels are non-dominated as soon as they are chosen in X, thus allowing the generation of non-dominated paths between s and t along the labeling process.

If again the aim is the determination of the minimal set of non-dominated paths at most one label is stored for each node in a tree level, therefore no dominated subpaths need to occur in non-dominated solutions and the next result follows.

Proposition 1 *Let $p^* \in P_N$ for the MinHop-MaxMin path problem, then there is $p \in P$ formed by non-dominated subpaths from s to any node such that $(h(p), u(p)) = (h(p^*), u(p^*))$.*

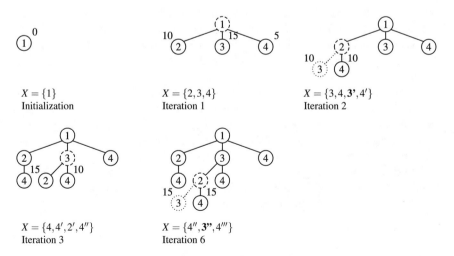

Fig. 6 MinHop-MaxMin maximal set of non-dominated paths from node 1 to node 4 in network G_2

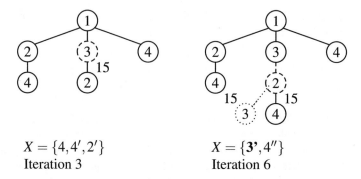

$$X = \{4, 4', 2'\} \qquad\qquad X = \{\mathbf{3'}, 4''\}$$
$$\text{Iteration 3} \qquad\qquad \text{Iteration 6}$$

Fig. 7 MinHop-MaxMin minimal set of non-dominated paths from node 1 to node 4 in network G_2

This result can be used to tighten the dominance test, given that a new label l_y can be discarded if there is l_x such that $\beta_x = \beta_y$ and

$$(\pi_x^h < \pi_y^h \text{ and } \pi_x^u \geq \pi_y^u) \text{ or } (\pi_x^h \leq \pi_y^h \text{ and } \pi_x^u > \pi_y^u),$$

that is, as X is manipulated as a FIFO list, if

$$\pi_x^u \geq \pi_y^u. \tag{13}$$

The same label will replace l_x whenever

$$\pi_x^h = \pi_y^h \text{ and } \pi_x^u < \pi_y^u. \tag{14}$$

Moreover, in this case the labels associated with a network node i correspond to a non-dominated path from s to i, and thus a non-dominated path from s to t is obtained whenever a label associated with t is chosen in X. The resulting method is not very different from Algorithm 2, but its pseudo-code is shown in Algorithm 4 for the sake of completeness.

When applying Algorithm 4 to the network G_2, the first iterations are similar to what was described earlier for Algorithm 3. Then, at iteration 3 the node 3 is scanned and only node $2'$ is added to the previous tree, because label $[2, 10, 3, 4]$ is equivalent to $l_{4'}$. The algorithm continues as described earlier and as depicted in Fig. 7. The final minimal set of non-dominated paths is $\{\langle 1, 4 \rangle, \langle 1, 2, 4 \rangle, \langle 1, 3, 2, 4 \rangle\}$.

The complexity of Algorithm 3 and 4 can be determined as for the MinHop-MinSum path problem. Therefore, the first is not polynomial in time whereas the second has a worst-case time complexity of $\mathcal{O}(mn)$.

Algorithm 4: MinHop-MaxMin minimal set of non-dominated paths determination

---

1 **for** $i \in N$ **do** $Last_i \leftarrow 0$
2 $nX \leftarrow 1; l_{nX} \leftarrow [0, +\infty, -, s]; X \leftarrow \{1\}$
3 $Last_s \leftarrow 1$
4 $\bar{P}_N \leftarrow \emptyset$
5 **while** $X \neq \emptyset$ **do**
6      $x \leftarrow$ first node in $X; X \longleftarrow X - \{x\}; i \leftarrow \beta_x$
7   **if** $i = t$ **then** $\bar{P}_N \leftarrow \bar{P}_N \cup \{x\}$
8   **for** $j \in N$ such that $(i, j) \in A$ **do**
9          $y \leftarrow Last_j$
10   **if** $y = 0$ or $(y \neq 0$ and $\pi_x^h + 1 > \pi_y^h$ and $\min\{\pi_x^u, u_{ij}\} > \pi_y^u)$ **then**
11             $nX \leftarrow nX + 1; l_{nX} \leftarrow [\pi_x^h + 1, \min\{\pi_x^u, u_{ij}\}, x, j];$ Insert nX at the end of X
12             $Last_j \leftarrow nX$
13   **else**
14   **if** $\pi_x^h + 1 = \pi_y^h$ and $\min\{\pi_x^u, u_{ij}\} > \pi_y^u$ **then** $l_y \leftarrow [\pi_y^h, \min\{\pi_x^u, u_{ij}\}, x, j]$

---

5 Computational Results

Computational experiments were carried out to evaluate the empirical performance of the methods described in the previous sections.

A first set of random networks with 1 000, 3 000, 5 000 and 7 000 nodes, dn arcs, for densities $d = 5, 10, 20, 30$, and uniformly integer cost (capacity) values generated in $[1, M]$, with $M = 100$ and $M = 10 000$, was considered. The results presented in the following were obtained over 30 different instances generated for each dimension of this data set.

The second set of instances is based on a simulation of a video traffic routing problem in undirected communication networks used in [6]. These networks have 1 000, 1 500, 2 000, 2 500 and 3 000 nodes and $4n$ arcs. Each node of the network corresponds to a point randomly chosen in a rectangular grid with dimension 400×240 and a mesh size unit of 10 Km. This grid simulates the United States of America geography. Each node has at least 2 and at most 10 neighbour nodes. Additionally, the generated networks contain at least a Hamiltonian path, to guarantee they are connected. The graphs are generated in three steps. In the first step a permutation of the order of the graph nodes is created. The second step consists of linking each pair of consecutive nodes in the permutation with an arc, thus obtaining a Hamiltonian path. Finally, in the last phase the remaining arcs are randomly assigned to pairs of nodes. Two values are associated with any arc, depending on the considered problem. Thus, given the arc (i, j) of the network, where i and j correspond to points (x_i, y_i) and (x_j, y_j) in the initial grid, respectively, the following values are generated:

- arc (i, j)'s delay, in milliseconds, given by

$$d_{ij} = \left(\frac{S^k_{\max}}{r_k} + \frac{S_{\max}}{R_{ij}} \right) + \frac{\ell_{ij}}{2c/3} = \left(\frac{53 \times 8}{1.5} + \frac{53 \times 8}{155.52} \right) \times 10^{-3} + \frac{\ell_{ij}}{200},$$

where $\ell_{ij} = \sqrt{(x_i - x_j)^2 + (y_i - y_j)^2}$ represents the Euclidean distance, in kilometers, between (x_i, y_i) and (x_j, y_j), $c = 300$ Km/ms is the speed of light and for any $(u, v) \in \mathscr{A}, R_{uv} = 155.52 \times 10^6$ bits/s is the bandwidth capacity of arc (u, v). Also:

- $r_k = 1.5 \times 10^6$ bits/s is the token generation rate of the leaky bucket (stochastic model associated with the nodes),
- $S^k_{\max} = S_{\max}$ is the maximum packet size of the flow k (in bits),

and $S_{\max} = 53 \times 8$ bits which is the size of an ATM cell.
- arc (i,j)'s available bandwidth, in Mb/s, that is a random value denoted by $b_{ij} \in \{0.52, 2.52, \ldots, 150.52\}$, which corresponds to a link capacity of 155.52 Mb/s.

For each number of nodes 10 distinct seeds were used to generate networks and for each of them the algorithm was tested by considering $\frac{n^2}{25\,000}$ origin-destination node pairs.

All the tests were executed on an Intel® Core™i7-5820K at 3.3 GHz, with 64 Gb of RAM, and using the compiler gcc 4.8.5 running over openSUSE Leap 42.2.

5.1   MinHop-MinSum Path Problem

In order to evaluate the methods proposed for finding the maximal set of MinHop-MinSum paths two programs were coded in C, namely a labeling algorithm where X is as FIFO list with a standard dominance test, designated F1, and Algorithm 1, designated A1. Similarly, when concerning only the determination of the minimal set of non-dominated paths a standard labeling algorithm using a FIFO, F2, and Algorithm 2, A2, have been implemented.

5.1.1   Random Instances

Tables 1 and 2 present the minimum, mean and maximum numbers of non-dominated paths from s to t, regarding the randomly generated networks for the cases of costs in $[1, 100]$ and in $[1, 10\,000]$, respectively. Even though t can have $n - 2$ different non-dominated labels in the worst-case, the results show the actual number of labels to be much smaller. In the entire test set the registered mean number of P_N elements is always between 2 and 5, increasing slowly with density. When $M = 10\,000$ the results are very similar in general, only with some higher minimum and mean number of non-dominated paths.

A comparison between the mean running times of standard versions of a labeling algorithm, F1 and F2, and the methods introduced for finding the maximal and the

Table 1 Number of MinHop-MinSum non-dominated paths in random instances with $M = 100$

	$n = 1\,000$			$n = 3\,000$			$n = 5\,000$			$n = 7\,000$		
	Min.	Mean	Max.	Min.	Mean	Max.	Min.	Mean	Max.	Min.	Mean	Max.
$d = 5$	1	2.5	4	1	2.3	5	1	2.9	5	1	3.1	6
$d = 10$	2	3.1	6	1	2.7	6	1	2.9	6	1	3.7	6
$d = 20$	1	3.7	6	1	3.4	6	1	3.5	5	2	3.7	6
$d = 30$	1	4.1	6	1	3.7	6	2	3.8	6	2	4.7	7

Table 2 Number of MinHop-MinSum non-dominated paths in random instances with $M = 10\,000$

	$n = 1\,000$			$n = 3\,000$			$n = 5\,000$			$n = 7\,000$		
	Min.	Mean	Max.	Min.	Mean	Max.	Min.	Mean	Max.	Min.	Mean	Max.
$d = 5$	1	2.5	4	1	2.5	5	1	3.0	5	1	3.1	6
$d = 10$	2	3.1	6	1	2.7	6	1	3.2	6	2	3.8	6
$d = 20$	1	3.7	6	1	3.4	6	1	3.6	6	2	4.0	6
$d = 30$	1	4.1	6	1	3.9	6	2	4.0	5	4	5.0	7

minimal sets of non-dominated paths, A1 and A2, is presented in Tables 3 and 4. The first of these tables shows mean values of the improvement obtained by the new algorithms when $M = 100$. The values on the second table refer to instances where $M = 10\,000$. In most cases the CPU time is approximately reduced by half, although in some small size instances (with low density when finding the minimal set) the results are better for the standard versions. However, for these dimensions all running times are very close to 0 s. The algorithmic performance is very similar for the two cost ranges, although the times are slightly greater for the wider range. Also, the speed improvement regarding the standard implementation was, in general, bigger when $M = 10\,000$ rather than when $M = 100$. It is still worth noting that, as expected, the minimal set determination is easier than that of the maximal set. The difference seems to increase with the instances dimension. The CPU times taken by any of the implemented algorithms were very small for small size instances. Figures 8 and 9 show the mean running times for the A1 and A2 codes. The first two plots concern the variation depending on the number of network nodes, while the other two depend on the network density. The plots show there is an increase of the running times with n as well as with d, both for A1 and A2 as the theoretical complexity bound suggests. Still, the determination of the maximal, and of the minimal, sets of MinHop-MinSum paths in the considered data set was made in short times. Both A1 and A2 were able to solve problems with 7 000 nodes and 210 000 arcs in less than 13.20 ms.

Table 3 Percentage mean CPU times improvement of MinHop-MinSum non-dominated paths in random instances with $M = 100$

100× (F1-A1)/F1				
	$d = 5$	$d = 10$	$d = 20$	$d = 30$
$n = 1\,000$	40.0	33.3	50.0	52.0
$n = 3\,000$	43.8	46.7	44.3	45.2
$n = 5\,000$	38.7	43.3	45.2	47.9
$n = 7\,000$	38.5	43.3	43.9	48.1
100× (F2-A2)/F2				
$n = 1\,000$	25.0	28.6	50.0	31.6
$n = 3\,000$	46.7	40.0	51.5	48.2
$n = 5\,000$	41.4	45.6	52.8	52.6
$n = 7\,000$	34.1	57.0	50.5	56.9

Table 4 Percentage mean CPU times improvement of MinHop-MinSum non-dominated paths in random instances with $M = 10\,000$

100× (F1-A1)/F1				
	$d = 5$	$d = 10$	$d = 20$	$d = 30$
$n = 1\,000$	40.0	33.3	55.0	50.0
$n = 3\,000$	25.0	45.2	43.5	53.8
$n = 5\,000$	50.0	55.1	37.8	48.3
$n = 7\,000$	31.4	45.6	51.3	46.3
100× (F2-A2)/F2				
$n = 1\,000$	40.0	44.4	42.9	44.4
$n = 3\,000$	42.9	42.3	47.1	50.5
$n = 5\,000$	53.3	48.1	52.7	53.3
$n = 7\,000$	56.8	50.5	48.7	54.7

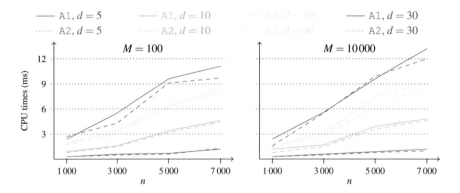

Fig. 8 Mean CPU times MinHop-MinSum non-dominated paths in random instances versus n

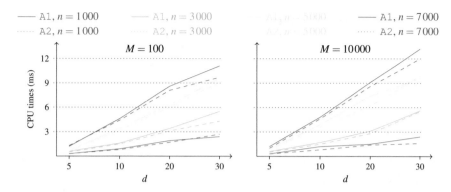

Fig. 9 Mean CPU times MinHop-MinSum non-dominated paths in random instances versus d

5.1.2 Communication Instances

Table 5 shows the minimum, mean and maximum numbers of non-dominated paths from s to t, obtained for the MinHop-MinSum problem on the communication network instances. For these problems the costs associated with the arcs were the delay values defined at the beginning of this section. Similarly to the previous results, the number of non-dominated paths ranged between 1 and 6, with means around 1.9 and slightly increasing with the number of nodes.

On Table 6 the running times of the methods F1 and F2 and of the methods A1 and A2, respectively, are compared. Those values show that an improvement of about 40% when using the new methods. In general this improvement is slightly bigger when finding the maximal set of MinHop-MinSum paths than when finding the minimal set of MinHop-MinSum paths.

Finally, the CPU times for algorithms A1 and A2 for these instances are depicted on Fig. 10. Finding the minimal set of non-dominated paths was easier than finding the maximal set of non-dominated paths. The plots also show a similar growth of the CPU times of both methods with the number of nodes of the network.

Table 5 Number of MinHop-MinSum non-dominated paths in communication instances

	$n = 1\,000$	$n = 1\,500$	$n = 2\,000$	$n = 2\,500$	$n = 3\,000$
Min.	1	1	1	1	1
Mean	1.9	1.9	2.0	2.0	2.0
Max.	5	5	6	6	6

Table 6 Percentage mean CPU times improvement of MinHop-MinSum non-dominated paths in communication instances

	$n = 1\,000$	$n = 1\,500$	$n = 2\,000$	$n = 2\,500$	$n = 3\,000$
100× (F1-A1)/F1	37.5	37.6	39.8	40.0	40.8
100× (F2-A2)/F2	39.2	34.2	35.6	37.2	38.2

Fig. 10 Mean CPU times MinHop-MinSum non-dominated paths versus n in communication instances

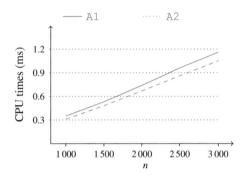

5.2 MinHop-MaxMin Path Problem

5.2.1 Random Instances

The procedure followed for this problem is analogous to the one used in the previous section. Two standard labeling algorithms using a FIFO were coded: one for finding the maximal and the other for finding the minimal sets of MinHop-MaxMin paths, F3 and F4, as well as Algorithm 3 and 4, A3 and A4, respectively.

The minimum, mean and maximum numbers of solutions in the tested instances with capacity values in [1, 100] are provided in Table 7. Similar values for instances with capacity values in [1, 10 000] are shown in Table 8. These values are greater than for the MinHop-MinSum problem. On average the maximal set of non-dominated paths has between 5.5 and 10.1 non-dominated solutions when $M = 100$, and between 5.5 and 11.3 non-dominated solutions when $M = 10\,000$, but both are still far from the upperbound $n - 2$. In general, for instances with $M = 10\,000$ the mean is slightly bigger than with $M = 100$.

Tables 9 and 10 show the relation between the original algorithms, F3 and F4, and the introduced methods, A3 and A4, for the minimal and the maximal sets determination, when $M = 100$ and when $M = 10\,000$, respectively. The improvement on running times is between 68 and 94% for finding all non-dominated paths and between 89 and 89% for finding the non-dominated objective values, regardless of the instance size. As expected these values are greater than the observed for the MinHop-MaxMin

Table 7 Number of MinHop-MinSum non-dominated paths in random instances with $M = 100$

	$n = 1\,000$			$n = 3\,000$			$n = 5\,000$			$n = 7\,000$		
	Min.	Mean	Max.	Min.	Mean	Max.	Min.	Mean	Max.	Min.	Mean	Max.
$d = 5$	1	5.5	15	1	6.5	11	2	6.7	12	2	7.8	17
$d = 10$	3	6.6	12	3	7.6	13	4	7.8	15	4	10.0	16
$d = 20$	4	6.5	8	3	9.5	17	3	7.3	12	6	8.9	12
$d = 30$	4	7.1	15	2	8.2	14	3	8.7	15	3	10.1	18

Table 8 Number of MinHop-MinSum non-dominated paths in random instances with $M = 10\,000$

	$n = 1\,000$			$n = 3\,000$			$n = 5\,000$			$n = 7\,000$		
	Min.	Mean	Max.	Min.	Mean	Max.	Min.	Mean	Max.	Min.	Mean	Max.
$d = 5$	1	5.5	15	1	6.7	11	2	7.1	13	2	8.3	16
$d = 10$	3	7.3	14	5	8.9	13	4	8.1	18	4	11.2	16
$d = 20$	3	6.9	13	3	10.7	20	3	8.4	14	6	10.7	12
$d = 30$	6	8.7	15	2	9.1	21	3	10.1	19	3	11.3	16

Table 9 Percentage mean CPU times improvement of MinHop-MaxMin non-dominated paths in random instances with $M = 100$

$100\times$ (F3-A3)/F3

	$d = 5$	$d = 10$	$d = 20$	$d = 30$
$n = 1\,000$	68.1	83.6	86.4	88.6
$n = 3\,000$	74.3	82.0	87.9	90.2
$n = 5\,000$	72.2	83.7	90.1	91.6
$n = 7\,000$	70.2	84.3	89.8	93.4

$100\times$ (F4-A4)/F4

	$d = 5$	$d = 10$	$d = 20$	$d = 30$
$n = 1\,000$	89.6	95.2	97.8	98.3
$n = 3\,000$	91.2	95.1	97.7	98.3
$n = 5\,000$	91.7	95.7	98.1	98.5
$n = 7\,000$	92.0	94.9	97.5	98.6

path problem, as much more weakly non-dominated labels now have to be stored. The general tendency is the same both for $M = 100$ and $M = 10\,000$.

The difference in performance when determining the maximal and the minimum sets of non-dominated paths is also evident on the plots in Figs. 11 and 12, which show the running times of A3 and A4 variation with n and with d, respectively. Solid lines, depicting A3 results, grow much faster than dashed lines, for A4. The growth seems to present a linear behaviour with n and with d. For larger instances it took in average 83.50 ms to find the maximal set of non-dominated paths and 20.30 ms to find the minimal set of non-dominated paths.

Table 10 Percentage mean CPU times improvement of MinHop-MaxMin non-dominated paths in random instances with $M = 10\,000$

100× (F3-A3)/F3				
	$d = 5$	$d = 10$	$d = 20$	$d = 30$
$n = 1\,000$	68.1	79.5	87.8	88.9
$n = 3\,000$	73.4	84.9	88.8	89.6
$n = 5\,000$	77.5	84.3	87.8	92.8
$n = 7\,000$	74.0	82.4	90.2	93.1
100× (F4-A4)/F4				
$n = 1\,000$	91.1	95.6	97.7	97.7
$n = 3\,000$	92.8	95.4	97.4	98.0
$n = 5\,000$	91.7	95.4	97.7	98.6
$n = 7\,000$	92.4	95.0	97.7	98.5

5.2.2 Communication Instances

For this set of problems the capacity values associated with the each arc were the available bandwidth values defined earlier in this section. According to Table 11 there were between 1 and 31 MinHop-MaxMin paths on the communication network instances, which, like before, is a higher number than for the MinHop-MinSum problem. The mean number of non-dominated paths seems to increase slowly with the number of network nodes.

On Table 12 the running times of the methods F3 and F4 and the methods A3 and A4 are compared. Those values show an improvement of around 80% when using the new methods for finding the maximal set of MinHop-MaxMin paths and around 94% when finding the minimal set of MinHop-MaxMin paths. In both cases the improvement increases with the number of nodes in the network.

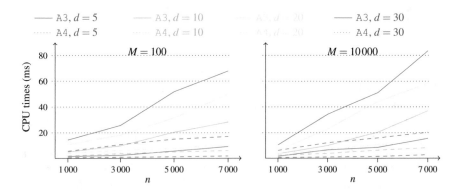

Fig. 11 Mean CPU times MinHop-MaxMin non-dominated paths in random instances versus n

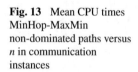

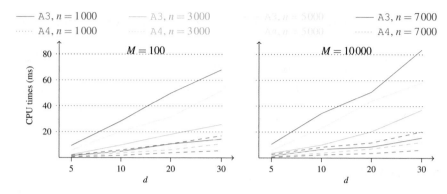

Fig. 12 Mean CPU times MinHop-MaxMin non-dominated paths in random instances versus d

Fig. 13 Mean CPU times MinHop-MaxMin non-dominated paths versus n in communication instances

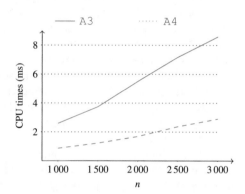

Figure 13 presents the mean CPU times of algorithms A3 and A4 for the communication instances. The results follow the tendencies observed for the MinHop-MinSum problem and the methods ran in a mean time of at most 0.6 ms.

Table 11 Number of MinHop-MaxMin non-dominated paths in communication instances

	$n = 1\,000$	$n = 1\,500$	$n = 2\,000$	$n = 2\,500$	$n = 3\,000$
Min.	1	1	1	1	1
Mean	6.9	7.0	7.3	7.4	7.6
Max.	21	24	24	29	31

Table 12 Percentage mean CPU times improvement of MinHop-MaxMin non-dominated paths in communication instances

	$n = 1\,000$	$n = 1\,500$	$n = 2\,000$	$n = 2\,500$	$n = 3\,000$
100× (F3-A3)/F3	76.6	78.9	79.9	80.8	81.2
100× (F4-A4)/F4	93.2	94.3	94.6	94.6	94.6

6 Conclusions

Labeling algorithms for bicriteria path problems minimizing the number of hops and either the path cost or the path capacity have been described, aiming at the computation of the maximal and the minimal sets of non-dominated paths. These methods make use of a breadth-first search tree by managing the set of labels as a FIFO and list node labels by non-decreasing order of the number of hops. Tuning the dominance tests according with this structure leads to non-polynomial algorithms for the problems of finding the maximal sets of non-dominated paths, for the MinHop-MinSum and the MinHop-MaxMin path problems. The minimal set versions of those algorithms has worst-case time of $\mathcal{O}(mn)$.

The new methods have shown an improvement between 25 and 57% for the MinHop-MinSum path problem running times and between 68 and 98% for the MinHop-MaxMin path problem over randomly generated instances. For a set of instances that simulate video traffic the improvement was between 34 and 41% in the first case and between 76 and 95% in the second. Additionally, numerical results indicate that only 13.20 ms are necessary to find the whole set of non-dominated paths in the first case, and 83.50 ms are needed in the second case, for instances with 7 000 nodes and 210 000 arcs. The determination of the non-dominated objective values was completed with 12.00 ms for the MinHop-MinSum and within 20.30 ms for the MinHop-MaxMin path problems over the same test bed. The biggest video traffic instances are defined over undirected networks with 3 000 nodes and 12 000 edges. For these instances the introduced algorithms were able to find the maximal, minimal, complete set of non-dominated paths for the MinHop-MinSum problem in less than 1.16, 1.05 ms. For the MinHop-MaxMin problem the same time bounds were of 8.58 ms, and 2.90 ms, respectively.

Future lines of research include comparing the introduced methods with other type of algorithms for bicriteria path problems, as well as the adaption of the introduced methods to problems with additional constraints, like the telecommunication problems mentioned in the introduction.

Acknowledgements This work was partially supported by the Institute for Systems Engineering and Computers at Coimbra – UID/MULTI/00308/2013, the Centre for Mathematics of the University of Coimbra – UID/MAT/00324/2013, and the grant SFRH/BSAB/113683/2015, funded by the

Portuguese Government through FCT/MEC and co-funded by the European Regional Development Fund through the Partnership Agreement PT2020.

References

1. R.K. Ahuja, T.L. Magnanti, J.B. Orlin, *Network Flows: Theory Algorithms and Applications* (Prentice Hall, Englewood Cliffs, NJ, 1993)
2. A. Balakrishnan, K. Altinkemer, Using a hop-constrained model to generate alternative communication network design. ORSA J. Comput. **4**, 192–205 (1992)
3. C. Bornstein, N. Maculan, M. Pascoal, L.L. Pinto, Multiobjective combinatorial optimization problems with a cost and several bottleneck objective functions: An algorithm with reoptimization. Comput. Oper. Res. **39**, 1969–1976 (2012)
4. J. Brumbaugh-Smith, D. Shier, An empirical investigation of some bicriterion shortest path algorithms. Eur. J. Oper. Res. **43**, 216–224 (1989)
5. G. Cheng, N. Ansari, Finding a least hop(s) path subject to multiple additive constraints. Comput. Commun. **29**, 392–401 (2006)
6. J. Clímaco, J. Craveirinha, M. Pascoal, A bicriterion approach for routing problems in multimedia networks. Networks **11**, 399–404 (2003)
7. T.H. Cormen, C.E. Leiserson, R.L. Rivest, C. Stein, *Introduction to Algorithms* (MIT Press, Cambridge, MA, 2001)
8. G. Dahl, L. Gouveia, On the directed hop-constrained shortest path problem. Oper. Res. Lett. **32**, 15–22 (2004)
9. X. Gandibleux, F. Beugnies, S. Randriamasy, Multi-objective shortest path problems with a maxmin cost function. 4OR - Quart. J. Belgian, Fr. Ital. Oper. Res. Soc. 4 47–59 (2006)
10. R. Guerin, A. Orda, Computing shortest paths for any number of hops. IEEE/ACM Trans. Netw. **10**, 613–620 (2002)
11. P. Hansen, Bicriterion path problems in *Multiple Criteria Decision Making: Theory and Applications*, ed. by G. Fandel T. Gal. Lectures Notes in Economics and Mathematical Systems, vol. 177,(Springer, Heidelberg, 1980), pp. 109–127
12. M. Iori, S. Martello, D. Pretolani, An aggregate label setting policy for the multi-objective shortest path problem. Eur. J. Oper. Res. **207**, 1489–1496 (2010)
13. E. Martins, On a multicriteria shortest path problem. Eur. J. Oper. Res. **16**, 236–245 (1984)
14. E. Martins, On a special class of bicriterion path problems. Eur. J. Oper. Res. **17**, 85–94 (1984)
15. L. Pinto, C. Bornstein, N. Maculan, The tricriterion shortest path problem with at least two bottleneck objective functions. Eur. J. Oper. Res. **198**, 387–391 (2009)
16. S. Randriamasy, L. Fourniã©, D. Hong. Distributed adaptive multi-criteria load balancing: Analysis and end to end simulation. *INFOCOM Poster and Demo Session* (2006)
17. W. Riedl, A complete characterization of jump inequalities for the hop-constrained shortest path problem. Discret. Appl. Math. **225**, 95–113 (2017)
18. A.J.V. Skriver, K.A. Andersen, A label correcting approach for solving bicriterion shortest-path problems. Computers and Operations Research **27**, 507–524 (2000)

Distance-Vector Algorithms for Distributed Shortest Paths Computation in Dynamic Networks

Gianlorenzo D'Angelo, Mattia D'Emidio and Daniele Frigioni

Abstract Computing and updating distributed shortest paths is a core functionality of today's communication networks. The solutions known in the literature are classified into two categories, namely *Distance-Vector* and *Link-State* algorithms. Distance-Vector algorithms usually require each node of the network to store the *distance* toward every other node in a data structure called *routing table*, thus requiring linear storage per node. Such a data structure is used to compute the *next hop* to be used to forward data toward any destination node of interest. This is usually done by solving very simple equations, thus requiring few computational time per node. The main drawback of Distance-Vector algorithms is that, in dynamic scenarios, they can suffer of the *looping* and *count-to-infinity* phenomena, though quite efficient countermeasures for such issues are known. Link-State algorithms, instead, require a node of the network to know and store the entire network topology, to compute its distance and next hop toward any destination. This is usually done by means of a centralized shortest-path algorithm, hence requiring quadratic storage and rather high computational effort per node. The main drawback of Link-State algorithms is that, notwithstanding they do not incur in *looping* and *count-to-infinity* problems, they perform quite poorly in dynamic scenarios, since nodes need to receive and store up-to-date information on the entire network topology after each change. In the last years, there has been a renewed interest in devising new lightweight distributed shortest-path solutions for large-scale Ethernet networks, where Distance-Vector algorithms are an attractive alternative to Link-State solutions when scalability and reliability are key issues or when the memory resources of the nodes

G. D'Angelo · M. D'Emidio
Gran Sasso Science Institute (GSSI), Viale Francesco Crispi, I-67100 L'Aquila, Italy
e-mail: gianlorenzo.dangelo@gssi.infn.it

M. D'Emidio
e-mail: mattia.demidio@gssi.infn.it

D. Frigioni (✉)
Department of Information Engineering, Computer Science and Mathematics,
University of L'Aquila, Via Vetoio, I-67100 L'Aquila, Italy
e-mail: daniele.frigioni@univaq.it

© Springer International Publishing AG, part of Springer Nature 2018
A. Adamatzky (ed.), *Shortest Path Solvers. From Software to Wetware*,
Emergence, Complexity and Computation 32,
https://doi.org/10.1007/978-3-319-77510-4_4

99

of the network are limited. In this chapter, we hence focus on Distance-Vector solutions by reviewing classic approaches and recent algorithmic developments in this category.

1 Introduction

The problem of computing and updating shortest paths in a distributed network whose topology dynamically changes over the time is a core functionality of today's communication networks, which has been widely studied in the literature. Broadly speaking, solutions found are classified into two categories, namely *Distance-Vector* and *Link-State* algorithms.

Distance-Vector algorithms usually require each node of the network to store (at least) the *distance* (i.e. the weight of a shortest path) toward every other node of the network and to store it in a data structure called *routing table*, thus, most of the times, requiring linear storage per node. Such a data structure is used, occasionally with some auxiliary data, to compute the *next hop* (i.e. the next node on a shortest path) to be used to forward data toward any destination node of interest. This is usually done by solving very simple equations, thus being very parsimonious solutions from the computational point of view per node. The majority of the known Distance-Vector solutions (see, e.g., [1–6] and references therein) are based on the classical *Distributed Bellman-Ford* (DBF) approach, introduced for the first time in Arpanet in the late 60s [7], and still used in some real-world networks, as part of the RIP protocol [8]. DBF has been shown to converge to the correct distances if the link weights stabilize and all cycles have positive lengths [9]. However, the convergence time can be very high (and possibly infinite) due to the well-known *looping* and *count-to-infinity* phenomena (tough quite efficient countermeasures for such issues are known). Furthermore, if the nodes of the network are not synchronized, even in the *static* case, i.e. when no change occurs in the network, the overall number of messages sent by DBF is, in the worst case, exponential with respect to the size of the network [10].

Link-State algorithms, as for example the *Open Shortest Path First* (OSPF) protocol, are widely used in the Internet [11] and require each node of the network to know and store the entire network topology to compute its distance to any destination, usually by running the centralized Dijkstra's algorithm [12]. Thus, they induce a space occupancy per node that is quadratic in the number of nodes of the network. Link-State algorithms do not incur in both looping and count-to-infinity phenomena. However, they perform quite poorly in dynamic scenarios, where each node needs to receive and store up-to-date information on the entire network topology after any change. This is achieved by broadcasting each modification affecting the network topology to all nodes [8, 11, 13], and by using a centralized algorithm for dynamic shortest paths, as for example those described in [14–16].

In the last years, there has been a renewed interest in devising new efficient and light-weight distributed shortest-path solutions for large-scale Ethernet networks (see, e.g., [17–23]), where Distance-Vector algorithms seem to be an attractive alternative to Link-State solutions either when scalability and reliability are key issues or when the memory resources of the nodes of the network are limited. This is the reason why, in this chapter, we focus on Distance-Vector solutions and both review classic approaches and overview the most recent and efficient solutions of the category.

The most important Distance-Vector algorithm in the literature is surely the *Diffuse Update Algorithm* (DUAL) [24], which is part of CISCO's widely used *Enhanced Interior Gateway Routing Protocol* (EIGRP) [25]. DUAL is more complex than DBF, and uses different data structures in order to guarantee freedom from looping and count-to-infinity phenomena. Another loop-free Distance-Vector algorithm, named *Loop Free Routing* (LFR), has been proposed in [26]. Compared with DUAL, LFR has the same theoretical message complexity but it uses an amount of data structures per node which is always smaller than that of DUAL. Moreover, in [26] LFR has been experimentally shown to be very effective in terms of both messages sent and memory requirements per node in some real-world networks of particular interest.

Recently a general technique, named *Distributed Computation Pruning* (DCP), has been introduced in [27]. It can be combined with any Distance-Vector algorithm in order to overcome limitations, such as high number of messages sent, high space occupancy per node, low scalability, or poor convergence. DCP has been designed to be efficient on networks following a power-law node degree distribution, which are often simply referred as *power-law networks*. Such class of networks is of particular practical relevance, since it includes some of the most important nowadays network applications. The Internet, the majority of modern wireless sensor networks, and social networks are examples of power-law networks. The main idea that DCP tries to exploit is that a power-law network with n nodes typically has average node degree much smaller than n (usually a small constant) and a high number of nodes with small degree (less than 3). Nodes with small degree often do not provide any useful information for the distributed computation of shortest paths, in the sense that there are many topological situations in which these nodes should neither perform nor be involved in any kind of distributed computation, as their shortest paths depend on those of higher degree nodes. In [27] the effectiveness of DCP has been shown via an extensive experimental evaluations conducted within OMNeT++ [28], a network simulator widely used in the literature. As input to the algorithms, instances of power-law networks similar to those considered in [17, 26] were used, that is the Internet topologies of the *IPv4 topology dataset* [29] of the *Cooperative Association for Internet Data Analysis* (CAIDA), which provides data and tools for the analysis of the Internet infrastructure, and the random topologies generated by the *Barabási-Albert* algorithm [30].

This chapter is organized as follows. In Sect. 2 we give all the necessary background and notation. In Sect. 3 we review DBF. In Sects. 4 and 5 we survey DUAL and LFR, respectively. In Sect. 6 we describe DCP, show how to combine it with both DUAL and LFR, and overview the experimental study of [27] which gives evidence of the practical effectiveness of DCP. Finally, in Sect. 7 we give some concluding remarks.

2 Background

In this section, we provide all the necessary background and notation that will be used through the chapter. As a general assumption, we consider the scenario where a network is made of processors which are connected through (bidirectional) communication channels and exchange data using a message passing model, in which:

- each processor can send messages only on its own communication channels, i.e. to processors it is connected with;
- messages are delivered to their destination within a finite delay;
- there is no shared memory among the processors;
- the system is *asynchronous*, that is a sender of a message does not wait for the receiver to be ready to receive the message. The message is delivered within a finite but unbounded time.

2.1 Asynchronous System

The asynchronous system considered in this chapter is based on that described in [31], which is briefly summarized below. The *state* of a processor v is the content of the data structure stored by processor v. The *network state* is the set of states of all the processors in the network plus the network topology and the channel weights. An *event* is the reception of a message by a processor or a change to the network state. When a processor p sends a message m to a processor q, m is stored in a buffer located at q. When q reads m from its buffer and processes it, the event "reception of m" occurs. Messages are transmitted through the channels in *First-In-First-Out (FIFO)* order, that is, messages arriving at processor q are always received in the same order as they are sent by p. An *execution* is a (possibly infinite) sequence of network states and events. A non-negative integer number is associated to each event, the *time* at which that event occurs. The time is a *global* parameter and is not accessible to the processors of the network. The time must be non-decreasing and must increase without any bound, if the execution is infinite. Finally, events are ordered according to the time at which they occur. Several events can happen at the same time as long as they do not occur on the same processor. This implies that the times related to a single processor are strictly increasing.

2.2 Graph Notation

We represent a network by an undirected weighted connected graph $G = (V, E, w)$, where V is a finite set of n nodes, one for each processor, E is a finite set of m edges, one for each (bidirectional) communication channel, and w is a weight function $w : E \rightarrow \mathbb{R}^+$ that assigns to each edge a real value representing the optimization parameter associated to the corresponding channel, such as, e.g. latency. Regarding the notation, given a graph $G = (V, E, w)$, we will denote by:

- (v, u) an edge of E that connects nodes $v, u \in V$, and by $w(v, u)$ its weight, respectively;
- $N(v) = \{u \in V : (v, u) \in E\}$ the set of neighbors of each node $v \in V$;
- $deg(v) = |N(v)|$ the degree of v, for each $v \in V$;
- $maxdeg = \max_{v \in V} deg(v)$ the maximum degree among the nodes in G.

Furthermore, we will use $\{u, \ldots, v\}$ to represent a generic path in G between nodes u and v and, given a path $P = \{u, \ldots, v\}$, we will use $w(P)$ to denote its *weight*, i.e. the sum of the weights associated to its edges. A path $P = \{u, \ldots, v\}$ is called a *shortest path* between u and v if and only if P is a path having minimum weight among all possible paths between u and v in G. Given two nodes $u, v \in V$, we will denote by $d(u, v)$ the topological *distance* between u and v, i.e. the weight of a *shortest path* between u and v. Finally, we will call $via(u, v)$ the *via* from u to v, i.e. the set of neighbors of u (there might be more than one) that belong to a shortest path from u to v. More formally, $via(u, v) \equiv \{z \in N(u) \mid d(u, v) = w(u, z) + d(z, v)\}$.

2.3 Dynamic Networks

In this chapter, we consider a set of common realistic assumptions that have been considered in the great majority of the works on distributed shortest paths. In particular, we focus on the case of *dynamic networks*, i.e. networks that vary over time due to change operations occurring on the processors or on the communication channels, respectively. We denote a sequence of update operations on the edges of the graph G representing the network by $\mathscr{C} = \{c_1, c_2, \ldots, c_k\}$. Assuming $G_0 \equiv G$, we denote by $G_i, 0 \leq i \leq k$, the graph obtained by applying c_i to G_{i-1}. Without loss of generality, we restrict our focus on the case where operation c_i either increases or decreases the weight of an existing edge in G_i, as insertions and deletions of nodes and edges can be easily modelled as weight changes (see, e.g., [17] for more details). Moreover, we consider the case of networks in which a change in the weight of an edge (either increase or decrease) can occur while one or more other edge weight changes are under processing. A processor v of the network might be affected by a subset of these changes. As a consequence, v could be involved in the *concurrent* executions related to such changes. We will use $w^t()$, $d^t()$, and $via^t()$ to denote a given edge weight, distance, or via in graph G_t, respectively.

2.4 Complexity Measures

In the remainder of the chapter, the performance of some of the considered algorithms will be measured in terms of two parameters, namely δ and Δ, which have been considered in several works on the matter (see, e.g. [1, 2, 24, 26] and reference therein) since they capture pretty well the amount of distributed computation that has to be carried out to update the shortest paths in dynamic networks, as a consequence of one or more update operations.

In more details, given a sequence $\mathscr{C} = \{c_1, c_2, ..., c_k\}$ of update operations, we define parameter $\sigma_{c_i,s}$ to represent, for each operation c_i and for each node s, the set of nodes that change either the distance or the via toward s as a consequence of c_i. More formally, for each operation c_i and for each node s, such parameter is defined as

$$\sigma_{c_i,s} = \{v \in V \mid d^{t_i}(v, s) \neq d^{t_{i-1}}(v, s) \text{ or } via^{t_i}(v, s) \neq via^{t_{i-1}}(v, s)\}.$$

If a node $v \in \bigcup_{i=1}^{k} \bigcup_{s \in V} \sigma_{c_i,s}$, then v is said to be *affected*. We denote by Δ the overall number of affected nodes, $\Delta = \sum_{i=1}^{k} \sum_{s \in V} |\sigma_{c_i,s}|$. Furthermore, given a generic destination s in V, $\sigma_s = \bigcup_{i=1}^{k} \sigma_{c_i,s}$ and $\delta = \max_s |\sigma_s|$. It follows that a node can be affected for at most δ different sources.

Finally, to properly analyse the behaviour of the solutions described in this chapter with respect to convergence time, we will consider the so-called FIFO network scenario which can be briefly summarized as follows. As shown in [32], the performance of a distributed algorithm in the asynchronous model depend on the time needed by processors to execute the local procedures of the algorithm and on the delays incurred in the communication among nodes. Moreover, these parameters influence the scheduling of the distributed computation and hence the number of messages sent. For these reasons, in order to correctly evaluate the performance of a distributed algorithm, the realistic case is considered, where the weight of an edge models the time needed to traverse such edge (the delay occurring on that edge if a packet is sent on it) and all the processors require the same time to process every procedure (the delay occurring on a processor if a procedure is performed on it), which is assumed to be instantaneous. In this way, the distance between two nodes models the minimum time that such nodes need to communicate. Then, the time complexity is measured as the number of steps performed by the processors, that is the number of times that a processor performs a procedure.

2.5 Distance-Vector Algorithms

In this section, we summarize the main characteristics of Distance-Vector algorithms that will be useful for their description and analysis. Distance-Vector algorithms based on shortest paths are usually able to handle concurrent updates, and share a set of common features which can be briefly summarized as follows. Given a weighted graph $G = (V, E, w)$, a generic node v of G executing a Distance-Vector algorithm:

- knows the identity of any other node of G, as well as the identity of its neighbors and the weights of its adjacent edges;

- maintains a routing table that has n entries, one for each $s \in V$, which consists of at least two fields:

  - the *estimated distance* $D_v[v, s]$ towards s, i.e. an estimation on $d(v, s)$, that will converge to the correct value in finitely many steps;
  - the *estimated via* (often also called *next hop*) $\text{VIA}_v[s]$, i.e. an estimation on one or more elements of $via(v, s) \equiv \{z \in N(v) \mid d(v, s) = w(v, z) + d(z, s)\}$;

- handles edge weight increases and decreases either all together, by a single procedure, or by two separate routines; in the former case (see, e.g., [24]), we will denote such unified routine by HANDLECHANGEW, while in the latter case (see, e.g., [17]), we will denote the two procedures by HANDLEINCREASEW and HANDLEDECREASEW, respectively;

- requests data to neighbors, regarding estimated distances, and receives the corresponding replies from them, through a dedicated exchange of messages (for instance, by sending a *query* message, like in [24], or by sending a *get.feasible.dist* message, like in [26]);

- propagates a variation, occurring on an estimation on the distance or on the via, to the rest of the network as follows:

  - if v is performing HANDLECHANGEW, then it sends out to its neighbors a dedicated notification message (from now on denoted by *update*); a node that receives this kind of message executes a corresponding routine, from now on denoted by HANDLEUPDATE;
  - if v is performing HANDLEINCREASEW (HANDLEDECREASEW, respectively) then it sends to its neighbors a dedicated notification message (denoted from now on by *increase* or *decrease*, respectively); a node that receives an *increase* (*decrease*, respectively) message executes a corresponding routine, from now on denoted by HANDLEINCREASE (HANDLEDECREASE, respectively).

It is known that a Distance-Vector algorithm can be designed to be free of looping or count-to-infinity phenomena by incorporating suitable sufficient conditions in the routing table update procedures. Three such conditions are given in [24]. The less restrictive, and easier to implement, of the conditions in [24] is the so-called SOURCE NODE CONDITION (SNC), which can be implemented to work in combination with a Distance-Vector algorithm if and only if such an algorithm maintains, besides the already mentioned routing table, a so-called *topology table*. The topology table of a node v has to contain enough information for v to be able to determine, for each $u \in N(v)$ and for each $s \in V$, the quantity $D_v[u, s]$, i.e. an estimation on the distance from u to s as it is known to v. Such values are then exploited by the SNC to establish whether a path is free of loops as follows. If, at time t, v needs to change $\text{VIA}_v[s]$ for some $s \in V$, then it can select as new via any neighbor $k \in N(v)$ satisfying both the conditions of the following *loop-free test*:

1. $D_v[k, s](t) + w^t(v, k) = \min_{v_i \in N(v)}\{D_v[v_i, s](t) + w^t(v_i, v)\}$, and
2. $D_v[k, s](t) < D_v[v, s](t)$,

Fig. 1 A graph G before and after a weight increase on the edge (s, v). Edges here are labeled with their weights

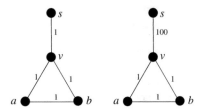

where $D_v[i, s](t)$ denotes, in this case, the estimated distance of neighbor $i \in N(v)$ as it is known to v at time t. If no such neighbor exists, then $\text{VIA}_v[s]$ does not change. If $\text{VIA}_G[s](t)$ denotes the directed subgraph of G induced by the set $\{\text{VIA}_v[s](t), \text{ for each } v \in V\}$, of the estimated vias, at time t, then the following result holds.

Theorem 1 ([24]) *Let G be a network whose $\text{VIA}_G[s](t_0)$ is loop-free at time t_0. If G undergoes a sequence of updates starting at a certain time $t' \geq t_0$ and SNC is used when nodes have to change their via, then $\text{VIA}_G[s](t)$ remains loop-free, for any $t \geq t' \geq t_0$.*

3 Distributed Bellmann-Ford

This section summarizes the main characteristics of the Distributed Bellmann-Ford (DBF) method. DBF requires each node v in the network to store the last known estimated distance $D_v[u, s]$ towards any other node $s \in V$, received from each neighbor $u \in N(v)$. In DBF, a node v updates its estimated distance $D_v[v, s]$ toward a node s by simply executing the iteration $D_v[v, s] := \min_{u \in N(v)}\{w(v, u) + D_v[u, s]\}$, when needed. As already mentioned in Sect. 1, it is known that DBF suffers of the well-known looping and count-to-infinity problems, which arise when a certain kind of link failure or weight increase operation occurs in the network. In Fig. 1, we show a classical topology where DBF counts to infinity. In particular, the left and right sides of such a figure show a graph G before and after a weight modification occurring on edge (s, v). In Fig. 2, we show the corresponding steps required by DBF to update both the distance and the via towards a distinguished node s, for each node of G, as a consequence of the change.

In detail, when the weight of edge (s, v) increases to 100, node v updates its distance and via towards s by setting $D_v[v, s]$ to 3 and $\text{VIA}_v[s]$ to node b. In fact, v knows that the distance from a (and b) to s is 2, while the weight of edge (v, s) is 100. Note that, v cannot know that the path from a to s with weight 2 is that passing through edge (v, s) itself. Now, we concentrate on the operations performed by nodes a and b. When node a (b, respectively) performs the updating step, it finds out that its new estimated via towards s is b (a, respectively) and its new distance is 3. In fact, according to a's information $D_a[v, s] = 3$ and $D_a[b, s] = 2$, therefore $w(a, b) + D_a[b, s] < w(a, v) + D_a[v, s]$. Subsequent updating steps (but the last one)

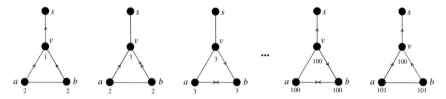

Fig. 2 The sequence of recomputations of $D_u[u, s]$ and $\text{VIA}_u[s]$ by a node u executing DBF. The value close to a node denotes its distance towards s while an arrowhead from x to y in edge (x, y) indicates that node y is the estimated via of x towards s

do not change the estimated via to s of both a and b, but only the estimated distances. For each updating step the estimated distances increase by 1 (i.e., by the weight of edge (a, b)). The counting stops after a number of updating steps that depends on the new weight of edge (s, v) and on the weight of edge (a, b). Note that, if edge (s, v) is deleted (i.e. his weight is set to ∞), the algorithm does not terminate.

In Fig. 3 we give an example of the execution of DBF on another (simple) network (which is part of an example in [24]) where a weight increase operation occurs and the algorithm does not count to infinity. In the figure, the value close to a node indicates its distance to node s and an arrowhead from x to y in edge (x, y) indicates that node y is the successor of x towards node s. An arrowhead from x to y close to edge (x, y) denotes that node x is sending a message to y containing the current distance from s to t, the value of such distance is reported close to the arrow.

At a certain point in time, edge (b, s) changes its weight from 2 to 10 (see Fig. 3a). When node b detects the weight increase, it updates the value of $D_b[b, s]$ to the minimum possible value, that is $D_b[b, s] = \min_{u \in N(b)}\{w(b, u) + D_b[u, s]\} = w(b, c) + D_b[c, s] = 4$. Then, node b sends $D_b[b, s]$ to all its neighbors (Fig. 3b). As a consequence of such messages, nodes a and c update $D_a[b, s]$ and $D_c[b, s]$, respectively, compute their optimal distances to s that are 4 and 5, respectively, and send them to their own neighbors (Fig. 3c). Nodes s and d only update $D_s[b, s]$ and $D_d[b, s]$, respectively. In Fig. 3d, node b updates $D_b[c, s]$ to 5 as a consequence of the message sent by c. As c was the successor node of b towards s, b needs to update $D_b[b, s]$ to $\min_{u \in N(b)}\{w(b, u) + D_b[u, s]\} = w(b, a) + D_b[a, s] = 5$. After this update, b sends $D_b[b, s]$ to its neighbors. Node d behaves similarly by updating its distance to s to 6. In Fig. 3e–g, the message sent by b is propagated to nodes c and d in order to update the distances from this nodes to s.

As a concluding remark of this section, we recall that, if the nodes of the network are not synchronized, even in the *static* case, i.e. when no change occurs in the network, it can be shown that overall number of messages sent by DBF is, in the worst case, exponential with respect to the number of nodes in the network as stated in the next theorem [33, Chap. 15].

Theorem 2 ([33]) *Let n be any even number, $n \geq 4$. Then there is a weighted graph G with n nodes, in which the DBF algorithm sends at least $\Omega(c^n)$ messages in the worst case, for some constant $c > 1$.*

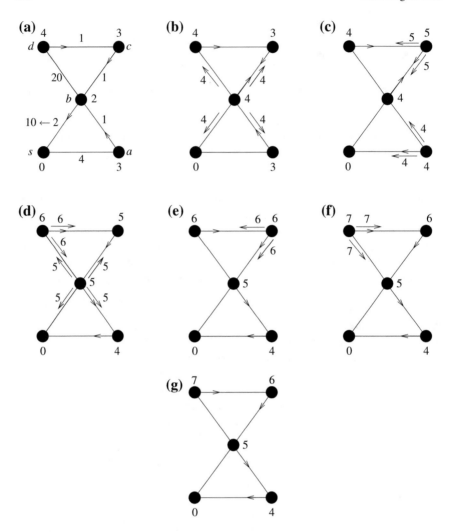

Fig. 3 Example of execution of DBF

4 Diffuse Update Algorithm

This section describes the main characteristics of the Diffuse Update Algorithm (DUAL), and analyses its complexity, under the FIFO networks assumption. We omit the proofs of correctness and formal initialization of the flag variables. The full details are presented in [24]. The main steps of the algorithm are executed every time a weight change $c_i \in \mathscr{C} = \{c_1, c_2, ..., c_k\}$ occurs on an edge (x_i, y_i).

4.1 Data Structures

DUAL is more complex than classical Distance-Vector algorithms, like DBF, and uses different data structures in order to guarantee freedom from looping and counting to infinity. For each node v and for each destination s, it stores a slightly modified routing table where the two fields: the first is a numerical value $D_v[v, s]$ that represents the *estimated distance*; the second field is a vertex $FS_v[s]$ that represents the *feasible successor* of v towards s. This latter field takes the place of the standard value of $VIA_v[s]$, and represents an estimation on $via(v, s)$ that is always guaranteed to induce a loop-free $VIA_G[s](t)$ at any time t [24]. In order to compute $FS_v[s]$, DUAL requires that each node v be able to determine, for each destination s, a set of neighbors called the Feasible Successor Set, denoted as $FSS_v[s]$. To this aim, each node v explicitly stores the topology table, which contains, for each $u \in N(v)$, the distance $D_v[u, s]$ from u to s. Then, it computes $FSS_v[s]$ by using the SNC sufficient condition. In more details, node $u \in N(v)$ is inserted in $FSS_v[s]$ if the estimated distance $D_v[u, s]$ from u to s is smaller than the *feasible distance* $FD_v[v, s]$ from v to s. If a neighbor $u \in N(v)$, through which the distance from v to s is minimum, is in $FSS_v[s]$, then u is chosen as feasible successor. Moreover, in order to guarantee mutual exclusion in case multiple weight change operations occur, each node v performing DUAL uses some auxiliary data structures: (i) an auxiliary distance $RD_v[v, s]$, for each $s \in V$; and (ii) a finite state machine to process these multiple updates sequentially. The state of the machine consists, for each $s \in V$, of three variables:

1. the query origin flag $O_v[s]$ (integer value from the set $\{0, 1, 2, 3\}$)
2. the state $ACTIVE_v[s]$ (an integer and a boolean entry)
3. the replies status flag $R_v[u, s]$ (a boolean value fro each neighbor $u \in N(v)$)

It follows that DUAL requires $\Theta(n \cdot maxdeg)$ space per node, as all the data structures stored by a node v are arrays of size n, with the exception of the topology table $D_v[u, s]$ and the replies status flag, which are permanently allocated and require $\Theta(n \cdot maxdeg)$ space.

4.2 Algorithm

The main core of DUAL is a sub-routine, named DIFFUSE-COMPUTATION (see Fig. 7), which is performed by a generic node v, every time $FSS_v[s]$ does not include the node $u \in N(v)$ through which the distance from v to s is minimum. The DIFFUSE-COMPUTATION works as follows: node v sends queries to all its neighbors with its distance through $FS_v[s]$ by using message *query* (see Line 2 of Fig. 7). Accordingly, v sets $R_v[u, s]$ to true (see Line 3 of Fig. 7), for each $u \in N(v)$, in order to keep trace of which neighbor has answered to the *query* message (the value is set to false when a corresponding *reply* message is received, see Line 2 of Fig. 9).

Event: Node v invokes procedure DISTANCEDECREASE(u, s).
Procedure: DISTANCEDECREASE(u, s)

1 **if** $O_v[s] = 1 \wedge \text{ACTIVE}_v[s] = true$ **then**
2 $\quad \lfloor \quad O_v[s] =: 0;$

3 **if** $O_v[s] = 3$ **then**
4 $\quad \lfloor \quad O_v[s] =: 2;$

5 $c := \text{FS}_v[s];$
6 $D_v[v, s] := D_v[c, s] + w(v, c);$
7 $\text{FD}_v[v, s] := D_v[v, s];$
8 $\text{RD}_v[v, s] := D_v[v, s];$
9 $D_{min} := \min_{k \in N(v)} D_v[k, s] + w(v, k);$
10 $S_{min} := \text{argmin}_{k \in N(v)} D_v[k, s] + w(v, k);$
11 $D_v[v, s] := D_{min};$ $\qquad\qquad\qquad$ // $D_v[S_{min}, s] < \text{FD}_v[v, s]$ trivially true
12 **if** $\text{ACTIVE}_v[s] = false$ **then**
13 $\quad \mid \quad \text{FS}_v[s] := S_{min};$
14 $\quad \mid \quad \text{RD}_v[v, s] := D_v[v, s];$
15 $\quad \mid \quad$ **foreach** $k \in N(v)$ **do**
16 $\quad \mid \quad \lfloor \quad$ Send $update(v, s, D_v[v, s])$ to $k;$

Fig. 4 Pseudo-code of procedure DISTANCEDECREASE

Event: Node v receives an $update(u, s, d)$ message from u
Procedure: UPDATE(u, s, d)

1 $D_v[u, s] := d;$
2 **if** $D_v[u, s] + w(u, v) < D_v[v, s]$ **then**
3 $\quad \mid \quad D_v[v, s] := D_v[u, s] + w(u, v);$
4 $\quad \mid \quad$ **if** $\text{ACTIVE}_v[s] = false$ **then**
5 $\quad \mid \quad \mid \quad \text{FD}_v[v, s] := D_v[v, s];$
6 $\quad \mid \quad \mid \quad \text{RD}_v[v, s] := D_v[v, s];$
7 $\quad \mid \quad \mid \quad \text{FS}_v[s] := u;$
8 $\quad \mid \quad \mid \quad$ **foreach** $k \in N(v)$ **do**
9 $\quad \mid \quad \mid \quad \lfloor \quad$ Send $update(v, s, D_v[v, s])$ to $k;$

10 **else**
11 $\quad \mid \quad$ **if** $\text{FS}_v[s] = u$ **then**
12 $\quad \mid \quad \lfloor \quad$ call DISTANCEINCREASE$(u, s);$

Fig. 5 Pseudo-code of procedure UPDATE

Event: Node v invokes procedure DISTANCEINCREASE(u, s).
Procedure: DISTANCEINCREASE(u, s)

1   **if** $O_v[s] = 1 \wedge \text{ACTIVE}_v[s] = true$ **then**
2   $\quad \lfloor \ O_v[s] =: 0;$

3   **if** $O_v[s] = 3$ **then**
4   $\quad \lfloor \ O_v[s] =: 2;$

5   $c := FS_v[s];$
6   $D_v[v,s] := D_v[c,s] + w(v,c);$
7   $FD_v[v,s] := D_v[v,s];$
8   $RD_v[v,s] := D_v[v,s];$
9   $D_{min} := \min_{k \in N(v)} D_v[k,s] + w(v,k);$
10 $S_{min} := \underset{k \in N(v)}{\text{argmin}}\ D_v[k,s] + w(v,k);$
11 **if** $D_v[S_{min}, s] < FD_v[v,s]$ **then**
12 $\quad$ $D_v[v,s] := D_{min};$
13 $\quad$ **if** $\text{ACTIVE}_v[s] = false$ **then**
14 $\quad\quad$ $FS_v[s] := S_{min};$
15 $\quad\quad$ $RD_v[v,s] := D_v[v,s];$
16 $\quad\quad$ **foreach** $k \in N(v)$ **do**
17 $\quad\quad\quad \lfloor$ Send $update(v, s, D_v[v,s])$ to $k;$

18 **else**
19 $\quad$ **if** $\text{ACTIVE}_v[s] = false$ **then**
20 $\quad\quad \lfloor$ call DIFFUSE-COMPUTATION$(s);$

Fig. 6 Pseudo-code of procedure DISTANCEINCREASE

Event: Node v invokes procedure DIFFUSE-COMPUTATION(s).
Procedure: DIFFUSE-COMPUTATION(s)

1   **foreach** $k \in N(v)$ **do**
2   $\quad$ Send $query(v, s, D_v[v,s])$ to $k;$
3   $\quad$ $R_v[k,s] := true;$

Fig. 7 Pseudo-code of procedure DIFFUSE-COMPUTATION

From this point onwards v does not change its feasible successor to s until the DIFFUSE-COMPUTATION terminates.

When a neighbor $u \in N(v)$ receives a $query$, it triggers the execution of procedure QUERY (see Fig. 8). The procedure tries to determine if a feasible successor toward s exists after the update. If so, it replies to the $query$ by sending message $reply$ containing its own distance to s (see Lines 11–18 of Fig. 8). Otherwise, u propagates the DIFFUSE-COMPUTATION toward the rest of the network. In details, it sends out queries and waits for the replies from its neighbors before replying to v's original $query$. To guarantee that each node is involved in one DIFFUSE-COMPUTATION phase at the time, for a certain $s \in V$, an appropriate finite state machine behaviour

Event: Node v receives a $query(u,s,d)$ from u.
Procedure: QUERY(u,s,d)

1 $D_v[u,s] := d$;
2 **if** $FS_v[s] \neq u$ **then**
3 $\quad$⌊ Send $reply(v,s,D_v[v,s])$ to u;

4 **else**
5 $\quad$ **if** $O_v[s] = 1 \wedge \text{ACTIVE}_v[s] = true$ **then**
6 $\quad\quad$⌊ $O_v[s] = 2$;

7 $\quad$ **if** $O_v[s] = 0 \wedge \text{ACTIVE}_v[s] = true$ **then**
8 $\quad\quad$⌊ $O_v[s] = 2$;

9 $\quad$ $D_{min} := \min_{k \in N(v)} D_v[k,s] + w(v,k)$;
10 $\quad$ $S_{min} := \operatorname{argmin}_{k \in N(v)} D_v[k,s] + w(v,k)$;
11 $\quad$ **if** $D_v[S_{min},s] < FD_v[v,s]$ **then**
12 $\quad\quad$ $D_v[v,s] := D_{min}$;
13 $\quad\quad$ **if** $\text{ACTIVE}_v[s] = false$ **then**
14 $\quad\quad\quad$ $FS_v[s] := S_{min}$;
15 $\quad\quad\quad$ $RD_v[v,s] := D_v[v,s]$;
16 $\quad\quad\quad$ Send $reply(v,s,D_v[v,s])$ to u;
17 $\quad\quad\quad$ **foreach** $k \in N(v)$ **do**
18 $\quad\quad\quad\quad$⌊ Send $update(v,s,D_v[v,s])$ to k;

19 $\quad$ **else**
20 $\quad\quad$ $D_v[v,s] := D_v[u,s] + w(u,v)$;
21 $\quad\quad$ **if** $\text{ACTIVE}_v[s] = false$ **then**
22 $\quad\quad\quad$ $O_v[s] := 3$;
23 $\quad\quad\quad$ $FD_v[v,s] := D_v[v,s]$;
24 $\quad\quad\quad$ $RD_v[v,s] := D_v[v,s]$;
25 $\quad\quad\quad$ call DIFFUSE-COMPUTATION(s);

Fig. 8 Pseudo-code of procedure QUERY

is implemented by variables $O_v[s]$ and $\text{ACTIVE}_v[s]$ (see, e.g. Lines 6 and 8 of Fig. 8). Changes to distances, feasible distances and successors are allowed only under specific circumstances. Moreover, an auxiliary variable $RD_v[v, s]$, representing an upper bound to $D_v[v, s]$ is used by each node v, for each $s \in V$, to answer to certain types of queries, under the same circumstances, in order to avoid loops. We refer the reader to [24] for an exhaustive discussion on the subject. In the same paper, the authors show that the DIFFUSE-COMPUTATION always terminates, i.e. that there exists, under the FIFO assumption, a time when a node receives messages $reply$ by all its neighbors. At that point, it updates its distance and feasible successor, with the minimum value obtained by its neighbors and the neighbor that provides such distance. This is done during the execution of procedure REPLY, which is invoked upon the reception of each REPLY message (see Fig. 9). At the end of a DIFFUSE-COMPUTATION execution, a node sends message $update$ containing the new computed distance to its

Event: Node v receives a $reply(u,s,d)$ from u.
Procedure: REPLY(u,s,d)

```
1   D_v[u,s] := d;
2   R_v[u,s] := false;
3   if R_v[k,s] = false ∀ k ∈ N(v) then
4       if {O_v[s] = 1 ∧ ACTIVE_v[s] = true} ∨ {O_v[s] = 3 ∧ FD_v[v,s] = ∞} then
5           D_min := min  D_v[k,s] + w(v,k);
                   k∈N(v)
6           S_min := argmin D_v[k,s] + w(v,k);
                     k∈N(v)
7           if D_v[S_min,s] < FD_v[v,s] then
8               D_v[v,s] := D_min;
9               FS_v[s] := S_min;
10              O_v[s] := 1;
11              ACTIVE_v[s] := false;
12              FD_v[v,s] := D_v[v,s];
13              RD_v[v,s] := D_v[v,s];
14              foreach k ∈ N(v) do
15                  if k = u then
16                      └ Send reply(v,s,D_v[v,s]) to k;
17                  else
18                      └ Send update(v,s,D_v[v,s]) to k;

19          else
20              if O_v[s] = 0 then
21                  └ O_v[s] = 1;
22              if O_v[s] = 2 then
23                  └ O_v[s] = 3;
24              c := FS_v[s];
25              D_v[v,s] := D_v[c,s] + w(v,c);
26              FD_v[v,s] := D_v[v,s];
27              call DIFFUSE-COMPUTATION(s);
```

Fig. 9 Pseudo-code of procedure REPLY

neighbors (see Line 18 of Fig. 9). As mentioned above, DUAL starts every time a node x_i detects a weight change operation c_i occurring on one of its adjacent edges, say (x_i, y_i). In what follows, the cases in which c_i is a weight decrease and a weight increase operation are considered separately.

Weight decrease. If c_i is a weight decrease operation on (x_i, y_i), node x_i first tries to determine whether node y_i can be chosen as new $FS_{x_i}[s]$ or not, for each $s \in V$, without performing DIFFUSE-COMPUTATION. In fact, since c_i can induce only decreases in the distances, SNC is trivially always satisfied by at least one neighbor, which is either the current $FS_{x_i}[s]$ or y_i itself. This is done by invoking procedure DISTANCEDECREASE (y_i, s) for all $s \in V$(see Fig. 4), and the same routine is performed, symmetrically, by node y_i. In any of the two cases, it propagates the change by sending *update* messages to its neighbors, with the aim of notifying either a change in the distance or in the distance and the feasible successor. Each node in the graph, which receives such *update* message, in turn, determines whether $FS_v[s]$ has to be updated or not in the same way, and possibly propagates the change (see Line 9 Fig. 5). Note that, as the FIFO case is under consideration, each node of the graph updates its data structures related to s at most once as a consequence of c_i. Hence, since there are $|\sigma_{c_i,s}|$ nodes that change their distances or feasible successors towards s as a consequence of c_i and since each node v in $\sigma_{c_i,s}$ sends at most *maxdeg update* messages, the number of messages, related to a source s, sent as a consequence of a weight decrease operation c_i is $O(maxdeg \cdot |\sigma_{c_i,s}|)$, while the number of steps required to converge is $O(|\sigma_{c_i,s}|)$.

Weight increase. If c_i is a weight increase operation, the only nodes that sends messages, as a consequence of operation c_i and w.r.t. a source s, are those in $\sigma_{c_i,s}$ and their neighbors. In particular, after c_i occurs on (x_i, y_i), node x_i tries, for each $s \in V$, to determine whether a feasible successor still exists or not, by checking if nodes in $FSS_v[s]$ still satisfy SNC. This is done by invoking procedure DISTANCEINCREASE (y_i, s) for all $s \in V$ (see Fig. 6). The same routine is performed, symmetrically, by node y_i. In the affirmative case, node x_i immediately terminates its computation and sends an *update* message to each $u \in N(x_i)$ with the updated value of distance (see Line 11 of Fig. 6). Since we are considering the FIFO case, by SNC we know that, in the above case, the path in $VIA_G[s]$ from u to s does not contain x_i. Then, it follows that also node u does not execute a DIFFUSE-COMPUTATION nor send *update* to x_i, as a consequence of c_i. In the negative case, i.e. node x_i performs a DIFFUSE-COMPUTATION, and sends *query* messages to all its neighbors (see Line 20 of Fig. 6), and possibly (depending on the presence of alternative paths) induces other nodes in $VIA_G[s]$ to perform DIFFUSE-COMPUTATION. When x_i receives all the *reply* messages, it chooses as new feasible successor a neighbor $u \in N(x_i)$ which, in turn, does not perform DIFFUSE-COMPUTATION nor send *update* messages to x_i, with respect to s, as a consequence of c_i.

The correctness of DUAL is given in the next theorem.

Theorem 3 ([24]) *Let $C = \{c_1, c_2, \ldots, c_k\}$ be a sequence of edge weight changes on G. If DUAL is used, then FS is loop-free at any time $t \geq t_0$. Moreover, at the end of the algorithm $FS_v[s] \in via^{t_k}(s, v)$ and $D_v[s] = d^{t_k}(v, s)$, for each pair of nodes s and v.*

In any of the above cases, node x_i sends $O(|N(x_i)|)$ *update* messages while only in the second case, it sends $O(|N(x_i)|)$ *query* messages and each of the nodes in $N(x_i)$ sends $O(1)$ *reply* messages. Note that each node $v \in \sigma_{c_i,s}$ behaves as x_i when it receives either *update* or *query* messages from $FS_v[s]$. As a consequence, it follows that the total number of messages sent by each node $v \in \sigma_{c_i,s}$ is $O(|N(v)|) = O(maxdeg)$ and that the overall number of messages related to the source s sent as a consequence of a weight increase operation c_i is $O(maxdeg \cdot |\sigma_{c_i,s}|)$ while the number of steps required to converge is $O(|\sigma_{c_i,s}|)$.

Since $\sum_{i=1}^{k} \sum_{s \in V} |\sigma_{c_i,s}| = \Delta$, it follows that the overall number of messages sent during a sequence of weight modifications $\mathscr{C} = \{c_1, c_2, ..., c_k\}$, in the realistic case, and for each possible source s, is given by $\sum_{i=1}^{k} \sum_{s \in V} O\left(maxdeg \cdot |\sigma_{c_i,s}|\right) = O(maxdeg \cdot \Delta)$, while the overall number of steps required by the algorithm to converge is $\sum_{i=1}^{k} \sum_{s \in V} O(|\sigma_{c_i,s}|) = O(\Delta)$.

This implies the next theorem.

Theorem 4 ([24]) DUAL *requires* $O(maxdeg \cdot \Delta)$ *messages,* $O(\Delta)$ *steps, and* $\Theta(n \cdot maxdeg)$ *space occupancy per node.*

4.3 Example of Execution

In Fig. 10 we present an example of execution of DUAL, which is inspired by an example given in [24].

The example focuses on the graph of Fig. 10a, and on destination s. In the figure, the value close to a node indicates its distance to node s, and an arrowhead from x to y in edge (x, y) indicates that node y is the successor of x toward node s. Messages *query*, *reply*, and *update* are denoted by Q, R, and U, respectively. The number in parentheses following R denotes the reported distance contained in the *reply* message. Nodes involved in a DIFFUSE-COMPUTATION are highlighted in white. At a certain point in time, edge (b, s) increases its weight from 2 to 10 (Fig. 10a). When node b detects the weight increase, it determines that it has no feasible successor as none of its neighbors has a distance smaller than its current distance, that is 2. Accordingly, it starts a DIFFUSE-COMPUTATION by sending a query to its neighbors (Fig. 10b). In Fig. 10c, node c forwards the *query* and continues the DIFFUSE-COMPUTATION, because it has no feasible successor, while node a finds a feasible successor which is node s itself as $0 < 3$ and sends a *reply* to b. When node d receives node b's *query*, it simply sends a *reply* because it has a feasible successor. However, it becomes involved in the DIFFUSE-COMPUTATION when it receives the *query* from node c (Fig. 10d). When node d receives all the replies to its *query* (Fig. 10e), it computes its new distance and successor (12 and c, respectively), and sends a *reply* to c's *query* (Fig. 10f). Nodes c and b operate in a similar manner when they receives all the replies to their respective queries (Fig. 10f–g). At this point, the DIFFUSE-COMPUTATION is terminated and node b sends messages *update* containing the new computed distance to notify its neighbors (Fig. 10h). Such messages are propagated to the entire network in order to update the distances according to paths to s induced by successors nodes (Fig. 10i).

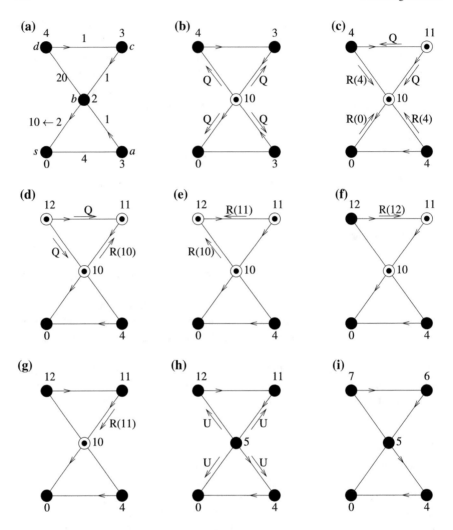

Fig. 10 Example of execution of DUAL

As a final observation of the section, notice that, the undesirable count-to-infinity phenomenon shown in Fig. 2 of Sect. 3, induced by the use of DBF, does not occur if DUAL is used, with SNC. For instance, at step 2, the SNC prevents node v to choose b as its successor, since the loop-free test fails. This triggers an execution of DIFFUSE-COMPUTATION, which is guaranteed to always produce an acyclic sub-graph induced by the feasible successors [24].

5   Loop Free Routing

This section describes the Loop Free Routing (LFR) algorithm of [26] and analyses its complexity, under the FIFO assumption. The algorithm consists of four procedures named UPDATE, DECREASE, INCREASE and SENDFEASIBLEDIST, respectively, shown in pseudocode in Figs. 11, 12, 13 and 14, respectively. The main procedures of the algorithm start every time a weight change $c_i \in \mathscr{C} = \{c_1, c_2, ..., c_k\}$ occurs on an edge (x_i, y_i).

5.1   Data Structures

Similarly to DUAL, LFR is more complex than the classical Distance-Vector algorithms, such as DBF, and uses a different set of data structures to realize a loop-free behaviour. In more details, it maintains, for each node v, a customized version of the standard routing table, that consists of two arrays $D_v[v, s]$ and $FS_v[s]$, that store the *estimated distance* and the so-called *feasible via*, respectively. This latter value represents a different kind of estimation on $via(v, s)$ that is always guaranteed to

---

Event: Node v receives an $update(u, s, D_u[u, s])$ message from u
Procedure: UPDATE$(u, s, D_u[u, s])$

1 **if** STATE$_v[s] = false$ **then**
2 $\quad$ **if** $D_v[v, s] > D_u[u, s] + w(u, v)$ **then**
3 $\quad\quad$ | $\quad$ DECREASE$(u, s, D_u[u, s])$;
4 $\quad$ **else**
5 $\quad\quad$ **if** $D_v[v, s] < D_u[u, s] + w(u, v)$ **then**
6 $\quad\quad\quad$ | $\quad$ INCREASE$(u, s, D_u[u, s])$;

Fig. 11 Pseudocode of procedure UPDATE$(u, s, D_u[u, s])$

Event: Node v invokes procedure DECREASE$(u, s, D_u[u, s])$
Procedure: DECREASE$(u, s, D_u[u, s])$

1 $D_v[v, s] := D_u[u, s] + w(u, v)$;
2 $UD_v[s] := D_v[v, s]$;
3 $FS_v[s] := u$;
4 **foreach** $k \in N(v) \setminus \{FS_v[s]\}$ **do**
5 $\quad$ | $\quad$ send $update(v, s, D_v[v, s])$ to k;

Fig. 12 Pseudo-code of procedure DECREASE$(u, s, D_u[u, s])$

Event: Node v invokes procedure INCREASE($u, s, D_u[u, s]$)
Procedure: INCREASE($u, s, D_u[u, s]$)

```
 1  if FS_v[s] = u then
 2      STATE_v[s] := true;
 3      allocate TEMPD_v[·][s];
 4      TEMPD_v[u][s] := D_u[u, s];
 5      UD_v[s] := TEMPD_v[u][s] + w(u, v);
 6      foreach v_i ∈ N(v) \ {FS_v[s]} do
 7          receive UD_{v_i}[s] by sending get.dist(v, s, UD_v[s]) to v_i ;
 8          store received value in TEMPD_v[v_i][s];
 9      D_min := min_{u∈N(v)}{TEMPD_v[u][s] + w(u, v)};
10      VIA_min := argmin_{u∈N(v)}{TEMPD_v[u][s] + w(u, v)};
11      if TEMPD_v[VIA_min][s] ≥ D_v[v, s] then
12          foreach v_i ∈ N(v) \ {FS_v[s]} do
13              receive UD_{v_i}[s] by sending get.feasible.dist(v, s, UD_v[s]) to v_i;
14              store received value in TEMPD_v[v_i][s]
15          D_min := min_{u∈N(v)}{TEMPD_v[u][s] + w(u, v)};
16          VIA_min := argmin_{u∈N(v)}{TEMPD_v[u][s] + w(u, v)};
17      deallocate TEMPD_v[·][s];
18      D_v[v, s] := D_min;
19      UD_v[s] := D_v[v, s];
20      FS_v[s] := VIA_min;
21      foreach k ∈ N(v) do
22          send update(v, s, D_v[v, s]) to k;
23      STATE_v[s] := false;
```

Fig. 13 Pseudo-code of procedure INCREASE

induce loop-free $\text{VIA}_G[s](t)$ at any time t [26]. In addition, for each $s \in V$, a node v executing LFR stores the following data structures: $\text{STATE}_v[s]$, which represents the state of node v with respect to source s (v is in *active* state and $\text{STATE}_v[s] = true$ if and only if it is performing procedure INCREASE or procedure SENDFEASIBLEDIST with respect to s); $\text{UD}_v[s]$ which represents the estimated distance from v to s through the current $\text{FS}_v[s]$ (in particular, if v is active $\text{UD}_v[s]$ is always greater than or equal to $D_v[v, s]$, otherwise they coincide).

To implement the topology table and, consequently, SNC, node v stores an array TEMPD_v which represents a *temporary* data structure. The allocation of memory for this data structure is made for a certain s only when needed, that is when v becomes active with respect to a certain s, and it is deallocated right after v turns back in passive state with respect to the same s. The entry $\text{TEMPD}_v[u][s]$ contains $\text{UD}_u[s]$, for each $u \in N(v)$, and hence TEMPD_v takes $O(maxdeg)$ space per node.

Event: Node v receives a $get.feasible.dist(u, s, \mathrm{UD}_u[s])$ message from u
Procedure: SENDFEASIBLEDIST$(u, s, \overline{D})$

```
1   if FS_v[s] = u and STATE_v[s] = false then
2       STATE_v[s] := true;
3       allocate TEMPD_v[·][s];
4       TEMPD_v[u][s] := D̄;
5       UD_v[s] := TEMPD_v[u][s] + w(u, v);
6       foreach v_i ∈ N(v) \ {FS_v[s]} do
7           receive UD_{v_i}[s] by sending get.dist(v, s, UD_v[s]) to v_i;
8           store received value in TEMPD_v[v_i][s];
9       D_min := min_{u∈N(v)}{TEMPD_v[u][s] + w(u, v)};
10      VIA_min := argmin_{u∈N(v)}{TEMPD_v[u][s] + w(u, v)};
11      if TEMPD_v[VIA_min][s] ≥ D_v[v, s] then
12          foreach v_i ∈ N(v) \ {FS_v[s]} do
13              receive UD_{v_i}[s] by sending get.feasible.dist(v, s, UD_v[s]) to v_i;
14              store received value in TEMPD_v[v_i][s];
15          D_min := min_{u∈N(v)}{TEMPD_v[u][s] + w(u, v)};
16          VIA_min := argmin_{u∈N(v)}{TEMPD_v[u][s] + w(u, v)};
17      send D_min to FS_v[s];
18      deallocate TEMPD_v[·][s];
19      D_v[v, s] := D_min;
20      UD_v[s] := D_v[v, s];
21      FS_v[s] := VIA_min;
22      foreach k ∈ N(v) do
23          send update(v, s, D_v[v, s]) to k;
24      STATE_v[s] := false;
25  else
26      if STATE_v[s] = true then
27          TEMPD_v[u][s] := D̄;
28      send UD_v[s] to u;
```

Fig. 14 Pseudo-code of procedure SENDFEASIBLEDIST

5.2 Algorithm

At any time $t < t_1$, before LFR starts, we assume that, for each pair of nodes $v, s \in V$, the values stored in $\mathrm{D}_v[v, s](t)$ and $\mathrm{FS}_v[s](t)$ are correct, that is $\mathrm{D}_v[v, s](t) = d^t(v, s)$ and $\mathrm{FS}_v[s](t) \in via^t(v, s)$. The description focuses on a distinguished node $s \in V$ and each node $v \in V$, at time t, is assumed to be passive with respect to that s.

The algorithm starts when the weight of an edge (x_i, y_i) changes. As a consequence, x_i (y_i, respectively) sends to y_i (x_i, respectively) message $update(x_i, s, \mathrm{D}_{x_i}[x_i, s])$ ($update(y_i, s, \mathrm{D}_{y_i}[y_i, s])$, respectively). Messages received at a node are stored in a queue and processed in FIFO order to guarantee mutual exclusion. If an arbitrary node v receives $update(u, s, \mathrm{D}_u[u, s])$ from $u \in N(v)$, then it performs procedure

UPDATE, which simply compares $D_v[v, s]$ with $D_u[u, s] + w(u, v)$ to determine whether v needs to update its estimated distance and or its estimated feasible via to s.

If node v is active, then the processing of the message is postponed by enqueueing it into the FIFO queue associated to s. Otherwise, we distinguish three cases, and discuss them separately, depending on the type of change in the estimated distance (or feasible via) that is induced by the message. In particular, if $D_v[v, s] > D_u[u, s] + w(u, v)$, then v performs procedure DECREASE, while if $D_v[v, s] < D_u[u, s] + w(u, v)$, then v performs procedure INCREASE. Finally, if node v is passive and $D_v[v, s] = D_u[u, s] + w(u, v)$ then it follows that there is more than one shortest path from v to s. In this case the message is discarded and the procedure ends.

DECREASE. When a node v performs procedure DECREASE, it simply updates D, UD and FS data structures by using the updated information provided by u. Then, the update is forwarded to all neighbors of v with the exception of $FS_v[s]$ which is node u (see Fig. 12).

INCREASE. When a node v performs procedure INCREASE(see Fig. 13), it first checks whether the update has been received from $FS_v[s]$ or not. In the negative case, the message is simply discarded while, in the affirmative case (only) v needs to change its estimation on distance and feasible via to s. To this aim, node v becomes active, allocates the temporary data structure $TEMPD_v$, and sets $UD_v[s]$ to the current distance through $FS_v[s]$. At this point, v first performs the so called LOCAL-COMPUTATION, which involves all the neighbors of v. If the LOCAL-COMPUTATION does not succeed, then node v initiates the so called GLOBAL-COMPUTATION, which involves in the worst case all the other nodes of the network. During the LOCAL-COMPUTATION, node v sends *get.dist* messages, carrying $UD_v[s]$, to all its neighbors, with the exception of u. A neighbor $k \in N(v)$ that receives a *get.dist* message, immediately replies with the value $UD_k[s]$, and if k is active, it updates $TEMPD_k[v][s]$ to $UD_v[s]$. When node v receives these values from its neighbors, it stores them in the array $TEMPD_v$, and it uses them to compute the minimum estimated distance D_{min} to s and the neighbor VIA_{min} which gives such a distance.

At the end of the LOCAL-COMPUTATION v checks whether a feasible via exists, by executing the loop-free test, according to the SNC. If the test fails, then v initiates the GLOBAL-COMPUTATION, in which it entrusts the neighbors the task of finding a loop-free path. In this phase, v sends $get.feasible.dist(v, s, UD_v[s])$ message to each of its neighbors. This message carries the value of the temporary estimated distance through its current feasible via. This distance is not guaranteed to be minimum but it is guaranteed to be loop-free. When v receives the answers to $get.feasible.dist$ messages from its neighbors, again it stores them in $TEMPD_v$ and it uses them to compute the minimum estimated distance D_{min} to s and the neighbor VIA_{min} which gives such a distance. At this point, v has surely found a feasible via to s and hence it deallocates $TEMPD_v$, updates $D_v[v, s]$, $UD_v[s]$ and $FS_v[s]$ and propagates the change by sending *update* messages to all its neighbors. Finally, v turns back in passive state and starts processing another message in the queue, if any.

A node $k \in N(v)$ which receives a $get.feasible.dist$ message performs procedure SENDFEASIBLEDIST. If $FS_k[s] = v$ and k is passive, then procedure - SENDFEASIBLEDIST behaves similarly to procedure INCREASE.

The only difference is that SENDFEASIBLEDIST needs to answer to the $get.feasible.dist$ message. However, within SENDFEASIBLEDIST, the LOCAL-COMPUTATION and the GLOBAL-COMPUTATION are performed with the aim of sending a reply with an estimated loop-free distance in addition to that of updating the routing table. In particular, node k needs to provide to v a new loop-free distance. To this aim, it becomes active, allocates the temporary data structure TEMPD$_k$, and sets $UD_k[s]$ to the current distance through $FS_v[s]$. Then, as in procedure INCREASE, k first performs the LOCAL-COMPUTATION, which involves all the neighbors of k. If the LOCAL-COMPUTATION fails, that is SNC is violated, then node k initiates the GLOBAL-COMPUTATION, which involves in the worst case all nodes of the network. At this point k has surely found an estimated distance to s which is guaranteed to be loop-free and hence, differently from INCREASE, it sends this value to its current via v as answer to the $get.feasible.dist$ message. Now, as in procedure INCREASE, node k can deallocate TEMPD$_v$, update its local data structures $D_v[v, s]$, $UD_v[s]$ and $FS_v[s]$, and propagate the change by sending $update$ messages to all its neighbors. Finally, v turns back in passive state and starts processing another message in the queue, if any.

The correctness of LFR is given in the next theorem.

Theorem 5 ([26]) *Let $C = \{c_1, c_2, \ldots, c_k\}$ be a sequence of edge weight changes on G. If LFR is used, then FS is loop-free at any time $t \geq t_0$. Moreover, at the end of the algorithm $FS_v[s] \in via^{t_k}(s, v)$ and $D_v[s] = d^{t_k}(v, s)$, for each pair of nodes s and v.*

Concerning the space complexity, LFR takes $O(n + maxdeg \cdot \delta)$ space per node, as all the data structures, stored by a node v, are arrays of size n, with the exception of TEMPD$_v[\cdot][s]$ which is allocated only when node v becomes active for a certain destination s, that is only if $v \in \delta_{c_i, s}$, and deallocated when v turns back in passive state for s, that is at most δ times. As each entry TEMPD$_v[\cdot][s]$ requires $O(maxdeg)$ space, the total space per node is $O(n + maxdeg \cdot \delta)$ in the worst case.

Concerning the message and time complexity, given a source s and a weight change operation $c_i \in \mathscr{C}$ on edge (x_i, y_i), the cases in which c_i is a weight decrease or a weight increase operation are considered separately. If c_i is a weight decrease operation, only nodes in $\sigma_{c_i, s}$ update their data structures and send messages to their neighbors. In detail, a node v can update its data structures related to s at most once as a consequence of c_i and, in this case, it sends $|N(v)|$ messages. Hence, v sends at most $maxdeg$ messages. Since there are $|\sigma_{c_i, s}|$ nodes that change their distance or via to s as a consequence of c_i, the number of messages related to the source s sent as a consequence of a weight decrease operation c_i is $O(maxdeg \cdot |\sigma_{c_i, s}|)$, while the number of steps required to converge is $O(|\sigma_{c_i, s}|)$.

If c_i is a weight increase operation, the only nodes which send messages with respect to operation c_i and source s are those in $\sigma_{c_i, s}$ and the neighbors of such nodes. In detail, each time that a node $v \in \sigma_{c_i,s}$ executes procedures INCREASE and SENDFEASIBLEDIST, it sends $O(|N(v)|)$ messages and each of the nodes in $N(v)$ sends $O(1)$ messages, for a total number of $O(|N(v)|) = O(maxdeg)$ messages sent. In the FIFO case, node v performs either procedure INCREASE or procedure SENDFEASIBLEDIST at most once with respect to c_i and s. It follows that each $v \in \sigma_{c_i,s}$ sends at most $O(maxdeg)$ messages. Therefore, the overall number of messages related to the source s sent as a consequence of operation c_i is $O(maxdeg \cdot |\sigma_{c_i,s}|)$ while the number of steps required to converge is $O(|\sigma_{c_i,s}|)$. Now, since $\sum_{i=1}^{k} \sum_{s \in V} |\sigma_{c_i,s}| = \Delta$, it follows that the overall number of messages sent during the whole sequence $\mathscr{C}$ and for each possible source s, is given by $\sum_{i=1}^{k} \sum_{s \in V} O\left(maxdeg \cdot |\sigma_{c_i,s}|\right) = O(maxdeg \cdot \Delta)$, while the overall number of steps required by the algorithm to converge is $\sum_{i=1}^{k} \sum_{s \in V} O(|\sigma_{c_i,s}|) = O(\Delta)$.

The next theorem follows from the above discussion.

Theorem 6   ([26]) LFR *requires* $O(maxdeg \cdot \Delta)$ *messages,* $O(\Delta)$ *steps, and* $O(n + maxdeg \cdot \delta)$ *space occupancy per node.*

5.3   Example of Execution

Fig. 15 shows an example of execution of LFR on the same graph of Fig. 10a, where the focus is on shortest paths towards node s. Given a node v, $FS_v[s]$ is represented by an arrow from v to $FS_v[s]$, and $D_v[v, s]$ and $UD_v[s]$ by a pair of values associated to v. Before LFR starts it is assumed that all nodes are in *passive* state, that is none of them is involved in a computation with respect to s. Passive nodes are represented as black circles, while active nodes by white circles. In the example, the algorithm starts when the weight of (b, s) increases from 2 to 10 (Fig. 15a). As a consequence b sends to s message $update(b, s, D_b[b, s])$, and s sends to b message $update(s, s, D_s[s, s])$, denoted as u(2) and u(0), respectively (Fig. 15b).

When a node v receives an *update* message with an increased distance to s, it checks whether it comes from $FS_v[s]$ and, in the affirmative case, it performs Procedure INCREASE, otherwise it discards the message. Hence, when s receives u(2), it immediately discards it and terminates, as $FS_s[s] \neq b$. On the other hand, when node b, receives u(0), since $FS_b[s] = s$ and $D_b[b, s] < D_s[s, s] + w(b, s)$, it needs to update its own routing table and hence it performs Procedure INCREASE. In detail, b first performs the LOCAL-COMPUTATION in which it tries to understand whether its routing table can be updated by using only the distances to s of its neighbors. To this aim, b switches to the *active* state, sets $\text{TEMPD}_b[s][s] = 0$ and $UD_b[s] = \text{TEMPD}_b[s][s] + w(b, s) = 0 + 10 = 10$, and it sends to all its neighbors, except $FS_b[s] = s$, a *get.dist* message carrying $UD_b[s]$ denoted as gD(10) (Fig. 15c). When a node $k \in N(b)$ receives gD(10), it immediately replies to b by sending $UD_k[s]$ (Fig. 15c). By using the replies of its neighbors, b computes $D_{min} = \min\{\text{TEMPD}_b[k][s] + w(b, k) \mid k \in N(b)\}$ and $VIA_{min} = \text{argmin}\{\text{TEMPD}_b[k][s] + w(b, k) \mid k \in N(b)\}$ and performs the loop-free test of

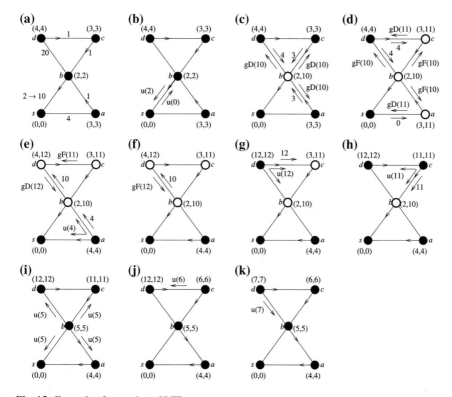

Fig. 15 Example of execution of LFR

the SNC, to check whether the provided distance corresponds to a loop-free path or not. Then b compares $\text{TEMPD}_b[\text{VIA}_{min}][s]$ with $D_b[b, s]$, which represents the last value of a loop-free distance computed by b. If the test succeeds, it follows that b has a feasible via to s. Then it turns back in passive state, updates its routing table and propagates the change to its neighbors. Otherwise, b performs the GLOBAL-COMPUTATION, where it computes a loop-free path by involving the other nodes of the graph. In this phase b sends to its neighbors a *get.feasible.dist* message (denoted as gF in the figure), bringing the most up to date estimated distance to s through the current successor $\text{FS}_b[s]$ of b. In this case, $D_{min} = D_a[a, s] + w(b, a) = 4$ and $\text{TEMPD}_b[a][s] = 3 > D_b[b, s] = 2$, hence b performs GLOBAL-COMPUTATION (Fig. 15d).

When a neighbor k of b receives *get.feasible.dist*, it performs Procedure - SENDFEASIBLEDIST. In detail, k first checks whether $\text{FS}_k[s] \neq b$ or $\text{STATE}_k[s] = true$. In this case, k immediately replies to b with $\text{UD}_k[s]$ (node d replies to b with 4 in Fig. 15d). If $\text{FS}_k[s] = b$ and $\text{STATE}_k[s] = false$ then node k, sets $\text{UD}_k[s] = \text{TEMPD}_b[b][s] + w(b, k)$ (for example node a sets $\text{UD}_a[s] = 10 + 1 = 11$ in Fig. 15d), performs first LOCAL-COMPUTATION and then GLOBAL-COMPUTATION (nodes a and c in Figs. 15d–f), in a way similar to Procedure INCREASE. To this aim, node a

performs LOCAL-COMPUTATION (Fig. 15d), which succeeds since $\text{TEMPD}_a[s][s] = 0$, and replies to b with 4 (Fig. 15e). Differently from a, the LOCAL-COMPUTATION of node c fails, and hence c performs GLOBAL-COMPUTATION as well, by sending gF(11) to d (Fig. 15e). When node d receives such message, it performs LOCAL-COMPUTATION by sending gD(12) to b which replies with $\text{UD}_b[s] = 10 \neq D_b[b, s] = 2$. Since the SNC is not satisfied at d, because

- $min\{\text{TEMPD}_d[z][s] + w(d, z) \mid z \in N(d)\} = 12$
- $argmin\{\text{TEMPD}_d[z][s] + w(d, z) \mid z \in N(d)\} = c$
- $\text{TEMPD}_d[c][s] = 11 > D_d[d, s] = 4$

then d performs GLOBAL-COMPUTATION by sending gF(12) to b, which immediately replies with 10, and updates $\text{TEMPD}_b[d][s] = 12$ (Fig. 15f).

At the end of this process, node d finds a new feasible via, node c, and replies to c with the corresponding minimum estimated distance $D_c[c, s] + w(d, c) = 12$. In addition, node d updates its routing table and propagates the change to its neighbors. When c receives the answer to the gF message from d (Fig. 15g) it behaves as d, by sending $D_b[b, s] + w(b, c) = 11$ to b, updating its routing table and propagating change to its neighbors.

When b receives all the replies to its gF messages (Fig. 15h), it is able to compute the new loop-free shortest path to s, to update $\text{UD}_b[s] = D_b[b, s] = 5$, to turn back in passive state and to propagate the change by means of *update* messages (Fig. 15i). The *update* messages induce the neighbors of b to update their routing tables as described above for b. As the value $\text{UD}_b[s]$, sent by b during the GLOBAL-COMPUTATION is an upper bound to $D_b[b, s]$, these *update* messages induce the neighbors to perform procedure DECREASE (Fig. 15j–k).

As a final observation of the section, notice that, the undesirable count-to-infinity phenomenon shown in Fig. 2 of Sect. 3, induced by the use of DBF, does not occur if LFR is used, with SNC, as well as for DUAL. In fact, for instance, at step 2, the SNC prevents node v to choose b as its successor, since the loop-free test fails after the LOCAL-COMPUTATION phase. This triggers an execution of GLOBAL-COMPUTATION, which is guaranteed to always produce an acyclic sub-graph induced by the feasible successors [26].

6 Distributed Computation Pruning

This section describes the Distributed Computation Pruning (DCP) technique introduced in [27]. The approach is not an algorithm by itself. Instead, it can be applied on top of any distance vector algorithm for distributed shortest paths.

Given a generic distance-vector algorithm **A**, the combination of DCP with **A** induces a new algorithm, which we denote by **A**-DCP. The DCP technique is designed to be efficient mainly in power-law networks, by forcing the distributed computation to be carried out only by a subset of few nodes of the network. In what follows, we first briefly summarize the main characteristics of such networks. Then, we present the technique that is able to exploit some of the mentioned characteristics.

6.1 Power-Law Networks

A *power-law network*, in the most general meaning, is a network exhibiting a power-law distribution of node degrees, thus having many nodes with low degree and few (core) nodes with very high degree. Such class of networks is very important from the practical point of view, since it includes many of the currently implemented communication infrastructures, like the Internet, the World Wide Web, some social networks, and so on. We refer the reader to [30] for more details on the subject. Practical examples of power-law networks, that will be considered in the remainder of the chapter, are the Internet topologies of the *CAIDA IPv4 topology dataset* [29], and the artificial instances generated by the *Barabási-Albert* model [30]. In Figs. 16 and 17 we show the power-law node degree distribution exhibited by typical CAIDA and *Barabási-Albert* networks, respectively. More details about these kinds of networks will be given in Sect. 6.7.

In the remainder of this section, we introduce some notation and definitions that are useful to capture scenarios typical of power-law networks We give some properties of shortest paths in these cases that then are exploited by DCP. Given a graph $G = (V, E, w)$, we classify nodes, edges and paths in G as follows. A node $v \in V$ is: *peripheral*, if $deg(v) = 1$; *semi-peripheral*, if $deg(v) = 2$; and *central* if $deg(v) \geq 3$. A peripheral or semi-peripheral node is *non-central*. A path $P = \{v_0, v_1, \ldots, v_j\}$ of G is *central* if v_i is central, for each $0 \leq i \leq j$. Any edge belonging to a central path is called *central edge*. A path $P = \{v_0, v_1, \ldots, v_j\}$ of G is *peripheral* if v_0 is central, v_j is peripheral, and all v_i, for each $1 \leq i \leq j - 1$, are semi-peripheral. In this case, v_0 is called the *owner* of P and of any node belonging to P. Any edge belonging to a peripheral path, accordingly, is called *peripheral edge*. Finally, a path $P = \{v_0, v_1, \ldots, v_j\}$ of G is *semi-peripheral* if v_0 and v_j are two distinct

Fig. 16 Power-law node degree distribution of a *CAIDA* graph with 8000 nodes and 11141 edges

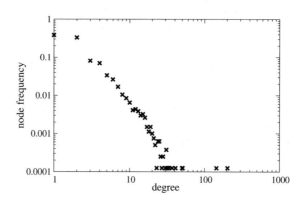

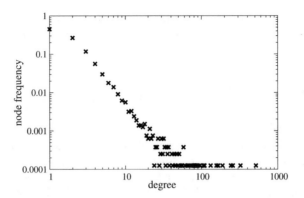

Fig. 17 Power-law node degree distribution of a *Barabási-Albert* graph with 8000 nodes and 12335 edges

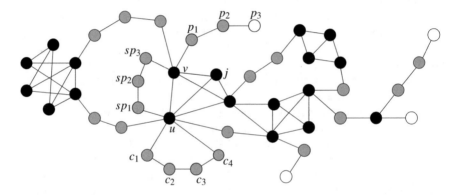

Fig. 18 A graph G and its corresponding nodes, edges and paths' classification. Central nodes are drawn in black, semi-peripheral nodes are drawn in gray, and peripheral node are drawn in white

central nodes, and all v_i are semi-peripheral nodes, for each $1 \leq i \leq j - 1$. Nodes v_0 and v_j are called the *semi-owners* of P and of any node belonging to P. Any edge belonging to a semi-peripheral path is called semi-peripheral edge. A further distinction for semi-peripheral paths occurs if $v_0 \equiv v_j$. In this case P is called a *semi-peripheral cycle*, and node $v_0 \equiv v_j$ is called the *cycle-owner* of P and of any node belonging to P. Each edge belonging to such a path is called *cyclic edge* and each node $u \neq v_0$ in P is called *cyclic node*.

In Fig. 18 we present examples of the above defined concepts. Central nodes are drawn in black, semi-peripheral nodes are drawn in gray, and peripheral node are drawn in white. The path $\{u, j, v\}$ is a central path, the path $\{v, p_1, p_2, p_3\}$ is a peripheral path whose owner is v, the path $\{u, sp_1, sp_2, sp_3, v\}$ is a semi-peripheral path whose semi-owners are u and v, while the path $\{u, c_1, c_2, c_3, c_4, u\}$ is a semi-peripheral cycle whose cycle-owner is u.

In what follows, we list some straightforward relationships that can be established between shortest paths involving central and non-central nodes of a given graph $G = (V, E, w)$.

Proposition 1 *Let $P = \{v, p_1, \ldots, p_j\}$ be a peripheral path of G whose owner is node v. Then, for each $x \in V \setminus \{p_1, \ldots, p_j\}$ and for each $1 \leq i \leq j$, we have that $d(x, p_i) = d(x, v) + w(v \rightarrow p_i)$, where $w(v \rightarrow p_i)$ is the weight of the unique sub-path of P connecting v to p_i.*

Proposition 2 *Let $S = \{u, sp_1, \ldots, sp_j, v\}$ be a semi-peripheral path of G whose semi-owners are nodes u and v. Then, for each $x \in V \setminus \{sp_1, \ldots, sp_j\}$ and for each $1 \leq i \leq j$, we have that $d(x, sp_i) = \min\{d(x, u) + w(u \rightarrow sp_i), d(x, v) + w(v \rightarrow sp_i)\}$, where $w(u \rightarrow sp_i)$ and $w(v \rightarrow sp_i)$ are the weights of the two sub-paths of S connecting sp_i to u and v, respectively.*

Proposition 3 *Let $C = \{u, c_1, \ldots, c_j, u\}$ be a semi-peripheral cycle of G whose cycle-owner is node u. Then, for each $x \in V \setminus \{c_1, \ldots, c_j\}$ and for each $1 \leq i \leq j$, we have that $d(x, c_i) = d(x, u) + \min\{w(P_i^1), w(P_i^2)\}$, where $P_i^1 = (u, c_1, \ldots, c_{i-1}, c_i)$ and $P_i^2 = (c_i, c_{i+1}, \ldots, c_j, u)$.*

6.2 The Technique

The DCP technique has been designed to exploit the aforementioned topological properties in order to reduce the communication overhead induced by distributed computations executed by Distance-Vector algorithms. In particular, it forces the distributed computation to be carried out by the central nodes only (which are few in real-world power-law networks). Non-central nodes, which are instead the great majority, play a passive role and receive updates about routing information from the respective owners, without taking part to any kind of distributed computation and by performing few trivial operation to update their routing data. Hence, it is clear that the larger is the set of non-central nodes of the network, the bigger is the improvement in the pruning of the distributed computation and, consequently, in the global number of messages sent by **A**-DCP. This observation will be supported through experimental means in Sect. 6.7.

6.3 Data Structures

In order to be implemented, DCP requires that a generic node of G stores some additional information with respect to those required by **A**. In particular, each node v needs to store and update information about adjacent non-central paths of G. To this aim, v maintains a data structure called CHAINPATH, denoted as CHP_v, which is an array containing one entry $\text{CHP}_v[s]$, for each central node s. $\text{CHP}_v[s]$ stores the list of all edges, along with their weights, belonging to those non-central paths that contain s. To build the CHAINPATH data structure, we assume that v knows the degree of all nodes of the network belonging to non-central paths. The following properties clearly hold:

- a central node obviously does not appear in any list of CHP_v;
- a peripheral node appears in exactly one list $CHP_v[s]$, where $s \in V$ is its owner;
- a semi-peripheral node appears in exactly two lists $CHP_v[v_0]$ and $CHP_v[v_{j-1}]$, if it belongs to a semi-peripheral path (v_0 and v_{j-1} are its semi-owners), while it appears in a single list $CHP_v[v_0]$, if it belongs to a semi-peripheral cycle (v_0 is its cycle-owner).

Hence, a node v, by using CHP_v is able to determine locally its type, and, in the case it is not central, it is also able to compute its owner, semi-owners, or cycle-owner. The space occupancy overhead per node due to the CHAINPATH can be estimated as follows:

- the CHAINPATH contains at most as many entries as the number of the central nodes;
- the sum of the sizes of all lists in the CHAINPATH is at most twice the number of non-central edges of G, since each non-central edge belongs to at most two chain paths;
- the number of non-central edges of G is $O(n)$, as they belong to paths in which every node has degree at most two.

Therefore, the worst case space overhead per node due to CHP_v is $O(n)$. Note that, despite this overhead, the use of DCP can induce a decrease in the space occupancy per node required by **A** for the following observations: (i) in most of the cases nodes do not ask and do not need to store information received from non-central nodes; (ii) computations which involve the whole network are performed only with respect to central destinations. Section 6.7 will give experimental evidence of this behavior. More details about the above statement are provided in the next section.

6.4 Description

The behavior of a generic algorithm **A**, when combined with DCP, can be summarized as follows. While in a classic routing algorithm every node performs the same code thus having the same behavior, in **A**-DCP central and non-central nodes are forced to have different behaviors. In particular, central nodes detect (and handle) changes concerning all kinds of edges, while peripheral, semi-peripheral, and cyclic nodes detect (and handle) changes concerning only peripheral, semi-peripheral, and cyclic edges, respectively. Changes affecting the distances towards/from peripheral, semi-peripheral, and cyclic nodes are handled by exploiting Propositions 1–3, as described below. Four different types of edge weight updates can occur on the network, i.e. (i) the weight of a central edge changes; (ii) the weight of a peripheral edge changes;

(iii) the weight of a semi-peripheral edge changes; (iv) the weight of a cyclic edge changes. In what follows, the generic algorithm **A**-DCP is described by considering the four cases separately.

Case (i) If the weight of a central edge (x, y) changes, then node x (y, respectively) performs the procedure provided by **A** for handling changes of this kind only with respect to central nodes. During this computation, if x (y, respectively) needs to know the estimated distances of its neighbors toward a central node s, it asks for it only to its central neighbors. If x (y, respectively) is the semi-owner of one or more semi-peripheral paths, it also asks for information to the other semi-owner of each semi-peripheral path, by means of a strategy we call TRAVERSE- PATH. In detail, node x (y, respectively) sends, for each semi-peripheral path it belongs to, a $sp.query(s)$ message to the corresponding semi-peripheral neighbor. The aim of this message is to traverse the semi-peripheral path in order to get the estimated distance, toward the considered node s, of the other semi-owner of the path. The $sp.query$ message contains only one field, i.e. the object s of the computation that has originated the message. When a semi-peripheral node receives a $sp.query(s)$ message from one of its two neighbors j, it simply performs a store-and-forward step and sends a $sp.query(s)$ message to the other neighbor $k \neq j$. The store-and-forward step is performed in a way that the ordering of the messages is preserved. A central node r that receives a $sp.query(s)$ message from one of its semi-peripheral neighbors u, simply replies to u with a $sp.reply(s, D_r[r, s])$ message, which carries the estimated distance of r towards s, which was requested by x (y, respectively). When a semi-peripheral node receives a $sp.reply(s, D_r[r, s])$ message from one of its two neighbors j, it simply performs a store-and-forward step and sends a $sp.reply(s, D_r[r, s])$ message to the other neighbor $k \neq j$. The strategy terminates whenever the central node x (y, respectively) receives $sp.reply(s, D_r[r, s])$: upon that event, x (y, respectively) stores $D_r[r, s]$ and uses it, if needed, while executing the procedure provided by **A** for the distributed computation of shortest paths.

Once x (y, respectively) has updated its own routing data toward a certain central node s, it propagates the variation to all its neighbors through a $gen.update(s, D_x[x, s])$ ($gen.update(s, D_y[y, s])$, respectively), which carries an updated value of $D_x[x, s]$ ($D_y[y, s]$, respectively). When a generic node v receives a $gen.update$ message from a neighbor u, it executes procedure GENERALIZEDUPDATE of Fig. 19 which, as first step, stores the current value of $D_v[v, s]$ in a temporary variable $D_v^{old}[s]$ (Line 1). Then, according to its status, the node performs different steps, which can be summarized as follows:

- if v is central, then it handles the change and updates its routing information toward the central node s (Line 3) by using the proper procedure of **A**, i.e. HANDLEUPDATE, HANDLEINCREASE, or HANDLEDECREASE, depending on the original structure of **A**, and forwards the change through the network accordingly (see Line 4).
- if v is a peripheral node whose owner is node r, then $D_v[v, s]$ is trivially updated by exploiting Proposition 1, i.e. by setting $D_v[v, s] = D_v[v, r] + D_v[r, s]$ (see Line 8).

Moreover, any specific data structure of $\mathbf{A}$ is accordingly updated, and $D_v[r, s]$ is propagated to the other neighbor of v.

- if v is a cyclic node whose cycle-owner is node r, then v exploits Proposition 3 and sets $D_v[v, s] = D_v[v, r] + D_v[r, s]$ (see Line 13). Moreover, since $D_v[r, s]$ is not changed, any specific data structure of $\mathbf{A}$ is accordingly updated, and $D_v[r, s]$ is propagated to the other neighbor of v.

- if v is a semi-peripheral node whose semi-owners are nodes r_1 and r_2, then the message carries the estimated distance from either r_1 or r_2 to s which can be used, according to Proposition 2, to update distances. In details, let us assume that the message carries $D_{r_1}[r_1, s]$, the other case is symmetric. We denote by u and z the neighbors of v which are closer (in terms of number of edges) to r_1 and r_2, respectively. If the distance from v to s is not affected by the change of $D_{r_1}[r_1, s]$, that is $D_{r_1}[r_1, s]$ increases but $\text{VIA}_v[s] \neq u$, then v simply discards the message. Otherwise, two cases may arise:

  - (i) if $D_{r_1}[r_1, s]$ is increased and $\text{VIA}_v[s] = u$, then node v updates (Line 22 of Fig. 19) $D_v[v, s]$ as the weight of the shortest between two paths: that formed by the shortest path from r_1 to s plus the path from r_1 to v, and that formed by the shortest path from z to s plus edge (z, v);
  - (ii) if $D_{r_1}[r_1, s]$ is decreased enough to induce a decrease also to $D_v[v, s]$, then v updates (Line 27 of Fig. 19) $D_v[v, s]$ as the weight of the path formed by the shortest path from r_1 to s plus the unique path from r_1 to v.

In both cases, any specific data structure of $\mathbf{A}$ is updated accordingly, and $D_{r_1}[r_1, s]$ is propagated to z. This behaviour mimics the distributed Bellman-Ford algorithm equipped with the split horizon heuristic [34, Section 6.6.3]: the information about the route for a particular node is never sent back in the direction from which it was received.

After updating the routing information toward the central node s, the node v calls the procedure PERIPHERYUPDATE, presented in Fig. 20, using s and $D_v^{old}[s]$ as parameters. This procedure first verifies whether the routing table entry of s is changed or not and, in the affirmative case (Line 1), it updates the routing information about the non-central nodes whose owner, semi-owner, or cycle-owner is s, if they exist, as follows:

- for each peripheral node z whose owner is s, node v sets $D_v[v, z]$ equal to the weight of the unique path from s to z plus the weight of the shortest path from v to s (Line 4).
- for each cyclic node z whose cycle-owner is s, node v sets $D_v[v, z]$ equal to weight of the shortest between the two possible paths from s to z plus the weight of the shortest path from v to s (Line 9).
- for each semi-peripheral node z such that one of the semi-owner nodes is s, node v performs procedure INNERSEMIPERIPHERYUPDATE (Line 12), if z and v lie on the same semi-peripheral path, and procedure OUTERSEMIPERIPHERYUPDATE (Line 14), otherwise. These procedures update the routing information toward z by exploiting the data stored in the CHAINPATH.

Event: Node v receives a *gen.update*$(s, \mathrm{D}.[\cdot, s])$ message from u
Procedure: GENERALIZEDUPDATE$(u, s, \mathrm{D}.[\cdot, s])$

1 $\mathrm{D}_v^{old}[s] := \mathrm{D}_v[v, s]$;
2 **if** v *is a central node* **then**   /* In this case $\mathrm{D}.[\cdot, s] \equiv \mathrm{D}_u[u, s]$ */
3   perform HANDLEUPDATE procedure of $\mathbf{A}$ wrt s;
4   **foreach** $k \in N(v)$ **do**
5   send *gen.update*$(s, \mathrm{D}_v[v, s])$ to k;

6 **if** v *is a peripheral node* **then**
7   let r be the owner node of v /* In this case $\mathrm{D}.[\cdot, s] \equiv \mathrm{D}_r[r, s]$ */
8      $\mathrm{D}_v[v, s] := \mathrm{D}_v[v, r] + \mathrm{D}_r[r, s]$;
9   update any specific data structure of $\mathbf{A}$;
10   send *gen.update*$(s, \mathrm{D}_r[r, s])$ to the other neighbor z of v;

11 **if** v *is a cyclic node* **then**
12   let r be the cycle-owner node of v /* In this case $\mathrm{D}.[\cdot, s] \equiv \mathrm{D}_r[r, s]$ */
13     $\mathrm{D}_v[v, s] := \mathrm{D}_v[v, r] + \mathrm{D}_r[r, s]$;
14   update any specific data structure of $\mathbf{A}$;
15   send *gen.update*$(s, \mathrm{D}_r[r, s])$ to the other neighbor z of v;

16 **if** v *is a semi-peripheral node* **then**
17   Let r_1 and r_2 be the semi-owner nodes of v /* In this case $\mathrm{D}.[\cdot, s] \equiv \mathrm{D}_{r_1}[r_1, s]$
    */
18   Let u and z be the neighbors of v which are closer to r_1 and r_2, respectively
19   **if** $\mathrm{D}_v[v, s] < \mathrm{D}_{r_1}[r_1, s] + \sum_{\{l,q\} \in \{v \to r_1\}} w(l, q)$ **then**
20   **if** VIA$_v[s] := u$ **then**
21   receive from z the value $\mathrm{D}_z[z, s]$, if not already stored by $\mathbf{A}$;
22             $\mathrm{D}_v[v, s] := \min \{\mathrm{D}_v[z, s] + w(v, z), \sum_{\{l,q\} \in \{v \to r_1\}} w(l, q) + \mathrm{D}_{r_1}[r_1, s]\}$;
23   **if** $\mathrm{D}_v[v, s] = \mathrm{D}_v[z, s] + w(v, z)$ **then**
24   VIA$_v[s] := \{z\}$;
25   send *gen.update*$(s, \mathrm{D}_{r_1}[r_1, s])$ to z;
26   **else**
27         $\mathrm{D}_v[v, s] := \sum_{\{l,q\} \in \{v \to r_1\}} w(l, q) + \mathrm{D}_{r_1}[r_1, s]$;
28   VIA$_v[s] := u$
29   send *gen.update*$(s, \mathrm{D}_{r_1}[r_1, s])$ to z;
30   update any specific data structure of $\mathbf{A}$;
31 call PERIPHERYUPDATE$(s, \mathrm{D}_v^{old}[s])$;

Fig. 19 Pseudo-code of procedure GENERALIZEDUPDATE

In detail, procedure INNERSEMIPERIPHERYUPDATE, described in Fig. 21, updates $\mathrm{D}_v[v, y]$ by comparing the weight of the only two paths that connect v and z. Note that, such two paths include the shortest paths between the semi-owners of v and v itself.

In particular, if we assume that S is the semi-peripheral path that includes v and z whose semi-owners are r_1 and r_2, node v computes the weight D_1 of the unique sub-path of S from v to z (Line 2), and determines the neighbor VIA$_1$ of v that belongs to such sub-path (Line 3).

Event: Node v invokes procedure PERIPHERYUPDATE$(s, D_v^{old}[s])$
Procedure: PERIPHERYUPDATE$(s, D_v^{old}[s])$

1 **if** $D_v[v,s] \neq D_v^{old}[s]$ **then**
2   **foreach** $z \in CHP_v[s]$ **do**
3   **if** P_z *is a peripheral path* $: z \in P_z \wedge v \notin P_z$ **then**
4              $D_v[v,z] := D_v[v,s] + \sum\limits_{\{l,q\}\in\{s\to z\}\subseteq P_z} w(l,q);$
5              $VIA_v[z] := VIA_v[s];$
6   **if** C_z *is a semi-peripheral cycle* $: z \in C_z \wedge v \notin C_z$ **then**
7   Let k_1 and k_2 be the neighbors of z
8              $D_v[s,z] := \min\limits_{N\in\{k_1,k_2\}} \{ \sum\limits_{\{l,q\}\in\{s\to N\to z\}\subseteq C_z} w(l,q) \};$
9              $D_v[v,z] := D_v[v,s] + D_v[s,z];$
10             $VIA_v[z] := VIA_v[s];$
11   **if** S_z *is a semi-peripheral path* $: z \in S_z \wedge v \in S_z$ **then**
12   INNERSEMIPERIPHERYUPDATE$(z);$
13   **if** S_z *is a semi-peripheral path* $: z \in S_z \wedge v \notin S_z$ **then**
14   OUTERSEMIPERIPHERYUPDATE$(z);$

Fig. 20 Pseudo-code of procedure PERIPHERYUPDATE

Then, if v belongs to the sub-path of S that connects z to r_1 (this detail can easily be deduced from the CHAINPATH), it computes the weight D_2 of the path formed by the sub-path of S from z to r_2 plus the shortest path from v to r_2 (Line 5). Note that such shortest path might contain node r_1. Otherwise, node v computes the weight D_2 of the path formed by the sub-path of S from z to r_1 plus the shortest path from v to r_1 (Line 8). Finally, node v sets $D_v[v, z]$ to the minimum weight of the above two possible paths (Lines 10 and 14, respectively) and, accordingly, updates $VIA_v[z]$.

Similarly, procedure OUTERSEMIPERIPHERYUPDATE, presented in Fig. 22, updates $D_v[v, z]$ by comparing the weight of the only two paths that connect v and z. Note that such paths include the shortest paths between the semi-owners of z and v. In detail, if S is the semi-peripheral path that contains z and r_1 and r_2 are the semi-owners of S, node v first computes two values D_{r1} and D_{r2} (Lines 2–3), equal to the weight of the path formed by the path between z and r_1 (z and r_2, respectively) plus the shortest path between v and r_1 (v and r_2, respectively), then sets $D_v[v, z]$ equal to the minimum of the weights of these two possible paths (Line 2) and, accordingly, updates $VIA_v[z]$.

Case (ii) If a weight change occurs on a peripheral edge (x, y_1), belonging to a peripheral path $P = \{r, \ldots, x, y_1, \ldots, y_n\}$ whose owner is r, then node x (y_1, respectively), handles the change by sending a *peri.change*$(x, y_1, w(x, y_1))$ message to each of its neighbors. In this case, the distance from each node of the network to x does not change, except for those nodes y_p with $p = 1, \ldots, n$ (which are topologically further than x from r). Each of these nodes, after receiving the *peri.change*$(x, y_1, w(x, y_1))$

Event: Node v invokes procedure INNERSEMIPERIPHERYUPDATE(z)
Procedure: INNERSEMIPERIPHERYUPDATE(z)

1   let S the semi-peripheral path containing z and v, and let r_1 and r_2 be the semi-owners of S
2   $D_1 := \sum\limits_{\{l,q\} \in \{v \to z\} \subseteq S} w(l,q)$;
3   $VIA_1 := \{j \in N(v) : j \in \{v \to z\} \subseteq S\}$;
4   **if** $v \in \{r_1 \to z\}$ **then**
5      $D_2 := D_v[v, r_2] + \sum\limits_{\{l,q\} \in \{r_2 \to z\} \subseteq S} w(l,q)$;
6      $VIA_2 := VIA_v[r_2]$;
7   **else**
8      $D_2 := D_v[v, r_1] + \sum\limits_{\{l,q\} \in \{r_1 \to z\} \subseteq S} w(l,q)$;
9      $VIA_2 := VIA_v[r_1]$;
10   **if** $D_1 \le D_2$ **then**
11      $D_v[v, z] := D_1$;
12      $VIA_v[z] := VIA_1$;
13   **else**
14      $D_v[v, z] := D_2$;
15      $VIA_v[z] := VIA_2$;

Fig. 21 Pseudo-code of procedure INNERSEMIPERIPHERYUPDATE

Event: Node v invokes procedure OUTERSEMIPERIPHERYUPDATE(z)
Procedure: OUTERSEMIPERIPHERYUPDATE(z)

1   Let S the semi-peripheral path containing z and let r_1 and r_2 be the semi-owners of S
2   $D_{r1} := D_v[v, r_1] + \sum\limits_{\{l,q\} \in \{r_1 \to z\} \subseteq S} w(l,q)$
3   $D_{r2} := D_v[v, r_2] + \sum\limits_{\{l,q\} \in \{r_2 \to z\} \subseteq S} w(l,q)$
4   $D_v[v, z] := \min\{D_{r1}, D_{r2}\}$
5   Let $b \in \{r_1, r_2\}$ be the node that provides $D_v[v, z]$
6   $VIA_v[z] := VIA_v[b]$

Fig. 22 Pseudo-code of procedure OUTERSEMIPERIPHERYUPDATE

message, first updates the CHP with the new value of $w(x, y_1)$ and then computes the distance to x and to all the other nodes s of the network by simply adding to $D_{y_p}[y_p, s]$ the weight change on edge (x, y_1). When a generic node v, different from nodes y_p, receives message $peri.change(x, y_1, w(x, y_1))$, it first verifies whether the update has been already processed or not, by comparing the new value of $w(x, y_1)$ with the one stored in its CHP. In the first case the message is discarded. Otherwise, it updates its CHP with the updated value $w(x, y_1)$ and its routing information only toward nodes y_p, as the shortest path toward x does not change. In particular, node v sets $D_v[v, y_p] = D_v[v, r] + D_v[r, y_p]$, where $D_v[r, y_p]$ is the weight of the peripheral path from r to y_p (note that, for each $v \in P$, $v \ne y_p$, $D_v[v, y_p]$ is computed by using only the information stored inside the CHP because it is equal to the weight of the peripheral path from v to y_p). Then, it propagates $peri.change(x, y_1, w(x, y_1))$ over the network by a flooding algorithm.

Case (iii) If the weight of a semi-peripheral edge (x, y), whose semi-owner nodes are r_1 and r_2, changes, then node x (y, respectively) sends two kinds of messages: a $semi.change(x, y, w(x, y))$, to each of its neighbors, and a $gen.update(s, \mathrm{D}.[\cdot, s]$ to x (y, respectively), for each central node s such that $\mathrm{VIA}_x[s] \neq y$ ($\mathrm{VIA}_y[s] \neq x$, respectively), where $\mathrm{D}.[\cdot, s]$ is the distance toward s of the semi-owner node of x (y, respectively) that belongs to the sub-path of the semi-peripheral path that does not include the edge (x, y). When a generic node v receives message $semi.change$, it first verifies whether the update has been already processed or not, by comparing the new value of $w(x, y)$ with the one stored in its CHP. In the first case the message is discarded. Otherwise, node v updates its CHP with the new value of $w(x, y)$ and it propagates $semi.change(x, y, w(x, y))$ over the network by a flooding algorithm. Moreover, if v is the semi-owner node of the semi-peripheral path P that includes edge (x, y), it also performs the procedure provided by **A** for the distributed computation of shortest paths with respect to central nodes. This step basically considers P as a single edge that connects the two semi-owner nodes of P itself, and induces such semi-owner nodes to behave like the weight of one of their adjacent edges has changed.

When a generic node v receives a $gen.update(s, \mathrm{D}.[\cdot, s]$ message from a neighbor u, it executes procedure GENERALIZEDUPDATE$(s, \mathrm{D}.[\cdot, s])$ of Fig. 19. After updating the routing information toward a central node s, node v calls the procedure PERIPHERYUPDATE presented in Fig. 20 using s and $\mathrm{D}_v^{old}[s]$ as parameters. The procedure works as in the case of a central edge weight change (see Case (i)).

Case (iv) If the weight of a cyclic edge (x, y) changes, both nodes x and y send a $cycl.change(x, y, w(x, y))$ message to each of their neighbors. Let r be the cycle-owner node of both x and y. When a generic node v receives message $cycl.change$, it first verifies whether the update has been already processed or not, by comparing the new value of $w(x, y)$ with the one stored in its CHP. In the first case the message is discarded. Otherwise, node v first updates its CHP with the updated value of $w(x, y)$ and propagates $cycl.change(x, y, w(x, y))$ over the network by a flooding algorithm. Then, two cases can occur: either node v belongs to the same semi-peripheral cycle of x and y or not.

- In the first case, node v first computes $d^\alpha = \mathrm{D}_v[v, s] - \mathrm{D}_v[v, r]$ and, hence, updates the routing information toward all the nodes of the semi-peripheral cycle, including r, by using the CHP data structure. Then, if $\mathrm{D}_v[v, r]$ changes, it updates the routing information toward all the other central nodes s of the network by setting $\mathrm{D}_v[v, s] = \mathrm{D}_v[v, r] + d^\alpha$. After updating the routing information toward a central node s, node v calls the procedure PERIPHERYUPDATE presented in Fig. 20 using s and $\mathrm{D}_v^{old}[s]$ as parameters. The procedure works as in the case of a central edge weight change (see Case (i)).

- In the second case, v computes, for each node z of the semi-peripheral cycle, the shortest path distance between z and its cycle-owner node r by using the CHP data structure. Finally, it assigns $D_v[v, z] = D_v[v, r] + D_v[r, z]$.

6.5 Combining DCP with DUAL

This section describes the combination of DCP to DUAL, denoted as DUAL-DCP. The main changes deriving by the application of DCP to DUAL can be summarized as follows.

If the weight of a central edge (u, v) changes, then node v verifies, only with respect to each central node s, whether $D_v[v, s] > D_v[u, s] + w(u, v)$ or not (note that the behaviour of node u is symmetric with respect to the weight change operation). In the first case, node v sets $D_v[v, s] = D_v[u, s] + w(u, v)$ and $FS_v[s] = u$ and propagates the change to all its neighbors. In the second case, node v first checks whether $FS_v[s] = u$ or not. If $FS_v[s] \neq u$, the node terminates the update procedure. Otherwise, node v tries to compute a new $FS_v[s]$. In this phase, if no neighbor of v satisfies SNC and node v needs to perform the DIFFUSE-COMPUTATION, it sends out *query* messages only to its central neighbors. Moreover, with the aim of knowing the estimated distance of each of the semi-owner of the semi-peripheral paths which node v belongs to, node v performs the TRAVERSE- PATH phase and sends $sp.query$ messages to each of its semi-peripheral neighbors. When node v receives all the replies to these messages, it updates its routing information towards s and propagates the change to all its neighbors. In all the cases, if the distance towards s changes, node v is able to update its routing information towards all the nodes in the non-central paths of s, if they exists.

If a weight change occurs on either a peripheral or a cyclic edge, then the nodes adjacent to the edge behave as described in Sect. 6. The difference from the generic case is that the involved nodes also update the topology table. If a weight change occurs on a semi-peripheral edge, differently from the general case, semi-peripheral nodes do not need to ask information to their neighbors, as DUAL permanently stores the topology table.

Note that, when DCP is combined with DUAL, certain nodes of the network can avoid to maintain some data structures of DUAL, as either the information stored in them can be inferred by using the CHAINPATH, or it is not needed due to the pruning mechanism of DCP. For instance, each node of the network executing DUAL-DCP, does not need to store the data structure of DUAL that implements the finite state machine with respect to non-central nodes, as no distributed computation can be initiated for this kind of nodes. Same considerations hold for the topology table. Other minor compressions in the space occupancy can be achieved by exploiting some relationships between the shortest paths from non-central nodes to their corresponding owner/semi-owners/cycle-owner.

6.6   Combining DCP with LFR

This section describes the combination of DCP to LFR, denoted as LFR-DCP. In what follows we summarize the main changes deriving by the application of DCP to LFR.

If the weight of a central edge (u, v) changes, then node v verifies, only with respect to central nodes $s \in V_c$, whether $D_v[v, s] > D_v[u, s] + w(u, v)$ or not (note that the behaviour of node u is symmetric with respect to the weight change operation). In the first case, node v sets $D_v[v, s] = D_v[u, s] + w(u, v)$ and $FS_v[s] = u$ and propagates the change to all its neighbors. In the second case, node v first checks whether $FS_v[s] = u$ or not. If $FS_v[s] \neq u$, the node terminates the update procedure. Otherwise, node v performs LOCAL-COMPUTATION, by sending $get.dist$ message to all its neighbors. If the LOCAL-COMPUTATION succeeds, node v updates its routing information and propagates the change. Otherwise, node v needs to perform the GLOBAL-COMPUTATION and it sends out $get.feasible.dist$ messages only to its central neighbors. Moreover, with the aim of knowing the estimated distance of each of the semi-owner of the semi-peripheral paths which node v belongs to, node v performs the TRAVERSE- PATH phase and sends $sp.query$ messages to each of its semi-peripheral neighbors. When node v receives all the replies to these messages, it updates its routing information towards s and propagates the change to all its neighbors. In all the cases, if the distance to s changes, node v is able to update its routing information towards all nodes in the non-central paths of s, if they exists. If a weight change occurs on a peripheral, semi-peripheral or a cyclic edge, then the nodes adjacent to the edge behave as described in Sect. 6.

Note that, also when DCP is combined with LFR, certain nodes of the network can avoid to maintain some data structures of LFR, as either the information stored in them can be inferred by using the CHAINPATH, or it is not needed due to the pruning mechanism of DCP. For instance, each node of the network executing LFR-DCP, does not need to allocate the temporary data structure TEMPD with respect to non-central nodes, as no node of the network can become active with respect to a non-central node. Moreover, this data structure, when allocated for some central node s, has a reduced size, equal to the number of central neighbors of s plus the number of semi-peripheral paths to which s belongs. Other minor compressions in the space occupancy and, in particular, in the data structures needed to coordinate the distributed computations, can be achieved also by exploiting some relationships between the shortest paths from non-central nodes to their corresponding owner/semi-owners/cycle-owner.

6.7   Practical Effectiveness of DCP

This section describes the results of the experimental study proposed in [27], which consider algorithms DUAL, LFR, DUAL-DCP and LFR-DCP, and show the practical effectiveness of the use of DCP. The experiments have been performed on a workstation equipped with a Quad-core 3.60 GHz Intel Xeon X5687 processor, with 12MB of internal cache and 24 GB of main memory, and consist of simulations within the OMNeT++ 4.0p1 environment [28]. The programs have been compiled with GNU g++ compiler 4.4.3 under Linux (Kernel 2.6.32).

6.7.1 Executed Tests

The experiments have been performed both on real-world and artificial instances of the problem, subject to randomly generated sequences of updates. In detail, we used both the power-law networks of the *CAIDA IPv4 topology dataset* [29], and the random power-law networks generated by the *Barabási-Albert* algorithm [30].

The CAIDA dataset is collected by a globally distributed set of monitors. The monitors collect data by sending probe messages continuously to destination IP addresses. Destinations are selected randomly from each routed IPv4/24 prefix on the Internet such that a random address in each prefix is probed approximately every 48 h. The current prefix list includes approximately 7.4 million prefixes. For each destination selected, the path from the source monitor to the destination is collected, in particular, data collected for each path probed includes the set of IP addresses of the hops which form the path and the Round Trip Times (RTT) of both intermediate hops and the destination. The power-law graphs of the *CAIDA* dataset have average node degree approximately equal to 2.5 and a number of nodes with degree smaller than 3 approximately equal to $3/4n$.

A Barabási–Albert topology is generated by iteratively adding one node at a time, starting from a given connected graph with at least two nodes. A newly added node is connected to any other existing nodes with a probability that is proportional to the degree of the existing nodes. The power-law graphs generated by the Barabási–Albert algorithm have average node degree approximately equal to 3 and a number of nodes with degree smaller than 3 approximately equal to $7/10n$.

Regarding CAIDA network instances, the files provided by the CAIDA consortium have been parsed to obtain a weighted undirected graph, denoted as G_{IP}, where a node represents an IP address in the dataset (both source/destination hosts and intermediate hops), edges represent links among hops, and weights of the edges are given by Round Trip Times. As the graph G_{IP} consists of almost 35000 nodes, it was not possible to use it for the experiments, as the amount of memory required to store the routing tables of all the nodes is $O(n^2 \cdot maxdeg)$ for DUAL. Hence, the tests have been performed on connected subgraphs of G_{IP}, with a variable number of nodes and edges, induced by the settled nodes of a breadth first search starting from a node taken at random. A subgraph of G_{IP} with h nodes is denoted with G_{IP-h}. Different tests have been generated, each test consisting of a subgraph of G_{IP} and a set of k edge updates, where k assumes values in $\{5, 10, \ldots, 200\}$. An edge update consists of multiplying the weight of a random selected edge by a percentage value randomly chosen in [50, 150%]. For each test configuration (a graph with a fixed value of k) 5 different experiments were performed (for a total amount of 200 runs) and the average values are reported.

Concerning Barabási–Albert instances, different tests have been randomly generated, where a test consists of a n nodes Barabási–Albert random graph, denoted
as G_{BA-n}, and a set of k edge updates, where k assumes values in $\{5, 10, \ldots, 200\}$.
Edge weights are non-negative real numbers randomly chosen in $[1, 1000000]$. Edge
updates are randomly chosen as in the CAIDA tests. For each test configuration (a
graph with a fixed value of k) 5 different experiments have been performed (for a
total amount of 200 runs) and average values are reported.

6.7.2   Analysis

Simulations have been ran on both CAIDA and Barabási–Albert instances with different number of nodes $n \in \{1200, 5000, 8000\}$. The results of the experiments
on the different instances are similar, hence only those on the bigger instances are
reported here, that is, $G_{IP-8000}$ (8000 nodes and 11141 edges), and $G_{BA-8000}$ (8000
nodes and 12335 edges), respectively. Notice that: $G_{IP-8000}$ has average node degree
equal to 2.8, a percentage of degree 1 nodes approximately equal to 38.5%, and a
percentage of degree 2 nodes approximately equal to 33%; $G_{BA-8000}$ has average
node degree equal to 3.1, a percentage of degree 1 nodes approximately equal to
45%, and a percentage of degree 2 nodes approximately equal to 26%.

In Fig. 23 the number of messages sent by DUAL and DUAL-DCP on $G_{IP-8000}$
are reported, while in Fig. 24 the number of messages sent by LFR and LFR-DCP
on $G_{IP-8000}$ are reported. These Figures show that the combinations of DUAL and
LFR with DCP provide a huge improvement in the global number of messages sent
on $G_{IP-8000}$. In particular, in the tests of Fig. 23 the ratio between the number of
messages sent by DUAL-DCP and DUAL is within 0.03 and 0.16 which means that
DUAL-DCP sends a number of messages which is between 3 and 16% that of DUAL.
In the tests of Fig. 24 the ratio between the number of messages sent by LFR-DCP
and LFR is within 0.10 and 0.26.

In Fig. 25 the number of messages sent by DUAL and DUAL-DCP on $G_{BA-8000}$
are reported, while in Fig. 26 the number of messages sent by LFR and LFR-DCP on

Fig. 23 Number of
messages sent by DUAL and
DUAL-DCP on $G_{IP-8000}$

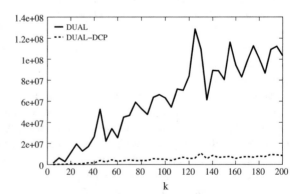

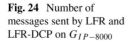

Fig. 24 Number of messages sent by LFR and LFR-DCP on $G_{IP-8000}$

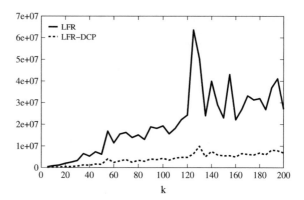

$G_{BA-8000}$ are reported. These Figures show that the use of DCP again gives a clear improvement in the global number of messages sent. In detail, in the tests of Fig. 25 the ratio between the number of messages sent by DUAL-DCP and DUAL is within 0.22 and 0.47. Similarly, in the tests of Fig. 26 the ratio between the number of messages sent by LFR-DCP and LFR is within 0.26 and 0.34. Notice that, the improvement provided by DCP in these artificial instances is smaller than in the real-world ones. This is due to the fact that the part of the distributed computation pruned by DCP is smaller in the Barabási-Albert networks with respect to the CAIDA networks, as they have: (i) a slightly higher average degree (ii) a wider range of the degree of the central nodes, that is a slightly larger standard deviation of the node degree. In fact, for instance, $G_{IP-8000}$ has an average degree equal to 2.8 and $maxdeg$ equal to 203 while $G_{BA-8000}$ has an average degree equal to 3.1 and $maxdeg$ equal to 515. Note also that this behaviour is more emphasized for LFR-DCP as LFR includes two sub-routines (called LOCAL-COMPUTATION and GLOBAL-COMPUTATION, respectively) where the worst case message complexity depends on the maximum degree, while DUAL uses a single sub-routine (namely the DIFFUSE-COMPUTATION) where the worst case message complexity depends on the same parameter.

To conclude the analysis, the space occupancy per node of each algorithm has been considered. The results are summarized in Table 1 where, both for $G_{IP-8000}$ and $G_{BA-8000}$, the maximum and the average space occupancy per node, in Bytes, of each algorithm is reported. Also the ratio between the space occupancy per node of the algorithms integrating DCP and that of the original algorithms is reported, for each test instance. Note that, since the space occupancy per node of LFR, and LFR-DCP depends on the number of weight change operations, median values for each of these algorithms are reported.

The experiments show that the use of DCP induces, in most of the cases, a clear improvement also in the space requirements per node. In particular, DUAL-DCP (LFR-DCP, respectively) requires a maximum space occupancy per node which is

Fig. 25 Number of
messages sent by DUAL and
DUAL-DCP on $G_{BA-8000}$

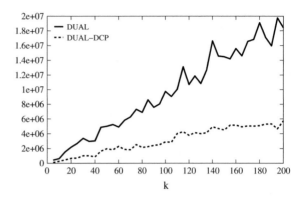

Fig. 26 Number of
messages sent by LFR and
LFR-DCP on $G_{BA-8000}$

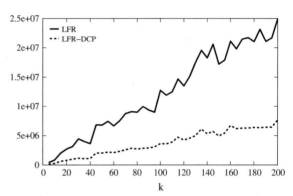

0.30 (0.72, respectively) and 0.29 (0.83, respectively) times that of DUAL (LFR, respectively) in $G_{IP-8000}$ and $G_{BA-8000}$, respectively. Notice that, the improvement is more evident in the case of DUAL, as its maximum space occupancy per node is by far higher than that of LFR. Concerning DUAL, this behaviour is confirmed also in the average case, where DUAL-DCP requires 0.81 and 0.92 times the average space occupancy per node of DUAL, in $G_{IP-8000}$ and $G_{BA-8000}$, respectively. On the contrary, the experimental data show that the average space occupancy per node of LFR-DCP is slightly greater than that of LFR and that the use of DCP induces an overhead in the average space occupancy per node which is equal to 53% and 77%, in $G_{IP-8000}$ and $G_{BA-8000}$, respectively. This is due to the fact that the average space occupancy of LFR is quite low by itself and that, in this case, the space occupancy overhead needed to store the CHAINPATH is greater than the space occupancy reduction induced by the use of DCP.

Table 1 Space occupancy per node of the considered algorithms

Graph	Algorithm	MAX		AVG	
		Bytes	Ratio	Bytes	Ratio
$G_{IP-8000}$	DUAL	8 320 000	1	311 410	1
	DUAL-DCP	2 517 680	0.30	252 625	0.81
$G_{IP-8000}$	LFR	549 170	1	192 871	1
	LFR-DCP	392 658	0.72	295 930	1.53
$G_{BA-8000}$	DUAL	20 800 000	1	323 350	1
	DUAL-DCP	6 130 860	0.29	300 081	0.92
$G_{BA-8000}$	LFR	520 289	1	192 619	1
	LFR-DCP	434 130	0.83	341 449	1.77

7 Conclusions

In the last years, there has been a renewed interest in devising new efficient light-weight distributed shortest paths solutions for large-scale Ethernet networks, where distance-vector algorithms are an attractive alternative to link-state solutions, when scalability and reliability are key issues or when the memory resources of the nodes of the network are limited.

In this chapter we have reviewed classic distance-vector approaches, and outlined the most recent and efficient solutions of this category. In particular, we have considered the classical *Distributed Bellman-Ford* (DBF) algorithm [7], the well-known *Diffuse Update Algorithm* (DUAL) of [24], the recent *Loop Free Routing* algorithm (LFR) of [26], and the *Distributed Computation Pruning* technique (DCP) of [27], which can be combined with distance-vector algorithms with the aim of overcoming some of their main limitations. We have analysed differences and similarities, pros and cons, of the various algorithm, and we have also provided a summary of the experimental results given in [26, 27], which show how the above mentioned algorithms behave in practice.

References

1. S. Cicerone, G. D'Angelo, G. Di Stefano, D. Frigioni, Partially dynamic efficient algorithms for distributed shortest paths. Theor. Comput. Sci. **411**, 1013–1037 (2010)
2. S. Cicerone, G.D. Stefano, D. Frigioni, U. Nanni, A fully dynamic algorithm for distributed shortest paths. Theor. Comput. Sci. **297**(1–3), 83–102 (2003)
3. P.A. Humblet, Another adaptive distributed shortest path algorithm. IEEE Trans. Commun. **39**(6), 995–1002 (1991)
4. G.F. Italiano, Distributed algorithms for updating shortest paths, in *Lecture notes in computer science on international workshop on distributed algorithms*, vol. 579, pp. 200–211 (1991)
5. A. Orda, R. Rom, Distributed shortest-path and minimum-delay protocols in networks with time-dependent edge-length. Distributed Comput. **10**, 49–62 (1996)

6. K.V.S. Ramarao, S. Venkatesan, On finding and updating shortest paths distributively. J. Algorithms **13**, 235–257 (1992)
7. J. McQuillan, Adaptive routing algorithms for distributed computer networks. Technical Report BBN Report 2831, (Cambridge, MA, 1974)
8. E.C. Rosen, The updating protocol of arpanet's new routing algorithm. Comput. Netw. **4**, 11–19 (1980)
9. D. Bertsekas, R. Gallager, *Data networks* (Prentice Hall International, 1992)
10. B. Awerbuch, A. Bar-Noy, M. Gopal, Approximate distributed bellman-ford algorithms. IEEE Trans. Commun. **42**(8), 2515–2517 (1994)
11. J.T. Moy, *OSPF: anatomy of an internet routing protocol*, (Addison-Wesley 1998)
12. E.W. Dijkstra, A note on two problems in connexion with graphs. Numerische Mathematik **1**, 269–271 (1959)
13. J. Wu, F. Dai, X. Lin, J. Cao, W. Jia, An extended fault-tolerant link-state routing protocol in the internet. IEEE Trans. Comput. **52**(10), 1298–1311 (2003)
14. F. Bruera, S. Cicerone, G. D'Angelo, G.D. Stefano, D. Frigioni, Dynamic multi-level overlay graphs for shortest paths. Math. Comput. Sci. **1**(4), 709–736 (2008)
15. D. Frigioni, A. Marchetti-Spaccamela, U. Nanni, Fully dynamic algorithms for maintaining shortest paths trees. J. Algorithms **34**(2), 251–281 (2000)
16. P. Narváez, K.-Y. Siu, H.-Y. Tzeng, New dynamic algorithms for shortest path tree computation. IEEE/ACM Trans. Netw. **8**(6), 734–746 (2000)
17. S. Cicerone, G. D'Angelo, G. Di Stefano, D. Frigioni, V. Maurizio, Engineering a new algorithm for distributed shortest paths on dynamic networks. Algorithmica **66**(1), 51–86 (2013)
18. G. D'Angelo, M. D'Emidio, D. Frigioni, Pruning the computation of distributed shortest paths in power-law networks. Informatica **37**(3), 253–265 (2013)
19. K. Elmeleegy, A.L. Cox, T.S.E. Ng On count-to-infinity induced forwarding loops in ethernet networks in *Proceedings of 25th IEEE conference on computer communications (INFOCOM2006)*, pp. 1–13 (2006)
20. A. Myers, E. Ng, H. Zhang, Rethinking the service model: Scaling ethernet to a million nodes, in *Proceedings 3rd workshop on hot topics in networks (ACM HotNets)*. (ACM Press, 2004)
21. S. Ray, R. Guérin, K.-W. Kwong, R. Sofia, Always acyclic distributed path computation. IEEE/ACM Trans. Netw. **18**(1), 307–319 (2010)
22. N. Yao, E. Gao, Y. Qin, H. Zhang, Rd: reducing message overhead in DUAL, in *Proceedings of1st international conference on network infrastructure and digital content (IC-NIDC2009)*, (IEEE Press, 2009) pp. 270–274
23. C. Zhao, Y. Liu, K. Liu, A more efficient diffusing update algorithm for loop-free routing, in *Proceedings 5th International Conference on Wireless Communications, Networking and Mobile Computing (WiCom2009)*, (IEEE Press, 2009) pp. 1–4
24. J.J. Garcia-Lunes-Aceves, Loop-free routing using diffusing computations. IEEE/ACM Trans. Netw. **1**(1), 130–141 (1993)
25. EIGRP, Enhanced interior gateway routing protocol. http://www.cisco.com/c/en/us/support/docs/ip/enhanced-interior-gateway-routing-protocol-eigrp/16406-eigrp-toc.html
26. G. D'Angelo, M. D'Emidio, D. Frigioni, A loop-free shortest-path routing algorithm for dynamic networks. Theor. Comput. Sci. **516**, 1–19 (2014)
27. G. D'Angelo, M. D'Emidio, D. Frigioni, D. Romano, Enhancing the computation of distributed shortest paths on power-law networks in dynamic scenarios. Theor. Comput. Syst. **57**(2), 444–477 (2015)
28. OMNeT++, Discrete event simulation environment. http://www.omnetpp.org
29. Y. Hyun, B. Huffaker, D. Andersen, E. Aben, C. Shannon, M. Luckie, K. Claffy, The CAIDA IPv4 routed/24 topology dataset. http://www.caida.org/data/active/ipv4_routed_24_topology_dataset.xml
30. R. Albert, A.-L. Barabási, Emergence of scaling in random networks. Science **286**, 509–512 (1999)
31. H. Attiya, J. Welch, Distributed Computing Wiley (2004)

32. E.W. Dijkstra, C.S. Scholten, Termination detection for diffusing computations. Informat. Process. Lett. **11**, 1–4 (1980)
33. N.A. Lynch, *Distributed algorithms*, Morgan Kaufmann Publishers (1996)
34. K. Pahlavan, P. Krishnamurthy, *Networking fundamentals: Wide* (Wiley, Local and Personal Area Communications, 2009)

Influenza Virus Algorithm for Multiobjective Energy Reduction Open Vehicle Routing Problem

Iraklis-Dimitrios Psychas, Eleni Delimpasi, Magdalene Marinaki
and Yannis Marinakis

Abstract We propose a Parallel Multi-Start Multiobjective Influenza Virus Algorithm (PMS-MOIVA) for the solution of the Multiobjective Energy Reduction Open Vehicle Routing Problem. The PMS-MOIVA could be categorized in the Artificial Immune System algorithms, as it simulates the process of annual evolution of influenza virus in an isolated human population. Two different versions of the algorithm are presented where their main difference is the fact that in the first version, PMS-MOIVA1, the algorithm focuses on the improvement of the most effective solutions using a local search procedure while in the second version, PMS-MOIVA2, the use of the local search procedure is applied equally in the whole population. In order to prove the effectiveness of the proposed algorithm a comparison is performed with the Parallel Multi-Start Non-dominated Sorting Genetic Algorithm II (PMS-NSGA II). The Multiobjective Energy Reduction Open Vehicle Routing Problem has two different objective functions, the first corresponds to the optimization of the total travel time and the second corresponds to the minimization of the fuel consumption of the vehicle taking into account the travel distance and the load of the vehicle when the decision maker plans delivery. A number of modified Vehicle Routing Problem instances are used in order to evaluate the quality of the proposed algorithm.

I.-D. Psychas (✉) · E. Delimpasi · M. Marinaki · Y. Marinakis
School of Production Engineering and Management,
Technical University of Crete, Chania, Greece
e-mail: ipsychas102@gmail.com

E. Delimpasi
e-mail: eldelimpasi@yahoo.gr

M. Marinaki
e-mail: magda@dssl.tuc.gr

Y. Marinakis
e-mail: marinakis@ergasya.tuc.gr

© Springer International Publishing AG, part of Springer Nature 2018
A. Adamatzky (ed.), *Shortest Path Solvers. From Software to Wetware*,
Emergence, Complexity and Computation 32,
https://doi.org/10.1007/978-3-319-77510-4_5

1   Introduction

In recent years, there is an increasing number of papers in two different fields of Vehicle Routing Problem variants, the first concerns Multiobjective Vehicle Routing Problems [18, 24, 36, 39] and the second concerns the optimization of energy or fuel consumption in the Vehicle Routing Problems. Considering the second variant the calculation of the tonne-kilometers (tonne-km or tkm) were used in order to calculate the "Fuel Efficiency" and the "CO_2 emissions" of a vehicle [19, 33, 34]. Leonardi's et al. [28] calculate the "Efficiency of the vehicle use" by a ratio tonne-kilometres/mass-kilometres and also they calculate the "CO_2 Efficiency" assuming that there are some other real route and environmental parameters that are multiplied with the "Efficiency of the vehicle use" in order to calculate the "CO_2 Efficiency" of a vehicle. Another parameter that can be taken into account for the calculation of fuel consumption is the parameter of speed [1, 2, 13, 23, 30, 44]. Xiao et al. [49] propose the Fuel Consumption Rate (FCR) for a VRP (FCVRP) in order to minimize the fuel consumption. A bi-objective Green Vehicle Routing Problem was proposed in [17] in order to minimize the total traveled distance and the CO_2 emissions. In [11] a bi-objective Pollution Routing problem's model is proposed where the first objective function minimizes the CO_2 emissions of a vehicle and the second objective function minimizes the driving time. For two more complicated multiobjective Energy VRPs please see [22, 35]. Two other energy Pick-up and Delivery VRP models are analyzed in [31, 45]. Some other CO_2 emissions minimizing models are presented in [7, 20, 21, 29, 46]. Also, CO_2 emissions could be minimized by creating shortest routes and by traveling with the best speed for the environment [42]. For a more extended review for the Energy and Green Vehicle Routing Problems please see [25, 32].

In general, the **Vehicle Routing Problem** (**VRP**) is a problem in which vehicles start from the depot and end to the depot after the servicing of a number of customers taking into account a number of constraints. For an overview of the VRP please see [26, 47].

Recently our research group published two papers, the first concerns the application of a new version of NSGA II, denoted as Parallel Multi Start NSGA II, for the solution of the Multiobjective Energy Reduction Vehicle Routing Problem [39] and the second concerns the application of a number of variants of the Differential Evolution algorithm for the solution of the same problem [40]. The latter algorithm is based on a modification of an algorithm that we have published for the solution of the Multiobjective Traveling Salesman Problem [41]. The novelty of the proposed research is twofold. Initially, a new formulation of a variant of the Vehicle Routing Problem is given, where the vehicles do not return to the depot, as in the Open Vehicle Routing Problem [3], and two objective functions are optimized simultaneously, the first corresponds to the optimization of the total travel time and the second objective function corresponds to the minimization of the fuel consumption of the vehicle taking into account the travel distance and the load of the vehicle when the decision maker plans delivery. The proposed model is denoted as Multiobjective Energy Reduction Open Vehicle Routing Problem (MEROVRP).

The other novelty of the chapter is the proposal of a new Artificial Immune System algorithm denoted as **Influenza Virus Algorithm (IVA)**. The algorithm simulates the process of annual evolution of influenza virus in an isolated *human population* with a constant number of people. Furthermore, it simulates the methods of *Reassortment* and *Mutation* that this virus uses in order to evolve into a host's body. Each solution represents a *carrier* of the virus. The application of an Artificial Immune System algorithm in a multiobjective environment has been studied from a number of researchers in the past [4, 5, 8]. The novelty of the proposed algorithm is that it is the first time, at least to our knowledge, that this natural process is simulated, transformed and presented as a Multiobjective Evolutionary Optimization algorithm [9]. This algorithm does not only creates new solutions by crossover using two selected parents from the previous generation as a simple genetic algorithm does, but also creates a number of clones of each solution and evolves them with two different methods, the Mutation that is an 2-opt application on the solution or Reassortment that is a crossover application on two solution. Using these two different methods the proposed algorithm gives the opportunity to the solutions to evolve more times and with different procedures than a genetic algorithm. For comparison reasons, we use the **Parallel Multi-Start NSGA II (PMS-NSGA II)** algorithm [39].

The structure of the chapter is as follows. In Sect. 2, the MEROVRP problem is described in detail and its formulation is presented. In Sect. 3, an analytical description of the proposed algorithms is presented. Finally, in Sect. 4, the computational results are presented and, then, concluding remarks and the future research are given in the last Section.

2 Multiobjective Energy Reduction Open Vehicle Routing Problem

In this chapter, a **Multiobjective Energy Reduction Open Vehicle Routing Problem (MERVRP)** is formulated. Two different objective functions are used where the first one concerns the minimization of the time needed for a vehicle to travel between two customers or a customer and the depot while the second one concerns the minimization of the fuel consumption when the decision maker plans routes with all the customers having only demands.

We used the same formulation as the one proposed in [39, 40] with the difference that the vehicle does not return to the depot after the servicing of the customers, as it happens in any variant of the Open Vehicle Routing Problem. For completeness and due to space limitations we present only the two objective functions and the constraints of the problem and the analysis of how this formulation resulted is given in [39, 40] with the difference that in the last two papers the vehicle returns to the depot after the completion of its routes.

In the first objective function, the minimization of the time needed to travel between two customers or a customer and the depot is given:

$$\min OF1 = \sum_{i=1}^{n}\sum_{j=2}^{n}\sum_{\kappa=1}^{m}(t_{ij}^{\kappa} + s_{j}^{\kappa})x_{ij}^{\kappa} \tag{1}$$

where t_{ij}^{κ} the time needed to visit customer j immediately after customer i using vehicle κ, s_{j}^{κ} is the service time of customer j using vehicle κ, n is the number of nodes and m is the number of homogeneous vehicles and the depot is denoted by $i = j = 1$. x_{ij}^{κ} denotes that the vehicle κ visits customer j immediately after customer i.

The second objective function is used for the minimization of the fuel Consumption (FC) taking into account the load and the traveled distance [39, 40]. We have to mention that in the second objective function we consider that the most loaded is the vehicle the more fuel it consumes:

$$\min OF2 = \sum_{j=2}^{n}\sum_{\kappa=1}^{m}c_{1j}x_{1j}^{\kappa}(1 + \frac{y_{1j}^{\kappa}}{Q}) + \sum_{i=2}^{n}\sum_{j=2}^{n}\sum_{\kappa=1}^{m}c_{ij}x_{ij}^{\kappa}(1 + \frac{y_{i-1,i}^{\kappa} - D_{i}}{Q}) \tag{2}$$

with the maximum capacity of the vehicle denoted by Q, the i customer has demand equal to D_i and $D_1 = 0$, x_{ij}^{κ} denotes that the vehicle κ visits customer j immediately after customer i with load y_{ij}^{κ} and $y_{1j}^{\kappa} = \sum_{i=1}^{n} D_i$ for all vehicles as the vehicle begins with load equal to the summation of the demands of all customers assigned in its route and c_{ij} is the distance from node i to node j.

The constraints of the problem are the following:

$$\sum_{j=2}^{n}\sum_{\kappa=1}^{m}x_{ij}^{\kappa} = 1, i = 1, \ldots, n \tag{3}$$

$$\sum_{i=1}^{n}\sum_{\kappa=1}^{m}x_{ij}^{\kappa} = 1, j = 2, \ldots, n \tag{4}$$

$$\sum_{j=2}^{n}x_{ij}^{\kappa} - \sum_{j=2}^{n}x_{ji}^{\kappa} = 0, i = 1, \ldots, n, \kappa = 1, \ldots, m \tag{5}$$

$$\sum_{j=2, j \neq i}^{n}y_{ji}^{\kappa} - \sum_{j=2, j \neq i}^{n}y_{ij}^{\kappa} = D_{i}, i = 1, \ldots, n, \kappa = 1, \ldots, m, \tag{6}$$

$$Qx_{ij}^{\kappa} \geq y_{ij}^{\kappa}, i = 1, \ldots, n, j = 2, \ldots, n, \kappa = 1, \ldots, m \tag{7}$$

$$x_{ij}^{\kappa} = \begin{cases} 1, & \text{if } (i, j) \text{ belongs to the route} \\ 0, & \text{otherwise} \end{cases} \tag{8}$$

Constraints (3) and (4) represent that each customer must be visited only by one vehicle; constraints (5) ensure that each vehicle that arrives at a node must leave from that node also. Constraint (6) indicate that the reduced load (cargo) of each vehicle after it visits a node is equal to the demand of that node. Constraints (7) are used to limit the maximum load carried by the vehicle and to force y_{ij}^{κ} to be equal to zero when $x_{ij}^{\kappa} = 0$ while constraints (8) ensure that only one vehicle will visit each customer.

3 Parallel Multi-Start Multiobjective Influenza Virus Algorithm (PMS-MOIVA)

In this Section the proposed algorithm is presented and analyzed in detail. Emphasis is given in the part of the algorithm denoted as Influenza Virus Algorithm and all the other parts that are common with our previous papers [39–41] are described in brief given the differences of their initial proposed versions and the appropriate reference where they were initially published.

3.1 Basic Parts of the Algorithm

The solutions are represented with the path representation of the tour. For example, if we have a solution with five nodes a possible path representation would be the "1 2 3 4 5". The node 1 is the depot. If a route does not start with the node 1 then we find it and we put it at the beginning of the route. For example a solution "2 3 1 5 4" is transformed to "1 5 4 2 3".

For the initialization of the population we use a method denoted as Parallel Multi Start Method [39, 40]. A brief description of the algorithm is as follows. More than one initial population are created (X populations). Each one of these populations is divided in K subpopulations with $w = W/K$ solutions each one, where W is the total number of solutions of the population and K is the number of objective functions. Thus, if the initial number of solutions W of one of the X populations is equal to 20, the number of objective functions (K) is equal to 2 and the number of population X is equal to 5, each one of the two subpopulation consists of $w = 10$ solutions and the total number of produced solutions (considering all the X populations) are equal to 100 ($W * X$). The initial members of each of the X (five in our example) populations are produced using, three different strategies, a Nearest Neighborhood procedure [27] (for the 20% of the populations, one in our example), a variant of the Variable Neighborhood Search (VNS) algorithm [14] as it was proposed in [39] (for the 40% of the populations, two in our example) and a variant of the Greedy

Table 1 How to produce the initial solutions for a two objective functions problem

		Number of individuals	Method used
Individuals in the first 40% of the populations			
W	w for OF1	1	VNS
		2 to w/3	Swap method
		(w/3) + 1 to 2w/3	2-opt method
		(2w/3) + 1 to w	Random solutions
	w for OF2	1	VNS
		2 to w/3	Swap method
		(w/3) + 1 to 2w/3	2-opt method
		(2w/3) + 1 to w	Random solutions
Individuals in the next 20% of the populations			
W	w for OF1	1	Nearest neighborhood
		2 to w/3	Swap method
		(w/3) + 1 to 2w/3	2-opt method
		(2w/3) + 1 to w	Random solutions
	w for OF2	1	Nearest neighborhood
		2 to w/3	Swap method
		(w/3) + 1 to 2w/3	2-opt method
		(2w/3) + 1 to w	Random solutions
Individuals in the last 40% of the populations			
W	w for OF1	1	GRASP
		2 to w/3	Swap method
		(w/3) + 1 to 2w/3	2-opt method
		(2w/3) + 1 to w	Random solutions
	w for OF2	1	GRASP
		2 to w/3	Swap method
		(w/3) + 1 to 2w/3	2-opt method
		(2w/3) + 1 to w	Random solutions

Randomized Adaptive Search Procedure (GRASP) [12] as it was proposed in [40] (for the rest of the populations, two in our example). All the other solutions of each of the populations are produced using a local search procedure based on VNS algorithm as it was analyzed in [40]. For more information please see Table 1.

3.2 Multiobjective Influenza Virus Algorithm

In real life, Influenza virus is a RNA virus. It consists of a genome that is enclosed in a host cell membrane. Influenza viruses are evolving by Mutation or by Reassortment into the hosts [16]. The process of Mutation produces an antigenic drift and the process of Reassortment produces an antigenic shift [6].

Antigenic drift

The Antigenic drifts are small changes in two antigens on the surface of the virus and cause the creation of strains derived from already existed strains. The new strain replaces the older strains when it enters into the population causing epidemics (each year, 5–20% of the population is infected with influenza viruses [15]) [48]. These new strains are drift variants. Antigenic drift happens every year. For this reason there is a lack of full immunity while the annual adjustment of the vaccine is essential.

Antigenic shift

The Antigenic shift is an effective fast change in viral genetics, which are sudden changes from one antigen to another. These large changes allow the virus to infect new species and to overcome quickly the protective immunity. For example the reassortment between avian strains and human strains can cause an antigenic shift. If a virus that infects people has completely new antigens, everyone would be infected and the new virus will cause a pandemic if the circumstances permit it [38]. Pandemics can start unpredictably every 10–40 years because of the appearance of a completely new strain to which the population has no natural immunity. In 1918 the Spanish influenza pandemic is estimated to have infected 50% of the world's population. The ability of this new strain to spread from person to person (contagiousness) as well as the urbanization and the high population density increase the risk of a rapid global spread of a pandemic due to a novel influenza virus.

The Influenza Virus Algorithm simulates the process of annual evolution of influenza in an isolated *human population* with a constant number of I individuals (humans). For each *human population* should be a *carrier* that will transmit the virus to a certain proportion of the *human population*. Thus, for an algorithm that optimize a population of solutions, as the proposed algorithm, we will have as many *carriers* as the number of the solutions. Also for every *carrier* there will be a *human population* that the *carrier* will transmit the virus to a certain proportion of it (of the population). In the proposed algorithm, the *carrier* corresponds to the problem's solution, the *infection* corresponds to the process of producing a new solution and one *year* corresponds to one iteration. The *Mutation* corresponds to a partial change of a solution (antigenic drift) and the *Reassortment* corresponds to the global change of a solution (antigenic shift).

In the proposed algorithm, the number of the initial solutions (*carriers*) is equal to W. Before the iterative procedure starts, for every solution (carrier) is determined randomly if it will be a *New Strain* or a *Pre-existing Strain* using an archive *strain*. If a solution has value 1 in the archive *strain*, then, this solution has been signed as a *New Strain*. On the other hand, if a solution has value 0 in the archive *strain*, then, this solution has been signed as a *Pre-existing Strain*.

Also, the personal best (*Carrier Best*) are initially set equal to the current solutions and the initial set of non-dominated solutions is calculated. As in the real life there are carriers of the virus in the population and there are, also, a number of healthy members of the population. In the proposed algorithm, every *population* has I healthy individuals. During a year, each carrier will infect a part of the *population* where it belongs to, thus, a table with one to, at most, $I/2$ clones of the solution (carrier) will be created (the number of the members of the healthy population that would be infected by each carrier). The number of the clones will be signed as $infect$. We consider that in every $infected$ individual occurs only one antigenic shift (Mutation) or one antigenic drift (Reassortment).

In each iteration and for every solution (carrier) the following procedure is applied. A variable den ($density$) is randomly generated (where $den \in (0, 1)$) which represents the annual $density$ of each $population$ of I individuals. Also, a variable con ($contagiousness$) is generated which represents the annual $contagiousness$ of each carriers virus i and is calculated by the following equations:

For the variant PMS-MOIVA1:

$$con_i = 1 - \frac{\sum_{k=1}^{K} \frac{value\ of\ the\ carrier\ (i)\ for\ the\ k\ objective\ function}{worst\ value\ of\ k\ objective\ function}}{K} \tag{9}$$

In this equation, the con tends to be equal to 1 as the fitness functions of a solution are improved. This equation gives the opportunity in better solutions to perform more mutations.

For the variant PMS-MOIVA2

$$con_i = \frac{\sum_{k=1}^{K} \frac{value\ of\ the\ carrier\ (i)\ for\ the\ k\ objective\ function}{worst\ value\ of\ k\ objective\ function}}{K} \tag{10}$$

In this equation, the con tends to be equal to 1 as the fitness functions of a solution deteriorate.

For example, for a carrier i of the proposed multiobjective problem if the $OF1_i$ is equal to 5 and the $OF2_i$ is equal to 100 and the larger values (considering the values of all carriers) is equal to 150 for the OF1 and is equal to 180 for OF2, then, considering the PMS-MOIVA2 con equation the con_i would be calculated as it follows:

$$con_i = \frac{(105/150) + (100/180)}{2} = 0.62 \tag{11}$$

If a solution (carrier) has been signed as a $New\ Strain$ ($strain = 1$), $con \geq 0.6$ and $den \geq 0.7$, then, this is a $Pandemic$ (the limits 0.6 and 0.7 were selected in order to simulate as faithfully as it could be the conditions of the real life) and the number $infect$ of the infected individuals of the $human\ population$ from that solution is calculated by the following equation:

$$infect = round(\frac{1}{2} * con * den * I) \tag{12}$$

By this way the value of the variant $infect$ will be always between 20 and 50% of the value of the variant I. Thus, the number of the clones of the i solution that will be thought as $Infecteds$ will be equal to $infect$.

In any other case, the situation is denoted as $Epidemic$ and the value of the variable $infect$ from the solution i is calculated by the following equation:

$$infect = round(\frac{1}{5} * con * den * I), \text{ if } infect > round(0.05 * I) \quad (13)$$

or

$$infect = round(0.05 * I), \text{ in any other case} \quad (14)$$

Based on the last two equations the value of the variant $infect$ will be always between 20 and 50% of the value of the variant I. Thus, for each carrier a table $Infecteds$ with $infect$ members, clones of the $carrier$, is calculated and each one of these solutions-clones has 90% to be mutated (mutation operator) and 10% to be replaced in the population by a new solution (reassortment operator—in real life this means that the human has been infected by more than one types of viruses). The Mutation operator is performed using a 2-opt in the solution of the table $Infecteds$ and creates a new solution and the Reassortment operator is performed, for the creation of a new solution, using a classic crossover operator using the solution-clone and a new randomly created solution $NewInf$. This new solution, that is created from the crossover, is signed as a $New\ Strain$. For more understanding please see Fig. 1. For the calculation of this new solution, each element of the solution-

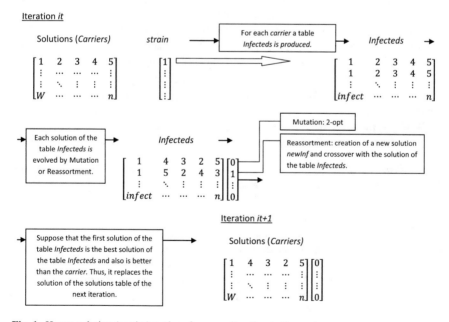

Fig. 1 How a solution (carrier) evolves from one iteration to the next

clone and the $NewInf$'s solution is transformed into a floating point in the interval [39, 40].

The equation for the calculation of the new solution ($infected$) is:

$$infected(i) = (1 - g) * NewInf(i) + g * carrier(i) \qquad (15)$$

where $g \in (0, 1)$ and is a random produced number and $i = 1, ..., n$ and $infected$ is the solution of the table $Infecteds$. After the crossover, the elements of the new solution (infected) are transformed back into the integer domain by assigning the smallest floating value to the smallest integer, the next highest floating value to the next integer and so on [39].

Then, the Pareto set of the table $Infecteds$ is calculated and one of the elements of this Pareto set is selected randomly as the *best infected* solution. This *best infected* from the Pareto set will become the new carrier of his *human population* for the next iteration. This procedure is continues for every solution (carrier) of the table *Carriers* until all solutions (carriers) are examined. Afterwards, a Variable Neighborhood Search (VNS) algorithm [39] ($vns_{max} = 20$ and $local_{max} = 10$) is applied in each *carrier*. The first iteration of the algorithm ends with the creation of the initial Pareto set with the non-dominated solutions of the initial population of the table *Carriers*.

The personal best solution for each *carrier* (*Carrier Best*) is updated using the following procedure. If a solution (carrier) in iteration it dominates its personal best solution, then, the personal best solution is replaced by the current solution. Otherwise, if the personal best solution dominates the current solution or the two solutions are not dominated between them, then, the personal best solution is not replaced. Then, the VNS method is applied at the personal best solutions of the table *Carriers Best* with both the vns_{max} and the $local_{max}$ equal to 10 [39]. At the end of each iteration, the set of the non-dominated solutions is updated and when all the processes of all populations have been completed, then, the Total Pareto Front is updated.

In Fig. 1, the basic steps of the method are presented and, then, a pseudocode of the proposed Parallel Multi-Start Multiobjective Influenza Virus algorithm is given.

Algorithm Parallel Multi-Start Multiobjective Influenza Virus
Do while the maximum number of Populations has not been reached:
Initialization
    Generation of the initial population
    Definition of the maximum number of iterations (years)
    Selection of the number of solutions W and of the value of the variable I
    Creation of the initial *Carriers* and define which solution
      will be signed as a *New Strain* and as a *Pre − existed Strain*
    Evaluation of each objective function for every solution
    Initialization of the *Carrier Best*
    Finding of the non-dominated set of the *Carriers*
Main Phase

Do while the maximum number of iterations has not been reached

  For each solution (carrier) **do**

   Evaluation of *con*

   $den = rand$

   If solution is a *New Strain* & $con \geq 0.6$ & $den \geq 0.7$ **then**

     Calculate *infect* using Eq. 12

   else

     Calculate *infect* using Eqs. 13 and 14

   Endif

   Creation of the *Infecteds*

   For each solution of the *Infecteds* **do**

     p=*rand*

     If p ≤ 0.9 **then**

       Call *Mutation*

       Define the new solution as a *Pre − existed Strain*

     else

       Call *Reassortment*

       Define the new solution as a *New Strain*

     Endif

     Evaluation of the objective function for every solution

   Endfor

   Initialization of the non-dominated set of the *Infecteds*

   Choose randomly the best solution of the *Infecteds*'s Pareto set

   Update of the solution (carrier) with the *best infected* solution

  Endfor

  Evaluation of the *Carriers* for each objective function

  Application of VNS to improve the solutions

  Update the personal best solutions (*Carrier Best*)

  Update of the non-dominated set

 Enddo

 Return Population's Pareto Front

Enddo

Return Total Pareto Front

4 Computational Results

The whole algorithmic approach was implemented in Visual C++. We used a data set of instances that was, initially, proposed in [39] for the solution of the Multiobjective Energy Reduction Vehicle Routing Problem and they are, also, suitable for the solution of the Multiobjective Energy Reduction Open Vehicle Routing Problem as the only difference between the two problems is that the vehicles do not return to the depot after the completion of the servicing of the customers. In order to compare the results of the proposed algorithm and to see the effectiveness of the procedure a variant of the NSGA II [10], the Parallel Multi-start NSGA II (PMS-NSGA II) is used as it was proposed in [39]. In order to evaluate the effectiveness of the procedures four classic measures were used, the range to which the front spreads (M_k) [50], the number of solutions of the Pareto front (L), the Δ measure which includes information about both spread and distribution of solutions [37], and the Coverage measure [50].

A number of different alternative values for the parameters of the algorithms were tested and the ones selected are those that gave the best computational results concerning both the quality of the solution and the computational time needed to achieve this solution and, also, taking into account the fact that we would like to test the algorithms with the same function evaluations. Thus, the selected parameters for all the algorithms are given in the following:

Parallel Multi-Start MOIVAs

- Number of carriers for each initial population: 100.
- Number of years: 500.
- $I = 100$.
- Number of initial populations: 10.

Parallel Multi-Start NSGA II

- Number of individuals for each initial population: 100.
- Number of generations: 500.
- Number of initial populations: 10.

After the selection of the final parameters, the two versions of the Parallel Multi-Start Multiobjective Influenza Virus Algorithm (PMS-MOIVAs) and the Parallel Multi-Start Non-dominated Sorting Genetic Algorithm II (PMS-NSGA II) were tested in ten instances (i.e., kroA100par3-kroB100par3, kroA100par3-kroC100par3, kroA100par3-kroD100par3, etc., as they are described in [39]). In the following tables the comparisons performed based on the four evaluation measures that are mentioned previously are given. In all Tables, kroA100par3 is denoted with A, kroB100par3 is denoted with B, and so on. If we have a combination of two instances, the instance is denoted by the combination of the two letters, for example kroA100par3-kroB100par3 is denoted with A−B in all Tables. In Table 2, the results of the first three measures are given while in Table 3, the results of the $Coverage$ measure are presented. In Fig. 2, four representative Pareto fronts are presented.

In general, it is preferred to find as many as possible non-dominated solutions (L measure), the expansion of the Pareto front to be as large as possible which shows that better solutions have been found in every dimension (M_k measure) and the spacing of solutions to be as smaller as possible which means that the non-dominated solutions are close between them (Δ measure). In the first table, the best value for each measure from the comparison of all algorithms is signed as bold while from the comparison of the two PMS-MOIVA algorithms the best values are underlined. On the other hand, in the second table, the best value for C measure from the comparisons of all the algorithms is signed as bold. In the C measure table, the notation of the algorithms

Table 2 Results of the first three measures for the three algorithms for the ten instances

	PMS-MOIVA1			PMS-MOIVA2			PMS-NSGA II		
Instances	L	M_k	Δ	L	M_k	Δ	L	M_k	Δ
A−B	40	**591.46**	**0.60**	<u>41</u>	591.17	0.67	**45**	597.21	0.66
A−C	38	<u>595.24</u>	**0.54**	<u>42</u>	591.09	0.63	**53**	**611.52**	0.65
A−D	47	551.27	<u>0.64</u>	54	**576.52**	0.65	**58**	559.72	**0.56**
A−E	<u>52</u>	564.61	0.63	45	<u>580.50</u>	**0.55**	**54**	**584.03**	0.63
B−C	46	571.82	<u>0.66</u>	48	<u>574.53</u>	0.81	**68**	**582.28**	**0.58**
B−D	<u>47</u>	570.51	0.66	44	**583.71**	**0.61**	**52**	571.63	0.76
B−E	42	<u>584.21</u>	0.66	<u>44</u>	579.86	**0.63**	**61**	**597.60**	0.72
C−D	<u>51</u>	559.04	**0.64**	45	**565.93**	0.72	**55**	559.23	0.69
C−E	35	580.33	0.63	<u>43</u>	**600.27**	<u>0.59</u>	**70**	597.95	**0.53**
D−E	<u>45</u>	<u>591.35</u>	0.67	<u>45</u>	582.55	<u>0.66</u>	**55**	**594.32**	**0.61**

Table 3 Results of the coverage measure

A−B	MOIVA1	MOIVA2	NSGA II	B-D	MOIVA1	MOIVA2	NSGA II
MOIVA1	0.00	**0.41**	**0.84**	MOIVA1	0.00	**0.43**	**0.83**
MOIVA2	0.38	0.00	**0.87**	MOIVA2	0.19	0.00	**0.79**
NSGA II	0.00	0.00	0.00	NSGA II	0.13	0.14	0.00
A−C	MOIVA1	MOIVA2	NSGA II	B-E	MOIVA1	MOIVA2	NSGA II
MOIVA1	0.00	**0.40**	**0.94**	MOIVA1	0.00	0.32	**0.98**
MOIVA2	0.34	0.00	0.87	MOIVA2	**0.52**	0.00	**1.00**
NSGA II	0.00	0.00	0.00	NSGA II	0.00	0.00	0.00
A−D	MOIVA1	MOIVA2	NSGA II	C-D	MOIVA1	MOIVA2	NSGA II
MOIVA1	0.00	**0.41**	**0.90**	MOIVA1	0.00	**0.44**	**1.00**
MOIVA2	0.40	0.00	**0.88**	MOIVA2	0.41	0.00	**0.98**
NSGA II	0.02	0.04	0.00	NSGA II	0.00	0.00	0.00
A−E	MOIVA1	MOIVA2	NSGA II	C-E	MOIVA1	MOIVA2	NSGA II
MOIVA1	0.00	**0.51**	**0.94**	MOIVA1	0.00	**0.60**	**0.99**
MOIVA2	0.42	0.00	**0.98**	MOIVA2	0.26	0.00	**0.89**
NSGA II	0.00	0.00	0.00	NSGA II	0.00	0.07	0.00
B−C	MOIVA1	MOIVA2	NSGA II	D-E	MOIVA1	MOIVA2	NSGA II
MOIVA1	0.00	**0.71**	**0.90**	MOIVA1	0.00	**0.51**	**0.85**
MOIVA2	0.20	0.00	**0.90**	MOIVA2	0.27	0.00	**0.87**
NSGA II	0.04	0.02	0.00	NSGA II	0.04	0.02	0.00

PMS-MOIVA# and PMS-NSGA II have been replaced with the notations MOIVA#
and NSGA II, respectively, in order to reduce the size of the Table.

In general the outcome of all evaluation measures could not give safe conclusions
concerning which algorithm performs better from the others. If we take into account

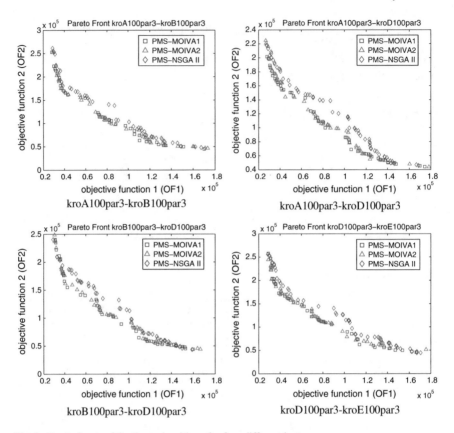

Fig. 2 Pareto fronts of the three algorithms for four different instances

only the results of the L measure (Table 2), then, the proposed algorithms PMS-MOIVA1 and PMS-MOIVA2 give inferior results compared to the PMS-NSGA II algorithm. On the other hand, if we take into account the coverage measure (Table 3), then, the proposed algorithms perform better than the PMS-NSGA II algorithm in all the instances. In the other two measures, all the algorithms give equally good results. However, if we, also, use the depiction of the results in Fig. 2, we can see that the two proposed algorithms perform better than PMS-NSGA II algorithm as the produced Pareto Fronts of the proposed algorithms always dominate the produced Pareto Fronts of PMS-NSGA II algorithm.

If we would like to compare the results of the two versions of the PMS-MOIVA algorithm, both algorithms perform equally well in the three out of the four measures (measures presented in Table 2). However, in the Coverage measure, PMS-MOIVA1 algorithm performs better than PMS-MOIVA2 in all but one instances. The same outcome could be seen from Fig. 2, although, as the solutions are very close between them, this outcome is not so obvious taking into account only the figure.

5　Conclusions and Future Research

In this chapter, a new algorithm based on the evolution of the Influenza Virus in different populations was presented. We have applied this algorithm in a multiobjective problem as we would like to test the effectiveness of the algorithm in a very difficult variant of the Vehicle Routing Problem. The results from the application in a multiobjective problem are positive and give us the strength to continue to the application to other multiobjective problems (not belonging to the variants of the Vehicle Routing Problem) and to single objective optimization problems (problems from the context of the Vehicle Routing Problem and more general problems).

References

1. J.M. Bandeira, T. Fontes, S.R. Pereira, P. Fernandes, A. Khattak, M.C. Coelho, Assessing the importance of vehicle type for the implementation of eco-routing systems. Transp. Res. Procedia **3**, 800–809 (2014)
2. T. Bektas, G. Laporte, The pollution-routing problem. Transp. Res. Part B **45**, 1232–1250 (2011)
3. J. Brandao, A tabu search algorithm for the open vehicle routing problem. Eur. J. Oper. Res. **157**, 552–564 (2004)
4. F. Campelo, F.G. Guimaraes, R.R. Saldanha, H. Igarashi, S. Noguchi, D.A. Lowther, J.A. Ramirez, A novel multiobjective immune algorithm using nondominated sorting, in *11th International IGTE Symposium on Numerical Field Calculation in Electrical Engineering* (2004), pp. 308–313
5. F. Campelo, F.G. Guimaraes, H. Igarashi, Overview of artificial immune systems for multiobjective optimization, in International Conference on Evolutionary Multicriterion Optimization, EMO 2007, vol. 4403 (LNCS, 2007), pp. 937–951
6. F. Carrat, A. Flahault, Influenza vaccine: The challenge of antigenic drift. Vaccine **39–40**, 6852–6862 (2007)
7. N. Charoenroop, B. Satayopas, A. Eungwanichayapant, City bus routing model for minimal energy consumption. Asian J. Energy Environ. **11**(01), 19–31 (2010)
8. C.A. Coello Coello, N.C. Cortes, An approach to solve multiobjective optimization problems based on an artificial immune system. in *1st International Conference on Artificial Immune Systems*, ed. by J. Timmis, P.J. Bentley (2002), pp. 212–221
9. C.A. Coello Coello, D.A. Van Veldhuizen, G.B. Lamont, *Evolutionary Algorithms for Solving Multi-Objective Problems*, (Springer, 2007)
10. K. Deb, A. Pratap, S. Agarwal, T. Meyarivan, A fast and elitist multiobjective genetic algorithm: NSGA-II. IEEE Trans. Evol. Comput. **6**(2), 182–197 (2002)
11. E. Demir, T. Bektas, G. Laporte, The bi-objective pollution-routing problem. Eur. J. Oper. Res. **232**, 464–478 (2014)
12. T.A. Feo, M.G.C. Resende, Greedy randomized adaptive search procedure. J. Glob. Optim. **6**, 109–133 (1995)
13. M. Figliozzi, Vehicle routing problem for emissions minimization. Transp. Res. Rec. J. Transp. Res. Board **2**, 1–7 (2011)
14. P. Hansen, N. Mladenovic, Variable neighborhood search: principles and applications. Eur. J. Oper. Res. **130**, 449–467 (2001)
15. S.A. Harper, J.S. Bradley, J.A. Englund, T.M. File, S. Gravenstein, F.G. Hayden, A.J. McGeer, K.M. Neuzil, A.T. Pavia, M.L. Tapper, T.M. Uyeki, R.K. Zimmerman, Seasonal influenza in

adults and children diagnosis, treatment, chemoprophylaxis, and institutional outbreak management: Clinical practice guidelines of the infectious diseases society of America. Clin. Infect. Dis. **48**, 1003–1032 (2009)

16. A.J. Hay, V. Gregory, A.R. Douglas, Y.P. Lin, The evolution of human influenza viruses. Philos. Trans. R. Soc. B Biol. Sci. **356**(1416), 1861–1870 (2001)

17. J. Jemai, M. Zekri, K. Mellouli, An NSGA-II algorithm for the green vehicle routing problem. in *International Conference on Evolutionary Computation in Combinatorial Optimization, Lecture Notes in Computer Science*, vol. 7245 (2012), pp. 37–48

18. N. Jozefowiez, F. Semet, E.G. Talbi, Multi-objective vehicle routing problems. Eur. J. Oper. Res. **189**, 293–309 (2008)

19. I. Kara, B.Y. Kara, M.K. Yetis, Energy minimizing vehicle routing problem. COCOA **2007**, 62–71 (2007)

20. C. Koc, T. Bektas, O. Jabali, G. Laporte, The fleet size and mix pollution-routing problem. Transp. Res. Part B **70**, 239–254 (2014)

21. C.A. Kontovas, The green ship routing and scheduling problem (GSRSP): A conceptual approach. Transp. Res. Part D **31**, 61–69 (2014)

22. R.S. Kumar, K. Kondapaneni, V. Dixit, A. Goswami, L.S. Thakur, M.K. Tiwari, Multi-objective modeling of production and pollution routing problem with time window: A self-learning particle swarm optimization approach. Comput. Ind. Eng. **99**, 29–40 (2016)

23. Y. Kuo, Using simulated annealing to minimize fuel consumption for the time-dependent vehicle routing problem. Comput. Ind. Eng. **59**(1), 157–165 (2010)

24. N. Labadie, C. Prodhon, A Survey on multi-criteria analysis in logistics: Focus on vehicle routing problems, *Applications of Multi-Criteria and Game Theory Approaches*. Springer Series in Advanced Manufacturing (2014), pp. 3–29

25. R. Lahyani, M. Khemakhem, F. Semet, Rich vehicle routing problems: From a taxonomy to a definition. Eur. J. Oper. Res. **241**, 1–14 (2015)

26. G. Laporte, The vehicle routing problem: An overview of exact and approximate algorithms. Eur. J. Oper. Res. **59**, 345–358 (1992)

27. E.L. Lawer, J.K. Lenstra, A.H.G.R. Rinnoy Kan, D.B. Shmoys, *The Traveling Salesman Problem: A Guided Tour of Combinatorial Optimization*, (Wiley, 1985)

28. J. Leonardi, M. Baumgartner, CO_2 efficiency in road freight transportation: Status quo, measures and potential. Transp. Res. Part D **9**, 451–464 (2004)

29. H. Li, T. Lv, Y. Li, The tractor and semitrailer routing problem with many-to-many demand considering carbon dioxide emissions. Transp. Res. Part D **34**, 68–82 (2015)

30. J. Li, Vehicle routing problem with time windows for reducing fuel consumption. J. Comput. **7**(12), 3020–3027 (2012)

31. C. Lin, K.L. Choy, G.T.S. Ho, T.W. Ng, A genetic algorithm-based optimization model for supporting green transportation operations. Expert Syst. Appl. **41**, 3284–3296 (2014)

32. C. Lin, K.L. Choy, G.T.S. Ho, S.H. Chung, H.Y. Lam, Survey of green vehicle routing problem: Past and future trends. Expert Syst. Appl. **41**, 1118–1138 (2014)

33. A. McKinnon, A logistical perspective on the fuel efficiency of road freight transport, in *OECD, ECMT and IEA: Workshop Proceedings*, Paris (1999)

34. A. McKinnon, Green logistics: The carbon agenda. Electr. Sci. J. Logistics **6**(3), 1–9 (2010)

35. J.C. Molina, I. Eguia, J. Racero, F. Guerrero, Multi-objective vehicle routing problem with cost and emission functions. Procedia Soc. Behav. Sci. **160**, 254–263 (2014)

36. N. Norouzi, R. Tavakkoli-Moghaddam, M. Ghazanfari, M. Alinaghian, A. Salamatbakhsh, A new multi-objective competitive open vehicle routing problem solved by particle swarm optimization. Netw. Spat. Econ. **12**(4), 609–633 (2012)

37. T. Okabe, Y. Jin, B. Sendhoff, A critical survey of performance indices for multi-objective optimisation. Evol. Comput. **2**, 878–885 (2003)

38. C.R. Parrish, Y. Kawaoka, The origins of new pandemic viruses: the acquisition of new host ranges by canine parvovirus and influenza A viruses. Ann. Rev. Microbiol. **59**, 553–586 (2005)

39. I.D. Psychas, M. Marinaki, Y. Marinakis, A parallel multi-start NSGA II algorithm for multi-objective energy reduction vehicle routing problem, in *8th International Conference on Evolutionary Multicriterion Optimization*, EMO 2015, Part I, LNCS 9018, ed. by A. Gaspar-Cunha, et al. (Springer International Publishing Switzerland, 2015), pp. 336–350

40. I.D. Psychas, M. Marinaki, Y. Marinakis, A. Migdalas, Non-dominated sorting differential evolution algorithm for the minimization of route based fuel consumption multiobjective vehicle routing problems. Energy Syst. (2016). https://doi.org/10.1007/s12667-016-0209-5

41. I.D. Psychas, E. Delimpasi, Y. Marinakis, Hybrid evolutionary algorithms for the multiobjective traveling salesman problem. Expert Syst. Appl. **42**, 8956–8970 (2015)

42. A. Sbihi, R.W. Eglese, Combinatorial optimization and green logistics, *4OR*, **5**(2), 99–116 (2007)

43. A.B. Shenderov, Avian influenza virus with pandemic potential: Suspected role of microbe/microbe and host/microbe interactions in change, adaptive evolution and host range shift. Microb. Ecol. Health Dis. **17**, 186–188 (2005)

44. Y. Suzuki, A new truck-routing approach for reducing fuel consumption and pollutants emission. Transp. Res. Part D **16**, 73–77 (2011)

45. N. Tajik, R. Tavakkoli-Moghaddam, B. Vahdani, S. Meysam Mousavi, A robust optimization approach for pollution routing problem with pickup and delivery under uncertainty. J. Manufact. Syst. **33**, 277–286 (2014)

46. A. Tiwari, P.C. Chang, A block recombination approach to solve green vehicle routing problem. Int. J. Prod. Econ. 1–9 (2002)

47. P. Toth, D. Vigo, *The Vehicle Routing Problem* (Monographs on Discrete Mathematics and Applications, Siam, 2002)

48. Y.I. Wolf, C. Viboud, E.C. Holmes, E.V. Koonin, D.J. Lipman, Long intervals of stasis punctuated by bursts of positive selection in the seasonal evolution of influenza A virus. Biol. Dir. 1–34 (2006)

49. Y. Xiao, Q. Zhao, I. Kaku, Y. Xu, Development of a fuel consumption optimization model for the capacitated vehicle routing problem. Comput. Oper. Res. **39**(7), 1419–1431 (2012)

50. E. Zitzler, K. Deb, L. Thiele, Comparison of multiobjective evolutionary algorithms: Empirical results. Evol. Comput. **8**(2), 173–195 (2000)

Practical Algorithms for the All-Pairs Shortest Path Problem

Andrej Brodnik and Marko Grgurovič

Abstract We study practical algorithms for solving the all-pairs shortest path problem. The Floyd-Warshall algorithm is frequently used to solve the aforementioned problem, and we show how it can be augmented to drastically reduce the number of path combinations examined. Very favorable results are shown via empirical tests that compare the new algorithm with known algorithms on random graphs. In addition to the all-pairs shortest path problem, we also investigate the highly related all-pairs bottleneck paths problem, and give an efficient average case algorithm. On top of that, we show how the bottleneck paths problem relates to the decremental transitive closure problem, and specifically how algorithms for the latter can be used to solve the former.

1 Introduction

Let $G = (V, E)$ denote a directed graph where E is the set of edges and $V = \{v_1, v_2, ..., v_n\}$ is the set of vertices of the graph. The function $\ell(\cdot)$ maps edges to (possibly negative) lengths. For a path π, we define its length to be the sum of the lengths of its edges: $\ell(\pi) = \sum_{(u,v) \in \pi} \ell(u, v)$. Additionally, we define $\forall (u, v) \notin E : \ell(u, v) = \infty$. From hereon we make the standard assumption that there are no cycles whose total lengths are negative, and without loss of generality, we assume G is strongly connected. To simplify notation, we define $m = |E|$ and $n = |V|$.

The research presented in this chapter was partially financially supported by the Slovenian Research Agency under N2-0053 and P2-0359.

A. Brodnik (✉) · M. Grgurovič
University of Primorska, Glagoljaška 8, 6000 Koper, Slovenia
e-mail: andrej.brodnik@upr.si

M. Grgurovič
e-mail: marko.grgurovic@famnit.upr.si

A. Brodnik
University of Ljubljana, Večna pot 113, 1000 Ljubljana, Slovenia

© Springer International Publishing AG, part of Springer Nature 2018
A. Adamatzky (ed.), *Shortest Path Solvers. From Software to Wetware*,
Emergence, Complexity and Computation 32,
https://doi.org/10.1007/978-3-319-77510-4_6

Furthermore, we define $d(u, v)$ for two vertices $u, v \in V$ as the length of the shortest path from u to v. It is also useful to define m^* as the number of edges (u, v) such that $d(u, v) = \ell(u, v)$. These are the edges that form the shortest path graph, and are the only edges necessary for its computation.

Finding shortest paths in such graphs is a classic problem in algorithmic graph theory. Two of the most common variants of the problem are the single-source shortest path (SSSP) problem and the all-pairs shortest path problem (APSP). In the SSSP variant, we are searching for paths with the least total length from a fixed vertex $s \in V$ to every other vertex in the network. Similarly, the APSP problem asks for the shortest path between every pair of vertices $u, v \in V$. In this chapter we will focus exclusively on the APSP variant of the problem, and without loss of generality, assume that we are not interested in paths beginning in v and returning back to v.

The asymptotically fastest APSP algorithm for dense graphs to date runs in $O(n^3 \log \log^3 n / \log^2 n)$ time [1]. For non-negative edge length functions and for sparse graphs, there exist asymptotically fast algorithms for worst case inputs [2–4], and algorithms which are efficient average-case modifications of Dijkstra's algorithm [5–7].

The APSP problem can easily be solved by n calls to an SSSP algorithm. There exist strategies that are more effective than simply running independent SSSP computations, such as the Hidden Paths Algorithm [8], the Uniform Paths algorithm [5], and most recently the Propagation algorithm [9]. The Propagation algorithm is more general than the former two, which are modifications of Dijkstra, in the sense that it works for any SSSP algorithm. Besides providing a speed-up for arbitrary SSSP algorithms, it also performs well in practice, as shown in [9].

As a truly all-pairs algorithm, Floyd-Warshall [10, 11] is frequently used to solve APSP. There exist many optimizations for the Floyd-Warshall algorithm, ranging from better cache performance [12], optimized program-generated code [13], to parallel variants for the GPU [14, 15]. One can also approach APSP through funny matrix multiplication, and practical improvements have been devised to this end through the use of sorting [16].

In spite of intensive research on efficient implementations of the Floyd-Warshall algorithm, there has not been much focus devoted to improvement of the number of path combinations examined by the algorithm. In Sect. 2, we will propose a modification of the Floyd-Warshall algorithm that combines it with an hourglass-like tree structure, which reduces the number of paths that have to be examined. Only those path combinations that provably cannot change the values in the shortest path matrix are omitted. The resulting algorithm is simple to implement, uses no fancy data structures and in empirical tests is faster than the Floyd-Warshall algorithm for random complete graphs on 256–4096 nodes by factors ranging from 2.5 to 8.5. When we inspect the number of path combinations examined however, our modification reduces the number by a staggering factor of 12–90.

In Sect. 4 we consider the all-pairs bottleneck paths (APBP) problem, which is highly related to the all-pairs shortest path problem. We show that an efficient algorithm whose bound depends on m^* can be obtained, and show how the APBP problem can be reduced to that of decremental transitive closure.

2   The Hourglass Algorithm

The Floyd-Warshall algorithm [10, 11] is a simple dynamic programming approach to solve the all-pairs shortest path problem. Unlike Dijkstra's algorithm, Floyd-Warshall can find shortest paths in graphs which contain negatively-weighted edges. In this section we will outline improvements that build on the base algorithm, but first we outline the pseudocode of the Floyd-Warshall algorithm in Algorithm 1. Intuitively, one might expect that the minimum operation in line 5, also sometimes referred to as relaxation, would not succeed in lowering the value of $W[i][j]$ every time. This is precisely what we aim to exploit: instead of simply looping through every node in line 4, we utilize the structure of shortest paths that we have computed up until now. This allows us to avoid checking many path combinations that the Floyd-Warshall algorithm inspects, but which provably cannot reduce the current value stored inside $W[i][j]$.

Algorithm 1 Floyd-Warshall Algorithm

1: **procedure** FLOYD-WARSHALL(W)
2:     **for** $k := 1$ to n **do**
3:         **for** $i := 1$ to n **do**
4:             **for** $j := 1$ to n **do**
5:                 $W[i][j] := \min(W[i][j], W[i][k] + W[k][j])$
6:             **end for**
7:         **end for**
8:     **end for**
9: **end procedure**

We will say a path $u \overset{k}{\rightsquigarrow} v$ is a k-shortest path if it is the shortest path between u and v that is only permitted to go through nodes $\{v_1, ..., v_k\}$. This means that $u \overset{n}{\rightsquigarrow} v$ would be the shortest path from u to v in the traditional sense. We denote the length of a path $u \overset{k}{\rightsquigarrow} v$ by writing $\ell(u \overset{k}{\rightsquigarrow} v)$, where the length is simply the sum of the lengths of all edges that are on the path.

The resulting algorithm is still a dynamic programming algorithm, but it now has a smaller pool of candidates to perform relaxation on, which makes it run faster. In Sect. 2.1, we show how to lower the number of candidates looped through in line 4 of Algorithm 1 by exploiting the tree structure of $k \overset{k-1}{\rightsquigarrow} j$ paths. In Sect. 2.2, we show how to exploit the structure of $i \overset{k-1}{\rightsquigarrow} k$ paths and further reduce the number of candidates in line 4. Both reductions are achieved by traversing a tree structure rather than looping through all nodes. These modifications yield two tree data structures, and joining them in the root yields an hourglass shaped data structure that combines the power of both.

2.1   The Single-Tree Algorithm

The simplest improvement involves the use of a tree, denoted as OUT_k, which is the shortest path tree containing paths that begin in node v_k and end in some node $w \in V \setminus \{v_k\}$, but only go through nodes in the set $\{v_1, ..., v_{k-1}\}$. In other words, these are paths of the form $v_k \overset{k-1}{\rightsquigarrow} w$ $\forall w \in V \setminus \{v_k\}$. Traversal of this tree is used to replace the FOR loop on variable j in line 4 of Algorithm 1. In order to reconstruct the shortest paths, the Floyd-Warshall algorithm needs to maintain an additional matrix, which specifies the path structure, but this additional matrix is otherwise not required for the functioning of the algorithm. In our algorithm, however, this information is essential, since the path structure is used during the algorithm's execution. We augment the Floyd-Warshall algorithm with a matrix $L[i][j]$ which specifies the penultimate node on the shortest path from i to j (i.e. the last node that is not j). This suffices for reconstructing the shortest path tree for all paths going out of k as follows: create n trees $\{T_1, ..., T_n\}$, now go through $j = 1$ to n and place T_j as the child of $T_{L[k][j]}$. This takes $O(n)$ time.

Assume that we have the $(k-1)$-shortest paths $i \overset{k-1}{\rightsquigarrow} j$ $\forall i, j \in V$ and we are trying to extend the paths to go through v_k, i.e. we want to compute $i \overset{k}{\rightsquigarrow} j$ $\forall i, j \in V$. First we construct OUT_k in $O(n)$ time. Now we can use the following lemma when extending the paths to go through v_k:

Lemma 1 *Let $v_x \in V \setminus \{v_k\}$ be some non-leaf node in OUT_k and let $v_y \neq v_x$ be an arbitrary node in the subtree rooted at v_x. Now let $v_i \in V \setminus \{v_k\}$ and consider a path $v_i \overset{k-1}{\rightsquigarrow} v_k \overset{k-1}{\rightsquigarrow} v_x$. If $\ell(v_i \overset{k-1}{\rightsquigarrow} v_k \overset{k-1}{\rightsquigarrow} v_x) \geq \ell(v_i \overset{k-1}{\rightsquigarrow} v_x)$, then we claim $\ell(v_i \overset{k-1}{\rightsquigarrow} v_k \overset{k-1}{\rightsquigarrow} v_y) \geq \ell(v_i \overset{k-1}{\rightsquigarrow} v_y)$.*

Proof By choice of v_y and v_x, we have $v_k \overset{k-1}{\rightsquigarrow} v_y = v_k \overset{k-1}{\rightsquigarrow} v_x \overset{k-1}{\rightsquigarrow} v_y$. Thus we want to show:

$$\ell(v_i \overset{k-1}{\rightsquigarrow} v_y) \leq \ell(v_i \overset{k-1}{\rightsquigarrow} v_k \overset{k-1}{\rightsquigarrow} v_x) + \ell(v_x \overset{k-1}{\rightsquigarrow} v_y).$$

Observe that $x < k$, since v_x is neither a leaf nor the root of OUT_k. Because $v_i \overset{k-1}{\rightsquigarrow} v_y$ is the $(k-1)$-shortest path and $x < k$ we have:

$$\ell(v_i \overset{k-1}{\rightsquigarrow} v_y) \leq \ell(v_i \overset{k-1}{\rightsquigarrow} v_x) + \ell(v_x \overset{k-1}{\rightsquigarrow} v_y).$$

Putting these together we get:

$$\ell(v_i \overset{k-1}{\rightsquigarrow} v_y) \leq \ell(v_i \overset{k-1}{\rightsquigarrow} v_x) + \ell(v_x \overset{k-1}{\rightsquigarrow} v_y) \leq \ell(v_i \overset{k-1}{\rightsquigarrow} v_k \overset{k-1}{\rightsquigarrow} v_x) + \ell(v_x \overset{k-1}{\rightsquigarrow} v_y).$$

Which is what we wanted to prove.

The algorithm then extends the $(k-1)$-shortest paths for each incoming node v_i by depth-first traversal[1] of OUT_k, starting with the root and avoiding the inspection of subtrees whose roots v_x did not yield a shorter path than $v_i \overset{k-1}{\rightsquigarrow} v_x$. Intuitively, one would expect this to exclude large subtrees from ever being considered. The pseudocode is given in Algorithm 2.

Algorithm 2 Single-tree Algorithm

1: **procedure** SINGLE-TREE(W)
2:    Initialize L, a $n \times n$ matrix, as $L[i][j] := i$.
3:    **for** $k := 1$ to n **do**
4:        Construct OUT_k.
5:        **for** $i := 1$ to n **do**
6:            Stack := empty
7:            Stack.push(v_k)
8:            **while** Stack $\neq$ empty **do**
9:                v_x := Stack.pop()
10:                **for all** children v_j of v_x in OUT_k **do**
11:                    **if** $W[i][k] + W[k][j] < W[i][j]$ **then**
12:                        $W[i][j] := W[i][k] + W[k][j]$
13:                        $L[i][j] := L[k][j]$
14:                        Stack.push(v_j)
15:                    **end if**
16:                **end for**
17:            **end while**
18:        **end for**
19:    **end for**
20: **end procedure**

Observe that the extra space required by the trees is merely $O(n)$, since we can reuse the same space. Constructing the tree takes $O(n)$ time which yields in total $O(n^2)$ time over the course of the entire algorithm.

2.1.1 Optimized Implementation

Instead of maintaining a stack and visiting nodes in the tree as in Algorithm 2, a much faster implementation is possible in practice. After building the tree OUT_k, we keep track of two permutation arrays: *dfs[]* and *skip[]*. The *dfs* array is simply the depth-first traversal of the tree, i.e. *dfs[x]* contains the x-th vertex encountered on a DFS traversal of OUT_k. For a vertex v_z, *skip[z]* contains the index in *dfs* of the first vertex after v_z in the DFS order that is not a descendant of v_z in OUT_k. Then, all we need to do is simply traverse *dfs* and whenever an improvement is not made, we jump to the next index via the *skip* array. It should be pointed out that the asymptotic time remains the same, as this is solely a practical optimization.

---

[1]Breadth-first traversal is also possible, of course.

2.2 The Hourglass Algorithm

We can augment Algorithm 2 with another tree. The second tree is similar to OUT_k, except that it is the shortest path "tree" for paths $w \overset{k-1}{\rightsquigarrow} v_k$ $\forall w \in V \setminus \{v_k\}$. Strictly speaking, this is not a tree,[2] but we can reverse the directions of the edges, which turns it into a tree with v_k as the root. We denote this tree as IN_k. Observe that if $v_a \neq v_k$ is a node in IN_k and v_b is a child of v_a in IN_k, then the $(k-1)$-shortest path from v_b to v_k goes through v_a, since the edges are reversed in the tree. Traversal of IN_k will be used as a replacement of the FOR loop on variable i in line 3 of Algorithm 1. In order to construct IN_k efficiently, we need to maintain an additional matrix $F[i][j]$ which stores the second node on the path from i to j (i.e. the first node that is not i). The construction of IN_k is now similar to what we had before: create n trees $\{T_1, ..., T_n\}$, then go through $i = 1$ to n and place T_i as the child of $T_{F[i][k]}$. This takes $O(n)$ time. Consequently, we have the following lemma:

Lemma 2 *Let $v_a \in V \setminus \{v_k\}$ be some non-leaf node in IN_k and let $v_b \neq v_a$ be an arbitrary node in the subtree rooted at v_a. Now let $v_j \in V \setminus \{v_k\}$ and consider a path $v_a \overset{k-1}{\rightsquigarrow} v_k \overset{k-1}{\rightsquigarrow} v_j$. If $\ell(v_a \overset{k-1}{\rightsquigarrow} v_k \overset{k-1}{\rightsquigarrow} v_j) \geq \ell(v_a \overset{k-1}{\rightsquigarrow} v_j)$, then we claim $\ell(v_b \overset{k-1}{\rightsquigarrow} v_k \overset{k-1}{\rightsquigarrow} v_j) \geq \ell(v_b \overset{k-1}{\rightsquigarrow} v_j)$.*

Proof Due to the choice of v_a and v_b we have: $v_b \overset{k-1}{\rightsquigarrow} v_k = v_b \overset{k-1}{\rightsquigarrow} v_a \overset{k-1}{\rightsquigarrow} v_k$. We want to show, that:
$$\ell(v_b \overset{k-1}{\rightsquigarrow} v_j) \leq \ell(v_b \overset{k-1}{\rightsquigarrow} v_a) + \ell(v_a \overset{k-1}{\rightsquigarrow} v_k \overset{k-1}{\rightsquigarrow} v_j).$$

Observe that $a < k$, since v_a is neither a leaf nor the root of IN_k. Thus we have:
$$\ell(v_b \overset{k-1}{\rightsquigarrow} v_j) \leq \ell(v_b \overset{k-1}{\rightsquigarrow} v_a) + \ell(v_a \overset{k-1}{\rightsquigarrow} v_j).$$

Putting these together we get the desired inequality:
$$\ell(v_b \overset{k-1}{\rightsquigarrow} v_j) \leq \ell(v_b \overset{k-1}{\rightsquigarrow} v_a) + \ell(v_a \overset{k-1}{\rightsquigarrow} v_j) \leq \ell(v_b \overset{k-1}{\rightsquigarrow} v_a) + \ell(v_a \overset{k-1}{\rightsquigarrow} v_k \overset{k-1}{\rightsquigarrow} v_j).$$

Observe that if we perform depth-first traversal on IN_k, we can temporarily prune OUT_k as follows: if v_a is the parent of v_b in IN_k and $v_a \overset{k-1}{\rightsquigarrow} v_j \leq v_a \overset{k-1}{\rightsquigarrow} v_k \overset{k-1}{\rightsquigarrow} v_j$, then the subtree of v_j can be removed from OUT_k while we are inspecting the subtree of v_a in IN_k, and later re-inserted. This is easy to do by using a stack to keep track of deletions. The pseudocode for the Hourglass algorithm is given in Algorithm 3. In practice, recursion can be replaced with another stack, and each node in the IN_k tree is then visited twice—the second visit would restore the subtrees that were removed from OUT_k by that node.

---

[2]The hourglass name comes from placing this structure atop the OUT_k tree, which gives it an hourglass-like shape, with v_k being the neck.

Algorithm 3 Hourglass Algorithm

1: **procedure** HOURGLASS(W)
2:    Initialize L, a $n \times n$ matrix, as $L[i][j] := i$.
3:    Initialize F, a $n \times n$ matrix, as $F[i][j] := j$.
4:    **for** $k := 1$ to n **do**
5:        Construct OUT_k.
6:        Construct IN_k.
7:        **for all** children v_i of v_k in IN_k **do**
8:            RECURSEIN($W, L, F, IN_k, OUT_k, v_i$)
9:        **end for**
10:    **end for**
11: **end procedure**
12: **procedure** RECURSEIN($W, L, F, IN_k, OUT_k, v_i$)
13:    Stack := empty
14:    Stack.push(v_k)
15:    **while** Stack $\neq$ empty **do**
16:        $v_x :=$ Stack.pop()
17:        **for all** children v_j of v_x in OUT_k **do**
18:            **if** $W[i][k] + W[k][j] < W[i][j]$ **then**
19:                $W[i][j] := W[i][k] + W[k][j]$
20:                $L[i][j] := L[k][j]$
21:                $F[i][j] := F[i][k]$
22:                Stack.push(v_j)
23:            **else**
24:                Remove the subtree of v_j from OUT_k.
25:            **end if**
26:        **end for**
27:    **end while**
28:    **for all** children $v_{i'}$ of v_i in IN_k **do**
29:        RECURSEIN($W, L, F, IN_k, OUT_k, v_{i'}$)
30:    **end for**
31:    Restore any subtrees we may have removed in line 24.
32: **end procedure**

The only extra space requirement of the Hourglass algorithm that bears any significance is the matrix F, which contains n^2 entries. It is important to note that the worst-case time complexity of the Hourglass (and Single-tree) algorithm remains $O(n^3)$. The simplest example of this is when all shortest paths are the edges themselves, at which point the tree structure is essentially flat and never changes.

2.3 Empirical Comparison

We now empirically examine how many path combinations are skipped by the Hourglass and Single-tree algorithms compared to the Floyd-Warshall algorithm. We performed two experiments, one on random complete graphs, and one on random sparse graphs. We measured the number of path combinations examined. Since the results are numbers that range from very small to very large in both cases, we display

Fig. 1 The percentage of path combinations examined by the two modifications of Floyd-Warshall, when compared to the original algorithm (which is always at 100%, not shown), for the input of complete graphs of various sizes

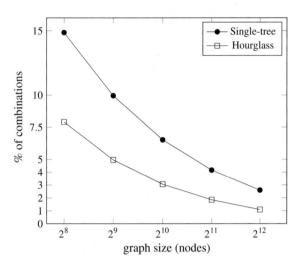

the results as a percentage of the Floyd-Warshall algorithm, which is always 100% in the plots, but is not drawn explicitly.

The input graphs were pseudorandomly generated. For complete graphs, this meant assigning each edge an independently uniformly distributed random length in the range $(0, 1)$. Sparse graphs were generated by starting with an empty graph on 1024 nodes and adding a desired number of edges, which were chosen independently according to the uniform random distribution, and assigned an independently uniformly distributed random length in the range $(0, 1)$. Edge lengths were represented using floating-point numbers in both cases.

The first experiment was for the input of random complete graphs of varying sizes. The results are shown in Fig. 1. The second experiment was for the input of a random graph of 1024 nodes whose number of edges varied from 10 to 80% where $100\% = n^2$. To make the comparison between Floyd-Warshall and the modified versions fairer in the second experiment, we augmented the Floyd-Warshall algorithm with a simple modification, that allowed it to skip combinations i, k where $W[i][k] = \infty$, which reduced the number of path combinations examined. The results of the second experiment are shown in Fig. 2.

In Fig. 1 we can see a significant reduction in terms of path combinations examined. This quantity dominates the algorithm's asymptotic running time and, as observed, decreases compared to the cubic algorithm when inputs grow larger. It might be possible to obtain sub-cubic asymptotic bounds in the average-case model, which is an open question. The experiments on sparse graphs in Fig. 2 show a reduction in path combinations examined as the graph becomes sparser, but the effect on the running time seems to be very minor.

Overall, the Single-tree algorithm is the simplest to implement and offers good performance. The Hourglass algorithm has the potential to be even faster, but would

Fig. 2 The percentage of
path combinations examined
by the two modifications of
Floyd-Warshall, when
compared to the original
algorithm (which is always
at 100%, not shown), for the
input of a graph with 1024
nodes and various edge
densities

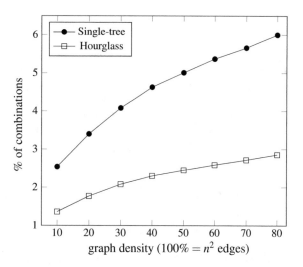

3   Empirical Comparison of APSP Algorithms

In this section, we analyze the results of empirical tests consisting of running five
shortest path algorithms on random graph instances and measuring their running
times.

3.1   Graphs and Algorithms

The experiments were conducted on the following two types of directed random
graphs.

Uniform random graphs: the edge length is uniformly randomly distributed
inside the interval $[0, 1)$. As these graphs grow denser, they start to favor the average-
case algorithms, since $m^* = \mathcal{O}(n \lg n)$ with high probability in complete graphs with
uniformly distributed random edge lengths [17].

Unweighted random graphs: edge lengths are set to 1. These graphs can be
viewed as a type of worst-case for the average-case algorithms, since $m^* = m$ always
holds, i.e. a direct edge is the shortest path between two nodes. It should be pointed
out, that breadth-first search (BFS) is extremely efficient in solving these instances
given its simplicity and $O(mn)$ running time (when solving APSP). However, since

likely require a better implementation. It is also worthwhile to note that the additional
space requirements for the Single-tree algorithm are very modest, as most applications would typically require storing the path reconstruction matrix regardless.

we consider these instances only as a worst-case of a more general shortest path problem, we did not include BFS in the comparisons.

In both cases, the graphs were constructed by first setting a desired vertex count and density. Then, a random Hamiltonian cycle is constructed, ensuring that the graph is strongly connected. Edges are added into the graph at random until the desired density is reached. Finally, algorithms are executed on the instance, and their running times recorded. We have explored densities ranging from $m = n^{1.1}$ to $m = n^2$, and vertex counts ranging from $n = 512$ to $n = 4096$. For each density and vertex count combination, we have generated 10 different random instances and averaged the running times of each algorithm.

Priority queues are integral to many shortest path algorithms. Pairing heaps were used in all experiments, since they are known to perform especially well in this capacity in practice. Unlike Fibonacci heaps, which have an $\mathcal{O}(1)$ amortized decrease key operation, the amortized complexity of decrease-key for pairing heaps is $\mathcal{O}(2^{2\sqrt{\lg\lg n}})$ [18]. The following algorithms have been compared:

Dijkstra [19]: solves all-pairs by solving multiple independent single-source problems. Using pairing heaps this algorithm runs in $\mathcal{O}(n^2 \lg n + mn2^{2\sqrt{\lg\lg n}})$.

Floyd-Warshall [10, 11]: classic dynamic programming formulation. Does not use priority queues and runs in $\mathcal{O}(n^3)$.

Propagation: the algorithm described in [9]. In essence, it is a more efficient way of using an SSSP algorithm to solve APSP. The underlying SSSP algorithm is Dijkstra's algorithm. Using pairing heaps this algorithm runs in $\mathcal{O}(n^2 \lg n + m^*n2^{2\sqrt{\lg\lg n}})$.

Single-tree: the algorithm from Sect. 2.1, with the optimizations outlined in Sect. 2.1.1.

Hourglass: the algorithm from Sect. 2.2.

The code has been written in C++ and compiled using `g++ -march=native -O3`. We have used the implementation of pairing heaps from the Boost Library, version 1.55. All tests were run on an Intel(R) Core(TM) i7-2600@3.40GHz with 8GB RAM running Windows 7 64-bit.

Results are shown as plots where the y axis represents time in milliseconds in logarithmic scale, and the x axis represents the graph edge density as $m = n^x$.

3.2 Uniform Random Graphs

The results for uniform random graphs are shown in Figs. 3, 4, 5 and 6.

The tests show that Propagation and Single-tree together outperform the other algorithms on all densities. As the size increases, Hourglass starts catching up to Single-tree, but the constant factors still prove to be too much for it to benefit from its more clever exploration strategy. The running time of Propagation depends on m^* instead of m, and $\frac{m}{m^*}$ in the uniform random graphs increases as the graphs grow

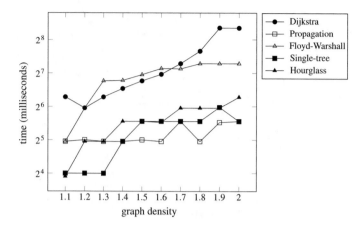

Fig. 3 512 vertices, uniform weights. The plot is quite erratic due to the extremely short running times

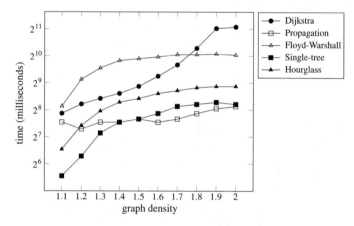

Fig. 4 1024 vertices, uniform weights. The general trend starts to form. Differences between the fastest three algorithms on the sparse instances are quite significant

denser, so it is expected that Dijkstra would be relatively slower the denser the graph is. It is quite surprising that the Single-tree and Hourglass algorithms are so efficient on sparse graphs, outperforming even Dijkstra, something that would seem incredibly difficult given its $O(n^3)$ worst-case time. This would suggest that its average-case time is significantly lower than its worst-case, but no theoretical bounds are known so far.

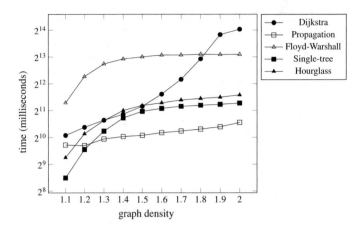

Fig. 5 2048 vertices, uniform weights. Floyd-Warshall's running time begins to increase drastically, as expected due to its cubic complexity. Differences between the fastest three algorithms on sparse instances start to decrease

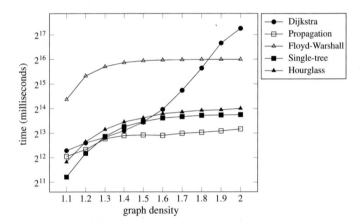

Fig. 6 4096 vertices, uniform weights. Propagation and Single-tree prove to be the fastest, with Single-tree outperforming Propagation on the sparser instances

3.3 Unweighted Random Graphs

The results for unweighted random graphs are shown in Figs. 7, 8, 9 and 10.

In these tests, Propagation performs quite poorly, but that is to be expected since $m = m^*$ in these graphs, resulting in no benefit from Propagation's more clever search strategy compared to Dijkstra. What is interesting is that the Single-tree and Hourglass algorithms are able to remain competitive with Dijkstra in spite of this, and

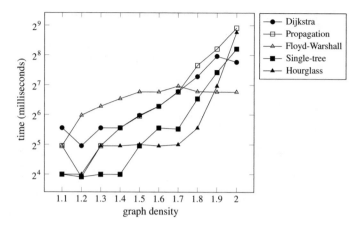

Fig. 7 512 vertices, unweighted. The plot is quite erratic due to the extremely short running times

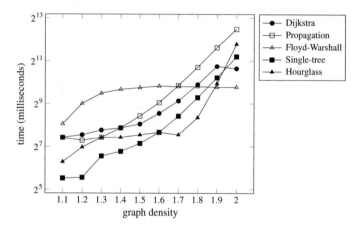

Fig. 8 1024 vertices, unweighted. A clearer picture begins to form, with Single-tree performing surprisingly well and being overtaken by Hourglass briefly as the graph grows dense

even outperforming it on the smaller graphs in some instances. It is worth mentioning that the n^2 case for unit graphs is somewhat pathological, as the instance is already solved since every vertex has a unit-length edge to every other vertex, which can be seen to cause a consistent dip in the running time in the case of Dijkstra.

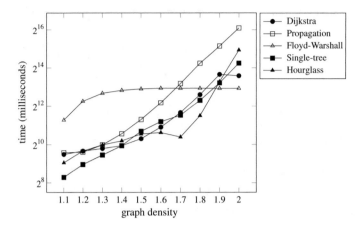

Fig. 9 2048 vertices, unweighted. Hourglass continues to perform best in the 1.7–1.8 range. Differences between the algorithms on the sparse instances begin to decrease, but Single-tree maintains good performance and is matched closely by Dijkstra

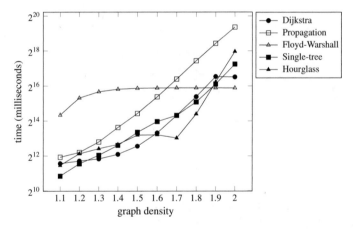

Fig. 10 4096 vertices, unweighted. Single-tree and Dijkstra remain closely matched, and Hourglass continues to dominate the 1.7–1.8 density range

4   All-Pairs Bottleneck Paths

The all-pairs bottleneck paths problem (APBP) is closely related to the all-pairs shortest path problem. The difference is only in the definition of the length of a path π, which is defined to be: $\ell(\pi) = \max_{(u,v)\in\pi} \ell(u, v)$. The length of edges is constrained to be non-negative. A solution to this problem is readily available by simply modifying the relaxation equation of shortest path algorithms to use maximum instead of addition. For example, modifying Dijkstra's algorithm in this way leads to a solution that runs in $O(mn + n^2 \lg n)$ using Fibonacci heaps [20]. A more efficient folklore modification of Dijkstra's algorithm is known to reduce this time down to

$O(mn)$. This folklore modification uses a bucket instead of a heap and sorts the edges by their lengths beforehand in $O(m \lg n)$ time. Then, shortest path lengths can be expressed as monotonically increasing integers from $1...m$ (referencing the edge lengths), and by using this bucket each SSSP computation takes $O(m)$ time in total. It should be pointed out, that in the case of undirected edges, we can solve APBP on the minimal spanning tree of G instead of on G itself, and still obtain the correct result. This can be a significant speed-up, since $m = n$ for any minimal spanning tree.

In this section, we will describe an algorithm that is more efficient, with an asymptotic running time of $O(m^*n + m \lg n)$. This algorithm will also allow us to state an interesting relationship between APBP and the dynamic transitive closure problem.

Given a graph $G = (V, E)$, the algorithm works by incrementally building the graph $G^* = (V, E^*)$ where $E^* \subset E$ are the edges (u, v) such that $d(u, v) = \ell(u, v)$. It accomplishes this by inserting edges into an initially disconnected set of vertices. The first step is to sort the set of edges E using their lengths. This can be done with any off-the-shelf sorting algorithm in $O(m \lg n)$ time.

Now we consider each edge in this sorted list from smallest to largest. Given an edge (u, v), check if v is reachable from u in G^*. If it is, ignore it and move to the next edge, and if it isn't, add (u, v) to G^*, and for every pair of vertices (w, q) that become reachable, set $d(w, q) = \ell(u, v)$. The algorithm finishes when we have considered every edge.

We summarize the algorithm in pseudocode as Algorithm 4.

Algorithm 4 APBP Algorithm

1: **procedure** APBP(V, E)
2:    Initialize D, a $n \times n$ matrix, as $D[i][j] := \infty$.
3:    $E^* := \emptyset$
4:    **for all** $(u, v) \in E$ from shortest to longest length **do**
5:        **if** $D[u][v] = \infty$ **then**
6:            $E^* := E^* \cup (u, v)$
7:            $D[u][v] := \ell(u, v)$
8:            **for all** (x, y) where $D[x][y] = \infty$ and $x \to y$ is reachable in $G^* = (V, E^*)$ **do**
9:                $D[x][y] := \ell(u, v)$
10:       **end for**
11:       **end if**
12:    **end for**
13: **end procedure**

Lemma 3 *For a graph $G = (V, E)$ the algorithm correctly computes $d(u, v)$:* $\forall u, v \in V$.

Proof By induction on the stage of the algorithm. Let $e_1, e_2, ..., e_n$ be the edges in sorted order, i.e. $\ell(e_1) \le \ell(e_2) \le \cdots \le \ell(e_n)$. Assume the algorithm is at a stage k, i.e. having examined the first $k - 1$ edges. For the case of $k = 1$, the shortest edge in the graph clearly forms the shortest path between the two vertices it connects.

For some case $n \geq k > 1$, let $e_k = (u, v)$ and consider first the case that u and v are already reachable in the current version of the graph G^*. That would imply that $d(u, v) \leq \ell(e_{k-1})$, due to the definition of the length of bottleneck paths, which means a shorter (or equal) path as e_k already exists, thus the edge can be safely omitted as it is redundant.

In the case u cannot yet reach v, then this is the shortest edge to connect the two vertices, and thus clearly $d(u, v) = \ell(e_k)$. For any additional vertex pairs (w, q) that become reachable after e_k is added into the graph, they clearly contain e_k on the path that connects them. Since all the other edges in the graph are shorter, by the definition of the length of bottleneck paths it holds that $d(w, q) = \ell(e_k)$, which completes the proof.

The running time of the algorithm depends heavily on how we check which previously unreachable vertex pairs have become reachable. The following simple approach works when adding some edge (u, v):

1. Gather all vertices that can reach u. This takes $O(n)$ time.
2. For each vertex that can reach u, start a breadth-first exploration of G^* from u, visiting only vertices that were previously not reachable.

Over the entire course of the algorithm, m^* edges are added to G^*, so the time for the first step is $O(m^*n)$. The second step is not more expensive than the cost of each vertex performing a full breadth-first exploration of G^* when it is fully built, thus at most $O(m^* + n)$ per vertex, amounting to $O(m^*n)$ in total. Overall, the cost is $O(m^*n)$.

Combining both times for the edge sorting and reachability checking, we arrive at the bound of $O(m^*n + m \lg n)$. It is worth pointing out that in the case of undirected graphs, G^* corresponds to the minimum spanning tree of G. This is interesting, because it means $m^* = O(n)$, so the running time of the algorithm becomes simply $O(n^2 + m \lg n)$ for undirected graphs. This remains true even if the *representation* is directed, i.e. each edge is simply repeated in both directions with the same length. In some limited sense, the algorithm is adaptive to the input graph.

4.1 Reduction to Decremental Transitive Closure

If instead of adding edges into the graph we consider the opposite scenario, that of removing edges (from largest to smallest) and checking when vertices are no longer reachable, we can reduce the problem to that of decremental transitive closure. In the latter problem, we are given a graph and a series of edge deletions, and the task is to maintain the ability to efficiently answer reachability queries. Relatively recent advancements in decremental transitive closure have led to an algorithm that has a total running time of $O(mn)$ under m edge deletions [21]. This immediately leads to an $O(mn)$ algorithm for all-pairs bottleneck paths. However, since transitive closure can be computed in $O(\frac{mn}{\lg n})$ time [22], a decremental algorithm that matches that running time could also lead to an $o(mn)$ *combinatorial* algorithm for APBP. While subcubic algebraic algorithms for APBP based on matrix multiplication exist [23], no $o(mn)$ combinatorial algorithm is known.

5 Discussion

In this chapter we have looked at practical algorithms for solving the all-pairs shortest path problem. It is typical of the more practically-minded APSP algorithms to rely on average-case properties of graphs, and most of them are modifications of Dijkstra's algorithm. However, the Floyd-Warshall algorithm is known to perform well in practice when the graphs are dense. To this end, we have suggested the Single-tree and Hourglass algorithms: modifications of the Floyd-Warshall algorithm that combine it with a tree structure, that allows it to avoid checking unnecessary path combinations. However, these two algorithms have no known average-case bounds, which would be an interesting topic for further research.

To compare practical performance, we have devised empirical tests using actual implementations. Since, as mentioned, the algorithms studied typically rely on average-case properties of graphs, we looked at both uniform random graphs and unweighted random graphs of varying density. The latter present a hard case for many of the algorithms and can highlight their worst-case performance, whereas the former are much more agreeable to the algorithms' assumptions. For the choice of algorithms we have included those known from past work, as well as the novel Hourglass and Single-tree algorithms. As it turns out, the new algorithms have proven to be quite efficient in the empirical tests that we have performed. The simpler Single-tree algorithm has ranked especially well alongside the Propagation algorithm, while at the same time it was more resilient when it came to worst-case inputs.

In addition, we have also briefly considered the case of all-pairs bottleneck paths, where we proposed a simple algorithm, the asymptotic running time of which can be parametrized with m^*. Additionally, we have shown ties to the decremental transitive closure problem, which might lead to faster algorithms for all-pairs bottleneck paths if faster algorithms for decremental transitive closure are found.

References

1. T.M. Chan, More algorithms for all-pairs shortest paths in weighted graphs. SIAM J. Comput. **39**(5), 2075–2089 (2010)
2. S. Pettie, A new approach to all-pairs shortest paths on real-weighted graphs. Theor. Comput. Sci. **312**(1), 47–74 (2004)
3. S. Pettie, V. Ramachandran, A shortest path algorithm for real-weighted undirected graphs. SIAM J. Comput. **34**(6), 1398–1431 (2005)
4. M. Thorup, Undirected single-source shortest paths with positive integer weights in linear time. J. ACM **46**(3), 362–394 (1999)

5. C. Demetrescu, G.F. Italiano, Experimental analysis of dynamic all pairs shortest path algorithms. ACM Trans. Algorithms **2**(4), 578–601 (2006)
6. D.R. Karger, Random sampling in cut, flow, and network design problems. Math. Oper. Res. **24**(2), 383–413 (1999)
7. Y. Peres, D. Sotnikov, B. Sudakov, U. Zwick, All-pairs shortest paths in $O(n^2)$ time with high probability, in *Proceedings of the 2010 IEEE 51st Annual Symposium on Foundations of Computer Science, FOCS '10* (IEEE Computer Societ, Washington, DC, USA, 2010), pp. 663–672
8. D.R. Karger, D. Koller, S.J. Phillips, Finding the hidden path: time bounds for all-pairs shortest paths. SIAM J. Comput. **22**(6), 1199–1217 (1993)
9. A. Brodnik, M. Grgurovic, Solving all-pairs shortest path by single-source computations: theory and practice. Discret. Appl. Math. **231**(Supplement C), 119–130 (2017) (Algorithmic Graph Theory on the Adriatic Coast)
10. R.W. Floyd, Algorithm 97: shortest path. Commun. ACM **5**(6), 345 (1962)
11. S. Warshall, A theorem on boolean matrices. J. ACM **9**(1), 11–12 (1962)
12. G. Venkataraman, S. Sahni, S. Mukhopadhyaya. A blocked all-pairs shortest-paths algorithm. J. Exp. Algorithmics, (8 Dec 2003)
13. S.C. Han, F. Franchetti, M. Püschel, Program generation for the all-pairs shortest path problem, in *Proceedings of the 15th International Conference on Parallel Architectures and Compilation Techniques, PACT '06* (ACM, New York, NY, USA, 2006), pp. 222–232
14. P. Harish, P.J. Narayanan, Accelerating large graph algorithms on the GPU using CUDA, in *Proceedings of the 14th International Conference on High Performance Computing, HiPC '07* (Springer, Berlin, Heidelberg, 2007), pp. 197–208
15. G.J. Katz, J.T. Kider, All-pairs shortest-paths for large graphs on the GPU, in *Proceedings of the 23rd ACM SIGGRAPH/EUROGRAPHICS symposium on Graphics hardware, GH '08* (Eurographics Association, Aire-la-Ville, Switzerland, Switzerland, 2008), pp. 47–55
16. J.J. McAuley, T.S. Caetano, An expected-case sub-cubic solution to the all-pairs shortest path problem in R (2009), arXiv:0912.0975
17. R. Davis, A. Prieditis, The expected length of a shortest path. Inf. Process. Lett. **46**(3), 135–141 (1993)
18. S. Pettie. Towards a final analysis of pairing heaps, in *46th Annual IEEE Symposium on Foundations of Computer Science, FOCS 2005* (IEEE Computer Society, Washington, DC, USA, Oct 2005), pp.174–183
19. E.W. Dijkstra, A note on two problems in connexion with graphs. Numer. Math. **1**, 269–271 (1959)
20. M.L. Fredman, R.E. Tarjan, Fibonacci heaps and their uses in improved network optimization algorithms. J. ACM **34**(3), 596–615 (1987)
21. J. Lacki, Improved deterministic algorithms for decremental transitive closure and strongly connected components, in *Proceedings of the Twenty-second Annual ACM-SIAM Symposium on Discrete Algorithms, SODA '11* (Society for Industrial and Applied Mathematics, Philadelphia, PA, USA, 2011), pp. 1438–1445
22. T.M. Chan, All-pairs shortest paths with real weights in $O(n^3 / \lg n)$ time, in *Proceedings of the 9th International Conference on Algorithms and Data Structures, WADS '05* (Springer, Berlin, Heidelberg, 2005), pp. 318–324
23. V. Vassilevska, R. Williams, R. Yuster, All-pairs bottleneck paths for general graphs in truly sub-cubic time, in *Proceedings of the Thirty-ninth Annual ACM Symposium on Theory of Computing, STOC '07* (ACM, New York, NY, USA, 2007), pp. 585–589

Computing Shortest Paths with Cellular Automata

Selim G. Akl

Abstract We describe cellular-automaton-based algorithms for solving two shortest path problems on arbitrary connected, directed, and weighted graphs with n vertices. The first problem is that of finding the shortest path from a given vertex to another given vertex of the graph. A two-dimensional cellular automaton, shaped as a triangle, with $O(n^2)$ cells, is used. The algorithm runs in $O(n)$ time. The second problem requires that all shortest paths between pairs of vertices be obtained. An $n \times n$ cellular automaton solves the problem in $O(n \log n)$ time.

1 Introduction

A cellular automaton, a biologically-inspired model of computation, is an arrangement of simple processors, or *cells* usually in a one or two dimensional geometric pattern [1, 7, 12]. Each cell receives data from the outside world and/or from its immediate neighbors and delivers output to the outside world and/or to its immediate neighbors. The cells operate in a synchronous fashion. In [2] it is shown how a cellular automaton can compute shortest paths in a rectangular lattice with weighted edges. The same computation is discussed in [9–11]. In this chapter we describe cellular-automaton-based algorithms for solving two problems pertaining to shortest paths in arbitrary graphs. Given a connected, directed, and weighted graph, as shown in Fig. 1:

Single-pair shortest path: It is required to compute the shortest path from one given vertex to another given vertex.

All-pairs shortest paths: For all ordered pairs of vertices, it is required to find the shortest path from the first vertex in the pair to the second.

S. G. Akl (✉)

School of Computing and Department of Mathematics and Statistics,
Queen's University, Kingston, ON K7L 3N6, Canada
e-mail: akl@cs.queensu.ca

© Springer International Publishing AG, part of Springer Nature 2018
A. Adamatzky (ed.), *Shortest Path Solvers. From Software to Wetware*,
Emergence, Complexity and Computation 32,
https://doi.org/10.1007/978-3-319-77510-4_7

Fig. 1 A connected,
directed, and weighted graph

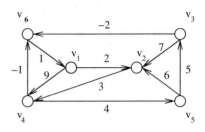

Finding shortest paths is a classical computation in computer science [5, 6, 8]. The algorithms in what follows are adapted for the cellular automaton model from ideas originally presented in [3, 4] in a different context.

2   Shortest Path Between Two Vertices

Dynamic programming is a powerful algorithmic method used in the solution of optimization problems in which a discrete function is to be minimized or maximized. The approach is to compute optimal solutions to subproblems of the main problem and then combine them to obtain a global optimal solution. This method has been applied successfully to a wide variety of problems, including scheduling problems, computing shortest paths in directed and weighted graphs, constructing binary search trees, and finding the longest common subsequence of two sequences. Our algorithms in this chapter use the dynamic programming approach.

Throughout the chapter, *time* is divided into discrete *time units*. A time unit is the time required by a cellular automaton cell to perform an arithmetic or logical operation. It is also the time required by a cellular automaton cell to receive a datum from the outside world, or to transfer a datum to a neighboring cell or to the outside world. The running time of an algorithm is measured and expressed in time units.

2.1   Preliminaries

Let $f(i, j)$ be a real-valued function of two integer variables i and j, $1 \leq i \leq j \leq n$. Initially,

$$f(i, j) = w(i, j) \text{ for } 1 \leq i < j \leq n, \text{ and } f(i, i) = 0 \text{ for } 1 \leq i \leq n.$$

Here, the values $w(i, j)$ are chosen as appropriate for the specific application at hand. The final value of $f(i, j)$ is to be obtained according to the rule

$$f(i, j) = \min_{i \leq k \leq j} (f(i, k) + f(k, j)) \text{ for } 1 \leq i < j \leq n.$$

This equation assumes that when $f(i, j)$ is to be computed for two given integers i and j, all values required for its computation, namely, $f(i, k)$ and $f(k, j)$ for $i \leq k \leq j$, have been previously obtained by the same rule and stored, and are hence available. Note also that for $k = i$ or $k = j$, the quantity $(f(i, k) + f(k, j))$ equals the given initial value $w(i, j)$.

Once $f(i, j)$ is computed, it, too, is stored and later used to compute subsequent values of f, for other values of i and j.

2.2 Shortest Path Problem Formulation

Suppose that G is a connected, directed, and weighted graph with n vertices $v_1, v_2, \ldots, v_n$. Further, let $w(i, j) \geq 0$ be the weight (also referred to as the *length*) of the directed edge (v_i, v_j), which connects vertex v_i to vertex v_j. If v_i is not directly connected by an edge to v_j, then $w(i, j) = \infty$. The weight (or length) of a path from v_i to v_j is the sum of the weights of the edges forming it. Under these conditions, $f(i, j)$, as defined above in Sect. 2.1, represents the length of a shortest path (i.e., a path with minimum weight) from v_i to v_j, which is allowed to go through intermediate vertices v_k, provided that $i < k < j$. Note that if the shortest path does indeed go through other vertices $v_h, v_l, \ldots, v_m$, then

$$w(i, h) + w(h, l) + \cdots + w(m, j) < w(i, j).$$

On a sequential model of computation, such as the Random Access Machine (RAM), for example, this problem can be solved in $O(n^2)$ time.

2.3 An Algorithm for a Triangular Cellular Automaton

We now show how the equation

$$f(i, j) = \min_{i \leq k \leq j} (f(i, k) + f(k, j))$$

can be computed on a two-dimensional cellular automaton, for $1 \leq i < j \leq n$. Specifically, we use a triangular arrangement.

Triangular cellular automaton. The automaton is shown in Fig. 2 for $n = 6$. In this arrangement, each cell $P(i, j)$ is connected to cells $P(i - 1, j)$ and $P(i, j + 1)$ by two links, namely, a simple (or *fast*) link and a buffer (or *slow*) link. The buffer link contains a secondary cell whose only purpose is to slow down the communication by one time unit. Thus:

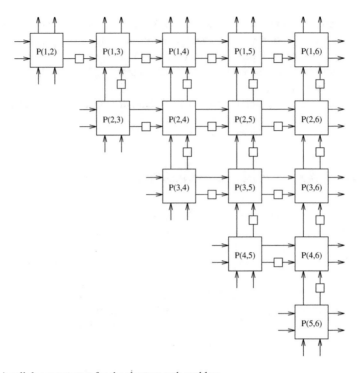

Fig. 2 A cellular automaton for the shortest path problem

1. If $P(i, j)$ sends a datum to $P(i - 1, j)$ and/or to $P(i, j + 1)$ on the fast link, the datum arrives at its destination during *the same* time unit.
2. If, on the other hand, the datum is sent on the slow link, it arrives during *the following* time unit.

Cell $P(i, j)$ will be in charge of computing $f(i, j)$ and will finish doing so at time unit $2(j - i)$. Once $f(i, j)$ has been computed, it is sent to $P(i - 1, j)$ and $P(i, j + 1)$ on the fast links. After leaving $P(i, j)$, a datum travels (simultaneously up and to the right) for $j - i$ time units on the fast links and then continues its motion on the slow links. Typically, a cell receives up to four inputs and produces up to four outputs, as shown in Fig. 3, where the small squares represent secondary cells. Each cell holds a variable F in an internal register. Initially, $F = w(i, j)$ for cell $P(i, j)$.

Scheduling data movement. Suppose that the algorithm begins during time unit 1 and that subsequent time units are numbered 2, 3, and so on. Let u denote the number of the current time unit at any point during the execution of the algorithm. When $u = 2(j - i)$, the final value of $f(i, j)$ has been computed by $P(i, j)$, and it is sent up and to the right on the fast links. Now, $f(i, j)$ stays on the fast links for $j - i$ time units. Therefore, it reaches $P(i, j + j - i)$ traveling right and $P(i - j + i, j)$ traveling up when

$$u = 2(j - i) + (j - i) = 3(j - i),$$

Fig. 3 Structure of a cell

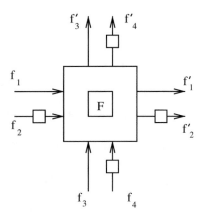

at which point it must switch to the slow links. Thus, a datum switches from a fast to a slow link as it goes through a cell $P(r, s)$ if $u = 3(s - r)/2$.

The Algorithm. Using the notation in Fig. 3, we see that each cell $P(i, j)$ performs the following computations whenever it receives input:

> **Step 1**: $F \leftarrow \min(F, f_1 + f_3, f_2 + f_4)$
> **Step 2**: if $u = 3(j - i)/2$
>   **then** (i) $f_2' \leftarrow f_1$
>   (ii) $f_4' \leftarrow f_3$
>   **else** (i) $f_2' \leftarrow f_2$
>   (ii) $f_4' \leftarrow f_4$
>   **end if**
> **Step 3**: if $u = 2(j - i)$
>   **then** (i) $f_1' \leftarrow F$
>   (ii) $f_3' \leftarrow F$
>   **else** (i) $f_1' \leftarrow f_1$
>   (ii) $f_3' \leftarrow f_3$
>   **end if**. ■

It is important to note that:

1. A cell $P(i, j)$ uses f_l, $1 \le l \le 4$, in the computation of F, provided that f_l was received during the current time unit. Similarly, $P(i, j)$ produces f_l as output, provided that it was received in the previous time unit.
2. Cells $P(i, i + 1)$ do not receive any input and begin operating when $u = 1$. They perform no computations and, when $u = 2$, produce F as $f(i, i + 1)$.
3. If a cell receives inputs f_1 and f_2 during time unit u, then during time unit $u + 1$ its outputs f_1' and f_2' are such that $f_1' = f_1$ and $f_2' = f_2$ (i.e., the input arriving on the fast input link is *not* switched to the slow output link). This is because if f_1 and f_2 are received by $P(i, j)$ during time unit u, then it must be the case that

$$u + 1 \neq 3(j - i)/2 \text{ and } u + 1 \neq 2(j - i).$$

4. If a cell's output f_2' is such that $f_2' = f_1$ during time unit $u + 1$, then the cell received no input f_2 during time unit u. Indeed, if $f_2' = f_1$ for $P(i, j)$ during time unit $u + 1$, then it must be the case that $u + 1 = 3(j - i)/2$. This property, together with the preceding one, establishes that a cell never has to place two values on its slow output link simultaneously.

5. If $F = f(i, j)$ is produced as output during time unit $u + 1$, then cell $P(i, j)$ received no input during time unit u. This follows from the fact that

$$\text{if } F = f(i, j) \text{ for } P(i, j) \text{ during time unit } u + 1,$$

then it must be the case that

$$u + 1 = 2(j - i).$$

This property ensures that a cell never has to place two values on its fast output link simultaneously.

6. Cell $P(i, j)$ finishes computing $f(i, j)$ when $u = 2(j - i)$. We prove this as follows. The number of links separating $P(i, i + 1)$ and $P(i, j)$ is

$$j - (i + 1) = j - i - 1.$$

Now $f(i, i + 1)$ is produced in time unit 2. It stays on the fast link one time unit, then switches to the slow links in time unit 3. It now traverses $j - i - 2$ slow links in $2(j - i) - 4$ time units and reaches $P(i, j)$ during time unit

$$2(j - i) - 4 + 3 = 2(j - i) - 1.$$

One time unit later $f(i, j)$ is produced.

Example 1 Let $n = 6$. We illustrate how $f(1, 6)$ is computed on the cellular automaton of Fig. 2, where

$$f(1, 6) = \min \ (w(1, 6), \ f(1, 2) + f(2, 6),$$
$$f(1, 3) + f(3, 6),$$
$$f(1, 4) + f(4, 6),$$
$$f(1, 5) + f(5, 6)).$$

The computation is illustrated in Fig. 4, in which only the top row of the triangular automaton is shown and u denotes the number of time units elapsed. The first values to emerge from their respective cells are $f(1, 2)$ and $f(5, 6)$, when $u = 2$. Subsequent computations are as follows:

$u = 3$: $f(1, 2)$ and $f(5, 6)$ are placed by $P(1, 3)$ and $P(4, 6)$ on the slow links.
$u = 4$: $f(1, 3)$ and $f(4, 6)$ are placed by $P(1, 3)$ and $P(4, 6)$ on the fast links.

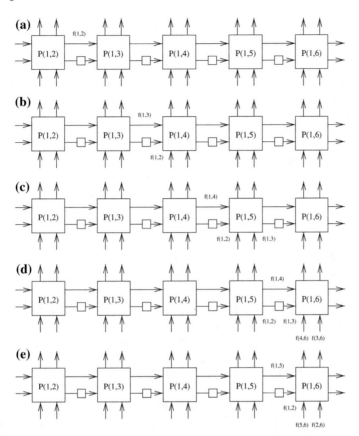

Fig. 4 Computing $f(1, 6)$: **a** $u = 2$; **b** $u = 4$; **c** $u = 6$; **d** $u = 7$; **e** $u = 8$

$u = 5$: $f(1, 3)$ and $f(4, 6)$ reach $P(1, 5)$ and $P(2, 6)$, respectively.

$u = 6$: $f(1, 4)$ and $f(3, 6)$ are placed by $P(1, 4)$ and $P(3, 6)$ on the fast links;
$\quad\quad\;\;$ $f(1, 3)$ and $f(4, 6)$ are placed by $P(1, 5)$ and $P(2, 6)$ on the slow links.

$u = 7$: $f(1, 4)$ and $f(1, 3)$ reach $P(1, 6)$ from the left;
$\quad\quad\;\;$ $f(4, 6)$ and $f(3, 6)$ reach $P(1, 6)$ from the bottom.

$u = 8$: $P(1, 6)$ computes $F \leftarrow \min\,(w(1, 6), f(1, 3) + f(3, 6), f(1, 4) + f(4, 6))$;
$\quad\quad\;\;$ $f(1, 5)$ and $f(1, 2)$ reach $P(1, 6)$ from the left;
$\quad\quad\;\;$ $f(5, 6)$ and $f(2, 6)$ reach $P(1, 6)$ from the bottom.

$u = 9$: $P(1, 6)$ computes $F \leftarrow \min\,(F, f(1, 5) + f(5, 6), f(1, 2) + f(2, 6))$.

When $u = 10$, $F = f(1, 6)$ is produced as output by $P(1, 6)$. ∎

2.4   Analysis

A problem of size n uses a triangular cellular automaton of $n(n-1)/2$ cells. Since the last output to be produced is $f(1, n)$, and this requires $2(n-1)$ time units, the algorithm has a running time of $t(n) = 2(n-1)$. It is important to note that this algorithm computes $f(i, j)$, the *length* of a shortest path from vertex v_i to vertex v_j. The path itself, that is, the actual sequence of intermediate vertices v_k, $i < k < j$, that form the path from v_i to v_j is obtained by keeping track of each index which achieves the minimum in the recurrence

$$f(i, j) = \min_{i \leq k \leq j} (f(i, k) + f(k, j)).$$

2.5   A Variant: No Need to Keep Track of Time

In the cellular automaton of Sect. 2.3, each cell needs to keep track of u, the number of time units elapsed, in order to "decide" whether it is time to produce the contents of F as output or to switch a value received on a fast link onto a slow link. We now show that by using $2(n-1)$ control signals, one per row and one per column, which travel at appropriate speeds from left to right and from bottom to top, respectively, each cell receiving these signals can determine what needs to be done (without having to store u).

The $(n-1)$ signals (one per row) traveling horizontally are the *switch* signals. They tell a cell when to switch an input from a fast to a slow link. Because the *switch* signal must reach $P(i, j)$ at time unit $3(j-i)/2$, it begins at time unit 3 at $P(i, i+2)$ and travels to the right on slow links connecting those cells for which $(j-i)$ is even, such that each link has three delays (and hence takes three time units to traverse).

The $(n-1)$ signals (one per column) traveling vertically are the *output* signals. They tell a cell when to produce the contents of F as output. Because the *output* signal must reach $P(i, j)$ at time unit $2(j-i)$, it begins at time unit 2 at $P(i, i+1)$ and travels upwards on slow links with one delay per link (similar to the slow links used for the data).

3   All-Pairs Shortest Paths

Suppose that we are given a directed and weighted graph G, with n vertices $v_1, v_2, \ldots, v_n$. The graph is defined by its weight matrix W, in which entry $w(i, j)$ represents the weight of edge (v_i, v_j), also referred to as its *length*, as was done in Sect. 2.2. However, unlike in Sect. 2.2, we assume more generally that W has positive, zero, or negative entries, as long as there is no cycle in G such that the sum of the weights of the edges on the cycle is negative.

The problem that we address here is known as the *all-pairs shortest paths problem* and is stated as follows: For every pair of vertices v_i and v_j in G, it is required to find the length of the shortest path from v_i to v_j along edges in G. Specifically, a matrix $\mathbf{D}$ is to be constructed such that d_{ij} is the length of the shortest path from v_i to v_j in G, for all i and j. Here, as before, the length of a path (or cycle) is the sum of the lengths of the edges forming it. It may be obvious now why we insisted that G have no cycle of negative length: If such a cycle were to exist within a path from v_i to v_j, then one could traverse this cycle indefinitely, producing paths of ever shorter lengths from v_i to v_j. Because G is directed, edges (v_i, v_j) and (v_j, v_i) are different, and a path from v_i to v_j is not the same as a path for v_j to v_i. Thus there are $n(n-1)$ distinct pairs of vertices in a directed n-vertex graph, and consequently $n(n-1)$ distinct shortest paths to be computed. On the RAM, this problem is solved in $O(n^3)$ time.

Two obvious ways are available to solve the all-pairs shortest paths problem, using the triangular cellular automaton of Sect. 2.2:

1. To employ a single triangular cellular automaton with $n(n-1)/2$ cells, repeatedly, $n(n-1)$ times, each time computing the shortest path for a different pair of vertices. This requires a running time of $n(n-1) \times 2(n-1)$. In this case, the total number of basic computations, such as additions, subtractions, and so on, expressed as the number of cells multiplied by the running time, is therefore $O(n^5)$.
2. Alternatively, to employ $n(n-1)$ distinct cellular automata, each equipped with $n(n-1)/2$ cells and operating independently from all the others to compute the shortest path for a different pair of vertices. Thus, by computing in parallel, the automata achieve a parallel running time of $2(n-1)$. The total number of basic computations, a measure of the work done collectively, is again $O(n^5)$.

In what follows we present a significantly more efficient solution to the all-pairs shortest path problem on a graph with n vertices, whose total number of computations is $O(n^3 \log n)$.

3.1 All Pairs Shortest Paths Cellular Automaton Algorithm

Let d_{ij}^k denote the length of the shortest path from v_i to v_j that goes through at most $k-1$ intermediate vertices. Thus, $d_{ij}^1 = w(i, j)$, that is, the length of the edge from v_i to v_j. In particular, if there is no edge from v_i to v_j, where $i \neq j$, then $d_{ij}^1 = w(i, j) = \infty$. Also, $d_{ii}^1 = w(i, i) = 0$. Given that G has no cycles of negative length, there is no advantage in visiting any vertex more than once in a shortest path from v_i to v_j. It follows that $d_{ij} = d_{ij}^{n-1}$, since there are only n vertices in G.

In order to compute d_{ij}^k for $k > 1$, we can use the recurrence

$$d_{ij}^k = \min_l (d_{il}^{k/2} + d_{lj}^{k/2}),$$

in which $k/2$ is rounded appropriately when needed. The validity of this relation is established as follows: Suppose that d_{ij}^k is the length of the *shortest* path from v_i to v_j and that two vertices v_r and v_s are on this shortest path (with v_r preceding v_s). It must be the case that the edges from v_r to v_s (along the shortest path from v_i to v_j) form a shortest path from v_r to v_s. (If a shorter path from v_r to v_s existed, it could be used to obtain a shorter path from v_i to v_j, which is absurd.) Therefore, to obtain d_{ij}^k, we can compute all combinations of *optimal subpaths* (whose concatenation is a path from v_i to v_j) and then choose the shortest one. The fastest way to do this is to combine pairs of subpaths with at most $k/2$ vertices each. This guarantees that a recursive computation of d_{ij}^k can be completed in $O(\log k)$ steps.

Let $\boldsymbol{D}^k$ be the matrix whose entries are d_{ij}^k, for $1 \leq i, j \leq n$. In accordance with the discussion in the previous two paragraphs, the matrix $\boldsymbol{D}$ can be computed from $D^1 = \boldsymbol{W}$ by evaluating $D^2, \boldsymbol{D}^4, \ldots, \boldsymbol{D}^m$, where m is the smallest power of 2 larger than or equal to $n - 1$ (i.e., $m = 2^{\lceil \log(n-1) \rceil}$), and then taking $\boldsymbol{D} = D^m$. In order to obtain $\boldsymbol{D}^k$ from $\boldsymbol{D}^{k/2}$, we use a special form of matrix multiplication in which the operations '+' and 'min' replace the standard operations of matrix multiplication—that is, '$\times$' and '+', respectively. Hence, if a matrix multiplication algorithm is available, it can be modified to generate $\boldsymbol{D}^m$ from $\boldsymbol{D}^1$. Exactly $\lceil \log(n-1) \rceil$ such matrix products are required.

The remainder of this section is devoted to showing how two matrices can be multiplied on a two-dimensional cellular automaton. An $n \times n$ cellular automaton for this purpose is shown in Fig. 5. In what follows various algorithms are described for multiplying two matrices on this automaton. Each algorithm, appropriately modified as described in the previous paragraph, can be used to compute the shortest path matrix $\boldsymbol{D}$. Each algorithm runs in $O(n)$ time. Since $O(\log n)$ such matrix products are required, the running time for computing all-pairs shortest paths for a graph with n vertices on an $n \times n$ cellular automaton is $t(n) = O(n \log n)$.

3.2   Matrix Multiplication on a Cellular Automaton

In the general case of matrix multiplication, we are given an $m \times n$ matrix A and an $n \times k$ matrix $\boldsymbol{B}$. It is required to compute an $m \times k$ matrix C equal to the product of A and B. The elements of $C = A \times B$ are given by

$$c_{ij} = \sum_{s=1}^{n} a_{is} \times b_{sj},$$

COLUMN

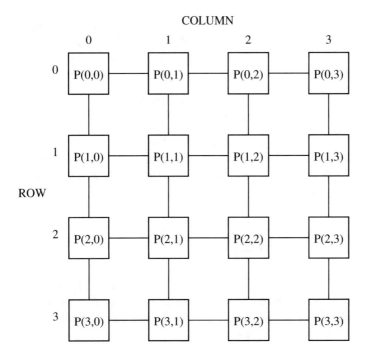

Fig. 5 Two-dimensional cellular automaton for all pairs shortest paths

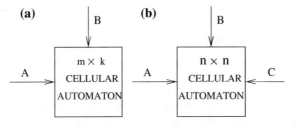

Fig. 6 Variants of data flow for matrix multiplication on a cellular automaton: **a** The elements of the product matrix, all initialized to 0, reside in the automaton at the beginning of the computation; **b** The elements of the product matrix, all initialized to 0, are fed into the automaton sequentially

for $1 \leq i \leq m$ and $1 \leq j \leq k$. We begin by describing an algorithm for performing this computation on a cellular automaton with m rows and k columns. Initially, cell $P(i, j)$ holds $c_{ij} = 0$. Matrices A and B are fed into the automaton as shown in Fig. 6a. We now show how to organize the input and subsequent computations so that, when the algorithm terminates, $P(i, j)$ contains the final value of c_{ij}.

Cellular automaton rows are numbered $1, 2, \ldots, m$ and columns $1, 2, \ldots, k$. Matrices A and B are fed into the cells in column 1 and row 1 as follows. Element a_{in} is the first element of row i of matrix A to enter $P(i, 1)$. Also, row i of A lags one time unit behind row $i - 1$, for $2 \leq i \leq m$. Similarly, element b_{nj} is the first element of column j of matrix B to enter $P(1, j)$. Also, column j of B lags one time unit behind column $j - 1$, for $2 \leq j \leq k$. This ensures that a_{is} meets b_{sj} in cell $P(i, j)$ at the right time, and the latter executes the following:

(a) $c_{ij} \leftarrow c_{ij} + (a_{is} \times b_{sj})$,
(b) sends a_{is} to $P(i, j + 1)$, provided $j < k$, and
(c) sends b_{sj} to $P(i + 1, j)$, provided $i < m$.

Elements a_{m1} and b_{1k} take $m + k + n - 2$ steps from the beginning of the computation to reach $P(m, k)$. Since $P(m, k)$ is the last cell to terminate, this many steps are required to compute the product. Assuming $m \leq n$ and $k \leq n$, the algorithm runs in $O(n)$ time.

3.3 A Variant: Three Input Streams

We now consider the case where matrices A, B, and C are fed to the cellular automaton as shown in Fig. 6b, with all the elements of C initialized to 0. Suppose that A and B are $n \times n$ matrices. We use a cellular automaton with n rows, numbered $1, 2, \ldots, n$ and $2n - 1$ columns, numbered $1, 2, \ldots, 2n - 1$.

(a) The elements of row i of A are fed to row i of the cellular automaton from the left, such that a_{i1} enters $P(i, 1)$ first, a_{i2} second, and so on. Two consecutive elements a_{ij} and $a_{i, j+1}$ are separated by two time units. Also, row i lags one time unit behind row $i - 1$, for $2 \leq i \leq n$.
(b) The elements of row i of C are fed to row i of the cellular automaton from the right, such that c_{i1} enters $P(i, 2n - 1)$ first, c_{i2} second, and so on. Two consecutive elements c_{ij} and $c_{i, j+1}$ are separated by two time units. Also, row i lags one time unit behind row $i - 1$, for $2 \leq i \leq n$.
(c) Let the diagonals of matrix B be numbered from 1 (bottom left) to $2n - 1$ (top right), such that the diagonal consisting of b_{n1} alone is diagonal 1, the diagonal consisting of $b_{n-1,1}$ and b_{n2} is diagonal 2, and so on. For $1 \leq j \leq 2n - 1$, the elements of diagonal j are fed to column j of the cellular automaton from the top, such that the top left element of diagonal j enters $P(1, j)$ first. Two consecutive elements of a diagonal are separated by two time units. Furthermore, b_{11} enters $P(1, n)$ when both a_{11} and c_{11} have reached that cell. In the following time unit, b_{21} and b_{12} enter $P(1, n - 1)$ and $P(1, n + 1)$, respectively. One time unit later, b_{31}, b_{22}, and b_{13} enter $P(1, n - 2)$, $P(1, n)$, and $P(1, n + 2)$, respectively. This continues until b_{nn} enters $P(1, n)$.

When a_{ik} meets b_{kj} and c_{ij} in a cell, the latter computes $c_{ij} \leftarrow c_{ij} + (a_{ik} \times b_{kj})$. The algorithm requires $O(n)$ time units.

3.4 The Case of Several Inputs

Assume that several pairs of matrices $(A_1, B_1), (A_2, B_2), \ldots, (A_q, B_q)$ are queued, waiting to be multiplied on the cellular automaton of Sect. 3.2 (or that of Sect. 3.3). We now show how the algorithm can be modified to allow this computation to be carried out in a *pipeline* fashion. This is accomplished as follows. The matrices $A_1, A_2, \ldots, A_q$ are fed to the $m \times k$ automaton in a pipeline fashion and move from left to right. Similarly, $B_1, B_2, \ldots, B_q$ are fed to the automaton in a pipeline fashion and move from the top down.

Once cell $P(i, j)$ has finished computing c_{ij} for one pair of matrices, it should produce it as output immediately before it becomes involved in computing the product of a new pair of matrices. This is because the product matrix may be needed for another computation once it is available and/or because $P(i, j)$ has storage room for only one c_{ij} at a time. If $P(i, j)$ is directly connected to an output device, then it can send c_{ij} to the outside world once it is computed.

Alternatively, suppose that only boundary cells in the cellular automaton are connected to output devices. In this case, the c_{ij} are produced as output in the following manner. Since each cell has a limited amount of local storage, it is important that c_{ij} be produced as output by $P(i, j)$ as soon as its computation is complete, and before $P(i, j)$ begins computing a new c_{ij} for the next pair of matrices. One way to produce the output is to send c_{ij} to $P(i, j - 1)$ as soon as its computation is complete, therefrom it travels leftwards to the output. Alternatively, c_{ij} can be sent to $P(i - 1, j)$ therefrom it travels upwards to the output.

3.5 A Second Variant: Resident Input

In this variant, an $n \times n$ cellular automaton is used the compute the product of two $n \times n$ matrices as follows. Initially, cell $P(i, j)$ stores elements a_{ij} and b_{ij} of the two matrices A and B, respectively. We describe an algorithm for computing $C = A \times B$ on this model, so that at the end of the computation $P(i, j)$ also holds c_{ij}. For ease of exposition, we present three versions of the algorithm.

1. The first, simplest but least efficient, version requires that one of the two matrices be transposed, and runs on a cellular automaton with so-called *wraparound* connections.
2. The second, more efficient but also slightly more complex, removes the need for initially transposing one of the two matrices, and instead uses the wraparound connections to simulate the transpose during the process of multiplying the two matrices.
3. Finally, we show how the wraparound connections can be discarded.

Therefore, we begin by describing how a matrix transpose is accomplished on the cellular automaton.

3.5.1   Matrix Transpose on a Cellular Automaton

Given an $n \times n$ matrix

$$B = \begin{pmatrix} b_{11} & b_{12} & \dots & b_{1n} \\ b_{21} & b_{22} & \dots & b_{2n} \\ & & \vdots & \\ b_{n1} & b_{n2} & \dots & b_{nn} \end{pmatrix},$$

it is required to compute the *transpose* of B; that is,

$$B^T = \begin{pmatrix} b_{11} & b_{21} & \dots & b_{n1} \\ b_{12} & b_{22} & \dots & b_{n2} \\ & & \vdots & \\ b_{1n} & b_{2n} & \dots & b_{nn} \end{pmatrix}.$$

In other words, every row in matrix B is a column in matrix B^T. The transpose of an $n \times n$ matrix can be computed on an $n \times n$ cellular automaton by assigning b_{ij} to cell $P(i, j)$ and then routing b_{ij} to cell $P(j, i)$, for all $1 \leq i, j \leq n$. We now show how this is done in each of the following two cases:

Case (a) The element b_{ij} carries along the indices of its destination cell, namely, j and i, as it travels from $P(i, j)$ to $P(j, i)$.

Case (b) The element b_{ij} carries no information whatsoever concerning i and j as it travels from $P(i, j)$ to $P(j, i)$.

Since the diagonal elements are not affected during the transposition, that is, elements b_{ii} of B^T, the data in the diagonal cells will stay stationary. Those below the diagonal are sent to occupy symmetrical positions above the diagonal. Simultaneously, the elements above the diagonal are sent to occupy symmetrical positions below the diagonal. Each cell $P(i, j)$ has three registers: One to store b_{ij} (and eventually b_{ji}), the second to store data received from the right or top neighbors, and the third to store data received from the left or bottom neighbors. When a datum traveling to the left (right) reaches a diagonal cell it switches its direction and moves down (up).

Routing in case (a) Suppose that element b_{ij} carries along (j, i) as it travels from $P(i, j)$ to $P(j, i)$. When b_{ij} reaches $P(j, i)$ the latter recognizes its indices in (j, i) and retains b_{ij} (other elements are forwarded by $P(j, i)$ in the direction in which they were moving).

Routing in case (b) Suppose that element b_{ij} carries no information about cell $P(j, i)$. The destination of b_{ij} is determined as follows: $P(j, i)$ retains the first element it receives from below (above) if $P(j, i)$ is above (below) the diagonal, and forwards all subsequent elements in the direction in which they were moving.

In either case, the algorithm runs in $O(n)$ time.

3.5.2 Using Wraparound Connections

Once again, for ease of exposition, we begin by temporarily assuming that the $n \times n$ cellular automaton in charge of multiplying matrices A and B, is such that the cells in each row and the cells in each column are connected to form a *ring*. The additional links are called *wraparound connections*. In this section we describe how a cellular automaton with these additional connections computes the product $C = A \times B$ of the two matrices as specified earlier, namely:

1. Cell $P(i, j)$ already stores elements a_{ij} and b_{ij} of A and B, respectively, at the beginning of the computation.
2. When the computation terminates, cell $P(i, j)$ also holds element c_{ij} of the product.

Each cell has four registers a, b, c, and d. Initially, the a and b registers of $P(i, j)$ store a_{ij} and b_{ij}, respectively. When the algorithm terminates, the c register of $P(i, j)$ holds c_{ij}. The algorithm is as follows (for convenience, numbering starts at 0).

Step 1: Compute the transpose of matrix B using the algorithm
of Sect. 3.5.1
Step 2: **for** $l = 0$ **to** $n - 1$ **do**
    for all cells **do in parallel**
      $d \leftarrow a \times b$
    end for
    for $i = 0$ **to** $n - 1$ **do in parallel**
      Compute the sum of all d registers in row i
      and place this sum in the c register of cell $P(i, (l + i) \bmod n)$
    end for
    for $j = 0$ **to** $n - 1$ **do in parallel**
      for $i = 0$ **to** $n - 1$ **do in parallel**
        Transfer the contents of the b register of every
        cell $P(i, j)$ to the b register of cell $P((i - 1) \bmod n, j)$
      end for
    end for
end for. ∎

The algorithm is simple; however, it runs in $O(n^2)$ time. An $O(n)$ time algorithm is as follows (for convenience, numbering starts at 1.)

Matrix multiplication on cellular automaton with wraparound connections

Step 1: (1.1) **for** $i = 1$ **to** n **do in parallel**
        Shift the contents of all a registers
        in row i cyclically to the right
        $n - i$ times
        end for
     (1.2) **for** $j = 1$ **to** n **do in parallel**
        Shift the contents of all b registers
        in column j cyclically down
        $n - j$ times
        end for
Step 2: **for** $i = 1$ **to** n **do in parallel**
     for $j = 1$ **to** n **do in parallel**
     $c_{ij} \leftarrow 0$
     for $k = 1$ **to** n **do**
        (2.1) $c_{ij} \leftarrow c_{ij} + (a \times b)$
        (2.2) Cyclically shift the contents of all
           a registers to the left once
        (2.3) Cyclically shift the contents of all b registers up once
     end for
     end for
     end for
Step 3: (3.1) **for** $i = 1$ **to** n **do in parallel**
        Shift the contents of all a registers in row i
        cyclically to the left $n - i$ times
        end for
     (3.2) **for** $j = 1$ **to** n **do in parallel**
        Shift the contents of all b registers in column j
        cyclically up $n - j$ times
        end for. ∎

3.5.3 Doing Away with the Wraparound Connections

Finally we demonstrate that the wraparound connections, while making the algorithm easy to describe, are not necessary, and can be discarded. The wraparound connections were used to circulate the elements of the two matrices so that every pair of numbers that must be multiplied meet in a cell. In the absence of the wraparound connections, their effect can be obtained by storing a second copy of each element of the two matrices. These "secondary" elements travel in a direction opposite to that of the "main" elements.

3.6 Analysis

Each of the matrix multiplication algorithms in Sects. 3.2, 3.3 and 3.5 computes the product of two $n \times n$ matrices in $O(n)$ time. Since the shortest paths algorithm of Sect. 3.1 consists of $O(\log n)$ matrix multiplications, its overall running time, using an $n \times n$ cellular automaton, is $O(n \log n)$. The total number of basic computations is $O(n^3 \log n)$. Once again we point out that this algorithm computes d_{ij}, that is, the *length* of a shortest path from v_i to v_j, for all i and j. In order to produce the path itself, that is, the sequence of vertices that appear on a shortest path from v_i to v_j, it is necessary to record each index that yields a minimum in the recurrence

$$d_{ij}^k = \min_l (d_{il}^{k/2} + d_{lj}^{k/2}).$$

4 Conclusion

It is rather counter intuitive at first glance that a highly organized and structured model of computation such as the cellular automaton be used to solve problems defined on a highly unstructured collection of data in an arbitrary graph. Yet, as shown in this chapter, the cellular automaton can be applied quite effectively to solve two shortest-path problems on graphs. This is a perfect example of the power of abstraction in the science of computation. An appropriate encoding and a clever algorithm is usually all it takes to do the job.

References

1. A.I. Adamatzky, *Identification of Cellular Automata* (Taylor and Francis, London, 1994)
2. A.I. Adamatzky, Computation of shortest paths in cellular automata. Math. Comput. Modelling. **23**, 105–113 (1996)
3. S.G. Akl, *The Design and Analysis of Parallel Algorithms* (Prentice Hall, Englewood Cliffs, New Jersey, 1989)
4. S.G. Akl, *Parallel Computation: Models and Methods* (Prentice Hall, Upper Saddle River, New Jersey, 1997)
5. G. Brassard, P. Bratley, *Algorithms: Theory and Practice* (Prentice Hall, Englewood Cliffs, New Jersey, 1988)
6. T.H. Cormen, C.E. Leiserson, R.L. Rivest, C. Stein, *Introduction to Algorithms* (MIT Press, Cambridge, Massachusetts, 2009)

 7. K. Preston Jr., M.J.B. Duff, *Modern Cellular Automata: Theory and Applications* (Springer, New York, 1985)
 8. R. Sedgewick, *Algorithms* (Addidon-Wesley, Reading, Massachusetts, 1983)
 9. Q. Sun, Z.J. Dai, A new shortest path algorithm using cellular automata model. Comput. Tech. Devel. **19**, 42–44 (2009)
10. M. Wang, Y. Qian, X. Guang, Improved calculation method of shortest path with cellular automata model. Kybernetes. **41**, 508–517 (2012)
11. X.J. Wu, H.F. Xue, Shortest path algorithm based on cellular automata extend model. Comput. Appl. **24**, 92–93 (2004)
12. S. Wolfram, *A New Kind of Science* (Wolfram Media, Champaign, Illinois, 2002)

Cellular Automata Applications in Shortest Path Problem

Michail-Antisthenis I. Tsompanas, Nikolaos I. Dourvas, Konstantinos Ioannidis, Georgios Ch. Sirakoulis, Rolf Hoffmann and Andrew Adamatzky

Abstract *Cellular Automata (CAs)* are computational models that can capture the essential features of systems in which global behavior emerges from the collective effect of simple components, which interact locally. During the last decades, CAs have been extensively used for mimicking several natural processes and systems to find fine solutions in many complex hard to solve computer science and engineering problems. Among them, the *shortest path problem* is one of the most pronounced and highly studied problems that scientists have been trying to tackle by using a plethora of methodologies and even unconventional approaches. The proposed solutions are mainly justified by their ability to provide a correct solution in a better time complexity than the renowned Dijkstra's algorithm. Although there is a wide variety regarding the algorithmic complexity of the algorithms suggested, spanning from simplistic graph traversal algorithms to complex nature inspired and bio-mimicking algorithms, in this chapter we focus on the successful application of CAs to shortest path problem as found in various diverse disciplines like computer

M.-A. I. Tsompanas · A. Adamatzky
University of the West of England, Bristol BS16 1QY, UK
e-mail: Antisthenis.Tsompanas@uwe.ac.uk

A. Adamatzky
e-mail: Andrew.Adamatzky@uwe.ac.uk

N. I. Dourvas · G. C. Sirakoulis (✉)
Laboratory of Electronics, Department of Electrical and Computer Engineering,
Democritus University of Thrace, Xanthi 67100, Greece
e-mail: gsirak@ee.duth.gr

N. I. Dourvas
e-mail: ndourvas@ee.duth.gr

K. Ioannidis
Information Technologies Institute Centre for Research and Technology Hellas,
6th km Charilaou-Thermi Road, 57001 Thermi, Thessaloniki, Greece
e-mail: kioannid@iti.gr

R. Hoffmann
Department of Computer Science, Technical University of Darmstadt,
Hochschulstrae 10, 64289 Darmstadt, Germany
e-mail: hoffmann@rbg.informatik.tu-darmstadt.de

© Springer International Publishing AG, part of Springer Nature 2018
A. Adamatzky (ed.), *Shortest Path Solvers. From Software to Wetware*,
Emergence, Complexity and Computation 32,
https://doi.org/10.1007/978-3-319-77510-4_8

199

science, swarm robotics, computer networks, decision science and biomimicking of biological organisms' behaviour. In particular, an introduction on the first CA-based algorithm tackling the shortest path problem is provided in detail. After the short presentation of shortest path algorithms arriving from the relaxization of the CAs principles, the application of the CA-based shortest path definition on the coordinated motion of swarm robotics is also introduced. Moreover, the CA based application of shortest path finding in computer networks is presented in brief. Finally, a CA that models exactly the behavior of a biological organism, namely the Physarum's behavior, finding the minimum-length path between two points in a labyrinth is given. The CA-based model results are found in very good agreement with the computation results produced by the in-vivo experiments especially when combined with truly parallel implementations of this CA in a Field Programmable Gate Array (FPGA) and on a Graphical Processing Unit (GPU). The presented implementations succeed to take advantage of the CA's inherit parallelism and significantly improve the performance of the CA algorithm when compared with software in terms of computational speed and power consumption.

1   Introduction

The shortest path problem has always been a hot topic in the study of graph theory, because of its wide application field, extending from operational research to the disciplines of geography, automatic control, computer science and traffic. According to its concrete applications, scholars in relevant fields have presented many algorithms, but most of them are solely improvements [27] based on Dijkstra's algorithm [7]. Shortest path problems can be solved in polynomial time by one of the many shortest path algorithms, such as Dijkstra [7] and Floyd-Warshall [15, 72], provided that edge lengths are deterministic, i.e. every feasible probability distribution, out of a given set, over all possible successor nodes assigns probability one to a single successor On the other hand, Cellular Automata (CAs) are models of physical systems, where space and time are discrete and interactions are local [69]. Prior and more recent works proved that CAs are very effective in simulating physical systems and solving scientific problems, because they can capture the essential features of systems where global behavior arises from the collective effect of simple components, which interact locally [1, 11, 53, 54, 73]. Furthermore, they can easily handle complex boundary and initial conditions, inhomogeneities and anisotropies [8, 62]. The last decades, a wide variety of CA applications have been proposed on several scientific fields, such as simulation of physical systems, biological modeling involving models for self-reproduction, biological structures, image processing, semiconductor fabrication processes, crowd evacuation, computer networks and quantum CAs [16, 17, 30, 31, 38, 45, 55, 56, 61, 65]. These problems are described in terms of CAs, spatially by an 1-d, 2-d or 3-d array of cells and a local rule, which is usually an arbitrary function that defines the new state(s) of its CA cell depending on the states of its neighbors. The CA cells can work in fully synchronous and parallel manner updating their own state. It is clear that the CA approach can be considered consistent with the modern notion of unified space time, where, in computer science, space corresponds to memory and time to processing unit. In analogy, in CA, memory (cell state) and processing unit (local rule) are inseparably related to a CA cell [49, 60]. Taking all the above into consideration, there is no surprise that CA have been also able to deal successfully with the shortest path problem providing coherent and computationally efficient solutions in a number of various scientific applications and fields as shown later in this chapter. In what follows, we will focus in some of the most pronounced CA based applications that present different confrontations and corresponding solutions concerning the shortest path problem.

2 The First Cellular Automata Approach in Shortest Path Problem

The first CA algorithm tackling the shortest path problem was proposed by Adamatzky [2] although, it can be claimed that the famous Lee algorithm [32] could be considered as the first CA alike approach (see the CA algorithm in the next Sect. 3 for unweighted cells). However, the CA algorithm proposed by Adamatzky is mainly studying a weighted graph, which is also oriented. The three most common variations of the problem, namely single source shortest path (S^3P), all pairs shortest path (APSP) and single source single destination shortest path (S^3DSP), were all faced by the proposed CA algorithm. The main aim of this work was to tackle in a parallel way the S^3P and APSP variants by implementing a CA with adjustable neighborhood radius. Solving a shortest path problem or identifying the most direct path in a network among two vertices (i.e. x and y) was approximated using a CA, by plotting the under study graph onto a rectangular mesh, where cells x and y represent the respective vertices. The proposed solution is reached when an excitation wave with starting point cell x, diffuses towards all directions and arrives at cell y.

CAs have simulated living processes, neural networks, cellular and animal populations, molecular liquids, membranes and excitable reaction-diffusion media, because they are the most material, perceptible and practical models [2]. The main feature of excitation in a medium is that signals can be propagated undamped over a long distance and the speed of wave propagation can be variable. In Adamatzky's work the speed of a wave is proportional to the weight of the edge connection between node x and node y. At the beginning the given graph is mapped onto the cellular array of CA. Source vertex x and destination vertex y of the graph are corresponding to cells x and y, respectively. The x cell is excited and the wave propagates in all directions around the lattice, modifying the states of cells. Computation is assumed finished when the wave reaches destination y.

The definition of a CA is a d-dimensional lattice L of n cells, cell states Q, neighborhood function u and transition function f. For every cell x a neighborhood function assigns a group of the closest cells $u(x)$. The local transition function f maps a set of neighborhood states into the set Q of cells states. Using all the above characteristics, the next state of cell x will be defined as the state of its neighborhood in the previous time step and the rule of the transition function in the following way: $x^{t+1} = f(u(x^t))$. Evolution of CA with initial configuration c is a series of transitions like: $c^0 \rightarrow c^1 \rightarrow c^2 \rightarrow \cdots c^t \rightarrow c^{t+1} \rightarrow \cdots$. In this (ours) work [2] a pointer p_x and a vector w_{xy} are used, which correspond to the direction from which the wave has propagated at the previous time step and the weight of the edge between nodes x and y, respectively. The set Q can take one of the following elements: $Q = \{+, \#, \bullet, 0, 1, 2, \ldots, v\}$. Pointer p_x takes values from a finite nonempty set $Y = \{1, 2, \ldots, k, \lambda\}$ where λ can be considered as the initial value of p_x. Every element w_{xy} of vector w_x, $y \in u(x)$, can be in one of the states of set $W = \{\infty, 0, \ldots, v\}$. When vertices (nodes) x and y are not connected with one another by an edge oriented from x to y then $w_{xy} = \infty$.

Let $G = \langle V, E \rangle$ be an oriented graph of n vertices arranged on the d-dimensional discrete lattice, every vertex $v \in V$ of which is connected with no more than k neighboring vertices $v_1, v_2, \ldots, v_k$ by input edges $v_1 v, v_2 v, \ldots, v_k v \in E$ of weights $w(v_1 v), w(v_2 v), \ldots, w(v_k v) \in \{0, 1, \ldots, v, \infty\}$. If some pair $v'v'' \in V$, $w(v'v'') = \infty$ is written when there is no edge $v'v''$ in set E (i.e. there is no edge between those vertices). There is a principal feature that graph G must have in order to be successfully mapped onto a cellular lattice: $k < n$. Let $p = (v_0, \ldots, v_m)$ be a shortest path from vertex v_0 to vertex v_m of graph G and $l(p)$ be a length of p. Then we have that $l(p) = min\{l(p') : p' = \{v_0, \ldots, v_m\}\}$.

First, the single source, single destination shortest path (S^3DSP) is considered. This can be used in S^3P and $APSP$ in a similar way. In the S^3DSP computation, at the beginning $x_s^0 = +$, where $t = 0$ is assumed and x_s and x_d are the source and destination nodes while $+$ is the wave of excitation. The computation stops when cell x passes in state $\#$ or every cell of L is in state $\#$ or $\bullet$ ($\bullet$ is a quiescent-like state). The second constraint is used to stop computation when there is no path from x_s to x_d. The virtual wave is moving in cellular lattice. The wave of states $+$ runs in all directions around the lattice from cell x_s until it is in x_d, or passes all the cells of L. The pointer p_x has an initial state λ as mentioned before. When the cell x takes the $+$ state then p_x changes its initial value and takes one from set $\{1, 2, \ldots, k\}$, which is the index of the neighbor from where the $+$ state has come from. In this way, the final path can be extracted easily from the final configuration of the CA by back-tracking over the pointers from cell x_d toward cell x_s. The transition function f works with the neighborhood function $u(x)$ and weights w_{xy} in the following way. Assuming that cell $x^t = \bullet$ and some of its neighbors from $u(x)$ are in state $+$ at time t. Cell x^t finds the neighbor $y^t = +$ that has the minimum weight value and x takes this value. Starting from state w_{xy}, cell x jumps in state 0, decreasing its current state on unit step at every time step $x^{t+1} = x^t - 1$. Cell x^t will take the state $+$ when $x^t = 0$ or there exists a neighbor $y^t = +$ and the weight of the edge between x^t and y^t is the minimum and $w_{xy}^t < x^t$. The transition from state 0 to state $+$ and from state $+$ to state $\#$ is occurring unconditionally.

When the simulation starts, pointers of all cells are in state λ. However, if cell x changes its state to w_{xy}, then for some neighbor y_j pointer p_x saves the index j of this

Fig. 1 The example of oriented graph G: arrows indicate orientation of edges; and intersection of two or more straight lines corresponds to a vertex of G; black and empty boxes are source vertex and destination vertex, respectively (adopted from [2])

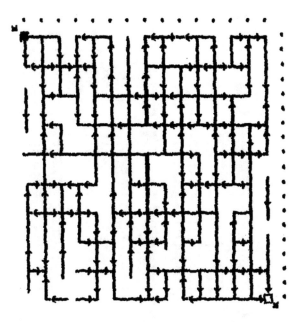

neighbor. During the decreasing of w_{xy} down to 0, pointer p_x can be modified when there is the condition $\forall y \in u(x) : y^t = + \wedge w_{xy} = min\{w_{xy'} : y' \in u(x) \wedge y'^t = +\}$ and $w_{xy}^t < x^t$. The state of pointer becomes constant after cell x departs from state 0. Concluding all these rules, x^{t+1} can take the following states in every case:

- #, when x^t is in states # or +.
- +, when $x^t = 0$.
- •, when $x^t = •$ and there in no neighbor $y^t = +$.
- w_{xy}, when $x^t = •$ or $x^t > 0$ and there is at least one neighbor with $y^t = +$ and the weight of the edge between x^t and y^t is the minimum of other edges in the neighborhood.
- $x^t - 1$, when $x^t > 0$ and there is no other neighbor $y^t = +$ that has a weight in the edge between x^t and y^t in order that $w_{xy} < x^t$.

An example can be shown in Fig. 1. The goal is the solution of a S^3DSP in a 2-d grid where the edges can take states ∞ or 0 and the source vertex is on the upper left and destination vertex on the bottom right. Q can take values $\{•, +, \#\}$, Y can take values $\{N, W, S, E, \lambda\}$ and W can take values $(w_{xN}, w_{xW}, w_{xS}, w_{xE})$. Symbols N, W, S, E are the indices for the northern, western, southern and eastern neighbors of the cell. The initial conditions are: $x_s^0 = +, \forall x \in L, x \neq x_s : x^0 := •$ and $p_x := \lambda$. The dynamic of CA evolution is shown in Fig. 2. The back-traced (a) and extracted (b) paths can be found in Fig. 3.

The longest path in G consists of $n - 1$ nodes. So, a CA of n cells, four neighbors and nine states that models a 2-d G graph of n nodes, some cut-off edges and edge

```
T=0                    T=1                    T=2                    T=3

+..............        -+.............        N-+............        NW-...........
...............        +.............         -+............        N-+...........
...............        .............          .............         .+............
...............        .............          .............         .............
...............        .............          .............         .............
...............        .............          .............         .............
...............        .............          .............         .............
...............        .............          .............         .............
...............        .............          .............         .............

T=4                    T=5                    T=6                    T=7

NWW............        NWW...........         NWW...........        NWW...........
NN-+...........        NNN-..........         NNNW..........        NNNW..........
+-+............        -N-+..........         ENN...........        ENNW..........
.+.............        +-............         -N.+..........        NN.-+.........
...............        .............          .............         .............
...............        .............          .............         .............
...............        .............          .............         .............
...............        .............          .............         .............
...............        .............          .............         .............

T=8                    T=9                    T=10                   T=11

NWW............        NWW...........         NWW...........        NWW..+........
NNNW...........        NNNW..........         NNNW.+........        NNNW.-........
ENNW...........        ENNW.+........         ENNW+-+.......        ENNW-S-.......
NN.N-+.........        NN.NW-........         NN.NWW........        NN.NWW+.......
...............        .............          .............         .............
...............        .............          .............         .............
...............        .............          .............         .............
...............        .............          .............         .............
...............        .............          .............         .............

T=12                   T=13                   T=14                   T=15

NWW..-.........        NWW..S........         NWW..S........        NWW..S.+......
NNNW.S.........        NNNW.S........         NNNW.S.+......        NNNW.S.-+.....
ENNWESW........        ENNWESW+......         ENNWESW-+.....        ENNWESWS-.....
NN.NWW-+.......        NN.NWWN-......         NN.NWWNW......        NN.NWWNW+.....
......+........        .....+-+......         ....+-N-+.....        ...+-ENW-+....
...............        ....+.........         .....+-+......        ....+-N-+.....
...............        .............          ......+.......        .......-+.....
...............        .............          .............         .............
...............        .............          .............         .............

T=16                   T=17                   T=18                   T=19

NWW..S+-+......        NWW..S-S-+....         NWW..SESS-....        NWW..SESSW....
NNNW.S.S-+.....        NNNW.S.SW-+...         NNNW.S.SWW-...        NNNW.S.SWWW...
ENNWESWSW......        ENNWESWSW+....         ENNWESWSW-+...        ENNWESWSWN-+..
NN.NWWNW-+.....        NN+NWWNWN-....         NN-NWWNWNS....        NNSNWWNWNS+...
..+-EENWW-+....        .+-EEENWW-....         +-EEENWWW.....        -EEEENWWWW....
....-ENN-+.....        ...NENNW-+....         .+.NENNWN-....        .-+.NENNWNN...
......N-+......        .....NN-......         ......NNN.+...        .....NNN+-+...
.......+.......        .....+-.......         .......-N.....        ......EN..+...
...............        .............          .............         .............

T=20                   T=21                   T=22                   T=23

NWW..SESSW.....        NWW..SESSW.+..         NWW..SESSW.-+..       NWW..SESSW.S-+.
NNNW.S.SWWW+...        NNNW.S.SWWW-+..        NNNW.S.SWWWS-+.       NNNW.S.SWWWSW-.
ENNWESWSWNN-+..        ENNWESWSWNNW-+.        ENNWESWSWNNWW-.       ENNWESWSWNNWWW.
NNSNWWNWNS-+...        NNSNWWNWNSN-+..        NNSNWWNWNSNN-+.       NNSNWWNWNSNNNW.
EEEEEENWWWW....        EEEEEENWWWW....        EEEEEENWWWW...        EEEEEENWWWW..+.
.N-.NENNWNN+...        .NW.NENNWNN-...        .NW+NENNWNNS..        .NW-NENNWNN...
..+...NNN-N-+..        ..-+..NNNENW...        ..N-+.NNNENW..        .NW-.NNNENW...
......EN.+-+...        ......EN.-N-..        ...+..EN.NNN..        ...-+.EN.NNN..
...............        .......+.+....         .........-+...        .........N-N-+.
...............        .............          .............        ..........+-+..

T=24                   T=25                   T=26                   T=27

NWW..SESSW.SS-+        NWW..SESSW.SSS-        NWW..SESSW.SSSW       NWW..SESSW.SSSW
NNNW.S.SWWWSWW.        NNNW.S.SWWWSW+        NNNW.S.SWWWSWW-       NNNW.S.SWWWSWWN
ENNWESWSWNNWWW+        ENNWESWSWNNWWW-       ENNWESWSWNNWWWS       ENNWESWSWNNWWWS
NNSNWWNWNSNNNW-        NNSNWWNWNSNNWWW       NNSNWWNWNSNNWWW       NNSNWWNWNSNNWWW
EEEEEENWWWW..-+        EEEEEENWWWW.N-        EEEEEENWWWW.+NW       EEEEEENWWWW.-NW
.NWSNENNWNNS.+.        .NWSNENNWNNS+-.       .NWSNENNWNNS-N.       .NWSNENNWNNSEN.
..NWW.NNNENW...        ..NWW.NNNENW.+.       ..NWW.NNNENW+-.       ..NWW.NNNENW-N.
...N-+EN.NNN.+.        ...NN-EN.NNN.-+       ...NNWEN.NNN.SS       ...NNWEN.NNN.SS
....+....NENW-+        ...-+..NENNW-        ....N-..NENWWW        ....NN..NENWWW
.......+-N-+.          ....+....-ENN-+       ...+-+..EENNN-       ...-N-+..EENNNN
```

Fig. 2 A CA evolution in the computation S^3DSP. Symbol—means state #. A state pointer for cell x is shown in the figure only if x was in state # at least two times (adopted from [2])

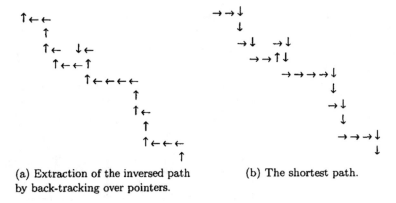

(a) Extraction of the inversed path   (b) The shortest path.
by back-tracking over pointers.

Fig. 3 The results of computation of S^3DSP (adopted from [2])

weights $\{\infty, 0\}$ can compute the shortest path in $O(n)$ computation time and APSP in $O(n^2)$, respectively.

Assuming a 2-d CA of n cells is used, each of which has four neighbors and $6 + 5n$ states modeling a 2-d rectangular grid where the edges of G graph can be of weight $\{0, \ldots, v, \infty\}$, the S^3DSP can be solved in $O(vn)$ upper time.

The same example is used as in Fig. 1, with set W taking values $\{0, 1, 2, \infty\}$. The graph with the edges' weight can be seen in Fig. 4. If the longest path consists of n vertices and the delay of state transition is v for any cell, then $O(vn)$ is an upper bound on the time for computation S^3P. If a CA uses k neighbors and $O(vk)$ states then the S^3DSP can be solved in $O((v + k)n)$ upper time. An S^3P can be extracted in $O(n^2)$ upper time. The evolution of the above example can be found in Fig. 5 and the result in Fig. 6.

After the initial introduction of the CA notion in the shortest path problem by Adamatzky [2], an increasing interest of the research community in the specific field was declared. In particular in 2004, more than a decade afterwards, Wu and Xue [74] tried to extend the CA model for shortest path calculation defining properly the cell state and providing suitable cellular evolution to dictate cell states evolution mainly focusing on the appropriate node order as resulted from the proposed CA algorithm. An update of the presented study was published a couple of years later by Li et al. [33] with the addition of the cell state turnover and Sun and Dai, based on this CA algorithm proposed the subtraction of the least surplus weight to advance once again the CA based algorithms for shortest path computation [58] as mentioned in Wang et al. paper [71]. In this last work, Wang et al. selected to adjust the cell state set by combining breeding and mature states, while trying to improve the resulting parallelism and at the same time, recording manner of cellular state turnover is modified to record all information sources [71].

Fig. 4 An oriented planar graph G every vertex of which has indegree 4. The edges are weighted with elements of set {0, 1, 2, ∞}. The left upper vertex is a source vertex and the right lower is a destination one (adopted from [2])

```
+12+0∞+∞∞∞+∞0+∞0+∞1+∞∞∞+∞∞∞+21+
  ∞   2   1   ∞   ∞   2   1   0   ∞   1
  ∞   2   2   2   ∞   ∞   1   1   1   ∞
+∞∞∞+20+2∞+∞2+∞1+0∞+10+∞∞∞+∞∞∞+
  2   ∞   1   0   1   0   0   0   1   ∞
  1   1   1   1   ∞   2   1   ∞   0   ∞
+∞1+∞∞∞+∞1+22+∞2+00+∞∞∞+∞1+1∞+
  2   1   0   ∞   2   0   0   ∞   ∞   ∞
  ∞   ∞   ∞   2   ∞   1   1   ∞   1   ∞
+∞2+1∞+2∞+02+∞∞∞+21+22+∞2+∞∞∞+
  0   ∞   0   ∞   ∞   ∞   2   1   ∞   ∞
  ∞   ∞   ∞   1   1   2   0   ∞   ∞   1
+∞0+2∞+20+∞0+00+22+02+12+11+
  ∞   2   ∞   ∞   ∞   1   0   ∞   ∞   0
  1   ∞   ∞   ∞   0   ∞   1   ∞   ∞   2
+∞0+00+22+∞2+∞0+∞∞∞+1∞+∞∞∞+1∞+
  ∞   ∞   ∞   ∞   2   ∞   1   ∞   2   ∞
  1   ∞   0   0   2   ∞   1   0   ∞   ∞
+∞∞∞+∞2+∞0+1∞+0∞+∞∞∞+∞∞∞+∞2+∞∞∞+
  ∞   ∞   ∞   ∞   0   ∞   0   0   ∞   ∞
  ∞   ∞   ∞   0   2   ∞   ∞   ∞   1   ∞
+10+∞∞∞+∞1+∞∞∞+∞0+∞0+∞0+∞∞∞+∞∞∞+
  2   0   ∞   ∞   0   2   0   2   ∞   ∞
  ∞   ∞   2   ∞   ∞   ∞   0   2   2   2
+21+11+∞∞∞+10+1∞+1∞+00+21+00+
  ∞   ∞   ∞   0   0   ∞   0   ∞   ∞   ∞
  ∞   2   ∞   ∞   ∞   ∞   0   ∞   2   ∞
+2∞+∞∞∞+0∞+1∞+∞0+22+21+20+00+
```

Moreover, in 2010, Wang [70] studied the shortest path solution on a three-dimensional surface using CAs. On the 2-d space, a straight line is the shortest distance between two points, but for a complex 3-d surface such as a path between mountain, the shortest walking path between the starting point and destination can not be a simple straight line. It is a more complex problem to find the shortest path on the complex 3-d surface. Such an approach has a considerable arbitrariness, and it is hard to find the best route. There are also obstacles like hills, rivers, lakes which block the routes from the source to the destination point. 3-d position data of the study area are imported, including the plane coordinates and elevation values of the starting point A and end point B. The problem is seeking the shortest path between A point and B point. The distance between A and B point was divided into n equal portions, and the vertical profile is made over each equal point, so that each vertical profile intersects three-dimensional surface, n profile curves are derived. Then, each profile curve is equal to m number of points according to horizontal distance, so that the path search problem of 3-d surface is transformed into a discrete optimization problem through gridding.

```
T=1            T=2            T=3            T=4            T=5

+.........     -2........     N1........     N+........     N-........
..........     ..........     ..........     ..........     .2........
..........     ..........     ..........     ..........     ..........
..........     ..........     ..........     ..........     ..........
..........     ..........     ..........     ..........     ..........
..........     ..........     ..........     ..........     ..........
..........     ..........     ..........     ..........     ..........
..........     ..........     ..........     ..........     ..........
..........     ..........     ..........     ..........     ..........


T=6            T=7            T=8            T=9            T=10

NW........     NW........     NW........     NW1.......     NW+.......
.1........     .+........     .-+.......     .N-.......     .NW.......
..........     ..........     .1........     .+1.......     .-+.......
..........     ..........     ..........     ..........     ..........
..........     ..........     ..........     ..........     ..........
..........     ..........     ..........     ..........     ..........
..........     ..........     ..........     ..........     ..........
..........     ..........     ..........     ..........     ..........
..........     ..........     ..........     ..........     ..........


T=11           T=12           T=13           T=14           T=15

NW-.......     NWS.......     NWS.......     NWS.......     NWS.......
.NW.......     .NW.......     .NW+......     .NW-2.....     .NWS1.....
.N-1......     .NN+......     .NN-2.....     .NNW1.....     .NNW+.....
..........     ..........     ...2......     ...1......     ...+......
..........     ..........     ..........     ..........     ..........
..........     ..........     ..........     ..........     ..........
..........     ..........     ..........     ..........     ..........
..........     ..........     ..........     ..........     ..........
..........     ..........     ..........     ..........     ..........


T=16           T=17           T=18           T=19           T=20

NWS.......     NWS.......     NWS.......     NWS..2....     NWS..1....
.NWS+.....     .NWS-1....     .NWSW+....     .NWSW-....     .NWSWW-...
.NNW-2....     .NNWW1....     .NNWW+....     .NNWW-+...     .NNWWW-...
..2-2.....     ..1N1.....     ..+N+.....     .1-N-1....     .+>NW+1...
...1......     ...+......     ..2-+.....     ..1N-+....     ..+NW-2...
..........     ..........     ..........     .....+....     ....-+....
..........     ..........     ..........     ..........     ...2......
..........     ..........     ..........     ..........     ..........
..........     ..........     ..........     ..........     ..........


T=21           T=22           T=23           T=24           T=25

NWS..+1...     NWS..-++..     NWS..S--..     NWS..SSS..     NWS..SSS..
.NWSW-+..     .NWSWWS-..     .NWSWWSW..     .NWSWWSW..     .NWSWWSW..
.NNWWWW...     .NNWWWW...     .NNWWWW...     .NNWWWW...     .NNWWWW...
.->NW-+...     .>>NWN-2..     .>>NWNN1..     .>>NWNNW+..     .>>NWNNN-2
.2-NWW1...     .1>NWW+...     .+>NWW-2..     .->NWWN1..     .>>NWWN+..
....N-....     ....NW....     ....NW1...     ....NW-...     ....NW-...
....1.....     ....+.....     ...1-.....     ...+N.....     ...-N.1...
..........     ..........     ...2......     ....1.....     ...++.....
..........     ..........     ..........     ..........     ..........


T=26           T=27           T=28           T=29           T=30

NWS..SSS..     NWS..SSS..     NWS..SSS..     NWS..SSS..     NWS..SSS..
.NWSWWSW..     .NWSWWSW..     .NWSWWSW..     .NWSWWSW..     .NWSWWSW..
.NNWWWW...     .NNWWWW...     .NNWWWW...     .NNWWWW...     .NNWWWW...
.>>NWNNW1.     .>>NWNNW+.     .>>NWNNN-.     .>>NWNNWW.     .>>NWNNWW.
.>>NWWN-2.     .>>NWWNN1.     .>>NWWNWN+.     .>>NWWNNW-1     .>>NWWNNW+
....NWN...     ....NWN...     ....NWN...     ....NWN...     ....NWN...
...>N.+...     ...>N.-...     ...>N.N...     ...>N.N+..     ...>N.N-2.
...--+....     ...NN-+...     ...NNW-+..     ...NNWW...     ...NNWWW..
..........     ..........     ......+...     .....1-+..     .....+N-1.
..........     ..........     ..........     ......+...     .....2-1..


T=31           T=32           T=33           T=34           T=35

NWS..SSS..     NWS..SSS..     NWS..SSS..     NWS..SSS..     NWS..SSS..
.NWSWWSW..     .NWSWWSW..     .NWSWWSW..     .NWSWWSW..     .NWSWWSW..
.NNWWWW...     .NNWWWW...     .NNWWWW...     .NNWWWW...     .NNWWWW...
.>>NWNNWW.     .>>NWNNWW.     .>>NWNNWW.     .>>NWNNWW.     .>>NWNNWW.
.>>NWWNNW-     .>>NWWNNWW     .>>NWWNWW     .>>NWWNWWW     .>>NWWNWWW
....NWN..2     ....NWN..1     ....NWN.2+     ....NWN.1-     ....NWN.+N
...>N.NS1.     ...>N.NS+.     ...>N.NS-.     ...>N.NSW.     ...>N.NSW.
...NNWWW..     ...NNWWW..     ...NNWW1.     ...NNWWW+.     ...NNWWW-.
....1-NW+.     ....+>NW-+     ...1->NWW-     ...+>>NWWW     ...->>NWWW
.....1N+..     .....+N-+.     ......-NW-+     ......>NWW-     .....>NWWW
```

Fig. 5 An evolution of CA which computed the $S^3 DSP$ on a graph with weighted edges. Symbol −means #. A state of the pointer for cell x is shown in figure only if x was in state # at least two times (adopted from [2])

Fig. 6 The shortest path
(adopted from [2])

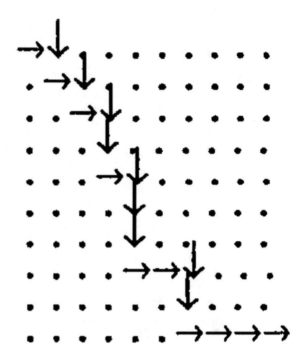

3   A Cellular Automata Algorithm Based on Lee's Algorithm

The Lee algorithm [32] is a well known fundamental routing algorithm that can find the shortest path between two points in a grid. The task was to find a CA similar to the Lee algorithm that uses a small number of states, independently of the grid size. Such an algorithm was found in [22] and then further described in [19]. It was developed when the expressive power of the cellular description language CDL [18, 20] was explored for different applications.

3.1   Lee Algorithm

A very well known approach to routing problems is the Lee algorithm [32]. Its purpose is to find an optimal path between two points on a regular grid. Each point of the grid is associated with a weight. The algorithm finds the path on the grid with the lowest sum of weights. By adjusting the weights of the grid points the user has some control over what is supposed to be an optimal path. Let us consider an example: The user simply searches for the shortest path between two points. In this case the user specifies the weight one for all grid points and the algorithm will find the path with

for all grid points i do
    acw(i) := infinity;
acw(starting point) :=0
steady := false
while not steady do
    steady := true;
    for all grid points i do
        MinNeighbour = min(acw(j)) where j in neighbours(i);
        acw(i) := MinNeighbour + weight(i) of this point;
        if acw(i) has changed then steady := false;

Fig. 7 First phase of the original Lee algorithm

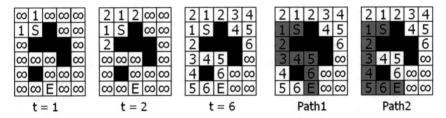

        t = 1   t = 2   t = 6   Path1   Path2

Fig. 8 Original Lee algorithm. A wave propagates from S to E. At each time step the cell's weight (one here) is added to the minimum of the neighbours' accumulated weights. When E is reached the path is backtraced along the neighbours with the minimal accumulated weights. Several alternative shortest paths are possible

the lowest number of grid points. This is the shortest path between the two chosen points S and E. In another example the user looks for a path that crosses the already existing paths as few as possible. In this case the user assigns a very high weight to all points of the existing paths and a very low weight to all other grid points. The algorithm will then find the path with as few crossings as possible.

The algorithm works in two phases, (1) wave propagation from S to E, and (2) building a path by backtracing from E to S.

In the first phase (Fig. 7) the accumulated weights (acw) for each node relative to the starting point are computed. All free cells are initialized to infinity. The accumulated weight for the starting point S is initialized to 0. Figure 8 shows for a sample grid the calculation of the accumulated weights. The weight of all grid points is one in this case. The wave reaches the end point at time step $t = 6$.

In the second phase the actual path is established. For this purpose the algorithm 'walks' back from the end point E to the starting point S. At each step the neighbour with the smallest accumulated weight is selected as part of the path.

Note, that there are several possibilities to build a path from the end point to the start point. At the end point you can either go up or left (Path1 and Path2 in Fig. 8, respectively) for this example.

Obstacles can be modeled by infinite weights at grid points. Since then the sum of the local weight and the accumulated weight of a neighbour will always be infinite, such grid points may never become part of the path.

At a first glance this algorithm looks like it perfectly fits onto a Cellular Automaton (CA). Unfortunately the number of states required to perform the algorithm is related to the longest path or more precisely to the largest accumulated weight that can occur. Thus we decided to develop a version of the algorithm [19, 22] which has a constant number of states. This version can only handle the shortest path problem with a unified weight at all grid points.

3.2 CA Based Lee Algorithm

The accumulated weights in the Lee algorithm are needed to find the shortest path. Instead of storing the accumulated weights we stored the direction in which we have to move in order to return back to the starting point. With these wave marks instead of the accumulated weights the algorithm requires only a constant set of states. Of course, we can not handle problems with arbitrary weights at the grid points.

We present the modified algorithm in CDL (Cellular Description Language). This language has been developed to describe CAs in a concise and readable way, independent of the target architecture. Compilers were built that could translate CDL into C functions for almost any software architecture and into Boolean equations for hardware synthesis, like the CEPRA family [21].

At the beginning of the first phase all cells are in the `free` state, except for the starting and end point. In the first phase, all cells check whether there is any cell in the neighbourhood that already has a wave mark. If a wave mark is found, the cell itself becomes a wave mark towards the already marked cell. This is performed in lines 25 and 26 of Fig. 9. The `one` function successively assigns all the elements of the groups `neighbours` and `wave` to the temporary variables `h` (a cell address) and `z` (a state). For each assignment the condition following the colon is checked. The evaluation of the `one` function stops if an assignment is found, that satisfies the condition, i.e. the corresponding neighbour is in state `start` or in any of the states `wave`. The first assignment is `h=N` and `z=wave_up`. The assignment to `z` is only used for its side effect to store temporarily the wave state corresponding to the neighbour being investigated. If the east neighbour is currently investigated this cell state will change to `wave_right` since it must point to this neighbour. Figure 10 at $t = 6$ shows a sample grid at the end of phase one. The wave marks are symbolized by small arrows. The black squares are obstacles. We had to introduce a special state to model obstacles, since we do not have weights at the grid points.

Phase one ends when the end point is reached. Now the path is built backward towards the start point along the wave marks (lines 29–34). If a cell is one of the `wave` states and it sees a neighbour cell that is a path towards this cell, then this cell becomes a path in the direction of its previous mark (Fig. 10, $t = 7$–13). This is done in the CDL program by adding four to the enumeration element; e.g. `wave_down`

```
(01) cellular automaton Lee_routing;
(02)
(03) const dimension = 2;    --2-dimensional CA
(04)       distance = 1;     --neighbourhood distance
(05)
(06)         --relative coordinates of cells: center, north, east, south, west
(07)       C=[0,0]; N=[0,1]; E=[1,0]; S=[0,-1]; W=[-1,0];
(08)
(09) type  celltype = (free, obstacle, Spoint, Epoint,
(10)                      --phase 1: wave marks
(11)                      wave_up, wave_right, wave_down, wave_left,
(12)                      --phase 2: path directions
(13)                      path_up, path_right, path_down, path_left,
(14)                      clear, Ready);
(15)
(16) group wave = {wave_up, wave_right, wave_down, wave_left};
(17)       path = {path_up, path_right, path_down, path_left};
(18)       neighbours = {N, E, S, W};
(19)
(20) var   h : celladdress;  --temporary, address of a neighbor
(21)       z : celltype;     --temporary, tested state (wave/path mark)
(22)
(23) rule
(24)   case *C of    --depending on contents * of cell with address C
(25)   free:    if one(h in neighbours & z in wave: *h in {start, wave}) then
(26)              *C := z; --set wave mark pointing to a neighbour in state start or wave
(27)   Epoint: if one(h in neighbours & z in path: *h in {start, wave}) then
(28)              *C := z; --set path mark pointing to a neighbour in state start or wave
(29)   wave:    if (*N = path_down) or (*E = path_left) or
(30)               (*S = path_up) or (*W = path_right) then
(31)              *C := *C + 4 --change wave mark into path mark
(32)            else
(33)            if one(h in neighbours: *h in {path, clear}) then
(34)              *C := clear; --clear unused path marks
(35)   clear:  *C := free;
(36)   Spoint: if one(h in neighbours: *h in path) then
(37)              *C := ready; --termination: start->ready
(38)   end;   --of case
```

Fig. 9 The shortest path algorithm described in the language CDL

becomes path_down. When the starting point is reached, its state changes from S to R. The ready state R signalizes the termination of phase two.

All unnecessary wave marks have to be cleared in order to allow subsequent routing passes. For this purpose all cells that see a neighbour cell which is a path not pointing towards this cell are cleared. Such a cell will never become part of the path. Also all cells are cleared that see a neighbour in the clear state. Since the building of the path moves along the shortest path, it is impossible that a cell in clear state could reach a cell which will be in the path but is not yet part of it. A cell in state clear will change to state free in the following time step.

The time complexity for the first and second phase is $O(p)$, where p is the path length. For a $n \times n$ square grid, the maximal path length is $2n - 1$ if there are no obstacles, and $O(n^2)$ if there are obstacles (spiral like path). The algorithm requires only 14 states and thus can be very easily realized in hardware.

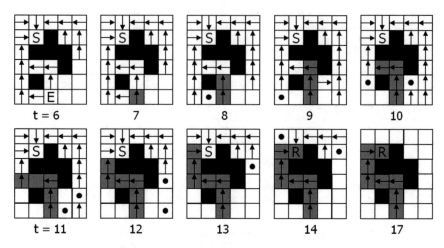

Fig. 10 A simulation of the algorithm. ($t = 6$) Wave marks at the end of phase one. ($t = 8$) A first unused wave mark changes into the state `clear` (•). ($t = 13$) Path is constructed. ($t = 14$) Starting point changes into the `ready` state (R). ($t = 17$) All unused wave marks are deleted

Based on the Lee algorithm, a CA algorithm with 14 states was designed that is able to route a shortest path between to nodes of a grid. The algorithm is independent of the grid size and needs only 14 states per cell. The time complexity is $O(p)$, where p is the path length.

4 Learning Automata

There are shortest path problems in which the lengths of the edges in a graph are allowed to be random. This makes the problem more difficult. A stochastic graph G is defined by a triple $G = \langle V, E, F \rangle$, where V represents the set of nodes, E specifies a set of edges and matrix F is the probability distribution describing the statistics of edge lengths. In stochastic graph G, a path π_i with length of n_i nodes and expected length of L_{π_i} from source node V_{source} to destination node V_{dest} is defined as an ordering $\{\pi_{i,1}, \pi_{i,2}, \ldots, \pi_{i,n_i}\}$ of nodes. $V_{\pi_{i,1}} = V_{source}$ and $V_{\pi_{i,n_i}} = V_{dest}$ are source and destination nodes, respectively and all the intermediates are nodes in path π_i. Assume that there are r distinct paths between V_{source} and V_{dest}. The shortest path between source node and destination node is defined as a path with minimum expected length. Stochastic shortest path problems can be grouped in two main classes: the first class aims to find a priori solution that minimizes the expected lengths, while the second one computes an online solution that allows decisions to be made at various stages.

In 2006, Beigy and Meybodi [3] introduced a network of Learning Automata (LAs), which were called Distributed Learning Automata (DLAs). More specifically, an automaton acting in an unknown random environment and improving its performance in some specified manner, is referred to as a learning automaton (LA). DLAs is a network of automata which collectively cooperate to solve a particular problem. A DLA can be modeled by a directed graph in which the set of nodes of a graph constitute the set of automata and the set of outgoing edges for each node constitutes the set of actions for corresponding automaton. When an automaton selects one of its actions, another automaton on the other end of edge, corresponding to the selected action, will be activated. They used this tool to propose some iterative algorithms and solve stochastic shortest path problem. In these algorithms, at each stage DLAs determine which edges to be sampled. This sampling method may result in decreasing unnecessary samples and hence decreasing the running time of algorithms. The automata approach to learning involves the determination of an optimal action from a set of allowable actions. Automaton selects an action from its finite set of actions, which serves as the input to the environment, which in turn emits a stochastic response at a specific time. The environment penalizes or rewards an action of the automaton with a penalty/reward probability. The state of the automaton is updated and a new action is chosen at the next time step.

In their algorithm a network of learning automata which is isomorphic to the input graph is created. Each node in the network represents a LA and the actions of this node in the LA is the outgoing edge of this node. Then, at the stage k, source automaton, which represents the source node in the graph, chooses one of its actions based on its action probability vector. If the action is a_m the automaton A_m is also activated on the other end of edge (s, m). The process of choosing an action and activating an automaton is repeated until the destination automaton is reached or for some reason moving along the edges of the graph is not possible or the number of visited nodes exceeds the number of nodes in the graph. After destination automaton is reached, the length of the traversed path is computed and then compared with a quantity called dynamic threshold.

Depending on the result of the comparison all the LAs (except the destination learning automaton) along the traversed path update their action probabilities. Updating is done in direction from source to destination or vice versa. If length of the traversed path is less than or equal to the dynamic threshold then all LAs along that path receive reward and if length of the traversed path is greater than the dynamic threshold or the destination node is not reached, then activated automata receive penalty. The process of traveling from the source LA to the destination LA is repeated until the stopping condition is reached which at this point the last traversed path is the path which has the minimum expected length among all paths from the source to the destination.

5 Shortest Path Based on Cellular Automata Algorithm for Computer Networks

As already mentioned in the Introductory Chapter the problem of finding the shortest path (SP) from a single source to a single destination in a graph arises as a sub-problem to many broader problems. In general, different path metrics are used for different application. For example, in communication systems, if each link cost is 1, then the minimum number of hubs is found. However, cost can also represent the propagation delay, the link congestion or the reliability of each link. In the latter case, if the individual communication links operate independently, then the problem can be stated as to find what path has the maximum reliability.

Here we focus on the computer networks and we present how the aforementioned problem is confronted and solved by the CAs approach. More specifically, Mardiris et al. [38] presented an interactive tool that offers automated modeling with the assistance of a dynamic and user friendly graphical environment, called Net_CA for modeling and simulation of computer networks based on CAs. More specifically, a 2-d NaSch [43] CA computer network model was developed and several computer networks were simulated, while algorithms for connectivity evaluation, system reliability evaluation and shortest path computation in a computer network [14] have also been implemented. The proposed system also produced automatically synthesizable VHDL code leading to the parallel hardware implementation of the above CA algorithms rendering Net_CA as a very reliable and fast simulator for wireless networks, ad hoc networks and, generally, for low connection reliability networks

In regards to the shortest path algorithm as expressed in CAs and applied for computer networks, let $G = (N, A)$ be a network, where N is the set of n nodes, $A \subseteq N \times N$ is the set of connections and L_i is the neighborhood of the node i. That is, each node i is mapped to a cell whose neighborhood is the set of nodes connected to it by its input connections. Associated with each connection $(p, q) \in A$ is a non-negative number; C_{pq} stands for the cost of connection from node p to node q. Non-existing connection costs are set to infinity. Let P_{st} be a path from a source node s to a destination node t, defined as the set of consecutive connected nodes: $P_{st} = \{s; n_1; n_2; \ldots; n_i; t\}$. The state of each node, at each time step t_s, is represented by a vector with two entries $V_i(t_s) = \{V_i^1(t_s), V_i^2(t_s)\}$: the first component is a pointer to the previous node in the path, while the second is the cost of the partial path up to node i. $V_i^1(t_s)$ is not necessary for evaluating the shortest path length, but it is used only for indicating the shortest path itself.

The evolution of CA algorithm is given by the following equation:

$$V_i^1(t_s) = \left\{ k, \left(V_i^2(t_s) + C_{k,i} \right) \right\} \tag{1}$$

The flowchart of the described CA algorithm is shown in Fig. 11. The minimum s–t path cost is the second component of the state vector of node t. In case this value is infinity, then there exists no s–t path.

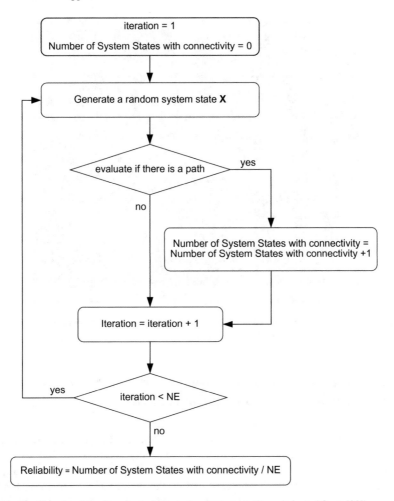

Fig. 11 The CA algorithm flowchart of shortest path computations (adopted from [38])

The results of the implementation of the presented shortest path computation algorithm to the proposed Net_CA system are depicted in Fig. 12. As before the user defines the starting topology. During the execution of the aforementioned algorithm each node is described by a pair of values, one for its number (name) according to the cost of the connection up to it, and the other for the minimum cost of connection of the starting node to the examined node. After the execution of the algorithm the shortest connectivity path between nodes s and t, if it exists, is colored yellow.

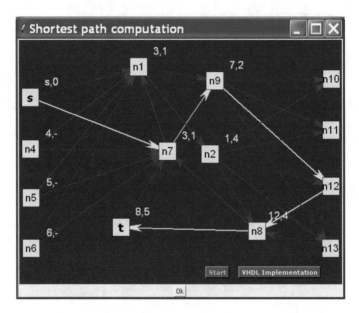

Fig. 12 The final simulation screen of the Net_CA system during the execution of the shortest path algorithm between nodes *s* and *t* (adopted from [38])

6 Shortest Path Definition Based on Cellular Structures for Coordinated Motion in Swarm Robotics

In robotics, the applied shortest path solvers must consider various constraints regarding both the environment and the utilized robot configurations. Robot navigation involves the determination of a continuous motion for a robot towards a goal location. Depending on the available information, the path planners must consider the presence of obstacles in order to avoid potential collisions. A priori knowledge of the configuration area status involves static obstacles which results to simple solutions of the path-planning problem. A global shortest path is extracted for the robot, which involves also collision avoidance scenarios. The complexity of such solutions increases in cases where no information is available in advance (dynamic environments) and therefore, they should extract the robot's motion in real-time. Robot's motions are recalculated at every time step in order to both avoid the detected objects and follow the defined shortest path. The periodic motions of a robot could also be affected by processes of coordination in multirobot systems. The required collective behavior that emerges from the interactions between the robots significantly impress the results of the path planner and its overall complexity. Cooperative robotic teams are extensively utilized for accomplishing additional tasks such as exploration [48], search and rescue [36] and formation control [46].

Several methods have been proposed for solving the shortest path problem in robot navigation, both for a single or multiple robots and for static or dynamic environments.

Visibility graphs have been exploited to identify Euclidean shortest paths among a set of polygonal obstacles in the plane [40]. The method applies a recursive process capable of solving dynamic navigation problems for a single robot. Collective behavior will dramatically increase the complexity of the approach due to recursiveness rendering the method improper for computing shortest paths in multirobot systems. Moreover, a modified version of potential fields has been proposed in order to consider both static and dynamic obstacles [42]. The method combines an Artificial Potential Field technique with a Bacterial Evolutionary Algorithm to reduce the extraction time of the optimal path. Despite the fast computation and the accurate results, the resulted time remains in high levels due to its dependency with the utilized robot configurations making its implementation on swarm robots unfeasible. Aiming at smoother transitions, heuristic based algorithms were introduced as potential path planning solvers. For example, an extended version of a D-star algorithm [57] was proposed in [13]. The method applies a bilinear interpolation function to compute the required motion fragments resulted by the vertex expansion. In order to reduce time complexity and define smoother transitions, a mathematical model inspired by Physarum polycephalum along with a heuristic rule function were proposed to solve the shortest path problem [75]. The method extract accurate results in limited time amounts nonetheless; it could not be implemented in low resources systems.

On the contrary, cell decomposition techniques display low time and computational complexity levels rendering the approaches proper for implementation in swarm robotic systems. The configuration area is partitioned into a lattice grid and every area cell is processed accordingly. For example, free space is retracted onto a Voronoi diagram while the evolution of a CA constructs its structure [68]. Moreover, a variant of the A-star algorithm, namely Theta-star, was extended in [47] where the acquired information is propagated along grid edges without constraining the paths to grid edges. The method handles accurately static objects however; the extracted paths for unknown environments are based on assumptions. Furthermore, numerous artificial intelligence algorithms were proposed as potential solvers of the shortest path problem in robotics. A fuzzy logic controller was proposed in [50] where obstacles of various shapes can be avoided and a single robot can follow the computed shortest path towards its final destination. A fuzzy-based cost function was also exploited by an ant colony optimization for the evaluation process of the potential solutions [51]. In addition, various types of artificial neural networks were utilized to extract optimum paths. A Guided Autowave Pulse Coupled Neural Network [59] and a Deep Convolutional Neural Network [23] were applied to create collision free trajectories for mobile robots. Despite their efficiency, special hardware resources and/or centralized control are required for their implementation to real robotic systems.

All the aforementioned methods display multiple limitations or specific drawbacks despite their efficiency in defining the required shortest paths for robot navigation. Their vast majority cannot include collective behavior since their complexity and their resource requirements can be increased significantly, even with proper implementation modifications. Several methods have been proposed to overcome such limitations and are specialized in computing the required paths in multiple robot teams. In general, methods that consider collaborations between the robot members present deviations regarding their complexity, which is related to the size of the robotic team and the collaborative tasks. The coordinated movement of a robotic team comprises one of the most widely studied research fields in swarm robotics. For example, a feedback law using Lyapunov type analysis was derived in [39] for a single robot thus, collision avoidance and tracking of the shortest path are accomplished. The method extends this result to the case of multiple nonholonomic robots. A coordinated control scheme based on leader-follower approach was also proposed to achieve the required formation maneuvers during the robots transit following their shortest routes [6]. First and second order sliding mode controllers were used for asymptotically stabilizing the vehicles to a time varying desired formation considering the optimum pathway. In addition, an improved rapidly exploring random tree (RRT) method was proposed in [35]. The modified RRT considers the kinematics of each mobile robot to extract the corresponding pathways while a dynamic priority strategy was introduced to avoid mutual collisions and retain the formation of the team. Except these approaches, various artificial intelligence based methods were also introduced in mutltirobot systems. A unified framework of a co-evolutionary mechanism and an improved genetic algorithm (GA) were introduced to compute the multiple paths of the team [52]. The improved GA converges to the optimum collision free paths while the co-evolution mechanism takes into full account the cooperation between the populations to avoid collisions between mobile robots. Finally, multiple shortest paths can also be defined with the use of artificial bee colony algorithms [34].

In general, most of the above methods can produce accurate results for multiple robots while retaining a formation nonetheless; their implementation in real systems is restricted. Most of these algorithms can only operate in simulation environments due to their resource requirements. Here a CA-based path planner is presented for robot teams, which also involves collective behaviors between the robot members in order to define the shortest routes for retaining their formation [24–26]. As already briefly commented CAs comprise a simple, yet efficient, computational tool that can be implemented in real systems of low cost miniature robots. CAs were successfully exploited as potential solution of the shortest path problem in [2] found earlier in this Chapter and could be denoted as a cell decomposition approach and proper for a single robot application. Marchese has also introduced the use of Spatiotemporal Cellular Automata (SCA) to define the desired shortest path [37]. Three level of maps are introduced where the first two maps reduce the problem of extremely large cell numbers. Limiting the search space to smaller areas and considering the interaction between the robots, motion planning is performed using the SCA. A simpler approach was introduced in [5] where the A-star algorithm was combined with CAs and tested successfully in real world planar environments. More specifically, the finite properties of the A-star algorithm were amalgamated with the CA rules to build up a substantial search strategy [4]. The corresponding algorithm's main attribute is that it expands the map state space with respect to time using adaptive time intervals to predict the potential expansion of obstacles.

In the following, a CA-based algorithm is introduced for dynamically extracting the required collision free pathways for every member of a robot team. The presented planner considers also the collective behavior that the robots must display in order to retain their formation. A swarm robotic team must cover a specific space in the configuration area while simultaneously each member must be able to detect and bypass every dynamic obstacle. For cases where a scatter formation is produced due to the existence of an obstacle, the team must be able to recover its initial formation following optimum paths via collaborations. The CA-based algorithm can extract the optimum pathways of every robot towards its final destination point while shortest paths are also computed for recovering the desired formation. In contrast with similar CA-based architectures, the proposed method does not require any type of central control making the system fully autonomous. In addition, the method is applicable to real systems comprised by miniature robots with low resource specifications since the next transit of ever robot depends on only its current location and the states of its adjacent robots. This flexibility and the method's efficiency were tested using different types of formations.

6.1 Proposed Method

As a cell decomposition approach, the configuration area where the teams operates is initially divided into a simple rectangular lattice of identical square cells. Both dimensions of the required lattice are expressed in cells and thus, they depend on the applied cell length. The latter is strictly related to the specifications of the distance sensor that is utilized from a robot in a real system. For the presented model, the desired covered distance in terms of cell numbers comprises the variable that determines the lattice size. Following the CA description, variable D is defined based on these requirements. Let z be the cell length and $x \times y$ cells the dimensions of the CA.

Since the dimensions are defined, the set of states should also be defined, meaning variable Q. According to the CAs definition, every cell can be denoted with only one discrete state at every evolution step based on the delineated set of states $Q = \{0, 1, \ldots, q\}$. The proposed model includes the use of multiple robots and so, the number of the possible states is relative to the number of the robot cells. Taking this notion one-step further, this robot state is exploited as an identifier for the collective behavior of the team. Assuming that the team includes r robot cells, the final set of states is comprised by three discrete subsets: CF denoting the absence of both obstacle and robot cells (free cells), CR denoting a robot cell and CO the presence of an obstacle. More specifically, every subset of state can be defined as $CF = 0$, $CR = 1, 2, \ldots, r$ and $CO = r + 1$. Essentially, in order to avoid the overlapping of the cell states, the Eq. 2 must be valid:

$$C_F \cap C_R \cap C_O = \{\emptyset\} \ \& \ C_F \cup C_R \cup C_O = Q \tag{2}$$

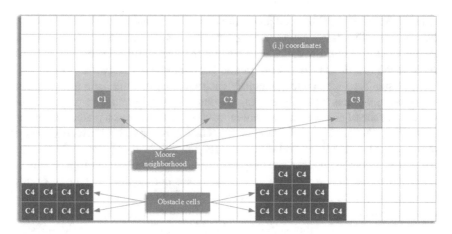

Fig. 13 Example setup of three robots

Due to the application, every robot cell must have a complete awareness of its surroundings in order to avoid properly the detected obstacles. Therefore, Moore neighborhood is exploited for every cell to be evolved accordingly (variable N neighborhood radius) with range equal to one. Figure 13 presents such a setup for a team of three robots.

The final variable of the quadruple F has to be defined, meaning the set of the transitions rules, in order to evolve the state of every cell. The applied local transition rules $F : Q^V \rightarrow Q$, considering the applied neighborhood, can be expressed as:

$$C_{x,y}^{t+1} = F(C_{x-1,y-1}^t, \ldots, C_{x,y}^t, \ldots, C_{x+1,y+1}^t) \tag{3}$$

or in a more compact, alternative formulation:

$$C_{(x,y)}^{t+1} = F\left[\sum_{i=-1}^{i=1} \sum_{j=-1}^{j=1} a_{ij} C_{(x+i,y+j)}^t\right] \tag{4}$$

where the a_{ij} are integer constants and thus the function F has a single integer as argument. Due to the discrete nature of the CAs, Cartesian coordinates are mostly utilized to characterize a specific cell. Nonetheless, the orientation of a robot must also be taken into consideration in order to achieve smoother transitions during the evolution. The corresponding parameter must follow the basic principles of a CA meaning the state of orientation should also be finite and integer. An additional parameter, θ^t, is inserted to the transitions rules at each evolution step displaying the following states $\theta^t = 0, 1, 2, 3, 4$ which are mapped to the values $\{-90, -45, 0, 45, 90\}$ expressed in degrees. Therefore, Eq. 4 is transformed as:

$$(C_{(x,y)}^{t+1}, \theta^{t+1}) = F\left[\sum_{i=-1}^{i=1}\sum_{j=-1}^{j=1} a_{ij}C_{(x+i,y+j)}^{t}, \theta^{t}\right] \tag{5}$$

The transition set of rules should consider both collision avoidance procedures and the collective behavior of the swarm robot cells. The avoidance of obstacles from a robot cell relies on the characterization of an adjacent cell in its Moore neighborhood as an obstacle cell. During its transit towards the final destination following the shortest path (straight line), the robot cell "checks" its contiguous area cell in order to define whether is comprised by free or obstacle cells. If a cell is occupied by an obstacle, the appropriate transition rules will be applied so that it could be bypassed. A small set of the corresponding transition rules are provided in Table 1. For the frontier cells, null boundary conditions are applied meaning that all virtual cells are always denoted as free cells.

In both scenarios (free space or present obstacles), the robot team must display collective behavior as one entity in order to retain or regain their formation. The proposed CA model involves all the appropriate procedures via the application of proper transition rules in order to define the required shortest paths for formation control. It is assumed that robot cells have the ability to exchange data regarding their position in the lattice and in the formation. A local relationship of master and slave is applied between the members of the team. Innermost robot cells are denoted as masters over their neighboring partners while the outermost as slaves. Master robot cells undertake to collect the required information from its slaves and the decide which transition rule should be applied. The latter could be either a command of moving towards one cell to the final destination point or a command of exchanging positions due to a scattered formation. Scattered formations are the result of a detected obstacle so the team members should collaborate to define the shortest paths aiming at recovering their initial structure. Depending on the required application, various transformation can be applied such as straight-line formation, triangular formation, etc. The only requirement for the formation control is to define the corresponding CA transition rules where the position in the lattice of every robot cell and additional checks based on their coordinates are required. Table 2 includes a set of such transition rules for straight-line formations where every cell should follow its optimum path (variable d_i denotes the desired path for i robot cell expressed as number of cells).

In case of a deviation, the corresponding robot cell will display a non-zero value between its d_i value and its current vertical coordinate. Thus, the team recognizes that the formation is scattered and the master robot cell apply the necessary transition rules to regain their formation following the extracted shortest paths by exchanging positions in the formation. The extracted optimum paths will eventually lead the team to converge to specific positions in the configuration area and at some point, their formation will be restored. Finally, the team will proceed as one unit towards the final destination by keeping their motion on the desired shortest route.

M.-A. I. Tsompanas et al.

Table 1 Example of transition rules for collision avoidance

Evolution time step t										Time step t + 1	
$C_{(x,y)}$	$C_{(x-1,y)}$	$C_{(x-1,y-1)}$	$C_{(x,y-1)}$	$C_{(x+1,y-1)}$	$C_{(x+1,y)}$	$C_{(x+1,y+1)}$	$C_{(x,y+1)}$	$C_{(x-1,y+1)}$	θ	$C_{(x,y)}$	θ
r	0	0	0	0	0	0	0	0	2	f	2
0	r	0	0	0	0	0	0	0	2	r	2
r	0	0	0	0	r+1	0	0	0	2	r	0
0	0	0	r	r+1	0	0	0	0	0	r	0
r	0	0	0	0	r+1	r+1	r+1	0	2	r	4
0	0	0	0	0	0	0	0	0	–	0	–
r+1	0	0	0	0	0	0	0	0	–	r+1	–

Table 2 Example of transition rules for formation control over a straight line

	Evolution time step t										Time step t + 1	
Case	$C_{(x,y)}$	$C_{(x-1,y)}$	$C_{(x-1,y-1)}$	$C_{(x,y-1)}$	$C_{(x+1,y-1)}$	$C_{(x+1,y)}$	$C_{(x+1,y+1)}$	$C_{(x,y+1)}$	$C_{(x-1,y+1)}$	θ	$C_{(x,y)}$	θ
$d_i - y_r = 0$	r	0	0	0	0	0	0	0	0	2	0	–
$d_i - y_r = 0$	0	r	0	0	0	0	0	0	0	2	r	2
$d_i - y_r > 0$	r	0	0	0	0	0	0	0	0	2	0	–
$d_i - y_r > 0$	0	0	0	0	r	0	0	0	0	2	r	2
$d_i - y_r < 0$	r	0	0	0	0	0	0	0	0	2	0	–
$d_i - y_r < 0$	0	0	0	0	0	0	r	0	0	2	r	2

6.2 Implementation in Real Swarm Robot Team

The main objective of the method is to display low computational and memory requirements so that it could be developed as a firmware and loaded on real robots. The simplicity of the developed CA renders the method suitable to achieve this task. To be fully functional, every utilized robot must be equipped with a proper hardware architecture including a microprocessor, distance sensors (e.g. IR), step motors and a communication interface (e.g. Bluetooth). No central control (e.g. base station) is required following the basic principles of the swarm robotics theory. Each robot of the team should be loaded with the method's implementation in order for the system to accomplish its goals. The team could also include different types of robots, forming a heterogeneous swarm, with the only restriction that all the robots exploit the same communication protocol. For testing purposes, without loss of generality, the method was tested on a three member's squad of miniature robots, called E-puck [41] (Fig. 14a). The e-puck robotic architecture comprises a fully open source platform providing full access to every of its modules. Equipped with all the aforementioned requirements for the method's implementation, it is a valid selection for swarm robotics applications.

The first stage of implementation is the determination of the lattice size that is pro-portional to the desired distance to be covered and the cell length. More specifically, cell length is strictly related to the proximity sensors' readings. In order to identify the presence of an obstacle, the sensors are enabled and based on the acquired data; an adjacent cell (which corresponds to an actual fragment of space) is denoted as free or as an obstacle cell. The IR sensors mounted on the e-puck (Fig. 14b) produce a scalable number, which represents the distance from an object. A higher value corresponds to a lower distance of an object and vice versa. Nonetheless, due to the

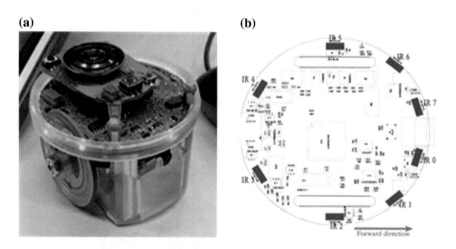

Fig. 14 **a** E-puck robot and **b** IR proximity sensors

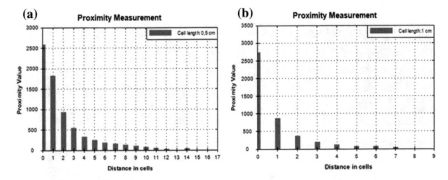

Fig. 15 Measures of the IR proximity sensors for different lengths: **a** 0.5 cm and, **b** 1 cm

nature of these sensors, their response is affected by the environment's ambient light leading to false positives (object detections) in large distances. On the other hand, in case of small cell lengths, the lattice is increased leading to higher memory resources and less accuracy in denoting a cell as obstacle due to the placement of the sensors on the robot. Therefore, multiple experiments were conducted in order to identify the proper cell size and ensure the required accuracy of the sensor's readings.

For this task, a special software was developed on a personal computer to help us acquire all the sensors' data and model their response. The software was connected with an e-puck robot via Bluetooth in order to acquire the required readings. At first, the smallest possible distance between the robot and an object was applied. At every time step, a backward motion was applied, covering a distance of one cell length and capturing the response of the sensor. When the maximum possible proximity (8 cm) was covered, the front sensors' (IR7 and IR0 of Fig. 14b) responses were transmitted back to the software for visualization and evaluation purposes. Multiple cell sizes were tested while Fig. 15 includes the sensors' responses for two cell lengths, 0.5 and 1 cm.

All actions executed by every robot can be summarized into two different subsets of execution. During the first stage, every robot acts as an individual and "scans" its adjacent environment in order to detect potential obstacles. The IR proximity sensors are enabled and according to their readings, the corresponding cells are denoted as free or obstacle cells. The presence of an obstacle is detected by comparing the acquired scalable values with the value that represents the cell length. For the tested environment, a cell length equal to 0.5 cm was used. In addition, two different formations were tested, namely a straight-line and a triangular formation. Snapshots of the entire procedure are provided in Fig. 16.

More specific, for the straight-line formation, all robots are deployed to their initial positions forming a straight line. As they cover the desired distance towards their final destination, all robots detect an obstacle that interfere their motion. Thus, obstacle avoidance transition rules are applied to bypass the obstacle. At the second stage, the central robot, which acts as a master, commands its right adjacent robot to exchange

(a)   **(b)**

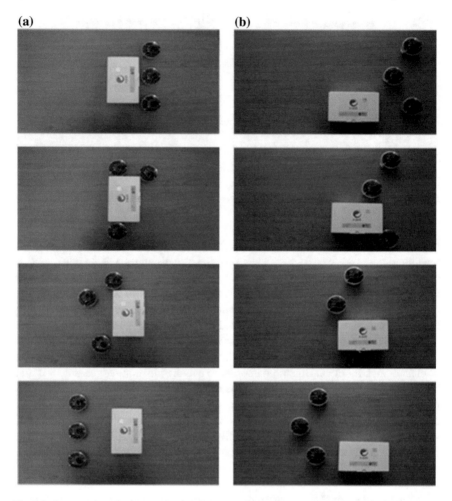

Fig. 16 Swarm robots from top to bottom in **a** straight line formation and, **b** triangular formation

their positions in order to recover the team's formation. Collective behaviors are executed and following the extracted shortest paths, the formation is retrieved with the minimum time cost. Finally, the team continues to their final positions covering the defined distance.

Similar process is followed for the triangular formation. All robots of the team are initially deployed forming a triangular formation. At some point, the left most robot detects an obstacle that must be avoided. The appropriate transition rules are applied to achieve this task. During that process, the central/master robot coordinates all the motions with the rest of the team members and decides that no position shifting is required. Until the robot avoids the box, the rest of the team freezes its motions and wait the discarded robot to regain its position to the formation. For this type of formation control, both vertical and horizontal coordinates in the lattice of every robot are exchanged. Since the formation is recovered, the team proceeds to its final destination as one entity, again via collaborations.

7   Physarum Polycephalum CA Model

Modern computers offer sufficient processing power to handle most of the analysis that several complex phenomena require. Physics, biology or chemistry can be characterized as complex phenomena. They are based on processes and systems using inhomogeneities, multiple interactions and complex constraints that even the modern computers cannot handle. CAs include all the necessary characteristics (handling of complex boundary and initial conditions, description of local interaction of a system with inhomogeneities and anisotropies that lead to global behavior, inherent parallelism) that makes them the appropriate tool to model and simulate natural phenomena.

A fungus, Physarum polycephalum, is such a system. Physarum Polycephalum is a large amoeba-like cell consisting of a dendritic network of tube-like structures (pseudopodia). It changes its shape as it crawls over a plain agar gel, and if nutrients is placed at two different points, it will extend pseudopodia that connect the two nutrient sources (FSs). Nakagaki et al. [44] showed that this simple organism has the ability to find the minimum-length solution between two points in a labyrinth. This resulted in an intensive period of research on this organism that exposed a great range of its computational abilities to spatial representations of various graph problems CAs are used extensively in this system because they have the ability to model the foraging behavior of plasmodium (physarum in its nutritious stage). Plasmodium spreads its pseudopodia and searches for chemo-attractants to lead it to nutrients that can devour and survive. It is very important for the survival of this life form to consume the least possible energy to find this chemo-attractants. This is the reason why the plasmodium creates tubes with minimum distance between food spots in a maze.

CAs is the most suitable paradigm to model such a structure [9, 10, 12, 28, 29, 63–67]. The maze can be modeled by creating a grid of cells with standard initial and boundary conditions. Plasmodium is not a unified mass but it is composed by many elementary parts that communicate and move. This local interaction leads in the movement of the whole plasmodium's mass. Each CA cell can model this elementary part of Physarum. The neighborhood of this cell will include walls of the maze, empty paths or other plasmodium's particles. This cell will interact with its environment, exchange stimuli and information and finally it will take a decision about the next direction of its movement. The evolution of this CA system leads to the final solution of the maze.

The maze used for the biological experiment (Fig. 17) was also used as an input for our algorithm. More specifically, Nakagaki et al. [44] took a growing tip of an appropriate size from a large plasmodium in a $25 \times 35\,\text{cm}^2$ culture trough and divided it into small pieces.

Fig. 17 The under study maze in correspondence to the one of [44]

Then, they positioned these in a maze created by cutting a plastic film and placing it on an agar surface. The plasmodial pieces spread and coalesced to form a single organism that filled the maze, avoiding the dry surface of the plastic film. At the start and end points of the maze, they placed agar blocks containing nutrient (ground oat flakes) and there were four possible routes between the start and the end points. The plasmodium pseudopodia reaching dead ends in the labyrinth shrank (Fig. 19a), resulting in the formation of a single thick pseudopodium spanning the minimum length between the nutrient-containing agar blocks (Fig. 19b). In our case, we artificially reconstructed the aforementioned maze taking into consideration the exact positions of the maze.

In order to simulate this biological experiment, the area is divided into a matrix of squares with identical areas and each square of the surface is represented by a CA cell. The type of neighborhood that was used in this CA model is the Moore neighborhood which means that we use the north, south, east, west, north-east, north-west, south-east and south-west neighbors. The state of the (i, j) cell at time t, defined as $C^t_{i,j}$ is equal to:

$$C^t_{i,j} = \{Topology_{i,j}, Chem^t_{i,j}, Dir^t_{i,j}, Phys^t_{i,j}, Pseudo^t_{i,j}\} \tag{6}$$

- $Topology_{i,j}$ is a variable which indicates the type of area of the corresponding (i, j) cell. The possible values of this variable are 0, 1, 2, 3 and indicate a free area, the spot of the initially placed FS, the spot of the initially placed plasmodium and the spot which represents a wall of the topology respectively.

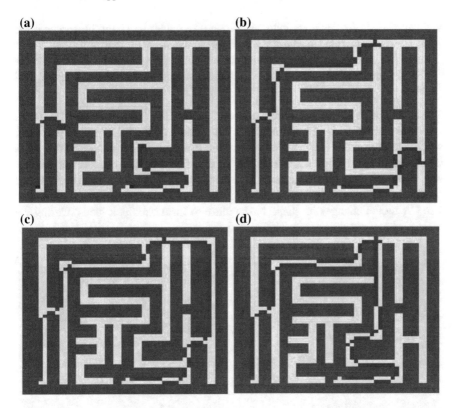

Fig. 18 The amoeba-like CA simulation results for the maze after **a** 500, **b** 1000, **c** 1200 and **d** 1500 time steps, respectively

- $Chem_{i,j}^t$ represents the concentration of chemo-attractants at time t in the area corresponding to the (i, j) cell. In order to calculate this variable for every cell, we make use of the concentration of the neighborhood to update the value of the central cell.
- $Dir_{i,j}^t$ is a variable that indicates the direction of the attraction of the plasmodium by the chemicals produced by the FS. For example, if the area around a corresponding cell has no chemo-attractants, the foraging strategy of the plasmodium is uniform and, thus, these parameters are equal to zero. If there is higher concentration of chemo-attractants in the cell at direction x from the one in direction y, then the parameter corresponding to direction x is positive and the parameter corresponding in the direction y is negative. This happens, in order to more accurately simulate the non-uniform foraging behavior of the plasmodium.
- $Phys_{i,j}^t$ indicates the volume of the cytoplasmic material of the plasmodium in the corresponding (i, j) cell. In order to calculate this variable for every cell, we make use of the neighbors' volumes.

- Finally, $Pseudo_{i,j}^t$ is a variable which can take values [0,1] and illustrates if the (i, j) cell is included in the final path of tubular network that is formed inside the plasmodium's body. This tubular network forms the shortest path between the FSs and the cell from where the plasmodium started to expand and it is our final solution.

The amoeba-like CA model simulation results after 500, 1000, 1200 and 1500 time steps are shown in Fig. 18. Compared to the results of the biological experiment, which are presented in Fig. 19, the algorithm can be considered successful. As is illustrated in Fig. 18, it takes 1200 time steps to find a solution that is not the best one. However, after 1500 time steps, it manages to solve the maze using the shortest possible route. It should be noted that in analogy to the real experiments, the amoeba-like CA model changes its shape in the maze to form one thick tube covering the

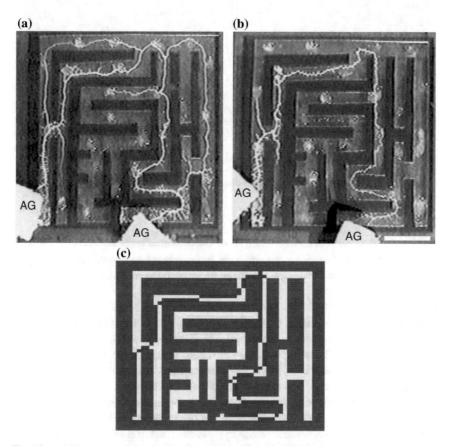

Fig. 19 a, b The maze solving by *P. polycephalum* after 4 and 8 h, respectively, as presented by Nakagaki et al. in [44], where yellow color, is the plasmodium and black are the maze 'walls'. **c** The final simulation results, after 1500 time steps, of the CA that mimics the *P. Polycephalum*'s behavior for the same maze. The situation in **b** is successfully reproduced

shortest distance between the FSs, so as to maximize its foraging efficiency, and therefore, its chances of survival.

The period of 1500 time steps (which correspond to about 45 s of real time on a PC) may seem like a long time, but compared to 8 h needed for the biological experiment, it is not a significantly long time period.

7.1   GPGPU Implementation

The term GPGPU (General-Purpose computing on Graphics Processing Units) refers to the use of the GPU processor as a parallel device for purposes other than graphic elaboration. More specific, GPU is often used in order to solve some complex computational problems that classical CPU cannot handle. This device has the ability to execute a great number of independent threads in parallel. So if a complex problem has an inherent parallel nature, an implementation in GPU is going to multiply the performance of its algorithm and the solution will be produced much faster. The GPU's architecture has a computational power that can exceed a teraFLOP and it is fully suitable for fine grain parallelism. The reason of the great success and enormous spread of the GPGPU application in the past few years, is CUDA programming model. The basic structure of CUDA is that it provides three key abstractions, namely the hierarchy with which the threads are organized, the memory organization and the functions that execute in parallel, called kernels.

In a CUDA application, some parts are performed in a parallel way and some other parts are performed in the classical serial way. The *device*, which is the name of the GPU in CUDA context, can be thought as an additional co-processor of the main CPU which is called *host* in the CUDA context. In order to take off the performance of our algorithm, we have to exploit the parts of the data that are made to work in parallel and execute them on the device as many time steps is necessary. In order to achieve this, we have to call one, two or more kernels which use thousands of threads.

One key problem towards the implementation in a GPU is the way the memory is organized and used. The threads can be organized and cooperate together by sharing a common fast *shared-memory* synchronizing in some points of the kernel within a so called *thread-block*. But the number of threads that a block can use is limited and at the most applications more parallelism is needed. CUDA gives the choice to launch kernels with a larger total number of threads by organizing block of threads together, by means of a grid of blocks. So it is possible to choose a cell of a CA, which can be thought as a particular data of the device memory, and associate it to a current thread of a kernel.

Threads can access different memory locations during execution. Generally, there are three types of memory used in CUDA applications, namely (*a*) the private memory, which is the memory its thread has for its own, (*b*) the shared memory, which is the memory being visible to all threads in a block and (c) the global memory, which is a larger memory on the device board but it is outside the computing chip. In this study, we make use of the global memory of the device. This memory is slower if compared with the shared memory but it can deliver a significantly higher memory bandwidth than the traditional CPU memory. It is measured that is about 20 times more efficient to access the global memory of the GPU than the CPU memory. As a result, when a CUDA application is designed, the minimum data transfers between CPU and GPU should take place.

The reason why GPGPU programming is used for CAs models can be explained easily when referring to the CAs' parallel nature. The local interaction of the neighbors that CAs methods propose is another fact that makes these implementations very suitable and very fast. These features make the CAs models ideal to be implemented in parallel computers. The basic idea when computing a CA model in GPU, which is also used in our implementation, can be described as follows: First, we compute the next state of all the cells in parallel. Afterward we use two memory regions to store the data. More specifically, we use one region for the $CA_{current}$, which indicates the CA states before the calculations and one for the CA_{next}, which in turns indicates the CA states after the calculations. Finally, the switching between the $CA_{current}$ and the CA_{next} in each time step takes place.

For this paper we store the CA data to the global memory of the device. The steps of the algorithm are: (a) Split the CA states and make use of a kernel for every one of them. In more detail, we make use of a kernel to hold the $Topology_{i,j}$, one kernel for the computation of the diffusion equation of the chemo-attractants, $Chem_{i,j}$, and their direction, $Dir_{i,j}$, one kernel for the computation of the diffusion equation of the mass of plasmodium, $Phys_{i,j}$ and finally one kernel that computes the $Pseudo_{i,j}$ to find the shortest path in the maze. (b) An initialization of the current state for all these kernels happens through a CPU-GPU memory copy operation (i.e. from host to the device global memory). (c) Every kernel runs in each time step and makes its calculations by using the information of the states of the other necessary kernels. For example, in order to calculate the $Pseudo_{i,j}$ we have to know which of the neighbors has the greater mass value. Therefore, we take this information from the kernel that executes the computation of the $Phys_{i,j}$. (d) At the end of each CA step, a device to device memory copy operation is used to update the new values in order to continue the calculations in the next CA step. (e) When the simulation is completed the final state of the automaton is being retrieved from the global memory of the device to the host through a GPU-CPU memory copy operation.

For the proposed GPU implementation of the presented slime mould CA based model we used the graphics card NVIDIA GT640. We used a 50×50 CA cells in order to synthesize the maze. The time needed for the presented solution to result is 2.47 s. In the serial code the time needed for the serial software implementation in MATLAB was approximately 45 s. Therefore, the increase in the performance in our implementation is about 18.2 times more than the one in MATLAB.

7.2 Hardware Implementation

Current FPGAs include logic density equivalent to millions of gates per chip and can implement very complex computations. CAs consist of a uniform n-dimensional structure, composed of many identical synchronous cells where both memory and computation are involved, thus matching the inherent design layout of FPGA Hardware. As a result memory and processing unit are closely related both in CAs cells and FPGA configurable logic blocks (CLBs). The structure of a cell consists of a combinational part connected with one or more memory elements in a feedback loop shape while the state of the memory elements is also defined by the inputs and the present state of these elements. For this implementation the design produced by using VHDL code has been analyzed and synthesized by Quartus II (32-bit version 12.1 build) FPGA design software of ALTERA Corporation.

Each CA cell is implemented by a hardware block called "PhysarumCell". Each "PhysarumCell" block is connected appropriately with its four neighbors (west, east, south and north). It uses the inputs from the neighbors and the previous state of itself to produce results that simulate the movement of the plasmodium. An "PhysarumCell" block has 22 inputs and 7 outputs.

After creating the lattice, the user has to provide only the topology of the experimental area by giving values to the 2-bit signals *topology* for each individual cell, namely the location of the FS and the location of the initial introduction of the plasmodium to the experimental area and the parameters for the diffusion equations.

The number of logic elements, registers and pins of the CA cell are presented in Table 3. Moreover, to illustrate the area needed for a fully interconnected system of a CA grid implementing the proposed bio-inspired model, the results of synthesizing a 10×10, a 15×15 and a 20×20 grid are illustrated in Table 4. The circuits are synthesized on several target devices and the results on the Stratix V 5SGXBB are presented here. The process ends in a few µs. In Table 4, it is shown that for almost every 150 CA cells there is an increment of 300,000 logic elements on average.

Table 3 FPGA hardware implementation details for one CA cell

Quartus II 32-bit Version	12.1 Build 243 01/31/2013 SP 1
Total logic elements	1,739
Total registers	45
Total pins	226

Table 4 FPGA hardware implementation for different topology sizes

	10×10	15×15	20×20
Total logic elements	161,162	370,447	666,060
Total registers	8,360	18,840	33,520
Total pins	317	692	1217

8   Conclusions

In this chapter the inseparable relationship between CAs and shortest path problem is depicted. CAs are a very powerful modeling tool that can capture the essential characteristics of this problem and produce effective results. They can manage the classical S^3P, S^3DSP and $APSP$ problems as presented in Sect. 2. They can learn and find solutions in a stochastic graph as presented in Sect. 4. They can also move to three-dimensional space and provide solutions to difficult territories as already shown in corresponding subsections. But their use is not only theoretical. They can be applied successfully for computer networks main problems as shown in Sect. 5. They are a very useful tool in robotics and their movement inside a maze section as demonstrated in Sect. 6. They can also describe and model very effectively and efficiently physical phenomena and living structures that have the ability to provide unconventional solution to shortest path problems as presented in Sect. 7. What is the reason that the complexity bounds are so good? It takes place because we used a very restricted form of a rectangular lattice to make the structure of a problem most resembling the architecture of computing device which solves this problem. Can CAs be applied in practice? Of course. The CAs algorithms are derived almost directly for biology and nature. For this reason, the implementation of these algorithms in massively parallel processors or neurocomputers is an event that already happens.

References

1. A.I. Adamatzky, *Identification of Cellular Automata* (Taylor & Francis, 1994)
2. A.I. Adamatzky, Computation of shortest path in cellular automata. Math. Comput. Modell. **23**(4), 105–113 (1996)
3. H. Beigy, M.R. Meybodi, Utilizing distributed learning automata to solve stochastic shortest path problems. Int. J. Uncertainty Fuzziness Knowl.-Based Syst. **14**(05), 591–615 (2006)
4. K. Charalampous, A. Amanatiadis, A. Gasteratos, Efficient robot path planning in the presence of dynamically expanding obstacles. Cell. Autom. 330–339 (2012)
5. K. Charalampous, I. Kostavelis, A. Amanatiadis, A. Gasteratos, Real-time robot path planning for dynamic obstacle avoidance. J. Cell. Automata **9** (2014)
6. M. Defoort, T. Floquet, A. Kokosy, W. Perruquetti, Sliding-mode formation control for cooperative autonomous mobile robots. IEEE Trans. Ind. Electron. **55**(11), 3944–3953 (2008)
7. E.W. Dijkstra, A note on two problems in connexion with graphs. Numer. Math. **1**, 269–271 (1959)
8. N.I. Dourvas, G.Ch. Sirakoulis, A.I. Adamatzky. Cellular automaton Belousov-Zhabotinsky model for binary full adder. Int. J. Bifurcat. Chaos **27**(06), 1750089 (2017)
9. N.I. Dourvas, M.-A.I. Tsompanas, G.Ch. Sirakoulis, *Parallel Acceleration of Slime Mould Discrete Models* (Springer International Publishing, Cham, 2016), pp. 595–617
10. N. Dourvas, M.-A.I. Tsompanas, G.Ch. Sirakoulis, P. Tsalides, Hardware acceleration of cellular automata physarum polycephalum model. Parallel Process. Lett. **25**(01), 1540006 (2015)
11. S. El Yacoubi, J. Was, S. Bandini (eds.), *Cellular Automata—12th International Conference on Cellular Automata for Research and Industry, ACRI 2016, Fez, Morocco, 5–8 Sept 2016. Proceedings*, volume 9863 of Lecture Notes in Computer Science (Springer, 2016)
12. V. Evangelidis, M.-A.I. Tsompanas, G.Ch. Sirakoulis, A.I. Adamatzky, Slime mould imitates development of roman roads in the Balkans. J. Archaeol. Sci.: Rep. **2**, 264–281 (2015)

13. D. Ferguson, A. Stentz, Using interpolation to improve path planning: the field d* algorithm. J. Field Robot. **23**(2), 79–101 (2006)
14. G. Fishman, A comparison of four Monte Carlo methods for estimating the probability of s–t connectedness. IEEE Trans. Rel. **35**(2), 145–155 (1986)
15. R.W. Floyd, Algorithm 97: shortest path. Commun. ACM **5**(6), 345 (1962)
16. I.G. Georgoudas, G. Koltsidas, G.Ch. Sirakoulis, I.Th. Andreadis, *A Cellular Automaton Model for Crowd Evacuation and Its Auto-Defined Obstacle Avoidance Attribute* (Springer, Berlin, Heidelberg, 2010), pp. 455–464
17. T. Giitsidis, G.Ch. Sirakoulis, Modeling passengers boarding in aircraft using cellular automata. IEEE/CAA J. Autom. Sinica **3**(4), 365–384 (2016)
18. C. Hochberger, R. Hoffmann, CDL—a language for cellular processing, in *Proceedings of the Second International Conference on Massively Parallel Computing Systems*, ed. by G.R. Sechi (1996), pp. 41–64
19. C. Hochberger, R. Hoffmann, Solving routing problems with cellular automata, in *Proceedings of the Second Conference on Cellular Automata for Research and Industry (ACRI '96)* (1996), pp. 89–98
20. C. Hochberger, R. Hoffmann, S. Waldschmidt, Compilation of CDL for different target architectures, in *Parallel Computing Technologies*, ed. by V. Malyshkin (1995), pp. 169–179
21. R. Hoffmann, K.-P. Völkmann, M. Sobolewski, The cellular processing machine CEPRA-8L. Math. Res. **81**, 179–199 (1994)
22. H. Hussain, Integration eines Compilers fur die Zellularsprache CDL in das XCellsim–System (Techn. Univ. Darmstadt, Comp. Science Dept., 1994)
23. T. Hwu, J. Isbell, N. Oros, J. Krichmar, A self-driving robot using deep convolutional neural networks on neuromorphic hardware, in *2017 International Joint Conference on Neural Networks (IJCNN)* (IEEE, 2017), pp. 635–641
24. K. Ioannidis, G.Ch. Sirakoulis, I. Andreadis, A path planning method based on cellular automata for cooperative robots. Appl. Artif. Intell. **25**(8), 721–745 (2011)
25. K. Ioannidis, G.Ch. Sirakoulis, I. Andreadis, Cellular ants: a method to create collision free trajectories for a cooperative robot team. Robot. Auton. Syst. **59**(2), 113–237 (2011)
26. K. Ioannidis, G.Ch. Sirakoulis, I. Andreadis, Cellular automata-based architecture for cooperative miniature robots. J. Cell. Autom. **8**(1–2), 91–111 (2013)
27. D.B. Johnson, A note on Dijkstra's shortest path algorithm. J. ACM **20**(3), 385–388 (1973)
28. V.S. Kalogeiton, D.P. Papadopoulos, I.P. Georgilas, G.Ch. Sirakoulis, A.I. Adamatzky, *Biomimicry of Crowd Evacuation with a Slime Mould Cellular Automaton Model* (Springer International Publishing, Cham, 2015), pp. 123–151
29. V.S. Kalogeiton, D.P. Papadopoulos, I.P. Georgilas, G.Ch. Sirakoulis, A.I. Adamatzky, Cellular automaton model of crowd evacuation inspired by slime mould. International Journal of General Systems **44**(3), 354–391 (2015)
30. M.G. Kechaidou, G.Ch. Sirakoulis. Game of life variations for image scrambling. J. Comput. Sci. **21**(Supplement C), 432–447 (2017)
31. K. Konstantinidis, A. Amanatiadis, S.A. Chatzichristofis, R. Sandaltzopoulos, G.Ch. Sirakoulis, Identification and retrieval of DNA genomes using binary image representations produced by cellular automata, in *2014 IEEE International Conference on Imaging Systems and Techniques (IST) Proceedings*, Oct 2014, pp. 134–137
32. C.Y. Lee, An algorithm for path connections and its applications. IRE Trans. Electron. Comput. **EC-10**(2), 346–365 (1961)
33. J. Li, B.H. Wang, P.Q. Jiang, T. Zhou, W.X. Wang, Growing complex network model with acceleratingly increasing number of nodes. Acta Physica Sinica **55**(8), 4051–4057 (2006)
34. J.-H. Liang, C.-H. Lee, Efficient collision-free path-planning of multiple mobile robots system using efficient artificial bee colony algorithm. Adv. Eng. Softw. **79**, 47–56 (2015)
35. S. Liu, D. Sun, C. Zhu, A dynamic priority based path planning for cooperation of multiple mobile robots in formation forming. Robot. Comput.-Integr. Manuf. **30**(6), 589–596 (2014)
36. A. Macwan, J. Vilela, G. Nejat, B. Benhabib, A multirobot path-planning strategy for autonomous wilderness search and rescue. IEEE Trans. Cybern. **45**(9), 1784–1797 (2015)

37. F.M. Marchese, Multi-resolution hierarchical motion planner for multi-robot systems on spatiotemporal cellular automata, in *Robots and Lattice Automata* (Springer, 2015), pp. 149–173
38. V.A. Mardiris, G.Ch. Sirakoulis, I.G. Karafyllidis, Automated design architecture for 1-D cellular automata using quantum cellular automata. IEEE Trans. Comput. **64**(9), 2476–2489 (2015)
39. S. Mastellone, D.M. Stipanovic, M.W. Spong, Remote formation control and collision avoidance for multi-agent nonholonomic systems, in *2007 IEEE International Conference on Robotics and Automation* (IEEE, 2007), pp. 1062–1067
40. S.K. Moghaddam, E. Masehian, Planning robot navigation among movable obstacles (NAMO) through a recursive approach. J. Intell. Robot. Syst. **83**(3–4), 603–634 (2016)
41. F. Mondada, M. Bonani, X. Raemy, J. Pugh, C. Cianci, A. Klaptocz, S. Magnenat, J.-C. Zufferey, D. Floreano, A. Martinoli, The e-puck, a robot designed for education in engineering, in *Proceedings of the 9th Conference on Autonomous Robot Systems and Competitions*, vol. 1 (IPCB: Instituto Politécnico de Castelo Branco, 2009), pp. 59–65
42. O. Montiel, U. Orozco-Rosas, R. Sepúlveda, Path planning for mobile robots using bacterial potential field for avoiding static and dynamic obstacles. Expert Syst. Appl. **42**(12), 5177–5191 (2015)
43. K. Nagel, M. Schreckenberg, A cellular automaton model for freeway traffic. Journal de Physique I **2**(12), 2221–2229 (1992)
44. T. Nakagaki, H. Yamada, Á. Tóth, Intelligence: Maze-solving by an amoeboid organism. Nature **407**(6803), 470 (2000)
45. L. Nalpantidis, G.Ch. Sirakoulis, A. Gasteratos, Non-probabilistic cellular automata-enhanced stereo vision simultaneous localization and mapping. Meas. Sci. Technol. **22**(11), 114027 (2011)
46. T.P. Nascimento, A.G.S. Conceiçao, A.P. Moreira, Multi-robot nonlinear model predictive formation control: the obstacle avoidance problem. Robotica **34**(3), 549–567 (2016)
47. A. Nash, S. Koenig, Any-angle path planning. AI Mag. **34**(4), 85–107 (2013)
48. C. Nieto-Granda, J.G. Rogers III, H.I. Christensen, Coordination strategies for multi-robot exploration and mapping. Int. J. Robot. Res. **33**(4), 519–533 (2014)
49. V.G. Ntinas, B.E. Moutafis, G.A. Trunfio, G.Ch. Sirakoulis, Parallel fuzzy cellular automata for data-driven simulation of wildfire spreading. J. Comput. Sci. **21**(Supplement C), 469–485 (2017)
50. A. Pandey, R.K. Sonkar, K.K. Pandey, D.R. Parhi, Path planning navigation of mobile robot with obstacles avoidance using fuzzy logic controller, in *2014 IEEE 8th International Conference on Intelligent Systems and Control (ISCO)* (IEEE, 2014), pp. 39–41
51. M.A. Porta Garcia, O. Montiel, O. Castillo, R. Sepúlveda, P. Melin, Path planning for autonomous mobile robot navigation with ant colony optimization and fuzzy cost function evaluation. Appl. Soft Comput. **9**(3), 1102–1110 (2009)
52. H. Qu, K. Xing, T. Alexander, An improved genetic algorithm with co-evolutionary strategy for global path planning of multiple mobile robots. Neurocomputing **120**, 509–517 (2013)
53. G.Ch. Sirakoulis, A.I. Adamatzky, *Robots and Lattice Automata* (Springer Publishing Company, Incorporated, 2014)
54. G.Ch. Sirakoulis, S. Bandini (eds.), *Cellular Automata—10th International Conference on Cellular Automata for Research and Industry, ACRI 2012*, Santorini Island, Greece, 24–27 Sept 2012. Proceedings, volume 7495 of Lecture Notes in Computer Science (Springer, 2012)
55. G.Ch. Sirakoulis, I. Karafyllidis, V. Mardiris, A. Thanailakis, Study of lithography profiles developed on non-planar SI surfaces. Nanotechnology **10**(4), 421 (1999)
56. G.Ch. Sirakoulis, I. Karafyllidis, D. Soudris, N. Georgoulas, A. Thanailakis, A new simulator for the oxidation process in integrated circuit fabrication based on cellular automata. Modell. Simul. Mater. Sci. Eng. **7**(4), 631 (1999)
57. A. Stentz et al., The focussed d* algorithm for real-time replanning. IJCAI **95**, 1652–1659 (1995)
58. Q. Sun, Z.J. Dai, A new shortest path algorithm using cellular automata model. Comput. Technol. Dev. **19**(2), 42–44 (2009)

59. U.A. Syed, F. Kunwar, M. Iqbal, Guided autowave pulse coupled neural network (GAPCNN) based real time path planning and an obstacle avoidance scheme for mobile robots. Robot. Auton. Syst. **62**(4), 474–486 (2014)

60. A. Tsiftsis, G.Ch. Sirakoulis, J. Lygouras, FPGA Processor with GPS for modelling railway traffic flow. J. Cell. Autom. **12**(5), 381–400 (2015)

61. A. Tsiftsis, I.G. Georgoudas, G.Ch. Sirakoulis, Real data evaluation of a crowd supervising system for stadium evacuation and its hardware implementation. IEEE Syst. J. **10**(2), 649–660 (2016)

62. M.-A.I. Tsompanas, A.I. Adamatzky, G.Ch. Sirakoulis, J. Greenman, I. Ieropoulos, Towards implementation of cellular automata in microbial fuel cells. PLoS ONE **12**, 1–16 (2017)

63. M.-A.I. Tsompanas, G.Ch. Sirakoulis, Modeling and hardware implementation of an amoeba-like cellular automaton. Bioinspir. Biomimetics **7**(3), 036013 (2012)

64. M.-A.I. Tsompanas, G.Ch. Sirakoulis, A.I. Adamatzky, *Cellular Automata Models Simulating Slime Mould Computing* (Springer International Publishing, Cham, 2016), pp. 563–594

65. M.-A.I. Tsompanas, G.Ch. Sirakoulis, A.I. Adamatzky, Evolving transport networks with cellular automata models inspired by slime mould. IEEE Trans. Cybern. **45**(9), 1887–1899 (2015)

66. M.-A.I. Tsompanas, G.Ch. Sirakoulis, A.I. Adamatzky, Physarum in silicon: the Greek motorways study. Nat. Comput. **15**(2), 279–295 (2016)

67. M.-A.I. Tsompanas, R. Mayne, G.Ch. Sirakoulis, A.I. Adamatzky, A cellular automata bioinspired algorithm designing data trees in wireless sensor networks. Int. J. Distrib. Sensor Netw. **11**(6), 471045 (2015)

68. P.G. Tzionas, A. Thanailakis, P.G. Tsalides, Collision-free path planning for a diamond-shaped robot using two-dimensional cellular automata. IEEE Trans. Robot. Autom. **13**(2), 237–250 (1997)

69. J. Von Neumann, *Theory of Self-Reproducing Automata* (University of Illinois Press, Champaign, IL, USA, 1966)

70. Y. Wang, Study for solving the path on the three-dimensional surface based on cellular automata method. Modern Appl. Sci. **4**(5), 196–200 (2010)

71. X.G.M. Wang, Y. Qian, Improved calculation method of shortest path with cellular automata model. Kybernetes **41**(3–4), 508–517 (2012)

72. S. Warshall, A theorem on boolean matrices. J. ACM **9**(1), 11–12 (1962)

73. J. Was, G.Ch. Sirakoulis, S. Bandini (eds.), *Cellular Automata—11th International Conference on Cellular Automata for Research and Industry, ACRI 2014*, Krakow, Poland, 22–25 Sept 2014. Proceedings, volume 8751 of Lecture Notes in Computer Science (Springer, 2014)

74. X.J. Wu, H.F. Xue, Shortest path algorithm based on cellular automata extend model. Comput. Appl. **24**(5), 92–3 (2004)

75. X. Zhang, Y. Zhang, Z. Zhang, S. Mahadevan, A. Adamatzky, Y. Deng, Rapid physarum algorithm for shortest path problem. Appl. Soft Comput. **23**, 19–26 (2014)

Checkerboard Pattern Formed
by Cellular Automata Agents

Rolf Hoffmann

Abstract Considered is a multi-agent system with agents, modeled by Cellular Automata. The agents have the task to form a checkerboard (CB) pattern on an $n \times n$ square field. The objective is to find the behavior (an algorithm) for the agents (realized by finite state automata, or finite state machines FSM), which can solve this task in shortest time, i.e. moving on a shortest path (space filling curve) for a single agent system. The target pattern class can be described by local matching patterns (3×3 templates). The degree of order is the number of template hits. Our goal is to find perfect CB patterns with a maximum degree of order. Firstly, a single-agent algorithm G1 with four states is designed, where the agent starts in a corner. The agent walks on a shortest path, but needs some additional steps to turn to a free direction when sensing a border. Secondly, FSM algorithms for multi-agent systems are evolved (found) by a Genetic Algorithm for an 8×8 field. Now the agents may start at any random position. Optionally an agent can emit a signal which can be sensed by another nearby agent. This signal was introduced to speedup the task. The evolved single-agent FSM algorithm uses another strategy than G1, a spiral-like trajectory. The fully packed system with 64 agents is the fastest, but it is also the most costly one, in terms of the product (time-steps $\times$ number of agents).

1 Introduction

1.1 The Problem

Given is a square field of $N = n \times n$ cells with border. k agents are moving around in the field. The agents' task is to construct a global configuration where a certain spatial pattern appears that belongs to a predefined pattern class. Each cell, except the border cells, contains a certain color $C \in \{-1, 0, 1\} \equiv \{bordercolor, white, black\}$. An

R. Hoffmann (✉)
Computer Science Department, Technical University Darmstadt,
Hochschulstr. 10, 64289, Darmstadt, Germany
e-mail: hoffmann@rbg.informatik.tu-darmstadt.de

© Springer International Publishing AG, part of Springer Nature 2018
A. Adamatzky (ed.), *Shortest Path Solvers. From Software to Wetware*,
Emergence, Complexity and Computation 32,
https://doi.org/10.1007/978-3-319-77510-4_9

agent can change the color of the site it is situated on, and the border color remains fixed. Initially the whole field is colored white, and the agents start at a special positions or are randomly distributed.

The objective is to find "intelligent" agents that are able to form a checkerboard pattern (CB pattern) in shortest time. All nodes of the grid have to be visited at least once (in order to color the cells) with the lowest possible frequency. If one agent is used only, the problem is similar to the problem of finding a shortest path. Space filling curves, like the one proposed by Peano [1], are possible solutions.

The agents' behavior shall be controlled by an embedded finite state machine (FSM). The capabilities of the agents shall be constrained, i.e. the number of control states, the action set and the amount of perceived information.

There are many applications for building patterns by agents, e.g. the forming of mechanical, chemical, biological or artificial structures, or the building of computational devices and communication networks. For example, nano-structures, functional polymers, and spin alignments, can be formed by nano-robots, or by beaming focused energy onto certain cells in order to change their physical state.

1.2   Using Agents

Problem solving with robots and agents has become more and more attractive [2–8]. Agents can be seen as moving processors in a spatial distributed computer system. Agents are "intelligent" and their capabilities can be tailored to the problem in order to solve it effectively, and often in an unconventional way. Important properties that can be achieved by agents are:

- **Generality and Flexibility**. The ability to solve the problem under a large variety of initial configurations or boundary conditions, e.g. by changing the size or the shape of the environment.
- **Scalability**. The ability to solve the problem with any number of agents. It is expected that the problem can be solved faster with more agents.
- **Tunability**. The ability to solve the problem faster or with a higher quality by increasing the agents' intelligence.
- **Fault-tolerance**. The ability to solve the problem with some degradation when parts of the system fail.
- **Adaptability**. The ability to cope with unexpected changes in the environment.

Owing to their intelligence, agents can be employed to design, model, analyze, simulate, and solve problems in the areas of complex systems, real and artificial worlds, games, distributed algorithms and mathematical questions.

Robots or agents controlled by a finite state machine (FSM) have a long history in computer science [9], sometimes they are simply called "FSMs", often with the property that they can move around on a graph or grid. For example, searching through the whole environment was addressed in [10, 11], routing in [12–15], all-to-all communication in [16], graph exploration in [17], and pattern formation in [18, 19].

1.3 Cellular Automata Agents

What is a Cellular Automata Agent (CA Agent)? Simply speaking, a CA agent is an agent that can be modeled within the CA paradigm. A simple example is the elementary CA rule 184 (Traffic Rule) in which the moving particles can be interpreted as agents. And what are the most important attributes of agents in our context?

1. **Self-contained** (an individual, complete in itself). In CA, this property can be realized by one cell, by a part of a cell, or by a group of cells.
2. **Autonomous** (not controlled by others). Agents operate on their own and control their actions and internal states. In CA, this property can be realized by the cell's state and the transition rule.
3. **Perceptive** (perceives information about the environment). In CA, this property is realized by reading and interpreting the states of the neighborhood.
4. **Reactive** (can react on the perceived environment). In CA, this property is realized by changing the cell's state taking into account the perceived information.
5. **Communicative** (can communicate with other agents). This property means that agents can exchange information, either indirectly through the environment (modeled as parts of the CA cells), or directly by observing other agents, or by listening to signals they are emitting.
6. **Proactive** (acts on its own initiative, not only reacting, using a plan). In CA, the cell's next state should not only depend on its neighbors' states but also on its own state. The number of states and inputs should not be too small in order to give the agent a certain intelligence to initiate changes and to deal in advance with difficult situations. And the agent's behavior is to a certain extent not foreseeable, it can be influenced by secret personal information or internal events. In CA, this can be accomplished by hidden states that cannot be observed by the neighbors, or by asynchronous internal triggers (e.g. by a random generator). As it is difficult to define proactivity in a strict way, it is a matter of viewpoint whether simple classical CA rules (like Game of Life, Traffic Rule) shall be classified as multi-agent systems or not.
7. **Local** (acts locally). Agents are small compared to the system size and can only act on their neighborhood. In CA, a rule is allowed to change the cell's own state only. Global effects arise from accumulated local actions.
8. **Mobile** (this feature is not required but often useful). Very often agents are able to move around in the environment. Then the points of activity and the rule applications are moving, too. When moving around, agents are often changing the environment.

Usually an agent performs *actions*. *Internal* actions change the agent's state, either visible or non-visible, whereas *external* actions change the state of the environment. The *environment* is composed of the ground environment (the playground), constant objects (like stones, boxes, obstacles), variable objects (like colors, markers, numbers), and other agents. Objects cannot change their state or location by themselves, only through the actions of agents.

In CA, an agent is not allowed to change the state of a neighboring cell. Therefore, if an agent wants to apply an external action to a neighboring cell, it can only issue a command that must be adequately executed by the neighbor. For example, if agent A sends a "kill" command to agent B, then agent B has to kill itself. This example shows that the CA modeling and description of changing the environment is indirect and does not appear natural. Other models like the "CA-w model" are helpful to simplify such descriptions.

1.4 CA and CA-w Models

In order to describe moving agents, moving particles or dynamic changing activities, the CA-w model (Cellular Automata with *write* access) was introduced [20]. This model allows to write information to a neighbor. This method has the advantage that a neighbor can directly be activated or deactivated, or data can be sent actively to it by an agent. Thus the movement of particles and agents can be described more easily.

The CA-w model is a restricted case of the more general, "*Global*" GCA-w [21–23]. In GCA-w any cell of the whole array can be modified whereas in the CA-w model only the local neighbors can be. Usually the cells of these models are a composition of (data, pointers). The neighbors are accessed via pointers, that can be changed dynamically like the data by an appropriate rule from generation to generation.

In order to avoid confusion between CA and CA-w, in this context the CA model can be attributed as "classical model" and the CA-w model as "implementation model" although both can be used for description and implementation.

The CA-w model can be mapped onto the CA model by increasing the neighborhood radius. If the radius of the CA-w model is R_1 for read and write access, then an equivalent behavior can be described in the CA model by using the radius $2R_1$ read access only.

A drawback is the possible occurrence of write conflicts. There are two solutions to handle conflicts:

- Use a conflict-resolving function, for example by applying a reduction operator (max, +, ...) or using a random or deterministic priority scheme.
- Avoid conflicts by algorithmic design, meaning that the parallel application of all rules will never cause a conflict.

The second solution is more elegant, leads to a more simple implementation, and many applications with agents can be described in this way.

1.4.1 Modeling the Agents' Mobility

How can an agent move from A to B? In the CA model, a couple of two consistent rules (*copy*-rule, *delete*-rule) must be performed (Fig. 1a): the first rule copies the

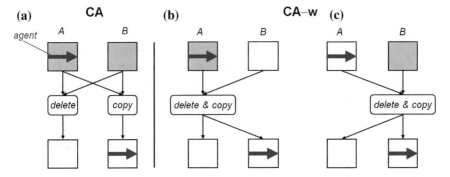

Fig. 1 CA model: **a** cell A deletes the agent and cell B copies it. CA-w model: **b** cell A deletes and copies the agent or **c** cell B deletes and copies the agent. Active cells executing a rule are shaded

agent from A to B, the second deletes it on A. Both rules have to compute the same moving condition, this means a redundant computation. There are two modeling options with CA-w to avoid this redundancy:

- Cell A (the agent) is responsible for the moving operation (Fig. 1b), it computes the moving condition and, if true, applies a rule that deletes itself on A and copies it to B.
- Cell B (the empty front cell) is responsible (Fig. 1c), it computes the moving condition and, if true, applies a rule that deletes the agent on A and copies it to B.

The second option was used for the software simulation of our problem. In this way, concurrent agents intending to move to the same empty site in front can easily be prioritized, and the moving condition needs only to be computed by the empty target cell.

1.5 Target Patterns and Degree of Order

How can a class of target patterns be defined? The idea is to use a set of small $m \times m$ matching patterns (or *templates*) and test them on each site (x, y) of the cell field. If a template fits on a site, then a hit (at most one) is stored at this site. Then the sum of all hits is computed which defines the *degree of order h*. This testing operation can be seen as a classical CA rule application.

The size of a template is not fixed, but it should be smaller than the whole field, and larger than one cell. A reasonable size is 3×3. The larger the templates are, the more sophisticated patterns can be generated.

Now we will define a very simple target pattern, the checkerboard pattern (*CB pattern*), which we are using throughout this chapter in order to demonstrate the whole process of pattern generation.

1.5.1 Checkerboard Pattern

A square cell field of $n \times n$ cells is considered. The CB pattern uses two colors (black and white) in a way that the colors alternate in each row and in each column. There are two possible solutions defining the class of target patterns, depending on the color at $(x, y) = (0, 0)$. The position x is counted from left to right and y from top to bottom. Two simple templates (Fig. 2a) define that class. The aimed global CB pattern has a maximum number of hits. The degree of order is $h = (n - 2)^2$ (Fig. 2a–e). The number of black and white cells are equal for n even and differ by one for n odd. In order to simplify the task and considerations the border length n is assumed to be even.

How can templates be found? There are three ways.

1. If some of the target patterns are given, they can be analyzed. Each site is inspected with its neighborhood according to the tentative template size. A new template is stored in the template set if it is part of the pattern in mind. So at

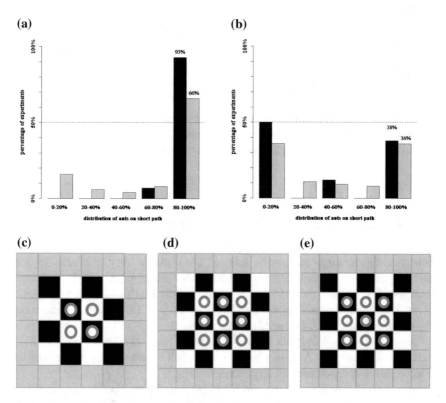

Fig. 2 Two templates define the class of CB patterns (**a**). There are two target patterns for a square cell field of $n \times n$ cells. The number of hits (degree of order) is $h = (n - 2)^2$. There are 4 hits for $n = 4$ (**b, c**) and 9 hits for $n = 5$ (**d, e**). Hits are depicted as dots

the end of the analyzing phase, the given patterns can be tiled (with a certain degree) by the collected templates. The tiling is performed in a way where the templates may overlap and the hits are as close to each other as possible. Then the given patterns and new similar ones (belonging to the same target class) can be generated by these templates.

2. Templates can be designed with respect to some patterns in mind which may be vague to a certain extent. Some most important aimed patterns or parts of them are analyzed as described before, and a first set of templates is fixed. Then patterns are generated using this first set. If the generated patterns are not satisfying (faulty patterns or missing patterns or unwanted patterns), the template set has to be updated until the generated patterns are satisfying. During this process new patterns may appear unintentional, which can be of interest and may be included in the set of target patterns. So, through this process, the designers creativity may be enhanced.

3. A game can be played in order to find new target patterns. The set of templates is generated exhaustively or at random satisfying some constraints (e.g. symmetry). Then the target pattern are observed and can be analyzed and selected by global properties (e.g. color distribution, fractal dimension). Thereby the influence of the templates on the patterns can be studied fundamentally, and interesting novel global patterns may appear.

2 Single-Agent System

2.1 The Specific Task

Only one agent is used. Initially it is placed in one of the four corners (Fig. 3). Its direction is chosen in a non-blocking way. For the starting position $(x, y) = (0, 0)$ the agent can move freely to the right (along the x-axis) until it meets the right border. The other starting positions are rotational symmetric.

What is the shortest path to visit all cells and thereby producing the desired CB pattern? In principle there is a simple solution: the agent moves first right until reaching the right border, then down one line, then to the left until reaching the left border, then down one line, then again to the right, and so forth until the agent reaches the bottom left corner where it can stop (Fig. 3a). Following the shortest path, every second cell is colored black, starting with a black cell at $(0, 0)$ or $(1, 0)$. The shortest path length for an agent with a global or large view is $t_{global} = N = n^2$. (In the special case that the last cell of the CB pattern is supposed to be white, the agent does not need to visit this last cell because it was already white initially.)

This path can easily be programmed by a supervisor who overviews the system globally. However, an agent can follow the optimal path only if its capabilities are powerful enough, i.e. its sensing, moving, turning and memory capabilities. What is the necessary local view to reach the optimum? The agent has to be able to view two

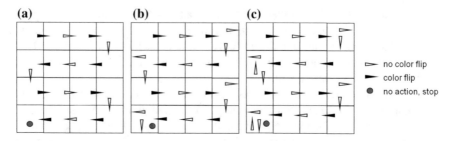

Fig. 3 Actions of a moving agent on a path forming a CB pattern. **a** Global view or large Neighborhood$_2$: agent walks on shortest path. **b** Neighborhood$_1$ (view one cell ahead): agent walks on fastest possible path. **c** Designed algorithm with Neighborhood$_1$

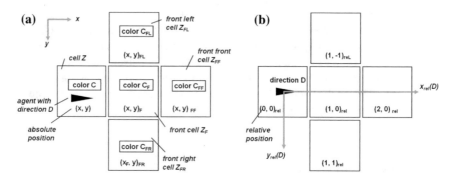

Fig. 4 **a** Potential neighbors in front of an agent dynamically depending on its direction. They can be identified by absolute coordinates. **b** Coordinates relative to the agent's direction are useful to describe the actual neighbors and the moving actions more easily

cells ahead (Z_F and Z_{FF}) and the left cell in front (Z_{FL}) in order to detect in time the borders and the bottom left corner (Fig. 4). Thus the needed dynamic neighborhood depending on the agent's direction D is

$NH_2 = Neighborhood_2(D) = \{(0, 0)_{rel}, (1, 0)_{rel}, (2, 0)_{rel}, (1, -1)_{rel}\}$.

Using this neighborhood, the optimal time (Fig. 3b) can be reached:

$t(NH_2) = t_{global}$.

Now we want to know the shortest time if the view is restricted to one cell ahead. Then the dynamic neighborhood consists of two cells only:

$NH_1 = Neighborhood_1(D) = \{(0, 0)_{rel}, (1, 0)_{rel}\}$.

The time becomes now a bit longer, because the necessary turn at a border needs 2 steps, first detecting the border, and second turning to a free cell in front. Thus the time is now

$t(NH_1) = N + 2(n/2) = n^2 + n$.

Now our objective is to find a real algorithm (a finite state automaton) that reaches or nearly reaches this time, using four states only.

2.2 Modeling

The CA or CA-w cell rule has to react on different non-uniform situations, such as: an agent is actively operating on the cell it is situated on, or a border cell is in front. Therefore the cell state is modeled as a record of several data items.

$CellState = (Color, Agent)$, where

$Color\ C \in \{-1, 0, 1\} \equiv \{border\ color, white, black\}$

$Agent = (active, Identifier, Direction, State)$

$active \in \{true, false\} \equiv \{1, 0\}$

$Identifier\ ID \in \{0, 1, ..., k - 1\}$, where k is the number of agents,

and for the single-agent-system $k = 1$

holds

$Direction\ D \in \{0, 1, 2, 3\} \equiv \{toN, toE, toS, toW\}$

$State\ s \in \{0, 1, ..., N_{State} - 1\}$,   where N_{State} is the number of states.

This means that each cell is equipped with a potential agent, which is either active or not. When an agent is moving from A to B, it is copied from A to B and the *active* bit of A is set to false.

The agent's hardware is depicted in Figs. 5 and 6. The function g is the main part of the algorithm controlling the agent. Inputs are the control state s, the current direction D, the color C and the color in front C_F. The direction has two functions: (1) it defines the cell in front which can be sensed, and (2) it defines the moving direction. Outputs are the new control state s_{new}, the new direction D_{new}, the new color C_{new}, the new active state, and the *move* request. The agent will move to the cell ahead if $move = 1$ and no blocking border is in front. The outputs can be distinguished into *internal* actions that influence the agent's state (active, s, D), *external* actions that influence the color (the environment), and *signals* (move) that are signaling certain conditions or sub states to the own agent or to potential agents in the neighborhood.

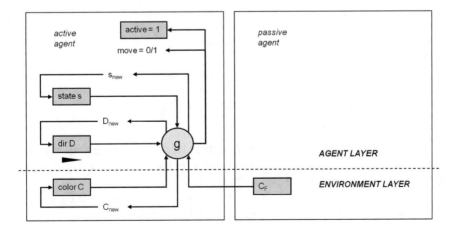

Fig. 5 Non-moving agent (staying in left cell)

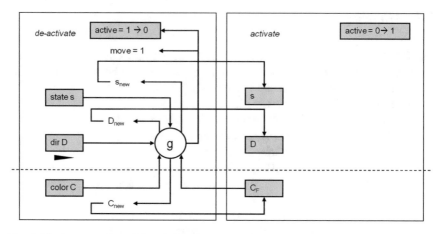

Fig. 6 Moving agent (from left to right cell)

The capabilities of the agents have to be defined before designing or searching for the agents' behavior. The main capabilities are: the perceivable data from the neighborhood, the actions an agent can perform, the capacity of its memory (number of possible control and data states) and its actual "intelligence" (useful pro- and reactive activity).

For our single-agent system the inputs, actions, and signals that an agent is able to perform are summarized in the following. The inputs are:

- its **own control state** s,
- its **own direction** D,
- the **cell's color** C the agent is situated on,
- the **color in front** C_F, including the detection of a border in front ($C_F = -1$).

The actions and signals are:

- **new state**: $s \leftarrow s_{new} \in \{0, \dots, N_{States} - 1\}$
- **turn**: $turn \in \{0, 1, 2, 3\}$
  The next direction is $D(t + 1) \leftarrow (D(t) + turn) \bmod 4$
- **flip color**: $flipcolor \in \{0, 1\}$
  The next color is $C(t + 1) \leftarrow (C(t) + flipcolor) \bmod 2$.
- **move**: $move \in \{0, 1\} \equiv \{wait, go\}$

All actions can be performed in parallel. There is only one constraint: when the agent's action is *go* and the situation is *blocked*, then the agent cannot move and has to wait, but still it can turn and change the cell's color.

Note that our agent's perception is very limited, it reacts only on on the color in front and the cell's own state. Therefore the task it not so easy to solve.

2.3 The Designed Algorithm

The designed agent's algorithm G1 to solve this task is depicted in Fig. 7. The agent walks on the shortest path (Figs. 3c and 8), but it needs at the right border an additional step and at the left border two additional steps to determine the next free cell. Thus the time to visit each cell is

$$t(G1) = N + \frac{n}{2} + 2\frac{n}{2} = N + 3\frac{n}{2}.$$

The time-complexity is $O(N)$. When the agent is free running and both colors CC_F are white, the states 0 and 5 are alternated, and the color is flipped when the state transition $0 \rightarrow 2$ takes place. When the right border is detected in state 5 the next state is 4, the color remains white, the *go* command is not executed because of the border, and the agent turns right. The next state transition is from 4 to 0. The color is not flipped and the agent moves forward and turns right. When the agent detects the left border, it first turns right and reaches state 4 again. There it detects a black cell in front, turns back in order to continue its walk, and reaches state 1. If it can move forward (not yet reached the bottom left corner) the color is flipped and the agent moves forward thereby turning left. When the agent detects the bottom left border, the agent remains in state 1 and stops all its actions. This is a clear condition that the algorithm has terminated.

2.4 Termination

We want to reflect upon the termination of multi-agent systems in our context of forming patterns. The goal is to form a certain pattern out of the defined pattern class with a certain degree of order. We will call the time to order a field $t_{success}(h_{target})$, if $h \geq h_{target}$.

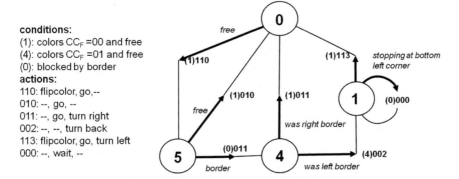

conditions:
(1): colors CC_F =00 and free
(4): colors CC_F =01 and free
(0): blocked by border
actions:
110: flipcolor, go,--
010: --, go, --
011: --, go, turn right
002: --, --, turn back
113: flipcolor, go, turn left
000: --, wait, --

Fig. 7 The agent's algorithm G1 performing the CB task, depicted as finite state automaton graph

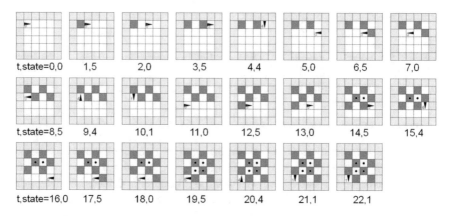

Fig. 8 Simulation of the agent's algorithm G1 on a 4 × 4 field. The agent has built a CB pattern at $t = 19$, and reaches state 1 at $t = 21$ and then stops all actions at $t = 22$

After this time, the degree of order may oscillate around the level h_{target}. Then, at time t_{stable}, the pattern may remain stable (reach a fix point) with a possibly other degree of order h_{stable}.

During the whole run, the agent may decide not to change the pattern anymore and (i) to stop all activities (here $move = wait$, no direction change, no external visible state change), or (ii) continue being alive (direction changing and/or moving), usually running in a cycle. The first case can be called "hard" stop, the second "soft" stop. If the agent stops hard/soft and the aimed degree of order is reached or is greater, then we call the multi-agent computation *hard/soft-terminated*.

3 Multi-agent System

3.1 The Task

Now we want to find algorithms for $k = 1, \ldots, N$ agents. The agents shall start at any random position and with any random direction. As it is quite difficult to design such algorithms, we will use a Genetic Algorithm to find them automatically. We restrict our approach to a field of size $N = n^2 = 8 \times 8$.

What is the shortest path for one agent to visit all cells and thereby producing the desired CB pattern? The agent can first find a corner as starting position and then use one of the algorithms discussed before. A more sophisticated algorithm would immediately start to color each second cell black in order not to visit already colored cells twice. We have already learned that the fastest algorithm would need a large neighborhood. As before we are interested in an algorithm with a few local neighbors. We would like to use the neighborhood NH_1 as before (sensing the cell in front only). But unfortunately we need to look two cells ahead in order to detect a moving conflict. Therefore we are using the neighborhood

$$NH_3 = Neighborhood_3(D) = \{(0, 0)_{rel}, (1, 0)_{rel}, (2, 0)_{rel}\}.$$

Optionally we will allow an agent E to emit a signal which can be detected by another agent A pointing to E. E must be a direct neighbor in front of A. The signal can be read by at most four agents pointing to E.

Now our objective is to find near optimal algorithms (finite state automata) with 6 states. We will also use the term *FSM* (finite state machine) to denote the agent's algorithm.

3.2 Modeling

The new agent's hardware (Fig. 9) is now slightly modified compared to the hardware before, and it is now called finite state machine (FSM). The function g (Figs. 5 and 6) is now partitioned into the sensor, the input mapping module, the state table and the "plus" operators used for the direction and color change. The *state table* is the *variable* (configurable) part, and the surrounding elements (wires, operators, memory elements) are *fixed* (pre-defined). The state table can also be seen as a program residing inside the FSM which is optimized by the Genetic Algorithm. The state table defines the actions (new state, turn, flip color, signal) to be performed. Inputs of the table are the control state s and several pre-defined input situations x.

The *sensor* gathers the information of the own cell and its neighborhood and transforms it to intermediate conditions which are then used by the input mapping module. The sensed data depends on the agent's direction. The sensed data is the

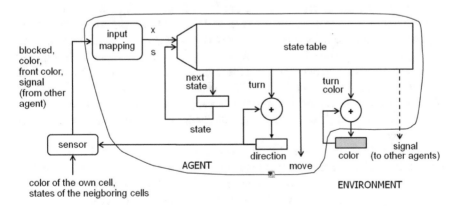

Fig. 9 An agent is controlled by a finite state machine (FSM). The state table defines the agent's next control state, its next direction, and whether to move or not. The table also defines the next color, and the signal to another agent

own color, the state of the front cell (color in front, agent in front), and the moving conflict situation.

The *input mapping module* reduces all possible input combinations to an index $x \in X = \{0, 1, \ldots, N_x - 1\}$ which is used in combination with the control state to select the actual line of the state table. The input mapping function was introduced in order to limit the size of the state table entries. It can be seen as a preprocessing operation related to the sensing capabilities. In principle the input mapping module is not really necessary, and then the intermediate conditions are directly part of the address which selects the actual line from the state table. But then the needed bit capacity of the state table would be much larger.

For multi-agent systems we have added a **signal** to the agent's hardware: an agent (A1) emits a $signal \in \{0, 1\}$ (output of the FSM table) that can be used by another agent (A2). Only agents A2 in the direct neighborhood of A1 and pointing to A1 can read this signal. The idea behind this signal is to enhance the system's performance by revealing a part of the agent's internal state to nearby agents.

In our multi-agent system an agent can now be blocked not only by a border but also by another agent in front. We are not distinguishing these two cases for the input mapping. If two or more agents meet head-on (with one free cell in between) want to move to the same free cell in front, this situation is a moving conflict. It is solved by giving priority to the agent with the lowest identifier (ID $= 0 \ldots k - 1$). Instead of using the identifier for prioritization, it would be possible to use other schemes, e.g. random priority, or a cyclic priority with a fixed or space-dependent base.

3.2.1   The Input-Mapping in Detail

The input mapping shown in Table 1 was used. Thirteen input situations x were distinguished. They select together with the current control state s the actual line of the state table. An agent perceives a border in front ($x = 0$), the color C and the color in front C_F when it can run freely ($x = 1 \ldots 4$), and in addition it can read the signal that another agent in front emits in case of blocking ($x = 5 \ldots 12$). The agents' FSMs will also be evolved for the case where agents emit no signals. Then no test for signals is performed any more, and the input values $x = 5 \ldots 8$ and $x = 9 \ldots 12$ coincide and are mapped to $x = 5 \ldots 8$. In a single-agent system the agent cannot be blocked by another agent and no conflict can appear, therefore only $x = 0 \ldots 4$ are relevant. In a fully packed system the agents cannot move, therefore $x = 1 \ldots 4$ are not relevant.

This mapping was designed by experience. Obviously, other input mappings can be defined, with another set of input situations (with more or less local information sources) and another number of x values. For instance the direction of the agents, or the detection of templates may enhance the performance of the agents.

Note that our agent's perception is very limited, it reacts only on

- own cell: color and control state,
- cell in front: color, signal (if blocking agent is there),

Table 1 The used input mapping. For each input situation a different index x (used for addressing the FSM table) is defined. NU = not used, NA = not available

x	Blocked	Color	Front color	Signal	
0	1	NU	−1	NA	Blocked by border
1	0	0	0	NA	Not blocked (agent can move)
2	0	1	1	NA	
3	0	1	0	NA	
4	0	0	1	NA	
5	1	0	0	0	Blocked (signal = 0)
6	1	0	1	0	
7	1	1	0	0	
8	1	1	1	0	
9	1	0	0	1	Blocked (signal = 1)
10	1	0	1	1	
11	1	1	0	1	
12	1	1	1	1	

- blocking: by border in front, by agent in front, or by conflict.

Under these restrictions, the agent's task is quite difficult to solve. Imagine that you are the agent, moving around in a dark room where you can observe only the color on the ground and in front, and sometimes you are detecting a border or an agent in front (emitting a binary signal)! It seems to be evident that the larger the agent's perception is with regard to the task to be solved, the easier it could be solved.

4 Evolving FSMs by a Genetic Algorithm

An ultimate aim could be to find an FSM that is optimal for all possible initial configurations on average. This aim is very difficult to reach because it needs an huge amount of computation time. Furthermore, it depends on the question whether all-rounders or specialists are favored. Therefore, in this work we searched only for *specialist* optimized for (i) a fixed field size of $N = n \times n$ ($n = 8$), (ii) a fixed number of agents $k = 1, 2, 4, 8, 16, 32, 64$, and (iii) 1000 initial random configurations (for training and evaluation).

The number of different FSMs which can be coded by a state table is $Z = (|s||y|)^{(|s||x|)}$ where $|s|$ is the number of control states, $|x|$ is the number of inputs and $|y|$ is the number of outputs. As the search space increases exponentially, we have restricted the number of states to $|s| = N_{states} = 6$, and the number of inputs to $|x| = 13$ (outputs of the input mapping). A relatively simple genetic algorithm similar to the one in [18] was used in order to find (sub) optimal FSMs with reasonable computational cost. A possible solution corresponds to the contents of the FSM's state table. For each input combination $(x, state) = j$, a list of actions is assigned: $actions(j) = (newstate(j), move(j), turn(j), flipcolor(j), signal(j))$.

The fitness is defined as the number t of time steps which is necessary to emerge successfully a target pattern with a given degree h_{target} of order, averaged over all given initial random configurations. *Successfully* means that a target pattern with $h \geq h_{target}$ was found. The fitness function t is evaluated by simulating the system with a tentative FSM_i on a given initial configuration. Then the mean fitness $t_{mean}(FSM_i)$ is computed by averaging over all initial configurations of the training set. t_{mean} is then used to rank and sort the tentative FSMs.

Evolved Finite State Machines

In general it turned out that it was very time consuming to find good solutions with a high degree of order, due to the difficulty of the agent's task in relation to their capabilities. In addition the search space is very large and difficult to explore. The total computation time on a Intel Xeon QuadCore 2 GHz was around 4 weeks to find the needed FSMs.

The best found FSM is denoted by

$$FSM(n, k, h_{target}),$$

for field size n, system with k agents, and a reached order of $h \geq h_{target}$. The reached order can also be given relatively as h_{rel} with a per cent suffix. If the order h_{target} is not explicitly given, then $h = h_{max}$ or $h_{rel} = 100\%$ resp. is assumed.

In order to evaluate the effect of the signal that an agent can emit, the FSMs were evolved *with* signal and *without* signal for $k \geq 2$. An FSM with signal is denoted by FSM^+.

Six FSM(8, k, 100%) without signal, and six FSM$^+$(8, k, 100%) with signal were evolved for $k = 2, 4, 8, 16, 32, 64$ (Figs. 15 and 16).

For $k = 1$ the signal is of no use, therefore only FSM(8, 1, 100%) = FSM(8, 1, 36) was evolved. Several FSM(8, 1, 100%) with the same best fitness were found by the Genetic Algorithm. Among them, one of them was selected which was also successful on other field sizes ($n = 4, 6, 8, 10$). The state table of this single-agent system is the following:

```
          /x=0        \  /x=1         \  /x=2         \  /x=3         \  /x=4        \
state      0 1 2 3 4 5    0 1 2 3 4 5    0 1 2 3 4 5    0 1 2 3 4 5    0 1 2 3 4 5
newstate   3 1 0 2 5 1    4 0 1 0 1 4    4 3 1 2 5 0    3 1 3 4 4 4    4 1 5 1 1 1
flipcolor  1 0 1 0 1 1    1 0 1 0 0 1    1 1 0 0 0 0    1 1 1 0 1 1    0 0 0 1 1 1
move       1 1 0 0 0 1    1 0 0 1 1 1    0 1 0 1 0 0    1 1 0 1 1 0    0 1 1 1 0 0
turn       3 2 2 1 1 3    0 0 0 0 1 0    2 1 1 2 3 0    2 1 2 3 0 2    3 2 2 2 1 1
```

5 Evaluations

5.1 Time-Steps of the Evolved FSMs for Fields of Size 8 × 8

The following measures were used in order to evaluate the performance of the multi-agent systems of field size 8 × 8 (Table 2):

- *successful fields*: the number of initial configurations (out of the given 1000 fields with random placement of the agents) which were successfully evolved over time to a CB pattern by the multi-agent system with FSM(8, k) or FSM$^+$(8, k).
- $t_{mean}, t_{min}, t_{max}$: The number of time-steps to form the pattern. The whole set of 1000 initial configurations which was used in the Genetic Algorithm was also used for evaluation, and mean, minimum and maximum time were computed.
- $speedup = t_{mean}(1)/t_{mean}(k)$: This is a well-known metric to evaluate how much faster a computer system works with k processors compared to one processor. For one processor an optimal sequential algorithm is used, and for $k > 1$, a parallel algorithm. Here we use another interpretation. Each agent is a (moving) processor. And for each k, a specific, near-optimal FSM (evolved by the Genetic Algorithm) is used.
- $cost\ per\ cell = t_{mean} \cdot k/N$: This measure is the sum of the time-steps that all agents need together, normalized to the number of cells. One can consider this

Table 2 Performance of the evolved FSMs: number of successful ordered fields; mean, minimum and maximal time; speedup = $t_{mean}(1)/t_{mean}(k)$; cost (number of all computational steps) per cell

	Agents k	1	2	4	8	16	32	64
FSM(8, k) without signal	Successful fields out of 1000	1000	998	1000	1000	1000	1000	1000
	Time t_{mean}	**203**		**314**	**174**	**88**	**77**	**71**
	t_{min}	102		55	45	32	27	14
	t_{max}	274		1541	606	374	400	341
	Speedup	1		0.65	1.17	2.31	2.64	2.86
	Cost per cell $t_{mean}\ k/64$	3.25		19.6	21.8	22	38.5	71
FSM$^+$(8, k) without signal	Successful fields out of 1000	1000	995	1000	1000	1000	1000	1000
	Time t_{mean}	**203**		**295**	**166**	**89**	**71**	**46**
	t_{min}	102		71	42	31	17	10
	t_{max}	274		1441	969	352	268	179
	Speedup	1		0.69	1.22	2.28	2.86	4.41
	Cost per cell $t_{mean}\ k/64$	3.25		18.4	20.1	22.2	35.5	46

measure to be the cost (in time-units) to be paid to all agents, in relation to the size of the field.

For $k = 1$, only FSM(8, 1) was evolved, because the signal is of no use, and therefore FSM = FSM$^+$. This FSM was selected among several equally best ones, in order to order the field independently of the field size. All possible initial configuration were used for training and testing. In fact there are (only) 100 different of them taking into account the rotational symmetry. The rotation rot(F) means, that a field F is rotated 90 degrees clock-wise. E.g., if an agent which is originally placed in the first quadrant (upper-left), then it is placed after rotation in the second quadrant (upper-right), and thereby the agent's direction is also rotated. There a four rotational equivalent fields: F, rot(F), $rot(rot$(F), $rot(rot(rot$(F))), and there are $(n/2)^2$ different positions that an agent can take (in the first quadrant)), with four different directions. Therefore there exist $4(n/2)^2$ different initial configurations with one agent. Note that the agent's actions are invariant under this rotation, therefore the new direction (output of the FSM table), was defined relative to the current direction.

For $k = 2$, FSM(8, 2) was found which was successful on 998/1000 fields, and the best found FSM$^+$(8, 2) was successful on 995/1000 fields only. Though the Genetic Algorithm was running for a very long time, no better algorithms were found. This rises the question if there exists a FSM(8, 2) with 6 states that is successful on any configuration. So further research has to be conducted on this topic.

For $k = 4, 8, 16, 32, 64$, the found algorithms were successful on all the 1000 fields (used for training and evaluation). Nevertheless it is important to note, that this test is not exhaustive, and the algorithms may fail for some special configurations. Indeed, it turned out for some tests, that the CB pattern cannot be formed perfectly if the agents are placed in a rotational symmetric way. One such case will be shown for a four-agent system in the next section.

Figure 10 shows how the time decreases with the number of agents until the system is fully packed. The positive effect of the signal is more noticeable when the communication becomes high for systems with a high density of agents.

The speedup for systems with several agents is relatively low, and it is even less than one with four agents. This means that four agents are doing worse that one only.

Fig. 10 Five Evolved FSMs for a different number of agents. Average time t_{mean} to order 1000 random fields of size 8×8 successfully

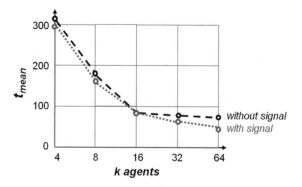

The single-agent system needs 3.25 time-steps on average per cell to form the pattern, whereas the fully packed system with signal needs 46 steps. Thus the fully packed system with signal is 4.41 times faster but $46/3.25 = 14$ times more costly.

The fully packed system is more similar to a classical CA because each cells contains an agent that cannot move, though it can turn its direction.

5.2 Simulation and Generality

Simulations for systems using FSM(8, k), k = 1, 4, 64, will be presented and discussed. And generality will be tested, how well FSM(8, 1) and FSM(8, 64) are performing for different field sizes n.

5.2.1 Single-Agent System

After evolving several FSM(8, 1) with equal fitness, they were tested exhaustively also on other field sizes: $n = 4, 6, 10$. The number of different initial configurations with one agent is $4(n/2)^2$. The selected FSM(8, 1) is successful for all these test cases (Table 3). The cost per cell increases linear with n, the function is $cost(n) = (n + 11)/6$, which can be derived from the given values. And the mean time is $t_{mean}(n) = N \cdot cost(n) = n^2(n + 11)/6$.

The time-complexity is $O(n^3) = O(N^{3/2})$. This algorithm is $O(n)$ slower, compared to the designed, restricted algorithm (Sect. 2.3), where the agent must start in a corner.

In Fig. 11 two simulations are shown, the fastest and the slowest. When the agent starts in the middle, it needs 102 time-steps to form the pattern. When the agent starts in the corner, it needs 274 time-steps. After reaching the order of $h_{max} = 36$, the agent continues its walk along the border, thereby changing slightly the pattern. ((t, h) = (274, 36), (275, 34), (276, 36), (277, 36), (278, 36), (279, 33), (280, 36), etc.).

What is the strategy of the agent? Following the simulation step by step for a 6 × 6 field (Fig. 12) one can see that the agent mainly follows a spiral path (counter-

Table 3 Single-agent FSM(8, 1) simulated for other field sizes

Size	4 × 4	6 × 6	8 × 8	10 × 10
# successful	Any 4 × 4	Any 4 × 9	Any 4 × 16	Any 4 × 25
t_{mean}	**40**	**100**	**203**	**350**
t_{min}	16	50	102	174
t_{max}	58	146	274	479
Cost per cell	**2.5**	**2.78**	**3.17**	**3.5**

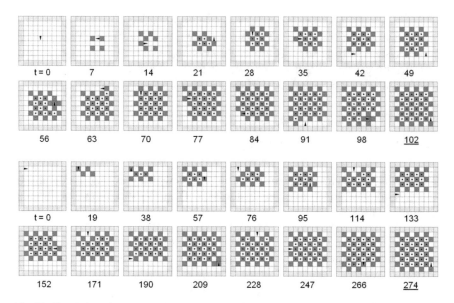

Fig. 11 Simulation of a 8 × 8 single-agent system with the best found FSM evolved for all possible 8 × 8 fields. The pattern is successfully formed at $t_{min} = 102$ (top) for an agent starting in the middle, and at $t_{max} = 274$ (bottom) for an agent starting in the top left corner

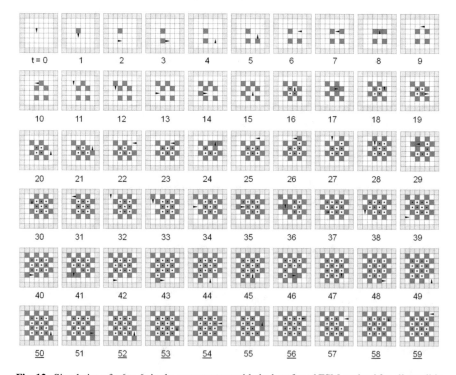

Fig. 12 Simulation of a 6 × 6 single-agent system with the best found FSM evolved for all possible 8 × 8 fields. The aimed pattern with degree 16 is formed at t = 50, 52, 53, 54, 56, 58, 59, etc

clockwise), enlarging the already visited area. On the way, the agent is performing tests and is slightly deviating from its main path. When the agent is blocked by the border, it runs along it. After having reached the maximum order at t = 50, the agent continues its walk along the border, thereby changing slightly the pattern (Fig. 12). The degree of order then oscillates between 16 and 13.

A question was, if there exists a dual algorithm working clockwise. The turn actions of the FSM table were manually changed (right and left turns were interchanged, 1 ↔ 3). Then this algorithm was tested and indeed, it showed the expected dual behavior.

5.2.2 Four-Agent System

The evolved four-agent algorithm FSM(8, 4) was successful on 1000 random initial configurations. But this not a proof that this FSM will be always successful. Especially sensitive are configurations that are rotational symmetric. Such a case was simulated (Fig. 13 (top)). Already snapshot at t = 30 shows that each agent builds a part of the pattern, but the parts do not fit together, and all corners are colored black and remain so. The agents are not able to break the symmetry and then to collaborate in order to build the aimed pattern. Furthermore, the system runs into a live-lock (the four last situations are cyclically repeated). In a live-lock the agents act in a way that there is no more progress in the system's global state towards the aimed pattern, and the system's global state runs in a cycle. An analogy is that two people meet head-on and each tries to step around the other, but they end up swaying from side to side, getting in each others way as they try to get out of the way.

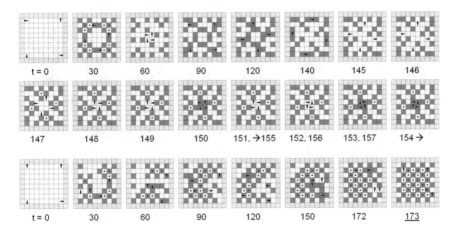

Fig. 13 The used FSM(8, 4) was successful on 1000 random configurations. (top) Starting from a symmetric configuration, the desired CB pattern could not be achieved. The system is running into a live-lock. (bottom) Starting from a slightly different asymmetric configuration, the CB pattern appears at t = 173

The next initially configuration was slightly modified, the agent in the upper-left corner is now directed downwards. Now the agents can solve the problem. The four corners are colored in black until t = 60, then the lower-right corner's color is changed into white at t = 90, and finally at t = 173 the pattern is established.

We can conclude, that some configurations, especially symmetric ones, may lead to live-locks in the multi-agent systems with the evolved FSM(8, k), k = 2, 4, 8, 16, 32, 64. Then, are the evolved FSMs useless? We are confident, that the FSMs are still useful, because they are functional with a high probability. In real life, we rely on many systems that are not working always perfectly. And there are some options to make systems more perfect and avoid live-locks: (i) break the symmetry by introducing some noise (e.g. change the agents' directions with a low probability), (ii) use slightly different FSMs for some agents, or (iii) use an asynchronous updating scheme.

I may thank a reviewer for his valuable comment on the break of the symmetries which I like to cite: *"In the deterministic setting, I think that it is possible to prove formally that some patterns are not solvable given certain initial configurations. E.g., the first configuration in* Fig. 13 *has the same symmetries of a Dihedral group* D_4 *whereas the target has less symmetries (that of a dihedral group* D_2*), then by showing that the symmetries of* D_4 *cannot be broken then the target cannot be reached. The case of the second starting configuration in* Fig. 13 *is different as it has no symmetries, then the target, with more symmetries, can be reached. Given a deterministic setting,*

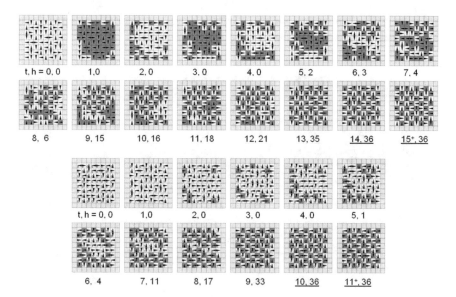

Fig. 14 Simulation of a 8 × 8 fully packed system with the best evolved FSM(8, 64) and FSM$^+$(8, 64). (top) Simulation for the field with shortest time (out of 1000) without using signals. (bottom) Simulation for the field with shortest time (out of 1000) using signals

the aim could be to exactly characterize those patterns that are solvable by certain starting configurations."

5.2.3   64-Agent System

The fully packed agent systems are fastest because all cells can be active simultaneously. The agents cannot move but change direction. The FSM without using signals

```
FSM(8,2)   /x=0 \/x=1 \/x=2 \/x=3 \/x=4 \/x=5 \/x=6 \/x=7 \/x=8 \
state      012345012345012345012345012345012345012345012345012345
newstate   211200401014431111315424410114201141050325523453001325
flipcolor  000110111000110111011010001110110001110101100101110001
move       001101101111101000111000110001100111010010010010010101
turn       322022101213312200113302022210332131212310332233201333

FSM(8,4)   /x=0 \/x=1 \/x=2 \/x=3 \/x=4 \/x=5 \/x=6 \/x=7 \/x=8 \
state      012345012345012345012345012345012345012345012345012345
nextstate  210241400014431150323432515011305251401444010101204225
flipcolor  001010101101111000010001000111100100110110001110011101
move       100001000100101010101011011100001010100110011011011100
turn       222113003011212230211303023013301231013133010122233110

FSM(8,8)   /x=0 \/x=1 \/x=2 \/x=3 \/x=4 \/x=5 \/x=6 \/x=7 \/x=8 \
state      012345012345012345012345012345012345012345012345012345
nextstate  405041015411431152315543302021034250320334115343104555
flipcolor  011001111111100000100000010111001010110001110101110011
move       010111111111010110111101000001110001111111101000000011
turn       222120331013112031003230321211312300001103113331222032

FSM(8,16)  /x=0 \/x=1 \/x=2 \/x=3 \/x=4 \/x=5 \/x=6 \/x=7 \/x=8 \
state      012345012345012345012345012345012345012345012345012345
nextstate  312215213532543221523304334034544204354301520444130542
flipcolor  000001011001111011000000000000110110100001100001111010
move       001001101101100111010110101001011111110110111011101000
turn       132223220231100211101313000221201330331301032210230031

FSM(8,32)  /x=0 \/x=1 \/x=2 \/x=3 \/x=4 \/x=5 \/x=6 \/x=7 \/x=8 \
state      012345012345012345012345012345012345012345012345012345
nextstate  221213131421543222031313333132431112422442214344250221
flipcolor  010001011111111010000000000000101110000100110000101010
move       001010101110001100011011101010110110111100001001110001
turn       202122201113323113301303120332031132303231011330303122

FSM(8,64)  /x=0 \/x=5 \/x=6 \/x=7 \/x=8 \
state      012345012345012345012345
nextstate  040422520044530555555344400034
flipcolor  000001011110000000011001111111
move        001111101100001011001000111110
turn        322203313312203122020130332300
```

Fig. 15 The state tables of the evolved finite state machines without signal

```
FSM+(8,2)  /x=0 \/x=1 \/x=2 \/x=3 \/x=4 \/x=5 \/x=6 \/x=7 \/x=8 \/x=9 \/x=10\/x=11\/x=12\
state      0123450123450123450123450123450123450123450123450123450123450123450123450123452345
newstate   5111424011140311103144202100113204105404350242533541242452532155521014151111411
flipcolor  0001101101011111110100100010100101111010001100000111100001101110000111001001101
move       1001111010110100100111101110010011110001110010010100000001101110010100011101
turn       1200201023112120001123022211100221120033201010300122303022233101111133330122310
signal     1011011110001001011110110001111011001001101000000011001111010011001010110111
```

```
FSM+(8,4)  /x=0 \/x=1 \/x=2 \/x=3 \/x=4 \/x=5 \/x=6 \/x=7 \/x=8 \/x=9 \/x=10\/x=11\/x=12\
state      0123450123450123450123450123450123450123450123450123450123450123450123450123452345
newstate   312140400524431150313433015011204300450324524152154110045133003042144353002324
flipcolor  0001011010011100110100010000111101011100100011000111101000010100000010000000010
move       011000000111010110111011110000111001011111010101000101011111011110001010110
turn       022113010011312031211333232211122323232202110003221321330101120311312220330200
signal     1001011111101111011110111001110011001011010111101110001111100010000010111100011
```

```
FSM+(8,8)  /x=0 \/x=1 \/x=2 \/x=3 \/x=4 \/x=5 \/x=6 \/x=7 \/x=8 \/x=9 \/x=10\/x=11\/x=12\
state      0123450123450123450123450123450123450123450123450123450123450123450123450123452345
newstate   4050410054114311513145433020210252553203342142421045550002220330420043131115310
flipcolor  0110001111111001001000000010111011110110001011101111111000001100001110001000000
move       1100111111110101100111101010001000011111000100000001010001001001010110111000100
turn       0221233303101120210032303211113023000011331033312223232302223302331021222200012
signal     110000000010000001001100111011011100101001100100010001111100001101010101111000
```

```
FSM+(8,16) /x=0 \/x=1 \/x=2 \/x=3 \/x=4 \/x=5 \/x=6 \/x=7 \/x=8 \/x=9 \/x=10\/x=11\/x=12\
state      0123450123450123450123450123450123450123450123450123450123450123450123450123452345
newstate   31225520353253232215233053340345442033443015204441205420320413433232014442052441
flipcolor  0000110010011010110000000000011011010000100001011000001100000111000011001
move       00100110110110011010110101010111011101011011011000000011110101110010101001100
turn       13222322022113021113131303022120133033130303221033003012202103311310331010301
signal     01000010011101011010100000001010100000100110000110101010001000100111000011110000
```

```
FSM+(8,32) /x=0 \/x=1 \/x=2 \/x=3 \/x=4 \/x=5 \/x=6 \/x=7 \/x=8 \/x=9 \/x=10\/x=11\/x=12\
state      0123450123450123450123450123450123450123450123450123450123450123450123450123452345
newstate   221214131311543221530303333032310111422442214344250221330312031131431525124125
flipcolor  00000011111011110110000001000001000001000000011000001101001010000100001000001000
move       00110110110101101010110100010100001001110000000011100011101010010100011010111111
turn       0021211011122311320130312022233013023321203013330033322100011212231330330022320
signal     01111110001001011100101000000100001110111100010100011100010000000000000000010
```

```
FSM+(8,64) /x=0 \/x=5 \/x=6 \/x=7 \/x=8 \/x=9 \/x=10\/x=11\/x=12\
state      0123450123450123450123450123450123450123452345
newstate   0354124000445155555451444055331215404422555540055154212
flipcolor  0010000111110000000001000010111100000000001000000000101111
move       111100101001001011011000011010011100001111101000011
turn       30220333331222311202313233230020321103022120311110331
signal     0111111011110010000010100000100001010101000110011100010
```

Fig. 16 The state tables of the evolved finite state machines with signal

shall be compared with the FSM$^+$ using signals. The initial configurations which led to the fastest pattern building were selected. For this example, the number of time-steps is $t = 14$ for FSM(8, 64) and $t = 10$ for FSM$^+$(8, 64) (Figs. 14, 15 and 16). Interesting is that both algorithms are successfully terminating (the pattern remains stable with $h = h_{max}$) while the agents continue to change direction. We have called this case *soft-termination* in Sect. 2.4.

How general is the system with respect to the field size? FSM(8, 64) and FSM$^+$(8, 64) were tested for $n = 4, 6, 8, 10$. The results of the simulations on 1000 random test fields are given in Table 4. FSM(8, 64) without signal was always successful on smaller systems, but only partly (997/1000) on the 10×10 system. FSM$^+$(8, 64) with signal was always successful on the 10×10 system, but not on smaller ones. This means that the algorithms evolved for the fully packed system are also successful on smaller or larger fields, but not for any initial configuration.

Table 4 Fully packed system evolved for FSM(8, 64), simulated on other field sizes

	Size	4×4	6×6	8×8	10×10
FSM(8, 64) without signal	# successful	1000	1000	1000	997
	t_{mean}	**26**	**46**	**71**	–
FSM(8, 64) with signal	# successful	970	995	1000	1000
	t_{mean}	–	–	**46**	**79**

6 Conclusion

The objective was to find algorithms (automata, finite state machines) for controlling the agents' behavior in order to form a checkerboard pattern in shortest time on an $n \times n$ field. The class of patterns was defined by two templates, small 3×3 local matching patterns. The degree of order is the number of template hits, for the CB pattern the maximum is $h_{max} = (n - 2)^2$. Firstly, a single-agent algorithm G1 with four states was designed, where the agent starts in a corner. The agent walks on a shortest path, forms the CB pattern, and then stops all activities. The time complexity is of order $O(n^2)$. Secondly, for 8×8 fields, several FSMs were evolved by a Genetic Algorithm with a different number of agents, also using a local communication signal between agents. Signals speedup the task, but only significantly if the density of agents is high. The agents are able to form successfully the aimed pattern with a maximum degree of order for 1000 random initial configurations. The evolved single-agent algorithm has a time-complexity of $O(n^3)$, where the agent can now start at any position. The fully packed system is the fastest, but it is also the most costly (cost = time-steps $\times$ number of agents). Although the evolved multi-agents systems did form a CB pattern for 1000 initial test configurations, there may exist special initial configurations which are not successful. It turned out that some totally symmetric initial configurations, cannot be solved successfully because of live-locks. Further research has to be conducted in order to find (1) more formally which patterns can / cannot be generated by certain multi-agent-systems, and (2) to find more general, efficient, and hard-terminating algorithms.

References

1. G. Peano, Sur une courbe, qui remplit toute une aire plane. Math. Ann. **36**(1), 157160 (1890)
2. M. Woolridge, N.R. Jenning, Intelligent agents: theory and practice. Knowl. Eng. Rev. **10**(2), 115–152 (1995)
3. S. Franklin, A. Graesser, Is it an agent, or just a program?: A taxonomy for autonomous agents, in *Proceedings of the ECAI'96 Workshop on Intelligent Agents III, Agent Theories, Architectures, and Languages*, ed. by J.P. Müller, M. Woolridge, N.R. Jennings. Springer (1997), pp. 21–35
4. J.H. Holland, *Emergence: From Chaos to Order* (Perseus Book, 1998)
5. M. Woolridge, *An Introduction to Multiagent Systems*, 2nd ed. (Wiley, 2002)

6. D. Pais, Emergent collective behavior in multi-agent systems: an evolutionary perspective. Ph.D. Dissertation, Princeton University, Princeton, NJ, 2012
7. F. Schweitzer, *Brownian Agents and Active Particles, Collective Dynamics in the Natural and Social Sciences*. Springer Series in Synergetics (Springer, 2003)
8. A.L. Rosenberg, Cellular ANTomata. Adv. Complex Syst. **15**(6) (2012)
9. C.L.E. Shannon, Presentation of a maze-solving machine, in *8th Conference of the Josiah Macy Jr. Foundation, Cybernetics* (1951), pp. 173–180
10. M. Blum, W. Sakoda, On the capability of finite automata in 2 and 3 dimensional space, in *18th IEEE Symposium on Foundations of Computer Science* (1977), pp. 147–161
11. M. Halbach, R. Hoffmann, Optimal behavior of a moving creature in the cellular automata model, in *8th International Conference, PaCT 2005, Proceedings*. LNCS, vol. 3606 (2005), pp. 129–140
12. P. Ediger, R. Hoffmann, Evolving probabilistic CA agents solving the routing task, in *16th International Workshop on Cellular Automata and Discrete Complex Systems, AUTOMATA, Nancy, France, 14–16 June 2010*
13. P. Ediger, R. Hoffmann, D. Désérable, Rectangular vs triangular routing with evolved agents. J. Cell. Autom. **8**(1–2), 73–89 (2013)
14. P. Ediger, R. Hoffmann, Routing based on evolved agents, in *23rd PARS Workshop on Parallel Systems and Algorithms, ARCS, Hannover, Germany*, vol. 2010 (2010), pp. 45–53
15. P. Ediger, R. Hoffmann, CA models for target searching agents. Autom. São José dos Campos. Brazil ENTCS **252**(2009), 41–54 (2009)
16. R. Hoffmann, D. Désérable, All-to-all communication with cellular automata agents in 2d grids—topologies, streets and performances. J. Supercomp. **69**(1), 70–80 (2014)
17. P. Fraigniaud, D. Ilcinkas, P. Guy, P. Andrzej, D. Peleg, Graph exploration by a finite automaton, in *MFCS 2004*. LNCS, vol. 3153 (2004), pp. 451–462
18. R. Hoffmann, How agents can form a specific pattern, in *ACRI Conference 2014*. LNCS, vol. 8751 (2014), pp. 660–669
19. R. Hoffmann, D. Désérable, Line patterns formed by cellular automata agents, in *ACRI Conference 2016*. LNCS, vol. 9863 (2016), pp. 424–434
20. R. Hoffmann, Rotor-routing algorithms described by CA-w. Acta Phys. Pol. B Proc. Suppl. **5**(1), 53–68 (2012)
21. R. Hoffmann, *The GCA-w Massively Parallel Model*, ed. by V. Malyshkin. LNCS, vol. 5698 (2009), pp. 194–206
22. R. Hoffmann, GCA-w: global cellular automata with write-access. Acta Phys. Pol. B Proc. Suppl. **3**(2), 347–364 (2010)
23. R. Hoffmann, GCA-w algorithms for traffic simulation. Acta Phys. Pol. B Proc. Suppl. **4**(2), 183–200 (2011)

Do Ants Use Ant Colony Optimization?

Wolfhard von Thienen and Tomer J. Czaczkes

Abstract Ant Colony Optimization (ACO) is a widespread optimization technique used to solve complex problems in a broad range of fields, including engineering, software development and logistics. It was inspired by the behaviour of ants which can collectively select the shorter of two paths leading to a food source. They are able to do so even without any single ant comparing the lengths of the two paths. Ants, like other eusocial insects, have no central authority to coordinate the sophisticated and complex work of their colony members. Coordination is achieved through self-organization, principles of which inspired the development of ACO algorithms. Here we discuss both the similarities and the considerable differences between the behaviour of real ant colonies and techniques used by ACO. We also describe some of the latest findings in ant research and how they may contribute to new ACO algorithms.

1 Introduction

1.1 Ant Colony Optimization

Ant Colony Optimization (ACO) refers to a family of optimization techniques that were inspired by the collective behaviour of ant colonies. ACO-techniques are used in different fields of software development, mathematics, engineering and logistics [22, 24]. They are especially useful for finding sufficiently good or near-optimal solutions for problems that are too complex to be solved simply by computational power. A typical example for such problems is the travelling salesman problem (TSP): A salesman must visit all the cities on his list once before he returns home. Let us imagine he lives in London and must visit Oxford, Manchester and Liverpool. It is very easy to solve this problem: one has to find all possible combinations of cities,

W. von Thienen (✉) · T. J. Czaczkes
Animal Comparative Economics Lab, Institute of Zoology, University of Regensburg,
Regensburg, Germany
e-mail: w@thienen.de

© Springer International Publishing AG, part of Springer Nature 2018 265
A. Adamatzky (ed.), *Shortest Path Solvers. From Software to Wetware*,
Emergence, Complexity and Computation 32,
https://doi.org/10.1007/978-3-319-77510-4_10

summarize the distances of each possible route, and compare them. In this case, there are only four cities involved. Thus, there are only three possible routes to compare and the salesman must do just a few simple calculations to find the shortest route. He can do so by simply adding the distances between the cities of each route. But what if he also wants to visit Plymouth, Southampton and Bristol? In this case, there are already 360 possible routes, which would take the salesman a long time to calculate without aid of a computer. If we add just four more cities, the number of routes will be close close to two million. If the salesman must visit 50 cities, the number of possible routes approximates the number of atoms in the known universe and no computer in the world would be able to calculate and compare all the different city combinations in an acceptable time. Such problems are called *NP-problems* (*nondeterministic polynomial*)[1]. In such problems, the number of possible solutions grows much faster than the number of problem instances.

In our technology-focused, interconnected society we must solve many such complex problems, and must do so quickly and efficiently. The TSP is just one example; other examples include the scheduling of aircraft landings, the routing of IP-packages through a communication network or planning the delivery routes of trucks.

1.2   Pattern Formation and Self-organization

In nature, complex problems are even more common. Living beings must find acceptable solutions for very complex problems in limited time since the number of parameters that influence a living system is extremely high, and competition is often fierce. Although the environment is extremely complex, nature has managed to provide organisms with strategies for finding solutions to a wide variety of problems. These strategies allowed living systems to develop from simple cells into complex multicellular beings, and allow organisms to react to all kinds of changing environmental conditions, organize their behaviour, survive and proliferate. Biologists have found that many of these strategies can be described by the concept of *self-organization*. Self-organization allows patterns to emerge in a system on a global level without the use of global information. Instead, patterns arise through interactions between the components of the system, using only local information [11, p. 8]. The stripes of zebras, the development of a complex organism from an egg, and the pattern of leaves and branches in a canopy—all of these and many others are well studied examples

---

[1] *Nondeterministic polynomial* means that an algorithm may be constructed which selects the different instances of the problem randomly (nondeterministic), constructs a solution for each instance and tests in polynomial time whether the solution solves the problem. A NP-problem is *complete* if any other NP-problem can be reduced to it in polynomial time. *Polynomial time* means that the time an algorithm needs to find a solution for a problem is no more than a polynomial function of the problem size. Although the test for each instance can be performed in polynomial time, there are no algorithms known that are able to solve a NP-complete problem in polynomial time. The reason is that the number of possible solutions grow much faster than the number of problem instances (from [57]).

of self-organization. Indeed, even evolution, the process that lies at the core of all living things, is a self-organizing process.

In this context, eusocial insects[2] such as ants, bees and termites have become important model organisms for biologists, as they offer a convenient way to study the rules that underlie self-organization. Although small, these animals are able to build very complex three dimensional structures orders of magnitude larger than the individual workers. Beehives, for example, consist of many thousands of hexagonal cells in which bees raise their brood and store their food. Termites mounds may rise several meters above the ground and are equipped with a sophisticated ventilation system [41]. Leaf cutter ants build vast subterranean "cities" extending up to seven meters underground, with tunnels up to 70 m long, inhabited by several millions ants, with a sophisticated tunnel system connecting chambers in which they farm fungi, organise waste disposal and ventilate the nest for climate control and air conditioning [35]. Less obvious, and much more important and impressive, is the ability of eusocial insects to coordinate a workforce of up to several million workers without central control. If we imagine a well-functioning city with thousands or even millions of citizens but without any central administration, we can catch a glimpse of how striking the self-organizing abilities of eusocial insects are.

1.3   Stigmergy

The key to understanding collective, self-organized decision-making by insects is the concept of *stigmergy*. Stigmergy describes a situation in which the output of a process serves as the input for the further development of the process. It was first defined in 1959 by the French biologist Grassé in relation to the construction process of the large, tower-like nest-structures of termites [29]. Since then, stigmergy was found to be an important principle of self-organization in all eusocial insects, allowing colonies to coordinate their collective behaviours.

Social insects such as ants can greatly increase their foraging efficiency by coordinating their foraging effort. For example, they could tune the number of workers they allocate to a food source depending on the amount of food available. How might ants achieve this? Ants deposit chemical substances on their way back to the nest after they have found a food source. These chemical substances, called *pheromones,* serve as a signal to other ants of the same species. While trail recruitment in ants can take many forms (for an overview see [19]), ACO typically considers the case of "mass recruitment". In this case, the chemical might mean *"follow this smell to food".* If another ant encounters the trail, either by smelling the pheromone at the nest entrance

---

[2]*Eusocial* means that individuals of different generations live together, cooperate in caring for the juveniles and only a subset of individuals reproduce (reproductive division of labor). The best known eusocial animals are ants, bees, wasps and termites, but rare examples of eusocial mammals (naked mole rats) and crustaceans exist. Eusociality is rare in the animal kingdom and only 2% of all insect species are eusocial. However, eusocial insects make up more than half of the biomass of all insects [36].

or by crossing the trail while searching for food, it is likely to follow the pheromone trail towards the newly discovered food source. Like the first ant, it will collect some food and lay pheromone while returning to the nest, reinforcing the trail. This will motivate more ants to follow the trail and themselves lay pheromone. As the trail gets stronger, the more likely ants are to react to it. As a result, a positive feedback loop is thus established that results in a constant flow of ants bringing food back into the nest. The number of ants at the food builds up exponentially. It is finally limited by pheromone evaporation and the crowding of the ants at the food source, since at a certain point the ants are unable to reach the food and therefore return without laying pheromone [64]. A further limiting factor is the amount of food available. When the food is exhausted, the ants cease laying pheromone, or begin searching for new food sources rather than return to the nest. The remaining pheromone slowly evaporates and the ants stop following the trail. The addition of a minor behaviour adds even more functionality: ants modulate the amount of pheromone they deposit depending on the quality of the food they have found [6]. Thus, the recruitment is stronger to higher-quality food sources, allowing ants to collectively select the best food source to exploit [7, 53].

1.4 Collective Decision Making in Ants

(modified from [57])

> By obeying algorithms, by using them as rules of thumb, each worker is able to make quick instinctive decisions in the midst of seeming chaos…With algorithms, the colony masters the problems natural selection has designed it to solve. The required information is distributed among the colony members. Thus, a distributed intelligence is greater than the intelligence of any one of the members, sustained by the increased pooling of information through communication. ([36], p. 58)

The above statement about ant colonies, made by the two famous sociobiologists and myrmecologists (ant scientists) E. O. Wilson and B. Hölldobler, provides not only a succinct description of how social organization works in ant colonies but also relates it to the way computer scientists think. Following principles are typical for the self-organized collective behaviour of ant colonies:

- There is no central authority [20].
- Individuals have only very limited knowledge about the state of the whole system and the environment [20].
- Individual ants act on a set of behavioural algorithms that are quite simple compared to the complex behaviour of the whole colony [35, pp. 315–315, 36, p. 57].
- Ants predominantly use pheromones deposited in the environment as signals for communication [34, 35, 63, p. 227].
- Changes in both the internal and/or external environment of the colony (e.g. internal: resource levels, colony demographics, external: food availability, tem-

perature) trigger positive and negative feedback loops, enabling structures or patterns to emerge [11, ch. 2]

- The colony as a whole reacts dynamically to changes in the environment [11, ch. 3], shifting the colony into a state that represents a good solution to the problems that the environment has posed upon the colony [36, p. 58].
- In many cases, colonies show a complete and abrupt transition from one stable state into another (*symmetry breaking* or *bifurcation*) [20].

Key to understanding self-organization in ants is to understand the way they communicate with each other using pheromones [35, p. 227]. In fact, ants use pheromones not only for marking trails to food sources and recruiting other ants towards them but for a vast array of different communication purposes such as warning others in case of danger, differentiating nestmates from non-nestmates, caste recognition, suppression of fertility in workers, and recognition of different development stages (for a general overview see [35, 36]). For the purposes of this chapter these can safely be ignored, but an understanding of how trail recruitment works is crucial, as this is what ACO is based on. Ants follow a pheromone trail by sensing the pheromone gradient around the trail and staying in a tunnel-like area of highest concentration [32]. They do not simply head towards the strongest pheromone concentration; rather, their decision is probabilistic. The higher the pheromone concentration an ant encounters, the higher the probability that it will respond to it. It is this probabilistic nature of trail following, which we must understand to truly perceive collective decision-making using trail pheromones.

1.5   The Deneubourg Model and Short Path Selection

(modified from [58, 59])

Deneubourg and colleagues showed in experiments on Argentine ants (*Linepithema humile*)[3] that ant colonies are able to collectively decide to use the shortest of two paths between their nest and a food source, and that a symmetry breaking occurs in that ants almost always focus on only one route to the food [28]. The Deneubourg group explained this behaviour based on four simple principles:

---

[3] Argentine ants (*Linepithema humile*) have spread from Argentina to become invasive to many ecosystems around the world. Unlike many other ant species, they show no or very little aggression towards members of other Argentine ant colonies, and they frequently share food, brood and workers between neighbouring colonies, which are often connected by trail systems. Since there is no clear distinction between the colonies, it is believed that they form large supercolonies which may spread over more than 3000 km [62]. Argentine ants have very poor vision, so are highly depend on pheromones [2]. They form extensive pheromone trails between their nest and their food sources and between the nests of different colonies [4, 37]. The ants deposit pheromone both on their way to and from the food [4] with up to four times more pheromone when returning to the nest [4]. This likely provides information about food quality similar to other ants, such as *Lasius niger* [6]. Very little is known about the absolute pheromone amounts the ants deposit on their trail [13, 58] and we are only able to measure relative pheromone concentrations by an indirect method based on the movements of their gaster when depositing pheromone [4].

Fig. 1 Shortest path
experiment (schematic)

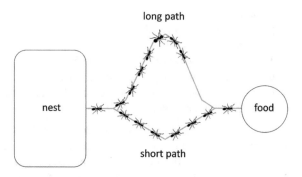

1. Argentine ants deposit pheromone on their trails in both directions. This assumption was already confirmed from observations (see above).
2. At a trail bifurcation, ants decide probabilistically which trail to follow depending on the pheromone concentration in two trails. This assumption was weakly confirmed by measurements of [33] on another ant species (*Lasius fuliginosus*).
3. The decision the ants take follows a simple mathematical choice function. We refer to this function as the *Deneubourg choice function (DCF*, see below). This function relates the pheromone concentration on each trail with the probability that an ant will take the one or the other trail. The DCF has only recently been rigorously tested [58].

$$pl = \frac{(k + c_L)^b}{(k + c_L)^b + (k + c_R)^b} \text{ and } pr = 1 - pl$$

pl—decision left branch c_L—pheromone concentration in the left branch, c_R—pheromone concentration in the right branch, b—exponent, k—constant.

4. An initial preference is amplified by positive feedback and finally results in the selection of a preferred solution by most of the ants (Figs. 1 and 2A).

The model explains the selection of the shortest of the two paths mainly by an initial time delay between the ants arriving at the food via the long path and those arriving via the short path. During this period, the long path carries no pheromone at all at the food-side and the returning ants prefer the short path on their way back to the nest and deposit pheromone on it. This generates a positive feedback loop in favor of the short path. Thus, in most experiments, the majority of ants follow the shortest path (see Fig. 2a). If both paths are of equal length, no time delay occurs. Stochastic differences between the paths occur at the beginning and positive feedback amplifies the path with an initially slightly higher concentration so that it is finally selected by most of the ants. Consequently, symmetry breaking occurs in most of the experiments—and in any one experiment either path is preferred by the majority of ants with equal probability (see Fig. 2b).

Based on these principles, other types of collective ant behaviour can be successfully explained, for example how ant colonies are able to select the higher quality

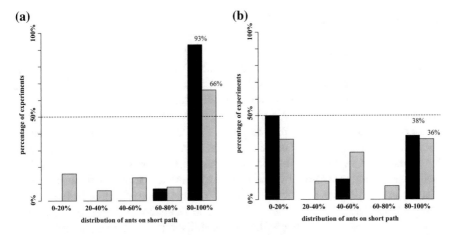

Fig. 2 Results of the original shortest path experiments, modified from [28]. The x-axis shows the distribution of ants that chose the short path (n = 500); the y-axis shows the number of experiments (n = 14 in **a**, n = 26 in **b**) or simulations (n = 1000) in percent. Black = original experimental results, grey = results of Monte Carlo simulations with parameters $b = 2$, $k = 20$

food source if food sources of different quality are presented to the colony [6, 7, 53] and how ant colonies can coordinate the division of labor by caste specific response thresholds [10, 27].

1.6 Extensions of the Deneubourg Model

1.6.1 U-Turns

One of the prerequisites for the Deneubourg model is the deposition of pheromone in both directions. Beckers [5] showed that the ant *Lasius niger*, like Argentine ants, is able to select the shortest path. However, in contrast to Argentine ants, *L. niger* do not lay pheromone on their first trip towards the food source. Therefore, an initial preference for the short path cannot build up and the model does not work. This discrepancy was solved by a significant modification of the Deneubourg model. It was found that the ants frequently made U-turns, and the probability of making a U-turn was greater on the long branch since the ants spend more time travelling the long branch. This leads to a preference of the short branch. It was also found that U-turning ants lay no pheromone and that the U-turn rate decreases with rising pheromone concentration. Both effects amplify the preference for the short branch. By extending the Deneubourg model with these effects, computer simulations could fully reproduce the short path selection rate of *L. niger* [5].

1.6.2   Short Path Selection in Labyrinths

When Argentine ants traverse a labyrinth to reach a food source, there are many possible routes that they may take. Due to the high number of possible routes, the initial time differences are less distinct than in the shortest path experiments and it can be expected that the ants' ability to select the shortest path will decrease.[4] However, experiments with Argentine ants gave a different result [56]. About 60% of the ants chose the shortest path and 20% the second shortest path. These results could be reproduced in computer simulations by assuming an extremely high value for the exponent of the DCF (b = 4), or by assuming that ants dislike changing their walking direction (see also [66]) and that they modulate pheromone deposition such that the ants lay twice as much pheromone on their way returning to the nest as coming from the nest. In contrast to the original Deneubourg model, the resistance to directional changes together with the probabilistic response to a pheromone was shown to be the main reasons for the system to be able to select the shorter paths in the labyrinth.

In a further experiment with Argentine ants, an established short route in a labyrinth was blocked. However, the Argentine ants were still able to find the remaining shortest route through the labyrinth [51]. This shows that they are able to dynamically react to a changing environment. Again, these results can be explained by factors such as ants avoiding changing their heading, or by the existence of a long-living exploratory pheromone besides a short-living trail pheromone, rather than by the amplification of initial time differences.

Vela-Pérez et al. [55] as well as Ramsch et al. [49] have developed models of ant foraging in labyrinths, which incorporate factors such as directional persistence or preference of small angle turns. These models also used the DCF as a decision rule. Vela-Pérez et al. [55] showed that the exponent $b > 1$ was critical for the system to show symmetry breaking and short path selection, and Ramsch et al. [49] used exponents between 2 and 4 in their simulations. Thus, while the DCF has clear limitations, it is still generally used as the basis for understanding collective route selection by ants.

---

[4]To test this, we extended the simulations of the shortest path experiments described in Fig. 2a (DCF with k = 20 and b = 2) by adding further virtual bridges arranged in line. In this way, we simulated different path length to the food, as in a labyrinth. We counted the number of ants taking the short or the long sections. While in the original setting with one bridge, 75% of the ants selected the short sections, this ratio dropped to 64% if three bridges were present, 62% if five bridges were present, and 57% with ten bridges. This demonstrates that the majority of ants still chose the short sections but the effect size drops the more paths are available, and that there was a considerable percentage of virtual ants that took the long sections (von Thienen unpublished data).

1.7 Ant Colony Optimization—ACO

Computer scientists were inspired by the findings of the Deneubourg group to develop a computer algorithm able to solve complex optimization problems [22] such as the traveling salesman problem and other NP-problems. Such an algorithm is not necessarily intended to find the ideal solution but rather a good approximation to the ideal solution in a time period short enough for practical purposes (see introduction). To do so, the ACO algorithm uses the same principles as the Deneubourg model, i.e., pheromone deposition, probabilistic response to different pheromone concentrations according to the DCF, and positive feedback. The best way to describe how it works is, again, to use the Travelling Salesman Problem (TSP) as an example. To solve the TSP, a number of virtual ants simultaneously and randomly visit the different cities until each ant has visited all cities. After completing their journeys, the ants mark their route with an amount of virtual pheromone that correlates with the inverse of the distances they travelled. This process is repeated in the next iteration, but this time, the virtual ants do not walk completely randomly. Rather, they slightly prefer the routes marked with pheromone. When repeating this process for many iterations (usually 1000), a path of the highest pheromone concentration emerges that usually marks the shortest path or is at least close to the shortest possible path. As a result, the algorithm converges to a solution that is close to the theoretical optimum.

Besides the TSP, ACO-algorithms have been applied to many different problems like data network routing [24, pp. 223–260], finding the best transport routes for trucks [24, pp. 155–159], scheduling of incoming airplanes at airports [9], machine scheduling in factories [8], data mining [46] and construction of phylogenetic trees [48] (for an overview see [23]). These are only some examples of ACO applications, and the list is steadily growing. ACO has become one of the most successful algorithms inspired by swarm intelligence, with over 3400 articles in scientific journals, dealing with ant colony optimization (according to a search in the Web Of Science Core Collection in March 2018).

The main components of the original ACO algorithms were the use of a virtual pheromone to mark good solutions, positive feedback to allow the algorithm to converge towards the optimal solution, and evaporation of the virtual pheromone to avoid the algorithm converging too early on a suboptimal solution.

Since its invention in 1996, the original ACO algorithm has undergone various improvements intended mostly to refine software algorithms rather than inspired by actual ant behaviour. Such improved algorithms included the *Elitist Ant System, Rank-Based ant system, Max-Min ant system* and *Ant Colony System (ACS)* [24, p. 73 ff.]. Nowadays, ACO is thus a meta-heuristic framework inspired by the Deneubourg model for finding good solutions to different kinds of optimization problems. In an *Elitist Ant System*, the best solution currently found is given a bonus by adding an extra amount of pheromone. This can be viewed as an elitist ant that had found the best current route and deposits an extra amount of pheromone on it. In a *Rank-Based Ant System*, the routes are ranked and marked with pheromone amounts corresponding to their rank. The *Max-Min Ant System* implements several different techniques. First, only the best route found is marked with pheromone. However, this may lead to a very early stagnation of the process. Therefore, a second mechanism is implemented by limiting the amount of pheromone to a certain range (minimum-maximum). In addition, the algorithm is reinitialized when stagnation occurs. In the *Ant Colony System*, negative feedback is implemented such that ants reduce the pheromone amount at a node they have crossed. This prevents the algorithm from running into stagnation.

2   Do Ants Use ACO?

2.1   There Is no Ant-God

The greatest difference between real ants and ACO is that all ACO algorithms currently used for practical applications have complete knowledge about the state of the system and perform operations on a global level. In fact, it was shown in the early stages of ACO-development that without such a global knowledge, the algorithm performed worse than other optimization techniques and was not suitable for practical purposes. Examples of such advanced ACO algorithms that use global knowledge are the above-mentioned Elitist Ant System, Rank-Based Ant System and Min-Max Ant System. Such advanced ACO algorithms perform centralized operations like calculating and comparing the lengths of routes found after each iteration, and using this knowledge to mark the best routes with additional pheromone.

Another form of global knowledge used by ACO is the information that defines the cost of going from one node of the solution space to another. This information is necessary to evaluate the quality of the solution found by ants. In case of the TSP, this is the distance between the cities. Usually, this information is available a priori and stored in a global table. However, it is not always necessary to provide such global information a priori. For instance, virtual ants may evaluate the distance by themselves while exploring the solution space and store it in their individual memory. They may also copy it to the global table or to a table at each node that can be read by other ants passing that node. However, since this has to be done for each ant, the cost in terms of performance is high and the approach is only advisable for applications in dynamic environments. A good example is AntNet, an ACO-algorithm for data network routing. The algorithm uses global routing tables and a globally synchronized time. It evaluates local heuristic information at each network node using network traffic by virtual ants and stores it in local tables at network nodes [23, 24, 44]. Thus, although operating in a distributed environment and performing decentralized operations, the algorithm still depends on global information. However, there are some similarities between the ways ants and ACO-algorithms use these kinds of global information: real ants are very capable of measuring distances and even keeping track of the directions they have walked [65], and many ant species have very good visual orientation [40]. But, by contrast to ACO, this information is only available to an individual ant due to experience and learning and not shared among the colony members. However, ants can modulate their pheromone deposition depending on their distance from the nest [21], which, together with other pheromone based information like food-quality, may be viewed as a kind of heuristic information similar to heuristic information used in ACO.

Thus, in contrast to ACO, there exists nothing like an ant-god that has global knowledge, constantly evaluates the best solution and directs the colony towards this solution. Instead, each ant has only a very limited knowledge and memory. It simply deposits pheromone along its path and reacts to pheromone in a probabilistic manner. This is sufficient for a behavioural pattern to emerge that is beneficial for the colony. Therefore, it is perhaps misleading to consider ACO a swarm-intelligence system. The algorithm depends on the global knowledge and only uses a stochastic process in the behaviour of its local agents to converge towards an optimal solution. However, for situations in which no central authority is available, the ACO-concept of depositing pheromone and reacting to it in a probabilistic manner may still be useful if it is combined with further techniques that incorporate knowledge about the environment, see for instance Meng et al. [45] who have developed an ACO algorithm for distributed robot swarms. Thus, while a major difference between ACO and real ant colony decision making is an increased centralization in ACO, this is mainly done to increase computational efficiency—an option which is not open to real ants.

2.2 Ants Do not Use the Deneubourg Choice Function

In biological modeling, it is important to find a balance between complexity and simplicity. Since biological systems are usually extremely complex, it is nearly impossible to develop models that would incorporate all possible factors. Instead, a model must focus on those factors that are essential for an understanding of the biological system investigated. The Deneubourg model was very successful in finding this balance and was very helpful in understanding ant behaviour. The main factors the model originally focused on were the probabilistic response of ants to pheromone—a simple choice function to give a mathematical description of the response—and the amplification of an initial preference of the short path caused by difference in time needed to travel both paths. At that point the model gave a sufficient explanation of the ants' behaviour in bifurcation experiments. Meanwhile research has continued and biologically important aspects not so far incorporated into the model must be considered.

2.2.1 Measuring the Dose-Response Relationship

Although experimental evidence existed for a general relationship between the ants' decision and the pheromone concentration in two other ant species [33, 54], no exact relationship between ants' decision and pheromone concentration had been measured for a broad range of concentrations, neither in Argentine ants nor in other ant species. In addition, it was not known whether the pheromone concentrations used in the experiments were realistic and occurred on natural trails. Moreover, the Deneubourg model did not distinguish between detection and discrimination of pheromones. The ability to detect a pheromone may be completely different from the ability to discriminate between two different pheromone concentrations. Further-more, the parameters of the DCF (k and b) lack a biological interpretation. Thus, a large part of the Deneubourg model—the exact relationship between pheromone concentration and path choice—was hypothetical. Even the original experimental results of the shortest path experiments published by the Deneubourg group were not fully reproduced by their simulations. The parameters of the DCF were chosen arbitrarily to give the best fit to the experimental data. However, in only 66% of the simulations did the majority of ants choose the shortest path, compared to 93% in the experiments (see Fig. 2a).

In von Thienen et al. [58] we attempted to test and parameterize the DCF. We measured the exact dose-response relationship in three ant species, one of which was the Argentine ant. We did this for detection as well as for discrimination. The results of our experiments were similar for the three ant species. As Deneubourg and his group assumed, the dose–response relationship followed an S-shaped curve (see Figs. 3 and 4). However, there were significant differences from the original assumptions. When we fitted the data to the DCF, the parameters were quite different from the parameters that the Deneubourg group had assumed. In the shortest path experiments the parameters were set to give a good fit to the experimental results with b = 2 and k = 20. Similar and even higher settings for the exponent were used in the model with U-turns and in the model of the labyrinth experiments, as well as in the model of Vela-Pérez et al. [55] (see above). In contrast to these assumptions, the values for the parameter b we deduced from experiment were b = 0.52 in detection experiments and b = 1.06 in discrimination experiments.

2.2.2 Weber's Law

In a further experiment, we tested how sensitive the discrimination abilities of the ants were to changes in absolute pheromone concentrations. The response of the ants followed the prediction of Weber's law. Weber's law is a key concept in understand-ing how animals (including humans) perceive the world. It states that the ability to distinguish two different physical stimuli only depends on their ratio and not on their absolute values (see Box 1). Weber's law can be expected to be of great biological importance to ants since the ability to discriminate between different concentrations should remain constant within a physiologically realistic range of concentrations. If this were not the case, the information encoded in the ratio of pheromone concentra-tions between two trails would be unstable due to changing environmental conditions such as humidity, temperature and evaporation. If we failed to confirm this stability, an important aspect of communication and information use in ant models could not work under realistic conditions. A similar finding in a different experimental setting

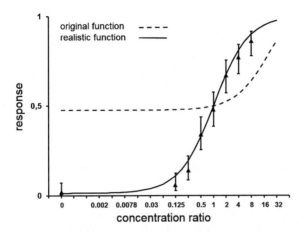

Fig. 3 Dose-response relationship (discrimination) for *Linepithema humile* (Argentine ant) fitted with the Deneubourg choice function. Symbols show the probability that the ants take the test trail (response) depending on the concentration ratio between test-and reference trail (with concentration = 1). Bars show the 95% confidence limits. The dashed line shows the original Deneubourg choice function used in the shortest path experiments with parameters k = 20 and b = 2. The solid line shows the Deneubourg choice function fitted to the data with parameters k = 0.02 and b = 1.06. The concentration ratio is given on a $\log_2$ scale

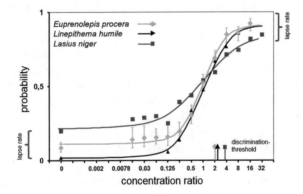

Fig. 4 Dose-response relationship (discrimination) for three ant species, fitted with the psychometric function. Symbols show the probability that the ants take the test trail (response) depending on the concentration ratio between test- and reference trail (with concentration = 1). Bars show the 95% confidence limits (only shown for *E. procera* for clarity reasons). The lines show the fitted psychometric functions. The concentration ratio is given on a $\log_2$ scale

(an open arena instead of bridges) was published before by Perna et al. [47]. They found that the more ants had passed by an arbitrary point in the arena following a certain direction, the more future ants passing that point would change their heading towards the direction in which most of the previous ants went. The relation between the change in heading and the ratio between the numbers of previous ants heading left or right followed Weber's law. The authors assumed that the pheromone concentration at a certain point is proportional to the number of ants that had passed that point. Thus, they concluded that the ants' response to differences in pheromone concentrations follows Weber's law too. Most importantly, the authors have shown that the Deneubourg model can only be consistent with Weber's law if the exponent is set to 1 (and the constant k $\neq$ 0). Since the Deneubourg model and its extensions only work with an exponent >1 (and k $\neq$ 0), it contradict Weber's law and thus is missing something critical.

2.2.3 Applying Psychophysical Theory to the Collective Behaviour of Ants

Although the Deneubourg model gives a general explanation of the ability of ants to collectively find the shortest path, the above arguments show that there are still important points missing in the model.

Most importantly, the response of an ant to pheromone is determined by the way it perceives the pheromone through its sensory organs, processes this information by its nervous system, and, finally, translates it into behaviour. Thus, a rigorous theory of how ants perceive pheromones and translate this information into a specific behaviour is needed to bring biological realism to our understanding of collective decision-making through pheromone trails. We found that a viable explanation could be provided by *psychophysical theory* (see Box 1). This theory describes the relation between the strength of a physical stimulus and its perception. It is based on Weber's law, which we found to be well followed in our experiments. This was the first indication that psychophysical theory might provide an appropriate explanation of perception mechanism, describing important aspects of ants colony behaviour. We have found in our measurements that the dose-response relationship of an ant colony can be described by psychophysical theory and follows a *psychometric function* (*PF*) (see Fig. 5). In contrast to the original Deneubourg model, this approach is not only based on an established theory of perception but, in addition, it allows us to define biologically meaningful parameters such as *detection and discrimination thresholds, information capacity* and *error rates* (*lapse rate*). These parameters describe characteristic differences between the three species we studied, and could be mapped to their specific ecological needs [58].

Box 1: Psychophysical theory,

(modified from [58–60])

Psychophysical theory was developed by Fechner [25] based on the works of Weber [61]. Weber sought to find a relationship between the strength of a physical stimulus and its sensory impressions. He discovered that the ability to discriminate between two stimuli depends on the ratio of the stimulus strengths, which, within certain limits, is independent of the absolute stimulus strengths (Weber's law). Fechner [25] discovered later that the sensory

(a) **(b)**

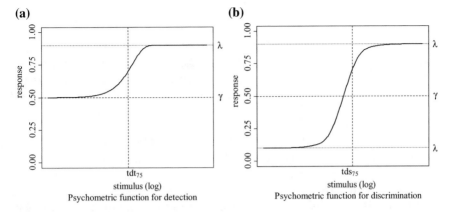

Fig. 5 Schematic examples of psychometric functions. In **detection experiments** (**a**) the psychometric function (PF) gives the probability that a stimulus is reported stronger (response) than a null-stimulus. In **discrimination experiments** (**b**) the PF gives the probability that a stimulus is reported stronger (response) than a constant stimulus, which is larger than zero. Lapse rate (λ, dotted horizontal line), guess rate (γ, dashed horizontal line), 75%-detection threshold (tdt$_{75}$, dashed vertical line), 75%-discrimination threshold (tds$_{75}$, dashed vertical line). Note that the 75%-threshold is the point at half the distance between guess rate and upper asymptote, thus, it may not be exactly at p = 0.75. Solid lines show the PF

impression of a physical stimulus is proportional to the logarithm of the stimulus strength, which is the reason why, for example, sound levels are measured on a logarithmic scale in units of decibel.

A mathematical framework has been developed that relates physical stimuli to sensory impressions. This framework incorporates the effect of noisy backgrounds that influence the ability to detect a signal and it gives clear mathematical definitions of sensory thresholds. One of the most useful mathematical tools is the *psychometric function* (*PF*), which describes the relationship between the probability of a positive response p to a stimulus and the stimulus strength x (see Fig. 5)

$$p(x) = \gamma + (1 - \lambda - \gamma) \cdot F(x)$$

x—stimulus strength, λ—guess rate, γ—lapse rate, $F(x)$—function describing the probability to detect a stimulus by the underlying sensory mechanism [39, p. 74]. For F(x) a probability distribution like the Weibull distribution is applied.

2.2.4 Testing the Models

We tested whether we could reproduce the original results of the shortest path experiments by using realistic parameter values that had been deduced from the experiments described above [59]. To do so, we repeated the simulations described by the Deneubourg group [28] with the realistic parameter values. Both the DCF and the PF could reproduce the experimental results when path lengths were equal. However, if the paths were of different length, neither the Deneubourg model with the DCF nor the model with the PF could reproduce the experimental results (see Fig. 6).

These results show that neither the DCF nor the PF form of the Deneubourg model can satisfactorily describe the collective behaviour of ants in a bifurcation experiment. As the first step toward finding the missing factors in the model, we tried to identify biologically reasonable mechanisms that could additionally amplify the initial differences between the paths. These conditions can be satisfied by the modulation of pheromone deposition since it is well documented that argentine ants deposit up to four times more pheromone on their way back to the nest compared to the way from the nest [4]. We incorporated this effect into our model. The results showed that the short path was selected significantly more often than the long path, and the PF showed a much higher rate of short path selection than the DCF (see

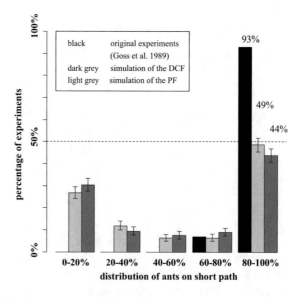

Fig. 6 (From [59]) Monte Carlo simulations of shortest path experiments with the DCF and the PF with realistic parameter values. Paths differ by a factor of two. Parameters for the DCF: $b = 1.06$, $k = 0.02$ (discrimination) and $b = 0.52$, $k = 0.02$ (detection). Parameters for the PF: $\gamma = 0.47$, $\lambda = 0.02$, $b = 0.87$ (discrimination) and $\gamma = 0.43$, $\lambda = 0.11$, $b = 0.6$, $tdt_{75} = 0.053$ (detection). The x-axis shows the distribution of ants that chose the short path ($n = 500$). The y-axis shows the number of experiments ($n = 14$) or simulations ($n = 1000$) in percent. Error bars give 95%-binomial confidence intervals of the simulations

Fig. 7 (From [59]) **a** Same
simulations as in Fig. 6, with
pheromone modulation. The
ants deposit four times more
pheromone on their way
back to the nest than in the
opposite direction. **b** Same as
in A but with the additional
assumption that the ants have
a higher responsiveness in
the initial phase. This was
achieved by increasing the
exponent (DCF) or the slope
(PF) four times in the first
100 iterations. Black =
original experimental results
modified from [28], dark
grey = simulation of the
DCF, light grey = simulation
of the PF

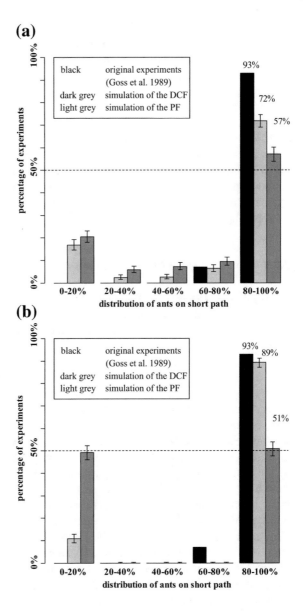

Fig. 7a). Further analysis showed that the lapse rate, a parameter that the DCF does
not have, was responsible for this difference. The lapse rate is a measure for the ants'
independence from pheromone trails [58].

Although the modulation of pheromone deposition gave better results for the short
path selection rate compared to experiments without modulation, the real ants still
performed much better than the virtual ants (93% vs. 72% in case of the PF). This

indicated that there were still some factors missing. We also tested several different biologically meaningful assumptions such as the combination of two pheromones with different evaporation rates, path memory, negative or positive feedback between pheromone concentration and pheromone deposition, higher pheromone deposition on the return path in the initial phase [59]. However, we could not fully reproduce the experimental results of the original shortest path experiments. We also tested the effect of U-turns because they had been shown to be important in short path selection in the case of *L. niger* ants [5] (see above). However, we could not confirm the result of Beckers [5] if we applied realistic parameter values (von Thienen, unpublished data).[5] Only by making the hypothetical assumption that ants show a higher responsiveness to pheromones for a short time after they have found food and in the beginning of the recruitment, we were able to fully reproduce the experiments by applying the PF (see Fig. 7b) [59].

2.2.5   The Ecological Meaning of the Deneubourg Model

Biologists are always interested in how a certain behaviour helps a species in its struggle to survive and how that behaviour was formed during the evolution of that species. In case of the shortest path experiments, we might ask how the ability to select the shortest path towards a food source gives the colony an advantage and increases its fitness sufficiently to be favored by evolution. Obviously, the ability to find the shortest path is in itself advantageous. It reduces predation risk due to shorter time periods outside the shelter of the nest, and also reduces the time and energy spent collecting food. However, to our knowledge, this advantage has never been demonstrated by field research and is therefore only hypothetical. It is also reasonable to ask whether these collective behaviours, which have been demonstrated in the lab, also occur in the 'real world'. Usually ants live in an open field where an infinite number of possible routes and obstacles between nest and food exist and form a sophisticated trail system connecting different nests and food sources with many intersections and nodes that had been formed over a longer time period. This trail system changes dynamically in response to the food supply outside the nest and changing environmental conditions [43]. Since in a network of trails many different paths exist that the ants may explore, it can be expected that the ability to select one of the shortest paths will be lower compared to a situation in which only two paths are available (See above footnote 4). In addition, food sources are discovered asynchronously and initial time differences due to different path lengths will therefore

---

[5]The model settings were similar to those described in Fig. 6 with the exception, that U-turns were incorporated as described by Beckers [5]. The U-turn settings were such that in average half of the ants made a U-turn while travelling a distance equal to the long path if no pheromone was present. The probability to take a U-turn decreased with the pheromone concentration according to the formula given by Beckers [5] with P_0 giving the U-turn-probability in the absence of pheromone set to 0.5, α set to 0.1 and C giving the pheromone concentration. The histogram of the short path selection was similar to that of the simulations without U-turns as shown in Fig. 6 from left to right: 29%–9%–7%–6%–49%.

only occur very rarely. Lastly, nature usually does not offer the sort of infinite supply of food present in laboratory experiments, which fuels such strong positive feedback cycles. Taken together, the 'real world' situation is worlds apart from the shortest path experiments with only two possible routes that are simultaneously explored.

With this in mind, it may happen that the results of the shortest path experiments simply represent an artefact of the special setup of the experiments, and may have little or no ecological meaning. Nevertheless, the experiments described above with Argentine ants in labyrinths and the models for their behaviour [51, 55, 56] were much more realistic than the original shortest path experiments and show that Argentine ants are able to find short paths in a more complex environment. However, the assumptions made in these extended Deneubourg models to explain the short path selection in labyrinths have not been tested with realistic parameter values for the DCF. Only the assumption of the modulation of pheromone deposition has so far been demonstrated to work in the model for shortest path selection with realistic parameter values (see above). Other assumptions, such as directional preference, combination of long- and short-living pheromones and preference of small angles have not been tested yet with realistic values of the exponent b in the DCF. Particularly problematic is the high value of the exponent, used in all of the extended Deneubourg models, which is unrealistic and contradicts Weber's law.

In contrast to models for shortest path selection, the Deneubourg model seems to work fine for the ants' ability to select the best food source if food sources of different quality are available, even when realistic parameter values are applied. The Deneubourg model could explain this behaviour very well for *L. niger* ants [6, 7, 53] (see introduction). This ability gives the ants a clear selective advantage and bears great biological meaning. We tested the model with realistic parameter values for the DCF and for the PF for *L. niger* as well as *L. humile*. We were able to confirm the results of Beckers et al. [6, 7, 53] for both functions and species.[6] Thus, in contrast to the shortest path experiments, the selection of food sources of different quality is consistent with the Deneubourg model and works with realistic parameter settings. Together with negative feedback factors like crowding at the food source and on the trail [18, 64], reduced pheromone deposition on marked trails [18] or the incorporation of the lapse rate [57, 58], the model might give good explanations of how ants are able to dynamically switch between different food sources and how

---

[6]The simulation settings were the same as described in Fig. 6 except that the paths were of equal length and led to two different food sources of different quality and that the returning ants deposited four times more pheromone on the right side with higher food quality than on the left side with lower food quality. *L. humile* and *L. niger* deposited pheromone in both directions with the exception that *L. niger* did not deposit pheromone on its first move to the food. For both ant species, we found a clear preference of the richer food source with following histogram similar to Fig. 6 from left to right:

DCF-*L. humile*: 11%–10%–8%–8%–63%—DCF-*L. niger*: 0%–0%–0%–100%–0%.
PF-*L. humile*: 1%–12%–17%–14%–56%—PF-*L. niger*: 0%–0%–0%–100%–0%.
Results are similar if pheromone is deposited only in one direction from food to nest.
([57], unpublished data).

they are able to optimize the foraging process if different food sources of different quality and distance are available.

2.3 Similarities Between ACO and Ant Trail Use

As we have seen, there exist deep differences between ACO and how ants actually use pheromone trails. However, besides the use of pheromones, there is one surprising parallel between how ACO works and at least one aspect of pheromone trail use by real ants. The pheromone-based solution search in ACO is very often coupled with local search to find local optima [24, pp. 92–96]. A similar function to localize searching is played by pheromone trails of many ant species. A good example is patrollers of the *Pogonomyrmex barbatus* ants. The nests of these ants are connected to different food sources by long-lasting trails. Each morning, the patrollers mark one of the trails, causing the foraging ants, which leave the nest, to follow this specific trail for the rest of the day [30]. Similar to ACO, ants combine different local search strategies, depending on the dispersion and type of food they are collecting [42]. They may for instance lay long-lasting trails to certain areas in which good food sites are located. Individual ants then independently search for food in these areas. Or a subset of foragers leave long-term trails to stable food-sources and engage in solitary search for short-living food sources scattered in the environment, like dead insects. Ants may also combine trails with long-lasting pheromone to stable food-sources like colonies of honeydew-secreting insects and trails with short-lasting pheromone leading to dead insects scattered in the area. Indeed, ants finding a resource near a well-established trail may even recruit other ants directly from the trail, without returning to the nest [26]. Similarly, as an army ant colony sweeps over an area, the trail system forms a fan-shaped structure, with a main trail representing the direction of the raid and many branching trails forming a constantly advancing front of foraging ants searching the local environment of the main trail for food [52].

3 Possible Further Improvements of ACO by Real Ant Behaviour

Over twenty years have passed since the invention of ACO and the original experiments of the Deneubourg group, and ACO-algorithms have become a successful tool for engineers and software developers. However, as has been stated before, the main improvements of ACO are strictly software-motivated extensions of the algorithm to improve performance or adapt it to specific problems. Here we will propose some possible improvements to ACO inspired by real ants that we, as biologists, have not tested yet, but that engineers and software developers might find interesting.

First, we will start with a disappointment. The use of a PF in ACO-algorithms most likely will not improve the algorithm. The PF is mathematically more complex than the DCF and therefore will not improve performance. In contrast, the DCF is a very simple function that can be calculated very fast. However, it should be kept in mind that the DCF does not follow Weber's law if it has exponents different from 1 and $k > 0$, which is common to ACO-algorithms. This leads to the effect that the solution space not being symmetrical with respect to absolute pheromone concentrations. In regions of low pheromone concentration the system behaves differently than in regions with high pheromone concentrations. To our knowledge, this effect has not been tested yet, and the implications for ACO algorithms have not been considered.

The concept of the lapse rate may be implemented in ACO. This might help avoid situations in which ACO algorithms reach stagnation before finding an optimum. This situation happens if all virtual ants follow the same path without coming close to the optimal solution. Usually this is avoided by adding a priori information to the algorithm about each node of the state-space, or by implementing local search algorithms that, for each ant, suggest optimal steps to take in the next iteration, or by letting the virtual pheromone evaporate [24, p. 70, pp. 215–216]. The concept of the lapse rate offers an alternate approach. By implementing the lapse rate concept into ACO algorithms, there is always a set of ants which more or less ignore the pheromone trails and randomly explore solutions beyond established pheromone marked routes. This could be achieved by defining a set of virtual ants that respond to pheromone weakly or not at all, or, perhaps more effectively, by allowing ants to be variable in their response to pheromones [12]. Inter-individual variability in social insect colonies can lead to increased effectiveness, but the costs of this variability are not yet well understood [38].

Czaczkes et al. [18] showed that ants deposit less pheromone if they are already walking on a pheromone-marked path. This is effectively a negative-feedback system, which makes strengthening of trails progressively harder, and gradually caps the maximum strength of a trail. By preventing trails from becoming too strong, this allows newly-discovered trails of a higher quality to successfully out-compete established trails, again preventing stagnation and trapping in a sub-optimal solution [14]. As mentioned before, the Ant Colony System (ACS) implements such a negative feedback by reducing the pheromone concentration on a node visited by an ant [24]. However, a more nuanced result with lower computational costs can be achieved by implementing a mechanism of local response to pheromone strength.

The modulation of pheromones described above, where ants deposit more pheromone as they return to the nest and encode food quality in it, has already been implemented in a similar way in the *Elitist Ant Systems* [24, p. 73]. In this case, the best of all currently found solutions is marked with an additional pheromone amount. However, the idea behind the Elitist Ant System is different from the situation observed in the nature since the evaluation of the best solution needs a central authority which the ants do not have. Instead, they incorporate directional and qualitative information into the system by depositing more pheromone when returning to the nest. Since classical problems solved by ACO, like the travelling salesman problem, do not require information about direction, the concept of pheromone modulation will not improve these algorithms. However, if the problem is such that directional information is important, the concept of pheromone modulation may be useful. For example, if a salesman prefers to visit cities in the east before he visits cities in the west (a priori definition), the virtual ants may deposit more pheromone when heading east, causing the algorithm to prefer cities in the east. Another variation may be that cities or group of cities may have different qualities (for instance sale rates). Routes to these cities might receive more pheromone, causing such cities to be preferred by the algorithm. This strategy might be good for a lazy salesman. He first visits the cities in which he can expect good business. If he has made enough sales, he may want to stop and go home, cancelling the less productive routes.

Memory and learning are important factors in ant behaviour and it had been shown that ants acquire route memory by visual learning which also modifies their response to pheromones [2, 3, 16, 17, 31, 60]. Several studies, [15] show that individual memories can improve collective decision-making process in ants. ACO algorithms already use the concept of individual memories to store a priori information or to store information collected during the exploratory phase. As an inspiration from real ants, we suggest the development of learning ACO algorithms. As an example, let us take a salesman who frequently repeats the same tour. While travelling he may modify the route, for instance, to avoid traffic congestions that result from road constructions works. After he has finished a tour, a second long lasting pheromone is added to the modified route segments that he has actually taken, increasing the attractiveness of these segments. In the next run of the ACO algorithm, this long lasting pheromone is still present from the first run and the marked segments will be preferred for the next journey of the salesman. In this way a kind of memory builds up that incorporates information of former tours into the construction process of new tours.

Our knowledge of how ants are able to collectively react to dynamically changing situations may inspire the development of ACO algorithms that react to changing situations in real time. Experiments and computer simulations show that the efficiency with which a group of ants exploit multiple food sources increases if individual experience and pheromone-based orientation are used together. We term this phenomenon *composite collective decision-making* [15]. The pheromone-based recruitment allows ants to quickly concentrate on the best food sources, and the individual memory component allows subsets of ants to specialize in less productive, but underexploited resources. Such a technique could be used to develop ant algorithms that use current infrastructure more efficiently. This may be of particular interest in intelligent transport system applications based on autonomous vehicles. ACO has been frequently used to develop algorithms for optimization of traffic flow (for instance [1]). These algorithms use the ACO-concepts to mark the virtual roads with pheromone depending on the traffic density, and then use that information to calculate the best route for a specific user. In a more advanced form, the algorithm is able to automatically find a balance between the goal of individual drivers to minimize their travel time and the overall goal to minimize the travel time of all drivers by optimizing the travel flow in the entire network. So far, such concepts remained the subject of theoretical studies based on simulations. However, in a modern context with big data and modern navigation devices able to communicate in real-time, they may be implemented in real-time traffic guidance systems that constantly evaluate the traffic situation, communicate with the navigation devices in cars and optimize the traffic flow. The algorithm for such systems could implement the concept of composite collective decision making by incorporating individual experience into each navigation device and using it together with the real-time data of the traffic situation encoded in virtual pheromone. For instance, the navigation device could remember the best found routes of the past of its user and thus fine tune the route calculation.

We have found that, apart from memory mechanisms, starvation can also alter the response to pheromone [60]. In dynamic situations this idea may be used to assign a time-dependent priority to each node of the state-graph. Let us suppose a travelling salesman has different time slots at which he has to visit his customers. To meet this requirement, he may update his scheme by a new ACO-run after each visit. In each run, the difference between the calculated arrival time at the remaining cities and the originally calculated arrival time is used to update the priority of the city and guide the system in such a way that cities with the greatest delay are visited next. In a similar way, Bencheihk et al. [9] calculated the scheduling of aircraft landing at an airport by ACO by penalizing the deviation from the scheduled time of arrival.

4 Conclusion

ACO is a very good example of how the complex and self-organizing behaviour of social insects inspired new optimizing solutions for complex technical and organizational problems. It is also a nice demonstration of how, by distilling the key aspects of a biological system and adding specific, non-biological improvements, the biomimetic approach can be very successful (see also [50]). ACO has only a few major aspects in common with real ant behaviour, namely the deposition of a virtual pheromone and the concentration–dependent, probabilistic response to the pheromone. This is to be expected, as ACO and collective ant behaviour developed and function under very different constraints. Real ants cannot use a central controller—there is no ant god—and so have developed a robust system which does not require one. ACO still requires central control, and time will tell whether fully decentralized ACO algorithms may generate practical solutions in acceptable time frames. In our opinion, applications for swarm robots, self-organizing factories, algorithms for social media, search algorithms, data mining, business intelligence, and applications in which many humans are simultaneously involved in real time (such as traffic) are interesting fields for further research on the application of fully decentralized ACO algorithms. At the same time, research on ant behaviour continues, and there are many things that we, as biologists, do not yet understand, and many unknown unknowns waiting to be discovered. We are still far from having a universal ant model that is able to incorporate at least the most important mechanisms of ants communication and their behavioural responses. We do not even have a commonly accepted model of how trail networks are formed, how trails are structured and what exactly happens when ants follow a trail and deposit pheromone. These behaviours vary significantly among different ant species, and will be strongly related to their specific ecological niche [19]. In many cases, we do not even know the exact pheromone composition of the pheromone signals that ants use. There are more than 14,000 ant species alive today [36, p. 11] and only very few of them have been studied in depth by biologists. And there are many other social insect groups such as bees, wasps or termites which add to our pool of potential inspiration. We have every reason to expect more surprising findings to inspire further improvements to ACO algorithms in the future.

Acknowledgements We would like to thank Gabriele Valentini and Tanya Latty for comments on earlier versions of this chapter. TC was funded by the Deutsche Forschungsgemeinschaft (grant number CZ 237/1-1)

References

1. D. Alves, J. von Ast, Z. Cong, B. De Schutter, R. Babûska, Ant colony optimization for traffic dispersion routing, in *Proceedings of the 13th International IEEE Conference on Intelligent Transportation Systems*, pp. 633–688 (2010)
2. S. Aron, R. Beckers, J.L. Deneubourg, J.M. Pasteels, Memory and chemical communication in the orientation of two mass-recruiting ant species. Insectes Soc. **40**, 369–380 (1993)
3. S. Aron, J.L. Deneubourg, J.M. Pasteels, Visual cues and trail-following idiosyncracy in Leptothorax Unifasciatus: an orientation process during foraging. Insectes Soc. **35**, 355–366 (1988)
4. S. Aron, J.M. Pasteels, J.L. Deneubourg, Trail-laying behaviour during exploratory recruitment in the argentine ant, Iridomyrmex humilis (Mayr). Biol. Behav. **14**, 207–217 (1989)
5. B. Beckers, Trails and U-turns in the selection of a path by the ants Lasius niger. J. Theor. Biol. 397–413 (1992)
6. R. Beckers, J.L. Deneubourg, S. Goss, Modulation of trail laying in the ant *Lasius niger* (Hymenoptera: Formicidae) and its role in the collective selection of a food source. J. Insect Behav. **6**, 751–759 (1993)
7. R. Beckers, J.L. Deneubourg, S. Goss, J.M. Pasteels, Collective decision making through food recruitment. Insectes Soc. **37**(3), 258–67 (1990)
8. J. Behnamian, M. Zandieh, S.M.T. Fatemi Ghomi, Parallel-machine scheduling problems with sequence-dependent setup times using an ACO, SA and VNS hybrid algorithm. Expert Syst. Appl. **36**, 9637–9644 (2009)
9. G. Bencheikh, J. Boukachour, A.E.L. Hilali Alaoui, Improved ant colony algorithm to solve the aircraft landing problem. Int. J. Comput. Theory Eng. **3**, 224–233 (2011)
10. E. Bonabeau, G. Theraulaz, J.L. Deneubourg, Quantitative study of the fixed threshold model for the regulation of division of labour in insect societies. Proc. R. Soc. Lond. B Biol. **263**, 1565–1569 (1996)
11. S. Camazine, J.-L. Deneubourg, N.R. Franks, J. Sneyd, G. Theraulaz, E. Bonabeau, *Self-Organization in Biological Systems*. Princeton University Press (2001)
12. D. Campos, F. Bartumeus, V. Mendez, J.S. Andrade Jr., X. Espadaler, Variability in individual activity bursts improves ant foraging success. J. R. Soc. Interface/R. Soc. **13** (2016)
13. D-H. Choe, D.B. Villafuerte, N.D. Tsutsui, Trail pheromone of the argentine ant, Linepithema humile (Mayr) (Hymenoptera: Formicidae). PLoS. One. **7**(9), e45016 (2012)
14. T.J. Czaczkes, How to not get stuck—negative feedback due to crowding maintains flexibility in ant foraging. J. Theor. Biol. **360**, 172–180 (2014)

15. T.J. Czaczkes, B. Czaczkes, C. Iglhaut, J. Heinze, Composite collective decision-making. Proc. R. Soc. B: Biol. Sci. **282** (2015)
16. T.J. Czaczkes, C. Gruter, L. Ellis, E. Wood, F.L. Ratnieks, Ant foraging on complex trails: route learning and the role of trail pheromones in Lasius niger. J. Exp. Biol. **216**, 188–197 (2013)
17. T.J. Czaczkes, C. Grüter, S.M. Jones, F.L.W. Ratnieks, Synergy between social and private information increases foraging efficiency in ants. Biol. Lett. **7**, 521–524 (2011)
18. T.J. Czaczkes, C. Gruter, F.L. Ratnieks, Negative feedback in ants: crowding results in less trail pheromone deposition. J. R. Soc. Interface/R. Soc. **10**, 20121009 (2013)
19. T.J. Czaczkes, C. Grüter, F.L. Ratnieks, Trail pheromones: an integrative view of their role in colony organisation. Annu. Rev. Entomol. **60**, 581–599 (2015)
20. C. Detrain, J.-L. Deneubourg, Self-organized structures in a superorganism: do ants "behave" like molecules? Phys. Live Rev. **3**, 162–187 (2006)
21. C. Devigne, C. Detrain, How does food distance influence foraging in the ant Lasius Niger: the importance of home-range marking. Insectes Soc. **53**, 46–55 (2006)
22. M. Dorigo, V. Maniezzo, A. Colorni, Ant system: optimization by a colony of cooperating agents. IEEE Trans. Syst. Man Cybern. Part B, Cybern.: Publ. IEEE Syst. Man Cybern. Soc. **26**, 29–41 (1996)
23. M. Dorigo, T. Stützle, Ant colony optimization: overview and recent advances, in *Handbook of Metaheuristics*, ed. by M. Gendreau, J.-Y. Potvin (Springer, 2010)
24. M. Dorigo, T. Stützle, *Ant colony optimization* (MIT Press, Cambridge, Mass, 2004)
25. G. Fechner, Elemente der Psychophysik. Breitkopf und Härtel (1860)
26. T.P. Flanagan, N.M. Pinter-Wollman, M.E. Moses, D.M. Gordon, Fast and flexible: argentine ants recruit from nearby trails. PLOS One **8**(8), e7088 (2013)
27. J. Gautrais, G. Theraulaz, J.L. Deneubourg, C. Anderson, Emergent polyethism as a consequence of increased colony size in insect societies. J. Theor. Biol. **215**, 363–373 (2002)
28. S. Goss, S. Aron, J.L. Deneubourg, J.M. Pasteels, Self-organized shortcuts in the Argentine ant. Naturwissenschaften **76**, 579–581 (1989)
29. P.P. Grassé, La reconstruction du nid et les coordinations interindividuelles chez *Bellicositermes natalensis* et *Cubitermes* sp. La théorie de la stigmergie: Essai d'interpretation du comportement des termites constructeurs. Insectes Soc. **6**, 41–80 (1959)
30. M.J. Green, D.M. Gordon, How patrollers set foraging direction in harvester ants. Am. Nat. **170**, 943–948 (2007)
31. C. Grüter, T.J. Czaczkes, F.L. Ratnieks, Decision making in ant foragers (*Lasius niger*) facing conflicting private and social information. Behav. Ecol. Sociobiol. **65**, 141–148 (2011)
32. W. Hangartner, Spezifität und Inaktivierung des Spurpheromons von Lasius fuliginosus Latr. und Orientierung der Arbeiterinnen im Duftfeld. Z Vgl Physiol. **57**, 103–136 (1967)
33. W. Hangartner, Orientierung von Lasius fuliginosus Latr. an einer Gabelung der Geruchspur. Insectes Soc. **16**, 55–60 (1969)
34. B. Hölldobler, The chemistry of social regulation: multicomponent signals in ant societies. Proc. Natl. Acad. Sci. U.S.A. **92**, 19–22 (1995)
35. B. Hölldobler, E.O. Wilson, *The Ants* (Harvard University Press, Cambridge, Mass, 1990)
36. B. Hölldobler, E.O. Wilson, *The Superorganism: The Beauty, Elegance, and Strangeness of Insect Societies*, 1st edn. (W.W. Norton, New York, 2009)
37. D.A. Holway, T.J. Case, Mechanisms of dispersed central-place foraging in polydomous colonies of the Argentine ant. Anim. Behav. **59**, 433–441 (2000)
38. R. Jeanson, A. Weidenmuller, Interindividual variability in social insects—proximate causes and ultimate consequences. Biol. Rev. Camb. Philos. Soc. **89**, 671–687 (2014)
39. F.A.A. Kingdom, N. Prins, *Psychophysics*, 1st edn. (Elsevier, London, 2010)
40. M. Knaden, P. Graham, The sensory ecology of ant navigation: from natural environments to neural mechanisms. Annu. Rev. Entomol. **61**, 63–76 (2016)
41. J. Korb, Termite mound architecture, from function to construction, in *Biology of Termites: A Modern Synthesis*, ed. by Y.R. David Edward Bignell, Nathan Lo (Springer Netherlands, 2011), pp. 349–373

42. M. Lanan, Spatiotemporal resource distribution and foraging strategies of ants (Hamnoptera: Formicidae). Myrmecol. News **20**, 53–70 (2014)
43. T. Latty, M.J. Holmes, J.C. Makinson, M. Beekman, Argentine ants (Linepithema Humile) use adaptable transportation networks to track changes in resource quality. J. Exp. Biol. (2017) (in press)
44. S. Liang, A.N. Zincir-Heywood, M.I. Heywood, The effect of routing under local information using a social insect metaphor, in *IEEE International Congress on Evolutionary Computation*, pp. 1438–1443 (2002)
45. Y. Meng, O. Kazeem, J. Muller, A hybrid ACO/PSO control algorithm for distributed swarm robots, in *Proceedings of the 2007 IEEE Swarm Intelligence Symposium (SIS 2007)* (2007)
46. R.S. Parpinelli, H.S. Lopes, A.A. Freitas, Data mining with an ant colony optimization algorithm. IEEE Trans. Evol. Comput. **6**, 321–332 (2002)
47. A. Perna, B. Granovskiy, S. Garnier, S.C. Nicolis, M. Labedan, G. Theraulaz, V. Fourcassie, D.J. Sumpter, Individual rules for trail pattern formation in Argentine ants (Linepithema humile). PLoS. Comput. Biol. **8**, e1002592 (2012)
48. M. Perretto, H.S. Lopes, Reconstruction of phylogenetic trees using the ant colony optimization paradigm. Genet. Mol. Res. **4**, 581–589 (2005)
49. K. Ramsch, C.R. Reid, M. Beekman, M. Middendorf, A mathematical model of foraging in a dynamic environment by trail-laying Argentine ants. J. Theor. Biol. **306**, 32–45 (2012)
50. F.L. Ratnieks (2008) Biomimicry: further insights from ant colonies?, in *Bio-Inspired Computing and Communication*, vol. 51
51. C.R. Reid, D.J.T. Sumpter, M. Beekman, Optimisation in a natural system: Argentine ants solve the Towers of Hanoi. J Exp Biol 214:50–58 (2011)
52. T.C. Schneirla, Raiding and other outstanding phenomena in the behavior of army ants. Proc. Natl. Acad. Sci. U.S.A. **20**, 316–321 (1934)
53. D.J.T. Sumpter, M. Beekman, From nonlinearity to optimality: pheromone trail foraging by ants. Anim. Behav. **66**(2), 273–80 (2003)
54. S.E. Van Vorhis Key, T.C. Baker, Trail-following responses of the Argentine ant, Iridomyrmex humilis (Mayr) to a synthetic trail pheromone component and analogs. J. Chem. Ecol. **8**, 3–14 (1982)
55. M. Vela-Pérez, M.A. Fontelos, J.J.L. Velásquez, Ant foraging and geodesic paths in labyrinths: analytical and computational results. J. Theor. Biol. **320**, 100–112 (2013)
56. K. Vittori, G. Talbot, J. Gautrais, V. Fourcassie, A.F. Araujo, G. Theraulaz, Path efficiency of ant foraging trails in an artificial network. J. Theor. Biol. **239**, 507–515 (2006)
57. W. von Thienen, A new approach in understanding pheromone based collective behavior of ants. phD thesis, Ludwig-Maximilians-Universität, Munich (2016)
58. W. von Thienen, D. Metzler, D.-H. Choe, V. Witte, Pheromone communication in ants: a detailed analysis of concentration dependent decisions in three species. Behav. Ecol. Sociobiol. **68**, 1611–1627 (2014)
59. W. von Thienen, D. Metzler, V. Witte, Modelling shortest path selection of the ant Linepithema humile using psychophysical theory and realistic parameter values. J. Theor. Biol. **372**, 168–178 (2015)
60. W. von Thienen, D. Metzler, V. Witte, How memory and motivation modulate the responses to trail pheromones in three ant species. Behav. Ecol. Sociobiol. **70**, 393–407 (2016)
61. E.H. Weber, De Pulsu, resorptione, auditu et tactu. Annotationes anatomicae et physiologicae, in ed. by Koehler. Leipzig (1834)
62. J.K. Wetterer, A.L. Wild, A.V. Suarez, N. Roura-Pacual, X. Espadaler, Worldwide spread of the Argentine ant, Linepithema humile (Hymenoptera: Formicidae). Myrmecol News. **12**, 187–194 (2009)
63. E.O. Wilson, A chemical releaser of alarm and digging behavior in the ant Pogonomyrmex badius (Latreille). Psyche **65**, 41–51 (1958)
64. E.O. Wilson, Chemical communication among workers of the fire ant Solenopsis saevissima (Fr. Smith). Anim. Behav. **10**, 134–164 (1962)

65. M. Wittlinger, R. Wehner, H. Wolf, The ant odometer: stepping on stilts and stumps. Science **312**, 1965–1967 (2006)
66. A.A. Yates, P. Nonacs, Preference for straight-line paths in recruitment trail formation of the Argentine ant, Linepithema Humile. Insectes Soc. 1–5 (2016)

Slime Mould Inspired Models for Path Planning: Collective and Structural Approaches

Jeff Jones and Alexander Safonov

Abstract Path planning is a classic and important problem in computer science, with manifold applications in transport optimisation, delivery scheduling, interactive visualisation and robotic trajectory planning. The task has been the subject of classical, heuristic and bio-inspired solutions to the problem. Path planning can be performed in both non-living and living systems. Amongst living organisms which perform path planning, the giant amoeboid single-celled organism slime mould *Physarum polycephalum* has been shown to possess this ability. The field of slime mould computing has been created in recent decades to exploit the behaviour of this remarkable organism in both classical algorithms and unconventional computing schemes. In this chapter we give an overview of two recent approaches to slime mould inspired computing. The first utilises emergent behaviour in a multi-agent population, behaving in both non-coupled and coupled modes which correspond to slime mould foraging and adaptation respectively. The second method is the structural approach which employs numerical solutions to volumetric topological optimisation. Although both methods exploit physical processes, they are generated and governed using very different techniques. Despite these differences we find that both approaches successfully exhibit path planning functionality. We demonstrate novel properties found in each approach which suggest that these methods are complementary and may be applicable to application domains which require structural and mechanical adaptation to changing environments.

J. Jones (✉)
Centre for Unconventional Computing, University of the West of England,
Bristol BS16 1QY, UK
e-mail: jeff.jones@uwe.ac.uk

A. Safonov
Centre for Design, Manufacturing and Materials, Skolkovo Institute of Science
and Technology, Moscow, Russia
e-mail: a.safonov@skoltech.ru

© Springer International Publishing AG, part of Springer Nature 2018
A. Adamatzky (ed.), *Shortest Path Solvers. From Software to Wetware*,
Emergence, Complexity and Computation 32,
https://doi.org/10.1007/978-3-319-77510-4_11

293

1   Introduction: Path Planning

Path planning is a classic problem in computer science and robotics [1, 2] which has been implemented in a wide variety of classical algorithms [3, 4]. These algorithms may be optimised in different ways, for example maximising execution speed or minimising memory resources. Path planning is also a canonical application task for unconventional computing substrates (see [5] for a thorough review). These novel computing schemes utilise the parallel spatial propagation of information within a physical medium and often employ a two-stage method involving diffusive propagation to discover all paths and a second stage to highlight or visualise the path between two particular points in an arena, such as a simple maze.

The giant amoeboid single-celled organism, true slime mould *P. polycephalum* is also adept at path planning as it is also famously known to solve mazes by adapting the transport networks which comprise its distributed body plan between nutrients placed at the start and end points of a maze. In the first approach [6] the slime mould was inoculated over the entire maze in separate pieces, before the organism re-fused it separate component parts and adapted its morphology to efficiently connect the nutrients. Slime mould has also been shown to solve mazes in a manner akin to the classical unconventional computing approach of growing towards and tracking the diffusing chemo-attractant gradient placed at the exit of the maze when the slime mould is itself inoculated at the start of the maze [7]. It appears that slime mould can not only solve mazes, but can do this task using two separate biological mechanisms, foraging and adaptation.

In this article we investigate two modelling approaches to approximating the behaviour of slime mould for path planning applications, the collective and structural approaches. The collective approach exploits collective behaviour in a multi-agent model of the slime mould plasmodium. The method utilises a population of identical mobile particles which sense the concentration of a diffusing virtual chemo-attractant factor. The collective approach can operate in both foraging and adaptive modes, which differ in how the particles are 'coupled' to their environment. The collective behaves as a virtual material in a spatially represented unconventional scheme where the initial problem configuration and final output 'solution' are both represented as spatial patterns.

In the non-coupled (foraging) mode, particles are inoculated at the start location of the maze and passively follow the diffusing wavefronts emanating from an attractant source placed at the exit of the maze. The foraging pattern of the slime mould is thus used to visualise the shortest path.

In the coupled (adaptation) mode particles are, like the original slime mould experiment, inoculated at all vacant spaces in the maze. Diffusive attractant sources are placed at the entrance and exit points of the maze. The particles not only follow the concentration gradient between the nutrient locations but also *modulate* this gradient (by depositing the same diffusive attractant) by the action of their own movement, resulting in an emergent collective cohesion of the population. Reducing the size of the population results in morphological adaptation as the collective adapts to the location of the diffusive stimuli and the obstacles within the maze. We demonstrate the similarities and differences between the two concepts, providing examples of how both methods can be used for path planning problems.

The two different modes of the collective approach also possess more subtle features which can be used in different aspects of path planning problems. For example, in the non-coupled approach, we demonstrate how variable path widths may be used to implement a novel quantitative method to assign path costs. Agents not only sense the timing of the diffusive wavefronts but also the relative *strength* of the competing fronts. In the coupled mode we demonstrate cases where multiple paths are required and the subsequent selection of a single path from multiple options. Collision-free paths may also be implemented via repulsion from the borders of the arena, and obstacle avoidance can be implemented by a repulsive field generated from obstacles as they are uncovered by the shrinking collective. These examples demonstrate how unconventional computing approaches can be used to implement different approaches to the same problem by exploiting and coupling different 'physical' properties of the underlying computing substrate.

The second modelling approach inspired by slime mould is the structural approach introduced by the authors in [8]. In these works the computing structures could be seen as growing on demand, and models for path planning develop in a continuum where an optimal distribution of material minimised internal energy. A continuum exhibiting such properties can be coined as a "self-optimising continuum". Slime mould of *P. polycephalum* well exemplifies such a continuum: the slime mould is capable of solving many computational problems, including mazes and adaptive networks [9]. Other examples of the material behaviour include bone remodelling [10], roots elongation [11], sandstone erosion [12], crack and lightning propagation [13], growth of neurons and blood vessels etc. Some other physical systems suitable for computations were also proposed in [14–17]. In all these cases, a phenomenon of the formation of an optimum layout of material is related to non-linear laws of material behaviour, resulting in the evolution of material structure governed by algorithms similar to those used in a topology optimisation of structures [18].

When applying dynamic approaches in solving the problems of biological growth or finding an optimal topology it is necessary to ensure a decrease in objective function over time. The systems energy is usually taken as an objective function. When using multi-agent modelling this is achieved through an adjustment of the collective morphology arising from the behaviours of individual agent particle interactions, and changes in the population size. When using the apparatus of ODE (Ordinary Differential Equations) a non-increase in objective function is usually ensured through meeting Lyapunovs condition determining a negative derivative of an objective function over its trajectory [18]. Therefore, both considered approaches used to model dynamic systems with decreasing internal energy describe a systems tendency to attractors representing morphology of slime mould networks.

Methods of topology optimisation of flow in a porous medium [19] were applied to simulate the growth of the slime mould [20]. We develop the ideas of material optimisation further and show, in numerical models, how path planning can be build in a conductive material self-optimise its structure governed by configuration of inputs and outputs.

We describe approaches to unconventional computing in Sect. 2. A brief overview of slime mould and slime mould computing follows in Sect. 3. The multi-agent approach is described in Sect. 4. Experiments using the non-coupled and coupled approaches are described in Sects. 6 and 7 respectively. In Sect. 8 we introduce topology optimisation aimed to solve a problem of a stationary heat conduction. Section 9 presents the simulation results for topology optimisation methods used in solution of test problems. We conclude in Sect. 10 by summarising the similarity and differences between the collective and structural approaches, along with suggestions for further research applications.

2 Unconventional Computing Substrates and Path Planning

Unconventional computing seeks to utilise the computing potential of natural physical systems to solve useful problems. Since these systems are localised in space, they typically use different mechanisms to classical approaches. In recent years physical propagation through space in chemical substrates has been used as a search strategy. Babloyantz first suggested that travelling wave-fronts from chemical reactions in excitable media could be used to approximate spatial problems [21]. Wave propagation in the Belousov-Zhabotinsky (BZ) chemical reaction was subsequently used to discover the path through a maze [22]. In this research a trigger wave was initiated at the bottom left corner of a maze and its propagating wave front recorded by time-lapse photography. Direction of wave propagation was calculated from the collective time-lapse information to give vectors which indicated the direction of the travelling wave. The path from any point on the maze to the exit (the source of the diffusion) was followed by tracking backwards (using the vector information) to the source.

Wave-front propagation differs from conventional diffusion and gradient formation because the diffusing fronts annihilate when they collide with an environmental barrier or another wave-front. As front propagation occurs at a fixed speed the annihilation zones indirectly encode information about distances travelled within the environment (shown, for example, in Fig. 5d). The propagation pattern generates a solution from any (and indeed *every*) point in the arena. Branching paths (for example around obstacles) are searched in parallel and the solution time is dependent on the spatial size (in terms of maximum path length) of the arena and the wave-front propagation speed. Although computationally efficient, a direct spatial encoding of the problem (arena, desired start and end points) must be stored, as opposed to a more compact graph or grid encoding in classical approaches.

Reading the output of the parallel calculations is not a simple approach using chemical substrates. Although the propagating wave solves the shortest path for all points in the arena, finding and tracking the desired path from start to end point requires separate processes. Different approaches have been attempted including image processing [23], using two wave-fronts in both directions [24], and hybrid chemical and cellular automata approaches [25]. More recently, a direct visual solution to path planning was devised in which an oil droplet (exploiting convection currents and surface tension effects) migrated along a pH gradient formed within a maze to track the shortest path through the maze [26].

3 Slime Mould Computing

The giant amoeboid Myxomycete organism true slime mould *P. polycephalum* has proven to be an ideal candidate for research into living unconventional computing substrates. *P. polycephalum* is a giant single-celled organism which can usually be seen with the naked eye (for a comprehensive guide, see [27]). During the plasmodium stage of its complex life cycle it adapts its body plan in response to a range of environmental stimuli (nutrient attractants, repellents, hazards) during its growth, foraging and nutrient consumption. The plasmodium is composed of a transport network of protoplasmic tubes which spontaneously exhibit contractile activity which is harnessed used in the pumping and distribution of nutrients. The organism is remarkable in that its complex behaviour is achieved without any specialised nervous tissue. Control of its behaviour is distributed throughout the simple material comprising the cell and the cell can survive damage, excision or even fusion with another cell.

The plasmodium of slime mould is amorphous in shape and ranges from the microscopic scale to up to many square metres in size. It is a giant single-celled syncytium formed by repeated nuclear division, comprised of a sponge-like actomyosin complex co-occurring in two physical phases. The gel phase is a dense matrix subject to spontaneous contraction and relaxation, under the influence of changing concentrations of intracellular chemicals. The protoplasmic sol phase is transported through the plasmodium by the force generated by the oscillatory contractions within the gel matrix. Protoplasmic flux, and thus the behaviour of the organism, is affected by changes in pressure, temperature, space availability, chemo-attractant stimuli and illumination [28–34]. The *P. polycephalum* plasmodium can thus be regarded as a complex functional material capable of both sensory and motor behaviour. Indeed *P. polycephalum* has been described as a membrane bound reaction-diffusion system in reference to both the complex interactions within the plasmodium and the rich computational potential afforded by its material properties [35].

Interest in slime mould computing was initiated by the work of Nakagaki et al. [6] who found that the slime mould could approximate the solution to a simple maze problem when the slime mould was inoculated as fragments covering the channels of a patterned maze. These fragments fused over a number of hours and, when the maze start and end points were covered by nutrient oat flakes, the plasmodium spontaneously adapted its transport network. Protoplasmic tubes were removed from redundant (dead end) paths and longer paths, leaving the transport network of the slime mould connected to the start and end points, thus finding a solution to the maze [6].

Subsequent research into the range of computational abilities of slime mould demonstrated that the plasmodium successfully approximates spatial representations of various graph problems. In [36] the authors examined the connectivity of the tube network when the plasmodium was presented with multiple sources of nutrients. They found that the plasmodium constructed networks that combined features of minimum path length (approximating the Steiner tree) and cyclic connectivity (giving resilience to random disconnection of a path). It has since been found that slime mould successfully approximates spatial representations of various graph problems including generation of Voronoi diagrams and collision-free path planning [37], Delaunay triangulation [38], spanning trees [35, 39, 40], proximity graphs [41], convex hulls and concave hulls [42]. These research examples all used the spatial foraging behaviour of the plasmodium to approximate graph problems which are conventionally solved using algorithmic approaches. Methods to control the propagation of the plasmodium using attractants, repellents and light irradiation were investigated by Adamatzky in [43–45].

The oscillatory phenomena and avoidance of light irradiation were exploited by Aono and colleagues for combinatorial optimisation problems [46–48], specifically small instances of the Travelling Salesman Problem, and found that the chaotic behaviour of the internal oscillations helped the plasmodium avoid deadlock situations, preventing the organism from becoming trapped in local minima—behaviour which is useful in terms of computational and biological search strategies. The behaviour of *Physarum* in response to strong long-distance attractant stimuli combined with short-distance repulsive stimuli was found to follow attractor cycles around simple stimuli and limit-cycle motion with more complex stimuli arrangements [49].

It is somewhat traditional in unconventional computing to validate the computational equivalence of a particular computing substrate with the components of classical computing devices [50–52]. It should be stressed that such research is motivated by exploring theoretical computational *potential*, rather than suitability. In [53], the authors demonstrated how a foraging plasmodium of *Physarum* could be used to construct simple logic gates. A similar approach based on the ballistic computing model was implemented using *Physarum* in [54]. The likelihood of extending this approach for more complex adding circuits was explored in simulation in [55] who found that foraging errors were compounded by small delays in signal timing at junctions, rendering the approach infeasible for larger adding circuits. More recently, photo-avoidance by *Physarum* was used to implement a range of logic gates [56]. The protoplasmic tubes of the *Physarum* plasmodium have also been shown to act as microfluidic logic gates under mechanical stimulation [57]. The relatively slow growth and propagation of the *Physarum* plasmodium limits its application for logical gates. However, Whiting recently demonstrated an approach whereby logic operations could be approximated by much faster changes in oscillatory streaming frequency [58]. The utilisation of different frequency responses with regard to arena size was used experimentally and in simulation in the proposal to simplify Adder circuits using a quantitative scheme [59].

Slime mould utilises its self-made protoplasmic network to transport nutrients within its cell body. The transport phenomena correspond to transportation networks formed by collectives in other living systems including fungi [60, 61], ants [62] and humans [63]. Since the plasmodial network is a single cell, constructed from 'bottom-up' principles, how does the structure of the plasmodium networks compare to other artificial transport networks which are typically constructed from hierarchical 'top-down' methodologies? The task is somewhat difficult as slime mould is only concerned with survival, rather than solving externally applied problems, however early research into the topic of nature-inspired transport networks using slime mould was performed by Adamatzky and Jones who found that *Physarum* networks closely approximated the major motorway network connecting the most populous UK urban areas [64]. The authors also found that the plasmodium effected an efficient response to simulated disastrous contamination of individual urban areas, implemented by diffusion of salts within the region. The plasmodium migrated away from contaminated regions to relatively unpopulated areas before re-establishing network connectivity when the damaged areas were contamination-free. This study was recently extended to include the major motorway networks in different countries [65], and an intriguing similarity between the historical evolution of human networks (for example, cattle droving trails, iron age trails, Roman roads, modern arterial routes) can be mirrored in the evolution of early stage fine-grained *Physarum* networks to later networks with thicker and more sparse connectivity [66]. The connectivity of *Physarum* networks was also compared with the regional rail system surrounding Tokyo by Tero et al. who, using a novel approach to represent environmental hazards using light irradiation, also found a similar correspondence between the human and plasmodial networks, in terms of distance and connectivity [67].

4   The Collective Approach: Multi-agent Slime Mould Computing

The multi-agent approach to slime mould computing was introduced in [68], consisting of a large population of simple mobile agents (a single agent and its controlling algorithm is shown in Fig. 1) whose collective behaviour was indirectly coupled via a diffusive chemo-attractant lattice. Each agent corresponds to a small subunit of the slime mould cell, capable of performing simple local movement. Agents sensed the concentration of a hypothetical 'chemical' in the lattice, oriented themselves towards the locally strongest source and deposited the same chemical during forward movement. By depositing the same chemical into the lattice, the particles are directly coupled to the lattice evolution, i.e. the particles are affected by changes in the lattice but also effect a direct change of the lattice configuration. The collective movement trails spontaneously formed emergent transport networks which underwent complex evolution, exhibiting minimisation and cohesion effects under a range of sensory parameter and scale settings. The overall pattern of the population

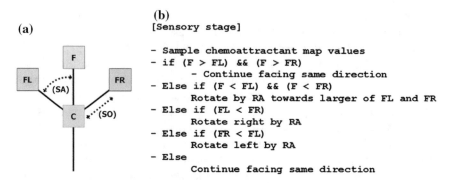

Fig. 1 Base agent particle morphology and sensory stage algorithm. **a** Illustration of single agent, showing location 'C', offset sensors 'FL','F','FR', Sensor Angle '*SA*' and Sensor Offset '*SO*', **b** simplified sensory algorithm

represented the *structure* of the *Physarum* plasmodium and the individual movement of particles within the pattern represented the *flux* within the plasmodium. The collective behaved as a virtual material demonstrating characteristic network evolution motifs and minimisation phenomena seen in soap film evolution (for example, the formation of Plateau angles, T1 and T2 relaxation processes and adherence to von Neumann's law [69]). A full exploration of the dynamical patterns were explored in [70] which found that the population could reproduce a wide range of Turing-type reaction-diffusion patterning (i.e. self-organised patterns emerging from local interactions, see [71] for an overview). The model was extended to include growth and shrinkage in response to environmental stimuli [72, 73]. In a comparison by image analysis and network analysis, the coarsening of the multi-agent networks were found to closely approximate the coarsening observed in the evolution of *P. polycephalum* transport networks [74].

5 Mechanisms of Collective Computation

The computational behaviour of the multi-agent approach is generated by the evolution of the virtual material over time and space. Although the type of pattern (and thus the type of material behaviour) can be influenced by parametric adjustment of the *SA* (Sensor Angle) and *RA* (Rotation Angle) values (Fig. 1a), the evolution is manifested most typically as a shape minimisation over time. At low *SA* and *RA* values the patterns are reticulate and adaptive, constantly changing their topology (for example, Fig. 2). As *SA* and *RA* values increase, the networks undergo minimisation, reducing the number of network edges and network nodes. Further increases result in labyrinthine patterns and the formation of circular minimal configurations. The Sensor Offset (*SO*) parameter acts as a scaling mechanism, altering the size of the patterns formed (see Fig. 3).

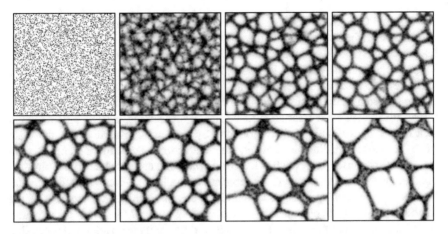

Fig. 2 Spontaneous formation and evolution of transport networks in the multi-agent model. Lattice 200×200, $\%p\,15$, SA $22.5°$, RA $45°$, SO 9, Images taken at: 2, 22, 99, 175, 367, 512, 1740 and 4151 scheduler steps

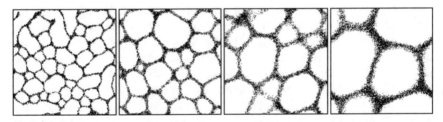

Fig. 3 Effect of Sensor Offset distance (SO) on pattern scale and granularity. Left to Right: Patterning produced with SO of 3, 9, 15, 25 pixels, Lattice 200×200. For all experiments: $\%p = 15$, SA $45°$, RA $45°$, Evolution stopped at 500 steps

The evolution of different patterns, however, does not in itself constitute computation. To perform useful computation we must be able to interact with the material, affecting its behaviour and evolution. This can be achieved by the placement of external attractant and repellent stimuli into the diffusive lattice. These stimuli, when projected into the lattice, form concentration gradients. The gradients constrain the natural minimisation of the material, which would otherwise typically condense the initial inoculation pattern into a minimal configuration. The final (or stable) state of the constrained material indicates that the computation has 'halted'. Because the virtual material is a spatially represented unconventional scheme, the initial problem configuration and final output 'solution' must both be represented as spatial patterns.

The approach follows the scheme suggested by Stepney [75] in which the physical properties of the unconventional computing substrate perform the computation (in this case by material adaptation) and the inputs to the computing substrate can be represented as external fields. In Stepney's scheme there is abundant scope for interfacing the physical substrate with classical computing devices. This interfacing of classical PCs to unconventional computing substrates can be used to program the substrate, read the current state of the substrate, extract data from the substrate and to halt the computation when the solution is found. For physical implementations these interfaces include magnetic fields, chemo-attractant gradients, light projection (for example, to project input patterns onto the chemical substrates) and video camera systems (to sample the current configuration). The interaction between the unconventional substrate and the external system may be very simple, such as the projection of a simple input stimulus pattern, followed by a recording sample of the final state of the computing medium. Alternately, the interaction may be more complex, such as a real-time closed-loop feedback system where the current state of the physical computing medium is sampled, interrogated by a classical computing device and fed back to the substrate. A good example of this more complex integration is the method by Aono and colleagues to control the migration of *Physarum* plasmodium within a stellate chamber, aided by video recording equipment and video projection equipment. Control of input illumination patterns (and thus control of the material substrate) was effected by a Hopfield-type algorithm running on a standard PC architecture (see [76] for details).

In the specific case of the multi-agent model of slime mould, the computing mechanism can be specified generically in the following scheme (specific implementations and differences will be described in later sections). The multi-agent population resides on a 2D diffusive lattice and the movement of each particle is indirectly coupled to their neighbours by deposition and sensing of chemo-attractant within the diffusive lattice. These interactions generate the emergent network formation and minimisation behaviour. This computing substrate is 'programmed' by inoculating the population at specific locations within the substrate, or as a particular pattern (Fig. 4). The evolution of the material is then constrained by placement of external spatial stimuli (chemo-attractant gradients, chemo-repellent gradients and simulated light irradiation). The computation proceeds with the morphological adaptation of the virtual material (constrained, to some degree, by the stimuli) and the final result is recorded as the stable pattern which the virtual material eventually adopts.

6   Non-coupled Collective Approach: Path Planning with Passive Particle Movement

For the first exploration of collective agent behaviour for path planning problems we examine the case where agents sense the concentration in the lattice but do not directly modify the lattice contents. This may be interpreted as a passive response by the agent population to the diffusive lattice. Initial experiments comprised a simple path choice between two channels with different lengths. The arena (Fig. 5a) contains a diffusion source (left circle) and the particles must be confined to start within a particular area of the path (right rectangle). The light grey area represents the area where waves are free to propagate and the surrounding darker grey areas are obstacles

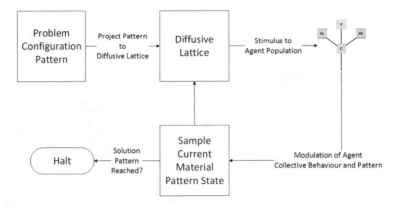

Fig. 4 Schematic illustration of unconventional computing approach using multi-agent model of slime mould

at which all waves will be annihilated. When particles reach the diffusion source they are immediately transported back to the initial start position area.

Figure 5b shows the partial propagation of the wavefront along the labyrinth. The wave is initiated at the source on the left side of the figure and splits when it encounters a junction, traversing both paths in parallel. Because the waves move at a constant velocity, the wavefront traversing the lower (shorter) path moves further along the labyrinth than the wave travelling the upper path. Before the wave has completed the labyrinth there is a period of initial confusion in the population (since there is no chemo-tactic path to follow), and random movement ensues with particles taking both the upper and lower path (Fig. 5c). Shortly afterwards the wave has completed its traversal (Fig. 5d) and the lower wave, arriving at the final junction first, travels the remaining distance to the particle source. When the two separate wave fronts meet, the particles choose the shorter path (Fig. 5e) because the wavefront travelling along the shorter path arrives at the agent source earlier than the wavefront from the longer path. The historical emergent trail record left behind by the particles (Fig. 5f) corresponds to the shortest path through the simple labyrinth.

Computation by diffusive propagation can be extended to more complex environments, as shown in Fig. 6. The wavefront is initiated at the exit (circled) and propagates backwards throughout the maze in parallel (Fig. 6).

When the wavefront reaches the particle start area (rectangle), the particles can then follow the path of strongest chemo-attractant and identify the shortest path through the maze (Fig. 7) by the historical record of the collective particle movement trails. Although the diffusion is a parallel process (for example, at branching points in the maze both branches are propagated simultaneously), the time taken for the wavefront to propagate through the maze is dependent on the total length (and tortuosity) of the maze.

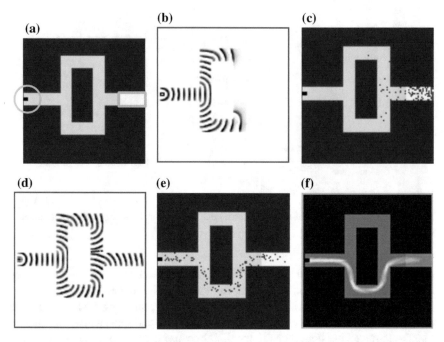

Fig. 5 Shortest path solution in a minimal complexity path choice problem. **a** Arena shape showing diffusion source (circle) and agent inoculation site (rectangle), **b** wavefront emanating from diffusion source, **c** initially random particle movement, **d** completion of diffusive propagation with wavefront collision point marked in blue, **e** agents follow highest concentration of attractant, **f** historical record of shortest path by agent population

6.1 Dynamical Response to Changing Problems

The particle population is also able to dynamically track changes to the environment without having to restart the algorithm. The iterative nature of the framework ensures that the population is collectively able to maintain a record of the shortest path via the trail map. An illustration of the response time-line to changes in the environment is shown in Fig. 8. The wave front is initiated at the source and begins to propagate through the maze. Particle movement is random at this point (Fig. 8a, 100 system steps). The particles follow the wavefront path and finds the path through the maze (Fig. 8b, 920 steps). The problem configuration is replaced with a longer, more complex maze (Fig. 8c, 1100 steps). There is population confusion as the new wavefront competes with the older collapsing front (Fig. 8d, 1500 steps). The new wavefront completes its traversal through the entire maze and particles follow the new path

(a) **(b)** **(c)**

(d) **(e)**

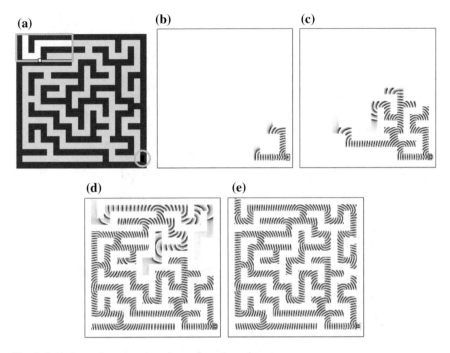

Fig. 6 Initiation and propagation of wavefront through a maze

(although the path is longer, the older and shorter path is no longer available). The particle positions indicate the new path but remnants of the older path persist in the trail pattern (Fig. 8e, 2100 steps). The trail erosion selection pressure ensures only the new path persists and the old path is forgotten (Fig. 8f, 4000 steps).

6.2 Quantitative *Field Propagation as a Cost Assignment*

One significant difference between the particle based approach and chemical implementations of RD computing is that in the particle approach, the wave propagation has quantitative as well as qualitative properties. In chemical approaches only the timing of the wave propagation is used for the computation, the wavefront which arrives first is the winner and the waves are annihilated on contact. In the particle based approach the amount of diffusing chemo-attractant also has an influence on particle path choice. In labyrinths with identical path widths, this property is not taken into consideration and only the timing of propagation is utilised. The quantitative aspect of the front propagation can be harnessed and exploited to assign path costs, as illustrated in Fig. 9.

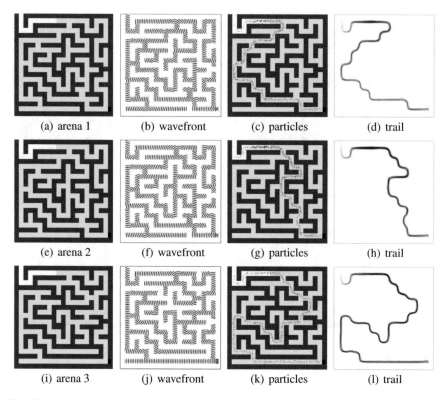

(a) arena 1   (b) wavefront   (c) particles   (d) trail

(e) arena 2   (f) wavefront   (g) particles   (h) trail

(i) arena 3   (j) wavefront   (k) particles   (l) trail

Fig. 7 Shortest path planning by non-coupled method in more complex environments. **a, e, i** Experimental arenas, **b, f, j** wavefront pattern, **c, g, k** particle paths, **d, h, l** collective historical record of particle paths through maze

Figure 9a shows a labyrinth with two possible paths, the right (narrow) side of which is considerably shorter. As befitting the shorter length, the wavefront on the right side arrives first at the particle start location (Fig. 9) and would be expected to be traversed by the particles. The narrow width of the right side path, however, acts to constrict the *amount* of chemo-attractant flowing through that path and the annihilation point where the two waves met is shifted to the right hand side by the strong competition of the flow in the longer left channel (Fig. 9c, circled region). For this reason the particles follow the path of greater flow and choose the longer side (Fig. 9d). When the width of the right side channel is changed to match that of the left side (Fig. 9e), the greater flow in the right side shifts the annihilation point back to the left (Fig. 9g, circled) and the particles preferentially choose the shorter right side path (Fig. 9h). When a simple erosion operator is used on the particle movement trail map the new path choice replaces the previous path choice over-time, in response

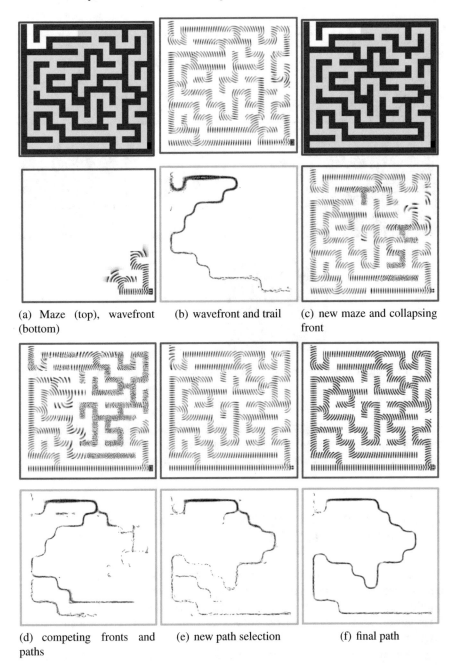

(a) Maze (top), wavefront (bottom)

(b) wavefront and trail

(c) new maze and collapsing front

(d) competing fronts and paths

(e) new path selection

(f) final path

Fig. 8 Evolution of diffusion field and shortest path trail pattern in a changing environment

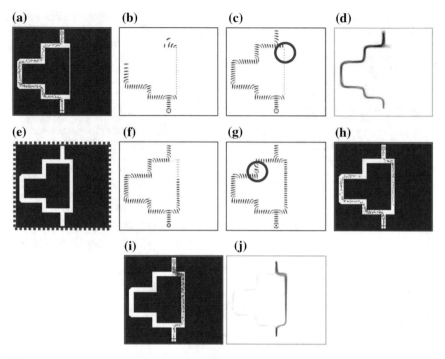

Fig. 9 Utilising the quantitative properties of front propagation to assign path costs via different channel widths

to path width changes (Fig. 9i, j). The width of the channels may therefore be used to assign additional costs to the problem routes, as with real life problem instances where, for example, a shorter path may be subject to more congestion.

7 Coupled Collective Approach: Path Planning with Cohesive Particle Collectives

The examples provided so far, using passive particle sensing of concentration gradients, reproduce the functionality of reaction-diffusion computing. In terms of slime mould behaviour, the approach approximates the foraging behaviour of *Physarum*. To replicate the adaptation behaviour of *Physarum* we exploit the collective cohesion property of the population when the particles modulate the concentration gradients. Exploitation of collective cohesion effects has been used to approximate Convex and Concave Hulls [77], combinatorial optimisation [78], spline curves [79] and the density classification problem [80]. For path planning a general approach was devised in [81] whereby a large population of the virtual plasmodium was inoculated within a lattice whose boundaries took the shape of a maze or robotic arena. Chemo-attractant was projected into the lattice at the desired start and end locations and the population size was reduced. The model plasmodium adapted its shape automatically as the population size reduced, conforming to the boundaries of the arena and forming a path between the start and end locations.

7.1 Maze Solution by Morphological Adaptation

Solution of a maze was the initial experimental finding which attracted the attention of the scientific community to the computational behaviour of *Physarum* [6, 82, 83]. The plasmodium solves the maze by morphological adaptation of its transport network after the plasmodium has completely covered the maze. It should be noted that the plasmodium network is only an approximation of the shortest path, and in the majority of runs with the living organism variants of the shortest path were found. The representation of the problem (initialising the plasmodium to completely cover all paths, then retracting redundant and longer veins) is also different to classical approaches to maze solving which typically search without knowing the complete maze configuration in advance, using a combination of depth and breath searching [84].

To assess the behaviour of the model plasmodium on maze solving a large virtual plasmodium was placed to completely cover all paths in a maze (Fig. 10a, an analogue of the original experimental design in [82]). The size of the virtual plasmodium was reduced over time using the growth and adaptation behaviour to maintain connectivity, and retraction of pseudopodia from dead-ends was observed (Fig. 10b–e). Unlike the real organism, however, all possible paths connecting the start to exit persisted. Attempts to 'force' the virtual plasmodium to choose the shorter of the paths were performed by removing particles selected at random. This resulted in the plasmodium shrinking to give thinner paths until connectivity was broken. However this method did not guarantee that the shortest path through the maze would persist. Gunji noted that protoplasmic flux in the *Physarum* plasmodium is not as idealised as that represented in the mathematical model by Tero et al. [85], noting that actual flow within the plasmodium is irregular, redundant and partially dependent on the shape of the organism itself [86].

How can the shape of the maze affect flux within the virtual plasmodium? The shape of the maze walls affect the tortuosity of the paths and evolution of the virtual plasmodium tends to enhance flux in paths where changes in direction are less frequent, thus favouring 'easier' as well as shorter paths. This may be partly due to the fact that the virtual plasmodium is not anchored firmly to the substratum, as in the case of the real plasmodium. The virtual plasmodium attempts to minimise the path choices by shifting the positions of the Steiner points. In an environment without obstacles the Steiner points are free to move and eventually competing paths merge to form a single path. In the maze, however, the walls provide obstacles to free

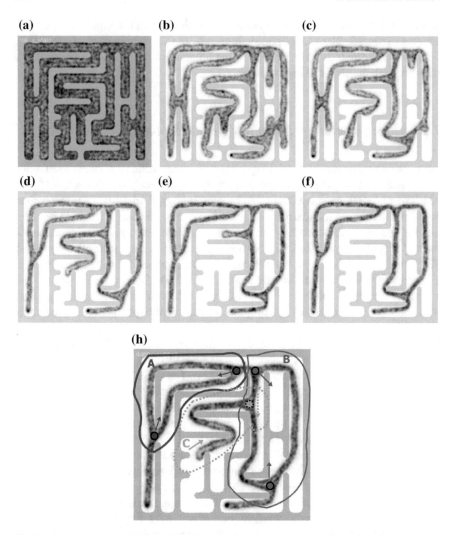

Fig. 10 Approximation of maze problem by virtual plasmodium. **a** population inoculated in entire habitable region of maze and nutrients (blue circles) projected at start and exit points, **b–e** retraction of virtual pseudopodia, **f** persistence of path combinations connecting start and exit, **g** regions of competing path flux

movement of the Steiner points. The competing paths, indicated by their respective Steiner points are indicated in Fig. 10g and separated by bounded regions. In the case of region 'C' the two paths above and below the point are effectively anchored, whereas the pseudopodium on the left is not anchored and shrinks back into the main vertical flow. The natural shrinkage direction of the Steiner points in regions 'A' and 'B' is indicated but movement in these directions is prevented by the maze walls. As diffusion does not cross maze walls (chemo-attractant is removed from wall regions) the walls 'pinch' the path thickness in these areas, reducing the attraction for paths which make significant contact with wall regions. Reducing the population size shrinks the width of the network paths but selection of shorter paths cannot be guaranteed and 'easier' (as opposed to shorter) paths were often selected. Reducing the frequency of chemo-attractant diffusion (for example, to every 50 scheduler steps instead of every step) mimicked a stronger adhesion of the path to the substratum whilst leaving individual particles free to decide path choices but did not reliably increase the path selection.

7.2 Collective Path Planning with Obstacle Avoidance

A number of variations on the cohesive adaptation approach are possible. In many applications a collision-free path may be required, for example if the desired path has to avoid close proximity to walls. To achieve this method by morphological adaptation we represented the walls of the arena as repellent sources (repellent sources project negatively weighted values into the diffusive lattice). We used the same arena as in earlier experiments, but with different start and end points (Fig. 11a, b). As the 'blob' of virtual plasmodium shrunk it formed the shortest path (following the walls) when repellent diffusion was not activated (Fig. 11c–f). When repellent diffusion from the arena walls was activated the virtual plasmodium still maintained its connectivity to the start and end points but also avoided the diffusing repellent values projecting from the walls of the arena (Fig. 11g). Further increasing the concentration of the repellent source increased the distance of the path from the walls (Fig. 11h).

To ensure only a single path is generated we devised a two-part repulsion mechanism. The first part of the mechanism occurs by initialising the blob to cover the *entire* arena (including the obstacles). This part in isolation would not solve the problem of multiple paths, however: if the obstacles repelled the blob immediately then the virtual plasmodium would simply flee the obstacle regions from all directions and multiple paths would still be retained. The second part of the mechanism ensures that only a single path is retained. The shrinkage process is performed more slowly and we generate repellent fields only from obstacles (more specifically, exposed *fragments* of large obstacles) that have been partially uncovered by the shrinkage of the blob.

The mass of the blob is thus shifted away from obstacles by their emergent repulsion field. Because the blob shrinks slowly inwards from the outside of the arena obstacles are slowly uncovered and the repulsion field further pushes the blob inwards until a single path connecting the source attractants is formed. The shrinkage and repulsion mechanism is illustrated in Fig. 12 where the arena (including obstacles) is completely covered by a large mass of particles comprising the virtual plasmodium

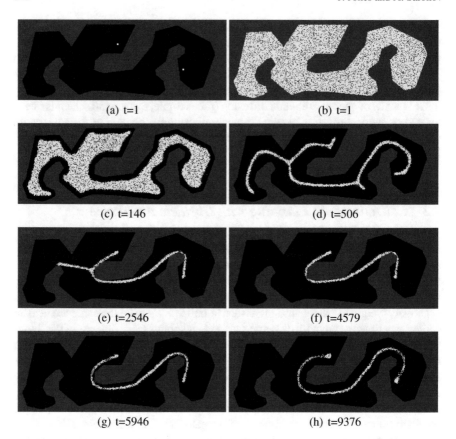

Fig. 11 Approximation of collision-free shortest path by morphological adaptation and repulsion. **a** 2D Arena defined by borders (grey), habitable regions (black) and start and end points (white), **b** initialisation of virtual plasmodium in habitable region, **c–f** shrinkage of blob causes adaptation of shape and attraction to points, forming the shortest path between the points, **g** repulsion field emitted from arena walls causes virtual plasmodium to avoid wall regions, forming a collision-free path, **h** increasing concentration of repulsion field causes further adaptation of the virtual plasmodium away from walls

(Fig. 12b). The blob shrinks inwards as the periphery of the blob is drawn inwards (Fig. 12c). When an obstacle is partially uncovered repellent is projected into the diffusive lattice at exposed obstacle fragments (Fig. 12d, arrowed). The blob at these regions is repelled and moves away from the exposed obstacle fragment. The shrinkage process continues and when a larger obstacle is partially exposed the repellent projected into the lattice again causes the blob to move away from this region

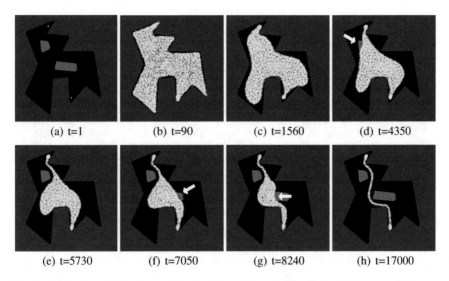

(a) t=1   (b) t=90   (c) t=1560   (d) t=4350

(e) t=5730   (f) t=7050   (g) t=8240   (h) t=17000

Fig. 12 Mechanism of shrinkage combined with repulsion at exposed obstacle fragments generates a single path. **a** Arena with habitable areas (black), inhabitable areas (dark grey), obstacles light grey and path source locations (white). **b** Blob initialised on entire arena, including obstacles, **c** gradual shrinkage of blob, **d** exposure of obstacle fragment generates repellent field at exposed areas (arrow), **e** blob moves away from repellent field of obstacle, **f** lower obstacle is exposed causing repellent field at these locations (arrow), **g** further exposure causes migration of blob away from these regions (arrow), **h** final single path connects source points whilst avoiding obstacles

(Fig. 12f, arrowed). Further exposure of this large lower obstacle causes the blob to continue to be repelled away (Fig. 12g) until eventually only a single path remains which connects the source attractants and threads between the obstacles (Fig. 12h).

In the presence of a large number of obstacles, the repulsion field emanating from newly-exposed obstacles acts to deform the shrinkage of the virtual plasmodium. The mass of particles is deformed both by the attractant stimuli from the start and end points of the path (Fig. 13d) and the gradual exposure of the obstacles as the blob shrinks. Figure 13 shows the deformation of the blob and also the changing concentration gradient field as the shrinkage continues, until only a collision-free path between the obstacles remains.

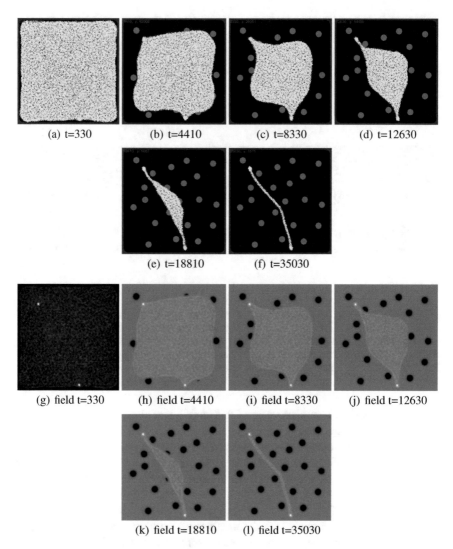

Fig. 13 Shrinkage and exposed repulsion method in complex obstacle field. (**a–f**) Uniform shrinkage of blob is distorted by attraction to start and end stimuli and repulsion from exposed obstacles, (**g–l**) visualisation of gradient field showing attractants at start and end stimuli (bright spots), blob (mid-grey mass) and repulsion field from exposed obstacles (dark circles). Greyscale gradient field images transformed by gamma correction ($\gamma = 0.6$) to improve clarity

8 The Structural Approach: Topology Optimisation

A topology optimisation in continuum mechanics aims to find a layout of a material within a given design space that meets specific optimum performance targets [87–89]. The topology optimisation is applied to solve a wide range of problems [90], e.g. maximisation of heat removal for a given amount of heat conducting material [91], maximisation of fluid flow within channels [92], maximisation of structure stiffness and strength [90], development of meta-materials satisfying specified mechanical and thermal physical properties [90], optimum layout of plies in composite laminates [93], the design of an inverse acoustic horn [90], optimisation of photonics-crystal band-gap structures [94], construction of logical gates [95].

A standard method of the topology optimisation employs a modelling material layout that uses a density of material, ρ, varying from 0 (absence of a material) to 1 (presence of a material), where a dependence of structural properties on the density of material is described by a power law. This method is known as Solid Isotropic Material with Penalisation (SIMP) [96]. An optimisation of the objective function consists in finding an optimum distribution of ρ: $\min_\rho f(\rho)$.

The problem can be solved in various numerical schemes, including the sequential quadratic programming (SQP) [97], the method of moving asymptotes (MMA) [98], and the optimality criterion (OC) method [90]. The topology optimisation problem can be replaced with a problem of finding a stationary point of an Ordinary Differential Equation (ODE) [18]. Considering density constraints on ρ, the right term of ODE is equal to a projection of the negative gradient of the objective function. Such optimisation approach is widely used in the theory of projected dynamical systems [99]. Numerical schemes of topology optimisation solution can be found using simple explicit Euler algorithm. As shown in [100] iterative schemes match the algorithms used in bone remodelling literature [101].

In this work the topology optimisation problem as applied to heat conduction problems [102]. Consider a region in the space Ω with a boundary $\Gamma = \Gamma_D \cup \Gamma_N$, $\Gamma_D \cap \Gamma_N = \emptyset$, separated for setting the Dirichlet (D) and the Neumann (N) boundary conditions. For the region Ω we consider the steady-state heat equation given in:

$$\nabla \cdot k\nabla T + f = 0 \text{ in } \Omega \tag{1}$$

$$T = T_0 \text{ on } \Gamma_D \tag{2}$$

$$(k\nabla T) \cdot \mathbf{n} = Q_0 \text{ on } \Gamma_N \tag{3}$$

where T is a temperature, k is a heat conduction coefficient, f is a volumetric heat source, and $\mathbf{n}$ is an outward unit normal vector. At the boundary Γ_D a temperature $T = T_0$ is specified in the form of Dirichlet boundary conditions, and at the boundary Γ_N of the heat flux $(k\nabla T) \cdot \mathbf{n}$ is specified using Neumann boundary conditions. The condition $(k\nabla T) \cdot \mathbf{n} = 0$ specified at the part of Γ_N means a thermal insulation (adiabatic conditions).

When stating topology optimisation problem for a solution of the heat conduction problems it is necessary to find an optimal distribution for a limited volume of conductive material in order to minimise heat release, which corresponds to designing a thermal conductive device. It is necessary to find an optimum distribution of material density ρ within a given area Ω in order to minimise the cost function:

$$\text{Minimize } C(\rho) = \int_{\Omega} \nabla T \cdot (k(\rho)\nabla T) \tag{4}$$

$$\text{Subject to } \int_{\Omega} \rho < M \tag{5}$$

In accordance with the SIMP method the region being studied can be divided into finite elements with varying material density ρ_i assigned to each finite element i. A relationship between the heat conduction coefficient and the density of material is described by a power law as follows:

$$k_i = k_{min} + (k_{max} - k_{min})\rho_i^p, \quad \rho_i \in \lfloor 0, 1 \rfloor \tag{6}$$

where k_i is a value of heat conduction coefficient at the i-th finite element, ρ_i is a density value at the i-th element, k_{max} is a heat conduction coefficient at $\rho_i = 1$, k_{min} is a heat conduction coefficient at $\rho_i = 0$, p is a penalisation power ($p > 1$).

In order to solve the problem (1)–(6) we apply the following techniques used in the dynamic systems modelling. Assume that ρ depends on a time-like variable t. Let us consider the following differential equation to determine density in i-th finite element, ρ_i, when solving the problem stated in (1)–(6):

$$\dot{\rho}_i = \lambda \left(\frac{C_i(\rho_i)}{\rho_i V_i} - \mu \right) \quad C_i(\rho_i) = \int_{\Omega_i} \nabla T \cdot (k_i(\rho)\nabla T)d\Omega \tag{7}$$

where dot above denotes the derivative with respect to t, Ω_i is a domain of i-th finite element, V_i is a volume of i-th element, λ and μ are positive constants characterising behaviour of the model. This equation can be obtained by applying methods of the projected dynamical systems [100] or bone remodelling methods [101, 103, 104].

For numerical solution of Eq. (8) a projected Euler method is used [99]. This gives an iterative formulation for the solution finding ρ_i [18]:

$$\rho_i^{n+1} = \rho_i^n + q[\frac{C_i(\rho_i^n)}{\rho_i^n V_i} - \mu^n] \tag{8}$$

where $q = \lambda \Delta t$, ρ_i^{n+1} and ρ_i^n are the numerical approximations of $\rho_i(t + \Delta t)$ and $\rho_i(t)$, $\mu^n = \frac{\sum_i C_i(\rho_i^n)}{\sum_i \int_{\Omega_i} \rho_{ev}d\Omega}$, ρ_{ev} is a specified mean value of density.

We consider a modification of Eq. (8):

$$\rho_i^{n+1} = \begin{cases} \rho_i^n + \theta \text{ if } \frac{C_i(\rho_i^n)}{\rho_i^n V_i} - \mu^n \geq 0, \\ \rho_i^n - \theta \text{ if } \frac{C_i(\rho_i^n)}{\rho_i^n V_i} - \mu^n < 0, \end{cases} \tag{9}$$

where θ is a positive constant.

Then we calculate a value of ρ_i^{n+1} using Eq. (9) and project ρ_i onto a set of constraints:

$$\rho_i^{n+1} = \begin{cases} \rho_{max} \text{ if } \rho_i^{n+1} > \rho_{max}, \\ \rho_i^{n+1} \text{ if } \rho_{min} \leq \rho_i^{n+1} \leq \rho_{max}, \\ \rho_{min} \text{ if } \rho_i^{n+1} < \rho_{min} \end{cases} \tag{10}$$

where ρ_{min} is a specified minimum value of ρ_i and ρ_{max} is a specified maximum value of ρ_i. A minimum value is taken as the initial value of density for all finite elements: $\rho_i^0 = \rho_{min}$.

The model can be described by the following parameters: ρ_{min} and ρ_{max} are minimum and maximum values of ρ_i, ρ_{ev} is a mean value of density, θ is an increment of ρ_i at each time step, p is a penalisation power, k_{max} is a heat conduction coefficient at $\rho_i = 1$, k_{min} is a heat conduction coefficient at $\rho_i = 0$.

The algorithm above is implemented in ABAQUS [105] using the modification of the structural topology optimisation plug-in, UOPTI, developed previously [106].

9 Topology Optimisation: Results

Calculations were performed using topology optimisation method for the finite element models of test problems (Fig. 14). The length scale of the models is 100 units. Square-shaped linear hexahedral plane elements of DC2D4 type (4-node, linear, see [105]) with a half unit length edges were used in calculations. The elements used have four integration points. The cost function value is updated for each finite element as a mean value of integration points for an element under consideration [105].

All parameters but ρ_{ev} are the same for all simulations: $\rho_{max} = 1$, $\rho_{min} = 0.0001$, $\theta = 0.01$, $p = 2$, $K_{max} = 1$, $K_{min} = 0.01$. For each test case, calculations were carried out with two values of the mean value of density ρ_{ev}: $\rho_{ev_1} = 0.05$, $\rho_{ev_2} = 0.1$.

Following boundary conditions is specified. Neumann boundary conditions are considered. Positive heat flux $Q_{Input} = 1$ is set at the entrance of the maze. Negative heat flux $Q_{Output} = -1$ is set at the exit of the maze. Adiabatic condition is set at the other boundaries.

Figures in this section show density distribution of the conductive material. The maximum values of ρ are shown by red colour, the minimum values by blue colour.

Figure 15 shows results of shortest path planning by topology optimisation approach for the minimal maze.

Figure 16 shows results of shortest path planning by topology optimisation approach for the complex maze.

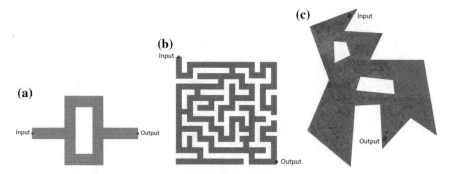

Fig. 14 Finite element models of test problems for topology optimisation. **a** Minimal maze; **b** complex maze; **c** arena with obstacles

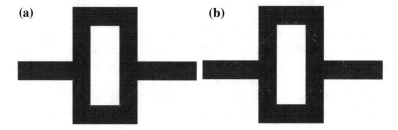

Fig. 15 Shortest path planning by topology optimisation approach for the minimal maze. **a** The mean value of density $\rho_{ev_1} = 0.05$; **b** the mean value of density $\rho_{ev_2} = 0.1$

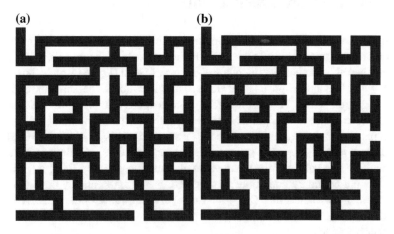

Fig. 16 Shortest path planning by topology optimisation approach for the complex maze. **a** The mean value of density $\rho_{ev_1} = 0.05$; **b** the mean value of density $\rho_{ev_2} = 0.1$

(a) t=60   (b) t=100

(c) t=120   (d) t=140

(e) t=150   (f) t=250

Fig. 17 Intermediate results of shortest path planning by topology optimisation approach for the complex maze with the mean value of density $\rho_{ev_1} = 0.05$

(a)   (b)   (c)

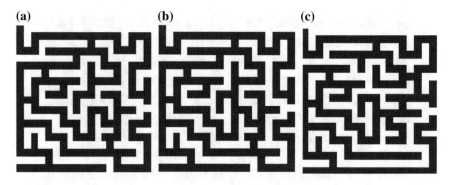

Fig. 18 Shortest path planning by topology optimisation approach for three variants of the complex maze. **a, b, c** The mean value of density $\rho_{ev_1} = 0.05$

(a)   (b)

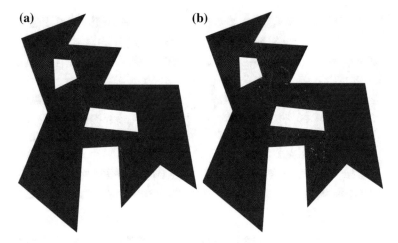

Fig. 19 Shortest path planning by topology optimisation approach for the arena with obstacles. **a** The mean value of density $\rho_{ev_1} = 0.05$; **b** the mean value of density $\rho_{ev_2} = 0.1$

Figure 17 shows intermediate results of shortest path planning by topology optimisation approach for the complex maze with the mean value of density $\rho_{ev_1} = 0.05$. Figure 18 shows results of shortest path planning by topology optimisation approach for three variants of the complex maze with the mean value of density $\rho_{ev_1} = 0.05$.

Figure 19 shows results of shortest path planning by topology optimisation approach for arena with obstacles.

Figures 15, 16 and 19 show the results for two values of the mean value of density ρ_{ev}: $\rho_{ev_1} = 0.05$, $\rho_{ev_2} = 0.1$. One can see the formation of two paths for the mean value of density $\rho_{ev_2} = 0.1$. On the other hand, for the mean value of density $\rho_{ev_2} = 0.05$ the optimal single path is observed. The Fig. 17 shows the selection of the optimal path among the two possible paths for the complex maze with the mean value of density $\rho_{ev_1} = 0.05$. The simulation results show an interesting fact that for a large amount of conductive material ($\rho_{ev_2} = 0.1$) algorithm indicates all possible paths. However, for a small amount of conductive material ($\rho_{ev_2} = 0.05$) algorithm indicates the optimal single path only. Supplementary videos can be found here [107].

10 Conclusions

We have demonstrated how unconventional computation of path planning problems may be performed directly in 2D space by morphological adaptation in a virtual material inspired by the adaptation of slime mould *P. polycephalum*. Unlike previous implementations of path planning problems in chemical substrates the method does not rely on a two-stage computation (one stage to perform the computation, another stage to highlight the path). The method computes a simple path with only two attractant sources. Multiple paths were represented by having more than two attractant sources and a single path was selected between two of these sources by removal of redundant sources. Collision-free paths were discovered by the simultaneous addition of repellent sources at arena boundaries. Obstacle avoiding paths were discovered using a mechanism whereby obstacles were represented by a gradual exposure of repellent sources. The contribution of this method is in the simplicity of the approach: the behaviour of the shrinking blob is distributed within the material itself and emerges from the simple and local interactions between the particles which comprise the blob. The path finding process is governed, to a large extent, by the spatial configuration of the arena and the obstacles within the arena. Since the blob initially occupies all of the space within the arena the path finding method may be described as subtractive—all redundant or inefficient paths are removed during the shrinkage process. This is achieved by withdrawal of pseudopodia (for example from dead-ends in the arena) and also by displacement of the blob by the repellent fields emitted from the gradually exposed obstacles. Unlike chemical-based approaches the method is not initiated at either the path start or end points but is initialised by shrinkage from the arena boundary. Diffusion from the source points still occurs but is merely used to anchor the blob material at these points and does not require propagation of the diffusion front throughout the entire arena. Likewise, repellent diffusion occurs from the boundary (for collision-free paths) and from obstacles (for obstacle-avoiding paths) but this diffusion also is only local and does not require propagation throughout the entire arena.

We also implemented path planning using topology optimisation of conductive material when solving stationary problems of heat conduction. In the simplest case of two sites in a plane with given heat fluxes the conductive material is distributed between the sites in a straight line. For the complex structure of the maze the conductive material is distributed between the sites along the optimal single path. The algorithm of optimal layout of the conductive material is similar to a biological process of bone remodelling. The algorithm proposed can be applied to a wide range of biological networks, including neural networks, vascular networks, slime mould, plant routes and fungi mycelium. These networks will be the subject of further studies.

Both collective and structural approaches exploit propagation of information within a 'physical' medium, albeit by very different methods (namely diffusion and emergent cohesion in the agent approach, and numerical calculation of heat flux within the structural approach). They demonstrate the efficiency of 'outsourcing' much of the computation to physical processes which also appears to be efficiently exploited by the *Physarum* plasmodium. Because slime mould is constantly remodelling its body plan in response to changing environmental stimuli, the two approaches described herein may also be amenable to tackling optimisation problems which are dynamic in their nature. In addition to the purely topological reorganisation, which is well studied in networks formed by *Physarum* and its models, we suggest that the two approaches described in this Chapter may be the bases of future explorations of slime-mould *mechanics*.

Acknowledgements This research was supported by the EU research project "Physarum Chip: Growing Computers from Slime Mould" (FP7 ICT Ref 316366).

Supplementary Materials

Topology Optimisation Approach

- Minimal maze, the mean value of density $\rho_{ev_1} = 0.05$: https://youtu.be/aXaca zRYt-8
- Minimal maze, the mean value of density $\rho_{ev_2} = 0.1$: https://youtu.be/Lf6IeCK wklQ
- Complex maze #1, the mean value of density $\rho_{ev_1} = 0.05$: https://youtu.be/ LCidzk607mE
- Complex maze #1, the mean value of density $\rho_{ev_2} = 0.1$: https://youtu.be/Fpnkzi CmS_U
- Complex maze #2, the mean value of density $\rho_{ev_1} = 0.05$: https://youtu.be/ 3tTkCK6US_Q
- Complex maze #3, the mean value of density $\rho_{ev_1} = 0.05$: https://youtu.be/ 6qNiXgHgFq4
- Arena with obstacles, the mean value of density $\rho_{ev_1} = 0.05$: https://youtu.be/ T4q_z_XphMQ
- Arena with obstacles, the mean value of density $\rho_{ev_2} = 0.1$: https://youtu.be/ pevR7eNZxhw

References

1. P. Raja, S. Pugazhenthi, Optimal path planning of mobile robots: a review. Int. J. Phys. Sci. **7**(9), 1314–1320 (2012)

2. C. Tam, R. Bucknall, A. Greig, Review of collision avoidance and path planning methods for ships in close range encounters. J. Navig. **62**(3), 455 (2009)
3. D.J. Zhu, J.-C. Latombe, New heuristic algorithms for efficient hierarchical path planning. IEEE Trans. Robot. Autom. **7**(1), 9–20 (1991)
4. J.H. Liang, C.H. Lee, Efficient collision-free path-planning of multiple mobile robots system using efficient artificial bee colony algorithm. Adv. Eng. Softw. **79**, 47–56 (2015)
5. A. Adamatzky, Physical maze solvers. All twelve prototypes implement 1961 Lee algorithm. In *Emergent Computation* (Springer, 2017), pp. 489–504
6. T. Nakagaki, Smart behavior of true slime mold in a labyrinth. Res. Microbiol. **152**(9), 767–770 (2001)
7. A. Adamatzky, Slime mold solves maze in one pass, assisted by gradient of chemo-attractants. IEEE Trans. NanoBioscience **11**(2), 131–134 (2012)
8. A. Safonov, J. Jones, Physarum computing and topology optimisation. Int. J. Parallel Emerg. Distrib. Syst. **32**(5), 448–465 (2017)
9. A. Adamatzky, *Advances in Physarum Machines: Sensing and Computing with Slime Mould*, vol. 21 (Springer, 2016)
10. P. Christen, K. Ito, R. Ellouz, S. Boutroy, E. Sornay-Rendu, R.D. Chapurlat, B. van Rietbergen, Bone remodelling in humans is load-driven but not lazy. Nat. Commun. **5** (2014)
11. B. Mazzolai, C. Laschi, P. Dario, S. Mugnai, S. Mancuso, The plant as a biomechatronic system. Plant Signal. Behav. **5**(2), 90–93 (2010)
12. J. Bruthans, J. Soukup, J. Vaculikova, M. Filippi, J. Schweigstillova, A.L. Mayo, D. Masin, G. Kletetschka, J. Rihosek, Sandstone landforms shaped by negative feedback between stress and erosion. Nat. Geosci. **7**(8), 597–601 (2014)
13. W. Achtziger, M.P. BendsOe, J.E. Taylor, An optimization problem for predicting the maximal effect of degradation of mechanical structures. SIAM J. Optim. **10**(4), 982–998 (2000)
14. J.F. Miller, S.L. Harding, G. Tufte, Evolution-in-materio: evolving computation in materials. Evol. Intell. **7**(1), 49–67 (2014)
15. A.J. Turner, J.F. Miller, Neuroevolution: evolving heterogeneous artificial neural networks. Evol. Intell. **7**(3), 135–154 (2014)
16. W. Banzhaf, G. Beslon, S. Christensen, J.A. Foster, F. Képès, V. Lefort, J.F. Miller, M. Radman, J.J. Ramsden, Guidelines: from artificial evolution to computational evolution. Nat. Rev. Genet. **7**(9), 729–735 (2006)
17. J.F. Miller, K. Downing, Evolution in materio: looking beyond the silicon box, in *Proceedings of the NASA/DoD Conference on Evolvable Hardware, 2002* (IEEE, 2002) pp. 167–176
18. A. Klarbring, B. Torstenfelt, Dynamical systems and topology optimization. Struct. Multidiscip. Optim. **42**(2), 179–192 (2010)
19. A.A. Safonov, Mathematical modeling for impregnation of reinforcing filler of fiberglasses during vacuum infusion. J. Mach. Manuf. Reliab. **39**(6), 568–574 (2010)
20. A. Safonov, J. Jones. Physarum computing and topology optimisation. Int. J. Parallel Emerg. Distrib. Syst. **32**(5), 448–465 (2017)
21. A. Babloyantz, J.A. Sepulchre, Front propagation into unstable media: a computational tool, in *Nonlinear Wave Processes in Excitable Media* (Springer, 1991), pp. 343–350
22. O. Steinbock, Á. Tóth, K. Showalter, Navigating complex labyrinths: optimal paths from chemical waves. Science **267**(5199), 868 (1995)
23. N.G. Rambidi, Biologically inspired information processing technologies: reaction-diffusion paradigm. Int. J. Unconv. Comput. **1**(2), 101–121 (2005)
24. K. Agladze, N. Magome, R. Aliev, T. Yamaguchi, K. Yoshikawa, Finding the optimal path with the aid of chemical wave. Phys. D Nonlinear Phenom. **106**(3–4), 247–254 (1997)
25. A. Adamatzky, B. de Lacy Costello, Reaction-diffusion path planning in a hybrid chemical and cellular-automaton processor. Chaos, Solitons & Fractals **16**(5), 727–736 (2003)
26. I. Lagzi, S. Soh, P.J. Wesson, K.P. Browne, B.A. Grzybowski, Maze solving by chemotactic droplets. J. Am. Chem. Soc. **132**(4), 1198–1199 (2010)
27. S.L. Stephenson, H. Stempen, I. Hall, *Myxomycetes: A Bandbook of Slime Molds* (Timber Press Portland, Oregon, 1994)

28. M.J. Carlile, Nutrition and chemotaxis in the myxomycete physarum polycephalum: the effect of carbohydrates on the plasmodium. J. Gen. Microbiol. **63**(2), 221–226 (1970)
29. A.C.H. Durham, E.B. Ridgway, Control of chemotaxis in Physarum polycephalum. J. Cell Biol. **69**, 218–223 (1976)
30. U. Kishimoto, Rhythmicity in the protoplasmic streaming of a slime mould, Physarum polycephalum. J. Gen. Physiol. **41**(6), 1223–1244 (1958)
31. T. Nakagaki, S. Uemura, Y. Kakiuchi, T. Ueda, Action spectrum for sporulation and photoavoidance in the plasmodium of Physarum polycephalum, as modified differentially by temperature and starvation. Photochem. Photobiol. **64**(5), 859–862 (1996)
32. T. Nakagaki, H. Yamada, T. Ueda, Interaction between cell shape and contraction pattern in the Physarum plasmodium. Biophys. Chem. **84**(3), 195–204 (2000)
33. A. Takamatsu, T. Fujii, I. Endo, Control of interaction strength in a network of the true slime mold by a microfabricated structure. BioSystems **55**, 33–38 (2000)
34. T. Ueda, K. Terayama, K. Kurihara, Y. Kobatake, Threshold phenomena in chemoreception and taxis in slime mold Physarum polycephalum. J. Gen. physiol. **65**(2), 223–34 (1975)
35. A. Adamatzky, B. de Lacy Costello, T. Shirakawa, Universal computation with limited resources: Belousov-zhabotinsky and Physarum computers. Int. J. Bifurc. Chaos **18**(8), 2373–2389 (2008)
36. T. Nakagaki, R. Kobayashi, Y. Nishiura, T. Ueda, Obtaining multiple separate food sources: behavioural intelligence in the *Physarum* Physarum plasmodium. R. Soc. Proc. Biol. Sci. **271**(1554), 2305–2310 (2004)
37. T. Shirakawa, Y.-P. Gunji, Computation of Voronoi diagram and collision-free path using the Plasmodium of Physarum polycephalum. Int. J. Unconv. Comput. **6**(2), 79–88 (2010)
38. T. Shirakawa, A. Adamatzky, Y.-P. Gunji, Y. Miyake, On simultaneous construction of voronoi diagram and delaunay triangulation by Physarum polycephalum. Int. J. Bifurc. Chaos **19**(9), 3109–3117 (2009)
39. A. Adamatzky, Physarum machines: encapsulating reaction-diffusion to compute spanning tree. Naturwissenschaften **94**(12), 975–980 (2007)
40. A. Adamatzky, If BZ medium did spanning trees these would be the same trees as Physarum built. Phys. Lett. A **373**(10), 952–956 (2009)
41. A. Adamatzky, Developing proximity graphs by Physarum polycephalum: does the plasmodium follow the toussaint hierarchy. Parallel Process. Lett. **19**, 105–127 (2008)
42. A. Adamatzky, Slime mould computes planar shapes. Int. J. Bio-Inspired Comput. **4**(3), 149–154 (2012)
43. A. Adamatzky, Routing Physarum with repellents. Eur. Phys. J. E Soft Matter Biol. Phys. **31**(4), 403–410 (2010)
44. A. Adamatzky, Manipulating substances with Physarum polycephalum. Mater. Sci. Eng. C **38**(8), 1211–1220 (2010)
45. A. Adamatzky, Steering plasmodium with light: dynamical programming of Physarum machine (2009), arXiv:0908.0850
46. M. Aono, M. Hara, Amoeba-based nonequilibrium neurocomputer utilizing fluctuations and instability, in *6th International Conference, UC 2007, LNCS, Kingston, Canada, 13–17 Aug 2007*, vol. 4618 (Springer, 2007), pp. 41–54
47. M. Aono, M. Hara, Spontaneous deadlock breaking on amoeba-based neurocomputer. BioSystems **91**(1), 83–93 (2008)
48. K. Ozasa, M. Aono, M. Maeda, M. Hara, Simulation of neurocomputing based on the photophobic reactions of Euglena with optical feedback stimulation. BioSystems **100**(2), 101–107 (2010)
49. A. Adamatzky, Simulating strange attraction of acellular slime mould Physarum polycephaum to herbal tablets. Math. Comput. Model. (2011)
50. M. Conrad, Information processing in molecular systems. Curr. Mod. Biol. (now BioSystems) **5**, 1–14 (1972)
51. N. Margolus, Physics-like models of computation. Phys. D **10**, 81–95 (1982)

52. M. Roselló-Merino, M. Bechmann, A. Sebald, S. Stepney, Classical computing in nuclear magnetic resonance. Int. J. Unconv. Comput. **6**(3–4), 163–195 (2010)
53. S. Tsuda, M. Aono, Y.-P. Gunji, Robust and emergent Physarum logical-computing. BioSystems **73**, 45–55 (2004)
54. A. Adamatzky, Slime mould logical gates: exploring ballistic approach (2010), arXiv:1005.2301
55. J. Jones, A. Adamatzky, Towards Physarum binary adders. Biosystems **101**(1), 51–58 (2010)
56. R. Mayne, A. Adamatzky, Slime mould foraging behaviour as optically coupled logical operations. Int. J. Gen. Syst. **44**(3), 305–313 (2015)
57. A. Adamatzky, T. Schubert, Slime mold microfluidic logical gates. Mater. Today **17**(2), 86–91 (2014)
58. J.G.H. Whiting, B.P.J. de Lacy Costello, A. Adamatzky, Slime mould logic gates based on frequency changes of electrical potential oscillation. Biosystems **124**, 21–25 (2014)
59. J. Jones, J.G.H. Whiting, A. Adamatzky, Quantitative transformation for implementation of adder circuits in physical systems. Biosystems **134**, 16–23 (2015)
60. D.P. Bebber, J. Hynes, P.R. Darrah, L. Boddy, M.D. Fricker, Biological solutions to transport network design. Proc. R. Soc. B Biol. Sci. **274**(1623), 2307–2315 (2007)
61. M. Fricker, L. Boddy, T. Nakagaki, D. Bebber, Adaptive biological networks. Adapt. Netw. 51–70 (2009)
62. T. Latty, K. Ramsch, K. Ito, T. Nakagaki, D.J.T. Sumpter, M. Middendorf, M. Beekman, Structure and formation of ant transportation networks. J. R. Soc. Interface **8**(62), 1298–1306 (2011)
63. D. Helbing, P. Molnar, I.J. Farkas, K. Bolay, Self-organizing pedestrian movement. Env. Plan. B **28**(3), 361–384 (2001)
64. A. Adamatzky, J. Jones, Road planning with slime mould: if Physarum built motorways it would route M6/M74 through newcastle. Int. J. Bifurc. Chaos **20**(10), 3065–3084 (2010)
65. A. Adamatzky, S. Akl, R. Alonso-Sanz, W. Van Dessel, Z. Ibrahim, A. Ilachinski, J. Jones, A. Kayem, G.J. Martínez, P. De Oliveira et al., Are motorways rational from slime mould's point of view? Int. J. Parallel Emerg. Distrib. Syst. **28**(3), 230–248 (2013)
66. E. Strano, A. Adamatzky, J. Jones, Physarum itinerae: evolution of roman roads with slime mould. Int. J. Nanotechnol. Mol. Comput. (IJNMC) **3**(2), 31–55 (2011)
67. A. Tero, S. Takagi, T. Saigusa, K. Ito, D.P. Bebber, M.D. Fricker, K. Yumiki, R. Kobayashi, T. Nakagaki, Rules for biologically inspired adaptive network design. Science **327**(5964), 439–442 (2010)
68. J. Jones, The emergence and dynamical evolution of complex transport networks from simple low-level behaviours. Int. J. Unconv. Comput. **6**(2), 125–144 (2010)
69. J. Jones, *From Pattern Formation to Material Computation: multi-agent Modelling of Physarum Polycephalum*, vol. 15 (Springer, 2015)
70. J. Jones, Characteristics of pattern formation and evolution in approximations of Physarum transport networks. Artificial Life **16**(2), 127–153 (2010)
71. H. Meinhardt, A. Gierer, Pattern formation by local self-activation and lateral inhibition. Bioessays **22**(8), 753–760 (2000)
72. J. Jones, Influences on the formation and evolution of physarum polycephalum inspired emergent transport networks. Nat. Comput. **10**(4), 1345–1369 (2011)
73. J. Jones, Mechanisms inducing parallel computation in a model of physarum polycephalum transport networks. Parallel Process. Lett. **25**(01), 1540004 (2015)
74. W. Baumgarten, J. Jones, M.J.B. Hauser, Network coarsening dynamics in a plasmodial slime mould: modelling and experiments. Acta Phys. Pol. B **46**(6) (2015). In–press
75. S. Stepney, The neglected pillar of material computation. Phys. D Nonlinear Phenom. **237**(9), 1157–1164 (2008)
76. M. Aono, Y. Hirata, M. Hara, K. Aihara, Amoeba-based chaotic neurocomputing: combinatorial optimization by coupled biological oscillators. New Gen. Comput. **27**(2), 129–157 (2009)

77. J. Jones, R. Mayne, A. Adamatzky, Representation of shape mediated by environmental stimuli in physarum polycephalum and a multi-agent model. Int. J. Parallel Emerg. Distrib. Syst. **0**(0), 1–19, 0
78. J. Jones, A. Adamatzky, Computation of the travelling salesman problem by a shrinking blob. Nat. Comput. **13**(1), 1–16 (2014)
79. J. Jones, A. Adamatzky, Material approximation of data smoothing and spline curves inspired by slime mould. Bioinspiration Biomim. **9**(3), 036016 (2014)
80. J. Jones, Embodied approximation of the density classification problem via morphological adaptation. Int. J. Unconv. Comput. **12**(2–3), 221–240 (2016)
81. J. Jones, A morphological adaptation approach to path planning inspired by slime mould. Int. J. Gen. Syst. **44**(3), 279–291 (2015)
82. T. Nakagaki, H. Yamada, A. Toth, Intelligence: maze-solving by an amoeboid organism. Nature **407**, 470 (2000)
83. T. Nakagaki, H. Yamada, A. Toth, Path finding by tube morphogenesis in an amoeboid organism. Biophys. Chem. **92**(1–2), 47–52 (2001)
84. V.J. Lumelsky, A comparative study on the path length performance of maze-searching and robot motion planning algorithms. IEEE Trans. Robot. Autom. **7**(1), 57–66 (1991)
85. A. Tero, R. Kobayashi, T. Nakagaki, Physarum solver: a biologically inspired method of road-network navigation. Phys. A: Stat. Mech. Its Appl. **363**(1), 115–119 (2006)
86. Y.-P. Gunji, T. Shirakawa, T. Niizato, M. Yamachiyo, I. Tani, An adaptive and robust biological network based on the vacant-particle transportation model. J. Theoret. Biol. **272**(1), 187–200 (2011)
87. M.P. Bendsoe, O. Sigmund, *Topology Optimization: Theory, Methods, and Applications* (Springer Science & Business Media, 2013)
88. B. Hassani, E. Hinton, *Homogenization and Structural Topology Optimization: Theory, Practice and Software* (Springer Science & Business Media, 2012)
89. X. Huang, M. Xie, *Evolutionary Topology Optimization of Continuum Structures: Methods and Applications* (Wiley, 2010)
90. M. Bendsoe, E. Lund, N. Olhoff, O. Sigmund, Topology optimization-broadening the areas of application. Control Cybern. **34**(1), 7 (2005)
91. A. Bejan, Constructal-theory network of conducting paths for cooling a heat generating volume. Int. J. Heat Mass Transf. **40**(4), 799–816 (1997)
92. T. Borrvall, J. Petersson, Topology optimization of fluids in Stokes flow. Int. J. Numer. Methods Fluids **41**(1), 77–107 (2003)
93. J. Stegmann, E. Lund, Discrete material optimization of general composite shell structures. Int. J. Numer. Methods Eng. **62**(14), 2009–2027 (2005)
94. H. Men, K.Y.K. Lee, R.M. Freund, J. Peraire, S.G. Johnson, Robust topology optimization of three-dimensional photonic-crystal band-gap structures. Opt. Express **22**(19), 22632–22648 (2014)
95. A. Safonov, A. Adamatzky, Computing via material topology optimisation. Appl. Math. Comput. **318**, 109–120 (2018)
96. M. Zhou, G.I.N. Rozvany, The COC algorithm, Part II: topological, geometrical and generalized shape optimization. Comput. Methods Appl. Mech. Eng. **89**(1–3), 309–336 (1991)
97. R.B. Wilson, *A Simplicial Method for Convex Programming* (Harvard University, Cambridge, MA, 1963)
98. K. Svanberg, The method of moving asymptotesa new method for structural optimization. Int. J. Numer. Methods Eng. **24**(2), 359–373 (1987)
99. A. Nagurney, D. Zhang, *Projected Dynamical Systems and Variational Inequalities with Applications*, vol. 2 (Springer Science & Business Media, 2012)
100. A. Klarbring, B. Torstenfelt, Dynamical systems, SIMP, bone remodeling and time dependent loads. Struct. Multidiscip. Optim. **45**(3), 359–366 (2012)
101. T.P. Harrigan, J.J. Hamilton, Bone remodeling and structural optimization. J. Biomech. **27**(3), 323–328 (1994)

102. A. Gersborg-Hansen, M.P. Bendsøe, O. Sigmund, Topology optimization of heat conduction problems using the finite volume method. Struct. Multidiscip. Optim. **31**(4), 251–259 (2006)
103. M.G. Mullender, R. Huiskes, H. Weinans, A physiological approach to the simulation of bone remodeling as a self-organizational control process. J. Biomech. **27**(11), 1389–1394 (1994)
104. W.M. Payten, B. Ben-Nissan, D.J. Mercert, Optimal topology design using a global self-organisational approach. Int. J. Solids Struct. **35**(3), 219–237 (1998)
105. Abaqus Inc. *Abaqus Analysis User Manual, Version 6.14*, 2014
106. A.A. Safonov, B.N. Fedulov, *Universal Optimization Software—UOPTI*, 2015
107. A.A. Safonov, *Youtube Channel of Alexander Safonov*, 2016

Physarum-Inspired Solutions to Network Optimization Problems

Xiaoge Zhang and Chao Yan

Abstract In this chapter, we introduce a mathematical model inspired by slime mould Physarum polycephalum, an amoeboid organism that exhibits phenomenal path-finding behavior. By comparing it to one of the classic shortest path algorithms—Dijkstra algorithm, we highlight and summarize the key characteristics that are unique in Physarum algorithm, namely flow continuity and adaptivity. Due to these features, the Physarum model responses autonomously to the changes of external environment, thereby converging to optimal solutions adaptively. Herein, we take advantage of its superior properties and develop various models to address several significant network optimization problems, including traffic flow assignment and supply chain network design. By comparing its performance with the state-of-the-art methods in terms of solution quality and running time, we demonstrate the efficiency of the proposed algorithms.

1 Introduction

Network, as an effective tool, has been widely used to characterize a large number of real-world systems, e.g., supply chain [1], manufacturing system [2], waterway network [3], traffic network [4, 5], airline network [6], and subway networks [7], to name a few. Typically, there are two basic components in a network: nodes and edges, in which the nodes are used to represent ad hoc entities depending on the specific context. For example, in a supply chain, a node can denote a supplier, a distribution center, a storage center, or a demand market. Whereas, in an airline network, a node may represent a hub airport, a destination city, or an origin city. With respect to

X. Zhang (✉)
Department of Civil and Environmental Engineering, School of Engineering,
Vanderbilt University, Nashville, TN 37235, USA
e-mail: xiaoge.zhang@vanderbilt.edu; zxgcqupt@gmail.com

C. Yan
Department of Electrical Engineering and Computer Science, School of Engineering,
Vanderbilt University, Nashville, TN 37235, USA
e-mail: chao.yan@vanderbilt.edu; authoryanchao@gmail.com

© Springer International Publishing AG, part of Springer Nature 2018
A. Adamatzky (ed.), *Shortest Path Solvers. From Software to Wetware*,
Emergence, Complexity and Computation 32,
https://doi.org/10.1007/978-3-319-77510-4_12

the edges, they are employed to characterize the activities or interactions among the entities. In particular, if there is an edge between two nodes in a transportation network, it entails that the two cities are connected to each other. Otherwise, there is no route between them. By representing a system using nodes and edges, we can capture a wide variety of activities that occur among different entities. By doing so, each given network topology corresponds to a possible system design.

Since the construction of the fundamental infrastructure systems is expensive, they must be rigorously scrutinized for cost-effectiveness. Mathematically, these problems can be cast as Network Design Problems (NDP) [8, 9]. Simply speaking, a network design problem is to identify an optimal subgraph from a graph subject to feasibility conditions [10–12]. Some well-known NDPs include traveling salesman problem (TSP) [13], minimum spanning tree problem (MSTP) [14], the shortest path problem (SPP) [15], the Steiner Tree Problem (STP) [16], and the bi-level network design problem [17], etc. These formulations characterize important matters in practical applications. For example, the traveling salesman problem is frequently encountered by many delivery firms, e.g., FedEx, DHL, and UPS, with the objective of minimizing the total fuel consumption as well as maximizing the delivery efficiency through optimizing the route for each truck in distributing the products with the constraint that each car must return to the distribution center after the delivery. Another instance is the transportation network design. In recent years, the bi-level network design has risen to be a significant model after the Braesss paradox [18] was identified. The Braesss paradox reveals that simply adding a new edge/link to a congested road traffic network may increase the overall journey time. From then on, the bi-level network design framework has been formulated and received increasing attentions over years. In this model, the passengers' path-choosing behavior acts as the lower-level model while the leader's decision making in terms of adding or removing certain edges from the traffic network plays as the upper model.

Since these problems have ubiquitous applications in practice, solving them efficiently has drawn increasing attentions. In the past decades, tremendous progress has been made towards finding either exact solutions or approximate solutions. The exact methods include branch and bound algorithm [19, 20], dynamic programming [21], and local search [22, 23], etc. The basic idea is to keep shrinking the region that contains a better solution. Unfortunately, when the magnitude of the problem instance increases, the search space grows exponentially and the time complexity for exact solutions becomes unacceptable in practice. Such critical deficiency has prevented the applications of these methods. Often, finding the optimal solution for a NP hard problem is non-tractable. Since most of the network design problems are NP-Hard, this further increases the difficulty to apply these algorithms. In recent years, there is a growing interest in the design and analysis of bio-inspired algorithms, e.g., Particle Swarm Optimization (PSO) [24], Genetic Algorithm (GA) [25], and Ant Colony Optimization (ACO) [26], due to their capability of providing approximately optimal solutions to complicated problems efficiently. All these methods are inspired by the phenomenon in the natural world. These algorithms are known as population-based optimizer: the initial individuals are generated in a probabilistic way, and they work together in the search process, in which each individual is assigned a value to measure its fitness in the whole population. The individuals of good fitness are selected from the current population to form a new generation. The same procedure continues until some predefined termination conditions are satisfied. Instead of searching the space exhaustively, these algorithms tackle the problems in a probabilistic manner, by which the near-optimal solutions are identified. Compared to the exact methods, one of the most significant properties of bio-inspired algorithms is their outstanding running time.

Among them, Physarum model has received increasing attentions in the past few years due to its prevailing capability in solving many practical network design problems, such as design of wireless sensor network [28], linear transportation problem [29] and route selection of delivery problem [30]. Physarum is a large, single-celled, amoeba-like creature. In most of its life, Physarum lives as a 'plasmodium', which is a single cell that contains many nuclei. The plasmodium searches for food by moving along like an amoeba and sending out a network of tendrils, through which the chemical signals and nutrients are transported [31]. Physarum does not have a brain, or even a nervous system, but nevertheless, they exhibit strong capability in making surprisingly sophisticated decisions. For example, Physarum is able to identify the shortest path between two nodes in a maze. The maze-solving behavior is demonstrated in Fig. 1. First, the slime mould extends its tendrils through every corner of the maze, essentially filling the entire maze (Fig. 1a). For more details, please refer to the video here: https://www.youtube.com/watch?v=czk4xgdhdY4. Two food sources are placed at the entrance (N_1) and the exit (N_2) of the maze. Then it tracks the tendrils that do not find food and leaves behind a trail of translucent slime that acts as external memory. Such trail enables the slime mould to recognize certain areas, which are referred to as dead ends, in this maze. The slime mould avoids the dead ends by decreasing the thickness of the tubes in these areas and grows exclusively along the shortest path from the source node to the destination node of the maze, as indicated in Fig. 1c.

In addition to identifying the shortest path in the labyrinth, Tero et al. also demonstrated that Physarum is able to build highly efficient networks by trading off among network cost, transport efficiency, and fault tolerance [32]. As shown in Fig. 2a, 36 food sources are placed in multiple positions to denote the geographical locations of major cities in the Tokyo area. The Physarum first grows from Tokyo and fills almost every corner of the available land space, as can be observed from Fig. 2b–d. Within certain period, Physarum optimizes the network structure by concentrating on the backbone tube network that connects the food sources, which can be seen from Fig. 2d, e. The ultimate network formed by Physarum is shown in Fig. 2f. Afterwards, Tero et al. compared the Physarum-formed network with the real Tokyo railway system using three metrics, namely: network cost, transport efficiency, and network robustness. The experiments revealed that Physarum is capable of constructing efficient network with comparable efficiency, fault tolerance, and cost to the real Tokyo railway network.

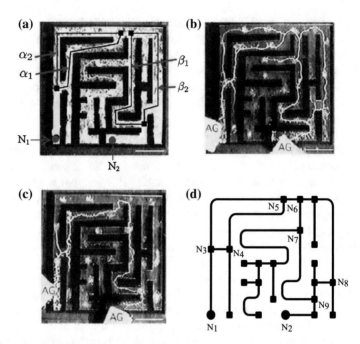

Fig. 1 **a–d** Illustrates Physarum's maze-solving process. **a** Initial state: the maze is filled by Physarum. **b** Intermediate state: the backbone path is formulated to connect the two food sources. **c** The path spanning the minimum length between the nutrient-containing agar blocks emerges. **d** A graphical representation of the maze: the source node N_1 and the sink node N_2 are indicated by solid circles and other nodes are shown by solid squares [27]

From then on, the Physarum model has been well-studied from the computational point of view, including route optimization [33, 34], 0–1 knapsack problem [35], network optimization [36–39], and load-shedding problem [40]. In this chapter, instead of providing a comprehensive review on the recent advances on the Physarum models, we focus on its progress and development in solving network optimization problems. Firstly, we present a generalized Physarum model, which has the capability to handle the route optimization problem with multiple sources and destinations in both directed and undirected networks. Secondly, we identify the principal differences between the Physarum model and other algorithms (i.e., Dijkstra algorithm, label setting algorithm) in finding the shortest paths. Thirdly, we leverage the core characteristics of the Physarum model to propose novel algorithms applied to different network optimization scenarios, where the edge weight is a function of the flow. Important problems, such as user equilibrium and system optimum problems in transportation networks, and supply chain network design, can be well solved.

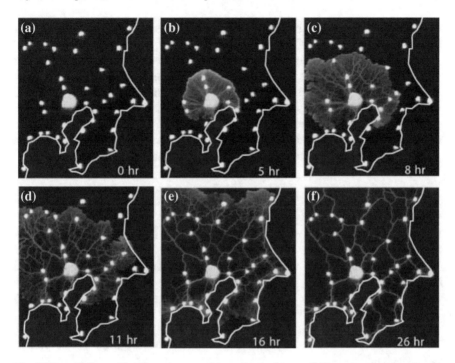

Fig. 2 Network formed by Physarum over time. **a** At t = 0, a small Physarum was placed at the location of Tokyo and it was bounded by the Pacific coast line. **b** At t = 5, the Physarum expanded from the initial food source gradually. **c** The Physarum spread progressively and colonized other food sources. **d** The Physarum yielded a network with a high similarity to the real rail network. **e** The backbone tubes interconnecting these food sources emerged. **f** The ultimate network formed by the Physarum [32]

2 Physarum Model

In 2000, Nakagaki et al. [41] demonstrated that the plasmodium of the slime mould *Physarum polycephalum* approximates the minimum-length path between two points in a given labyrinth. Before describing the maze-solving process of Physarum, we first recall some basic properties of *Physarum polycephalum*. Physarum is a large amoeba-like cell, and it consists of a dendritic network of tube-like structures. One of the most important features in Physarum is to adapt its shape when it crawls over a plain agar gel. If the food is placed at two different points, Physarum formulates the pseudopodia to connect the two points in a smart manner.

Figure 1 illustrates Physarum's maze-solving process. Initially, the maze is filled by Physarum, as illustrated in Fig. 1a. The two solid red circles, labelled as N_1 and N_2, indicate two separated food sources. As can be observed, there are four candidate routes to connect the two food sources, namely, α_1, α_2, β_1, and β_2. Figure 1b demonstrates that the backbone network only containing α_1, α_2, β_1, and β_2 is formulated and all other tubes in the dead end degenerate. Finally, as shown in Fig. 1c, the path spanning the minimum length between the food sources emerges while all the other segments disappear.

A mathematical model used to describe Physarum's shortest path finding behavior was developed in [27]. Here, we adopt this particular maze to illustrate this mathematical model. Suppose we represent the maze by a graph, as shown in Fig. 1d, in which two special nodes N_1 and N_2 are designated as the source node and sink node, respectively. All other nodes are labeled as $N_3, N_4, \ldots$ etc. The edge/link between nodes N_i and N_j is A_{ij}. The variable Q_{ij} is used to denote the flux through the tube A_{ij} from node N_i to N_j. Then the flux Q_{ij} can be formulated as:

$$Q_{ij} = \frac{D_{ij}}{L_{ij}}(p_i - p_j) \tag{1}$$

where p_i is the pressure at node N_i, D_{ij} is a conductivity (or diameter) of the tube A_{ij}, and L_{ij} is its length.

According to the flow conservation law, we have that the inflow and outflow must be balanced:

$$\sum_{\forall (i,j) \in A} Q_{ij} = 0 \quad (j \neq 1, 2) \tag{2}$$

With respect to the source node N_1 and the sink node N_2, the following two equations hold:

$$\begin{aligned} \sum_i Q_{i1} + I_0 &= 0 \\ \sum_i Q_{i2} - I_0 &= 0 \end{aligned} \tag{3}$$

where I_0 is the flux flowing from the source node. Since the total flux is fixed throughout the process in the experiment, thus I_0 is a constant.

To model the adaptive behavior of the slime mould, we assume that the conductivity D_{ij} changes over time according to the flux Q_{ij}. Thus, the evolution process of $D_{ij}(t)$ can be described as follows:

$$\frac{d}{dt}D_{ij} = y(|Q_{ij}|) - \gamma D_{ij} \tag{4}$$

where γ is the decay rate of the tube. This equation implies that the conductivity tends to vanish if there is no flux along the edge, while it is enhanced by the flux. It is natural to assume that y is a monotonically increasing continuous function satisfying $y(0) = 0$.

Then the network Poisson equation for the pressure is derived from Eqs. (1)–(3) as follows:

$$\sum_i \frac{D_{ij}}{L_{ij}}(p_i - p_j) = \begin{cases} -1 & \text{for} \quad j = 1, \\ +1 & \text{for} \quad j = 2, \\ 0 & \text{otherwise.} \end{cases} \tag{5}$$

By setting $p_2 = 0$ as the basic pressure level, all p_i's can be determined uniquely by solving the equation system (5), and each $Q_{ij} = D_{ij}/L_{ij}(p_i - p_j)$ is also obtained.

Herein, we assume that $y(Q_{ij}) = |Q_{ij}|$ because Physarum can converge to the shortest path regardless of the initial distribution of conductivities when $y\left(|Q_{ij}|\right) = |Q_{ij}|$, and $\gamma = 1$ [27]. With the flux calculated, the conductivity can be derived, where Eq. (6) is used instead of Eq. (4), adopting the functional form $y(Q) = |Q|$.

$$\frac{D_{ij}^{h+1} - D_{ij}^{h}}{\delta t} = |Q_{ij}^{h}| - D_{ij}^{h+1}. \tag{6}$$

where δt is a time mesh size and the upper index h indicates a time step.

In fact, the above equations model a dynamic evolution system, where the conductivity, D_{ij}, and flow, Q_{ij}, are the functions of time t. The conductivity D_{ij} evolves according to the adaptation equation (6), where variables Q_{ij} and p_i are determined by solving the network Poisson equation in Eq. (5) characterized by the value of D_{ij} and L_{ij} at each moment. In equilibrium ($D_{ij}^{h+1} = D_{ij}^{h}$), the conductivity along each link is equal to its flow. In non-equilibrium, the conductivity, D_{ij}, increases or decreases if the absolute value of the flow is larger than or smaller than the conductivity, respectively.

During the evolution of the system, some tubes grow or remain while others disappear. The system solves the shortest path problem while the conductivities of the edges on the shortest path converge to one, and the conductivities on the edges outside the shortest path converge to zero. In 2012, Bonifaci et al. [42, 43] proved that the mass of Physarum will eventually converge to the shortest N_1–N_2 path of the network, independent of the structure of the network or of the initial mass distribution.

3 A Generalized Physarum Model

Several crucial issues arise when we exploit the Physarum model to tackle the network optimization problems. Since the links in the maze are undirected, when Physarum finds the shortest path in the maze, its tubes can explore in either orientation along the segment. Whereas, many constraints are imposed with the traffic networks in reality, e.g., in the transporation networks, many roads are one-way. Hence, the first issue is how to extend the current Physarum model to handle the pathfinding problem in general cases. Another issue is that the original model can only find the shortest path in the network with one source node and one sink node, which is different from the realistic scenarios. For instance, in the supply chain, there are many suppliers providing the homogeneous product to various demand markets distributed in different locations. To address these issues, we develop a generalized Physarum model in this section.

3.1 Pathfinding in Directed Graph

One of the prevailing ways to represent the structure of a graph is the adjacency matrix [44]. The elements of the adjacency matrix indicate whether pairs of vertices are adjacent or not in the graph. Let $G(V, E, L)$ be a graph, where V denotes the set of vertices, E denotes the set of edges, and L represents the set of edge weights. Suppose $|V| = n$, then its adjacency matrix can be represented as:

$$L = \begin{bmatrix} 0 & L_{12} & L_{13} & \cdots & L_{1n} \\ L_{21} & 0 & L_{23} & \cdots & L_{2n} \\ L_{31} & L_{32} & 0 & \cdots & L_{3n} \\ \vdots & \vdots & \vdots & \ddots & \vdots \\ L_{n1} & L_{n2} & L_{n3} & \cdots & 0 \end{bmatrix}$$

where L_{ij} denotes the weight along the edge (i, j). In particular, if graph G is undirected, then its adjacency matrix is symmetrical. Thus, we have $L_{ij} = L_{ji}$, for $\forall (i, j) \in E$. Whereas, if graph G is directed and there is no edge from node j to node i, then we have $L_{ji} = \infty$.

From this standpoint, we can represent a directed link as a bidirectional one by using two individual variables L_{ij} and L_{ji}. By considering the orientation of each edge, we update Eq. (5) as:

$$\sum_i \left(\frac{D_{ij}}{L_{ij}} + \frac{D_{ji}}{L_{ji}} \right) (p_i - p_j) = \begin{cases} -1 & \text{for } j = 1 \\ +1 & \text{for } j = 2 \\ 0 & \text{otherwise.} \end{cases} \tag{7}$$

where D_{ij} is the conductivity of link A_{ij}.

If there is only one edge between node i and j, say $i \to j$, then $L_{ji} = \infty$, $\frac{D_{ji}}{L_{ji}} = 0$, and the equation degenerates to the same form as Eq. (5), and vice versa. If the link is bidirectional, then $\frac{D_{ij}}{L_{ij}} = \frac{D_{ji}}{L_{ji}}$, and we can update Eq. (5) as:

$$\sum_i \frac{2 * D_{ij}}{L_{ij}} (p_i - p_j) = \begin{cases} -1 & \text{for } j = 1 \\ +1 & \text{for } j = 2 \\ 0 & \text{otherwise.} \end{cases} \tag{8}$$

The only difference between Eqs. (5) and (8) is the constant term 2. Since the system evolves according to Eq. (6), the introduction of the constant term has no effect on its convergence. According to the evolution equation (6), the convergence condition still holds when $D_{ij} = Q_{ij}$. Since we bring in a constant term, the pressure of each node p_i' shrinks to half of its original value p_i calculated from Eq. (5). In the following iterations, the flow along each link shrinks by half because $Q_{ij} = \frac{D_{ij}}{L_{ij}}\left(p_i' - p_j'\right) = \frac{1}{2}\frac{D_{ij}}{L_{ij}}\left(p_i - p_j\right)$. When the system converges, the conductivities of the edges forming the shortest path converge to 0.5 while the conductivities of the edges outside the shortest path converge to 0.

In addition, the Physarum model can be readily represented by an electrical network. To be specific, the hydrostatic pressure p_i at node i Physarum is equivalent to the potential of the corresponding electrical network. R_{ij} can be modeled as $R_{ij} = \frac{L_{ij}}{D_{ij}}$. Let γ_{ij} be the current along the link (i, j), then we have:

$$\gamma_{ij} = \frac{U_{ij}}{R_{ij}} = \frac{p_i - p_j}{\frac{L_{ij}}{D_{ij}}} = \frac{D_{ij}}{L_{ij}}\left(p_i - p_j\right) = Q_{ij} \tag{9}$$

in which U_{ij} is the potential difference between i and j. It is observed that Eq. (9) exactly models the protoplasmic flow through the tubes.

In an electrical network, if there is a potential difference in an link (i, j) from i to j, then an electrical current γ_{ij} will flow from i to j according to Ohm's law. Consider the network in Fig. 3, the shortest path from node s to node t needs to be found. For the sake of simplicity, let the length of all the links be 1. s and t are the source and sink nodes, respectively. There are two alternate paths to connect s with t: $s \rightarrow 1 \rightarrow 3 \rightarrow t$, and $s \rightarrow 2 \rightarrow 4 \rightarrow t$. By making full use of the symmetry of the network, we rank the pressure of each node as: $p_s > p_1 = p_2 > p_3 = p_4 > p_t$.

Suppose we add another directed link $(3, 2)$ into the network. Since $p_2 > p_3$, the current tends to flow from node 2 to node 3, which is opposite to the link's direction. To solve this problem, we implement a checking procedure to guarantee the flow direction is consistent with the link orientation. Specifically, we record the direction of each link. If there is a directed edge from node i to node j, and the pressure p_j at node j calculated from Eq. (8) is larger than the pressure p_i at node i, the flux tends to flow from node j to node i, which is opposite to the link's orientation. In this case, $Q_{ij} = \frac{D_{ij}}{L_{ij}}\left(p_i - p_j\right) < 0$, which means the flow moves from node j to i, and we let the flow Q_{ij} be zero. In other words, we forbid the flow through the network when its direction is opposite to the edge's orientation. The general flowchart of the algorithm is summarized in Algorithm 1.

Fig. 3 One comprehensive example

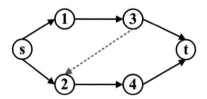

Algorithm 1: Physarum Algorithm for Directed Networks

Data: $G = (L, s, t)$, where L is an adjacency matrix of graph G, s is the starting node, and t
    is the ending node.

Result: The shortest path from s to t.

1   $D \longleftarrow (0, 1]$ $Q \longleftarrow 0$ $p \longleftarrow 0$

3   **while** *the termination condition is not met* **do**

5 | $p_t \leftarrow 0$

6̸7 | Calculate the pressure according to equation (8)

$$\sum_i \left(\frac{D_{ij}}{L_{ij}} + \frac{D_{ji}}{L_{ji}} \right) (p_i - p_j) = \begin{cases} -1 & \text{for } j = 1 \\ +1 & \text{for } j = 2 \\ 0 & \text{otherwise} \end{cases}$$

8 | $Q_{ij} \leftarrow D_{ij} \times (p_i - p_j)$

9 | **if** $L_{ij} \neq \infty$ and $p_i < p_j$ **then**

10 | $\lfloor$ $Q_{ij} = 0$

11     $\lfloor$ $D \leftarrow Q + D$

3.2 Network Optimization with Multiple Sources and Destinations

In a transportation network, there might be multiple source-sink pairs, whereas the original Physarum model is limited to a network with only one source node and one sink node. As a result, we modify Eq. (8) as follows to address the network optimization problem with multiple sources and sinks:

$$\sum_i \left(\frac{D_{ij}}{L_{ij}} + \frac{D_{ji}}{L_{ji}} \right) (p_i - p_j) = \begin{cases} -d^{s,t}, & \forall (s, t) \in \mathbb{OD}, \\ +d^{s,t}, & \forall (s, t) \in \mathbb{OD}, \\ 0, & \text{otherwise.} \end{cases} \tag{10}$$

where $\mathbb{OD}$ denotes the set of all OD pairs and $d^{s,t}$ denotes the travel demand between the origin node s and destination node t.

Given multiple sources and sinks, the proposed Physarum algorithm chooses the shortest path to transport the demand. Equation (10) guarantees that the flow is transported along the shortest paths, which has already been proved by Damian et al. [45].

4 Unique Features in Physarum Model

In this section, we identify the principal differences between Dijkstra algorithm and Physarum algorithm when they find the shortest path. By doing this, we highlight the unique characteristics in the Physarum model and summarize the benefits brought by such features.

4.1 Dijkstra Algorithm

Dijkstra algorithm was originally developed in 1956 by Edsger Dijkstra [46]. It is one of the most widely used approaches for finding the shortest path in a graph. In general, Dijkstra algorithm creates a tree of shortest paths from the starting node to all the other nodes in the graph. Herein, we briefly describe the procedures of Dijkstra algorithm in finding the shortest path. Let $G(V, E, L)$ be a graph, where V denotes the set of nodes, E represents the set of edges, and L is the set of link weights. For a given source node s, Dijkstra algorithm initializes the distance from all the nodes to the source node s as $+\infty$, and for each vertex $v \in V$, we maintain an attribute $v \cdot d$, and it is the upper bound of the distance from node v to the source node s, which is also referred to as *shortest-path estimate*. Then a set S is used to keep track of the nodes whose final shortest-path weights from the source node s have already been determined, and another set Q is used to represent the nodes that are not visited yet. Next, the algorithm repeatedly selects the vertex $u \in V - S$ with the minimum distance from node s, add u to S, and relaxes all edges leaving u. The general procedures of Dijkstra algorithm is summarized in Algorithm 2.

In such algorithm, $v \cdot \pi$ denotes the predecessor of node v in the shortest path, and the function **MIN-DIST** denotes identifying the node in set Q that has the minimum distance from node s. For the completeness of this chapter, we use one example to depict the basic procedures of Dijkstra algorithm. Figure 4a shows a weighted, directed graph with seven nodes, in which s and t play as the source and the destination nodes, respectively. The numbers along each edge indicate the weight of that particular link, and the value above each vertex represents the label of each node. Our goal is to find a path with the minimum total cost to connect the source node s and the ending node t. First of all, we initialize the distance from all the nodes to the source node as ∞ (see Fig. 4b), the set S is initialized as an empty set, as illustrated in lines 1–5 of Algorithm 2. Next, we pick the vertex with the minimum distance value (node s) from node s, remove it from set Q, and add it to set S. After the vertex s is included, we examine the edges that leave from node s. The edge $(s, 1)$ gives a path cost of 11 so we update the distance value of node 1 as 11. Likewise, we change the distance of node 2 from ∞ to a smaller value 8, as illustrated in Fig. 4c.

Now, among the nodes 1 and 2, node 2 has the smallest distance value. Thus, we remove it from set Q, and add it to set S. It can be observed that the edge $(2, 3)$ has a weight 2, and it indicates we can get from node s to node 3 for a cost of $8 + 2 = 10$. Hence, we update the distance from node s to node 3 as 10. In a similar way, we examine the other edges that leave node 2, and conduct the same operations. As a result, the distance from node 1 and node 5 to node s is updated as 9 and 14 (see Fig. 4d), respectively. The smallest value on a white vertex is node 1, so we add node 1 to the set S and remove it from the set Q as well (see Fig. 4e). In what follows,

---

Algorithm 2: Dijkstra algorithm for finding the shortest path

---

Data: A directed graph $G = (L, s, t)$, where L is an adjacency matrix of graph G, s is the starting node, and t is the ending node.

Result: The shortest path from s to t.

```
// Initialize the distance from source node s to all the
   other nodes
```

1: **for** each vertex $v \in G.v$ **do**

2:     $v.d = +\infty$

3:     $v.\pi = NULL$

4: **end for**

5: $S = \Phi$

6: $Q = G.V$

```
// Relaxtion
```

7: **while** $Q \neq \Phi$ **do**

8:     $u = \text{MIN-DIST}(Q)$

9:     $S = S \cup u$

10:     **for** each vertex $v \in G.adj[u]$ **do**

11:         **if** $v.d > u.d + L(u, v)$ **then**

12:             $v.d = u.d + L(u, v)$

13:             $v.\pi = u$

14:         **end if**

15:     **end for**

16: **end while**

---

we re-examine the distance values at the unvisited neighbor of node 9. The edge $(1, 3)$ has weight 6, and we can get node 3 with a cost of $9 + 6 = 15$, but it is more than the current shortest-path estimate 10. Thus, we do not update node 3's shortest-path estimate in this case. The other edge $(1, 4)$ has weight 5, and we can get node 4 from node s with a cost of $9 + 5 = 14$, which is less than its current shortest-path estimate ∞. Therefore, the shortest-path estimate of node 4 is updated (see Fig. 4e). Now, the minimum distance occurs at vertex 3. As a result, we remove it from set Q and add it into set S. Likewise, we check the links that leave node 3, and update the shortest-path estimates of adjacent nodes (node 4 and node t) as 12 and 18, respectively. Afterwards, node 12 becomes the node with minimum total weight, and it is removed from set Q and added into set S (see Fig. 4g). But it does not change the shortest-path estimates of the node in set Q. Eventually, node t is removed from set Q and added into set S (see Fig. 4h), and Dijkstra algorithm terminates because the queue Q is empty. The found shortest path from node s to node t is $s \rightarrow 2 \rightarrow 3 \rightarrow t$.

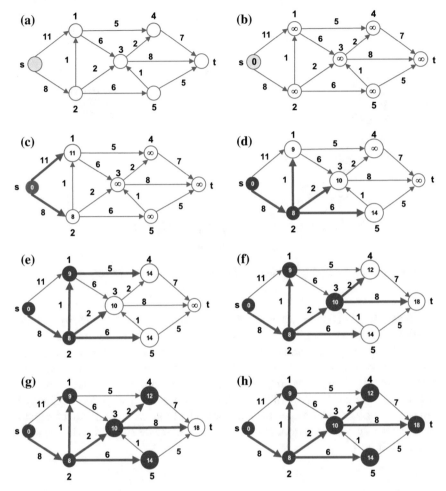

Fig. 4 The execution of Dijkstra algorithm. The source node s is the leftmost vertex. The shortest-path estimates appear within the vertices. The black vertices are in the set S, and the white vertices are in the set Q. **a** An illustrative example. **b** The network state before the first iteration of while loop of lines 7–16 in Algorithm 2. **c–h** The network state after each successive iteration of the while loop

4.2 Physarum Algorithm

As indicated in Algorithm 1, the first step in Physarum algorithm is to initialize all the parameters, including the conductivity D, the link flow Q, and the node pressure p. The second step is to derive the node pressure from the network flow conservation equation (10). Afterwards, the flow along each link is updated according to Eq. (1). Once the flow is obtained, the conductivity along each edge in the next iteration is calculated based on Eq. (6). The process continues until the termination condition

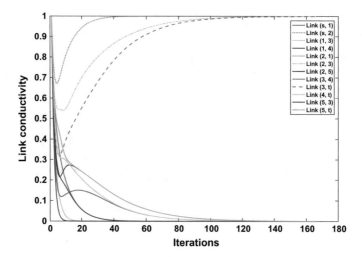

Fig. 5 An example illustrating Physarum model's shortest path finding process

is satisfied. In this chapter, when the difference of the conductivity along each link between two consecutive iterations is less than a prescribed threshold (10^{-3}), the algorithm halts. Figure 5 demonstrates the process of Physarum in finding the shortest path from node s to node t for the directed graph shown in Fig. 4a. It can be seen that the conductivity along all the links is initialized as the same value. With the increase of iterations, the conductivity of certain links converges to 1 while the conductivity along all the other links degenerates to zero. The links with conductivity equal to one form the shortest path from the source node s to the destination node t, which is $s \rightarrow 2 \rightarrow 3 \rightarrow t$.

4.3 Primary Differences

Obviously, the path found by Physarum algorithm is consistent with the one identified by the Dijkstra algorithm. However, there are huge differences in the procedures of the two algorithms in finding the shortest path. First of all, Dijkstra algorithm approaches the shortest path by updating the shortest-path estimate. During each iteration, Dijkstra algorithm adds one vertex to the set S and updates the shortest-path estimate to the latest. The same process continues until all the nodes are added to the set S. Thus, Dijkstra algorithm needs to maintain the set Q and S to keep track of which node is visited, which one is not. Whereas, Physarum algorithm does not maintain the same data structure that is used in Dijkstra algorithm, i.e., set S and Q, and the shortest-path estimate. In contrast, Physarum algorithm employs the network Poisson equation (see Eq. (8)) as a central unit to determine the pressure of each node with the consideration of link cost and conductivity, and the computed node pressure are then used to calculate the link flow in the next iteration. Another key underlying mechanism in the model is the positive feedback expressed in Eq. (6): larger conductivity results in larger flux, and this in turn increases the conductivity. When the above two mechanisms are combined together, it makes the Physarum model to converge to the shortest path, as proved in Ref. [42].

Secondly, the different underlying mechanisms in the two algorithms result in totally different pathfinding process. As can be seen from Fig. 4, the shortest path finding in Dijkstra algorithm can be seen as a discrete process, while in Physarum algorithm, the search process for the shortest path is continuous, as can be observed from Fig. 5. The continuous nature in the Physarum algorithm equips the algorithm with some significant advantages over the Dijkstra algorithm. Specifically, the Dijkstra algorithm performs the optimization in a discrete manner by iterating over all the nodes one by one. When the costs on some links related to the visited nodes happen to change, the Dijkstra algorithm needs to revisit those nodes and reconstruct the shortest path for the nodes affected by such a change. Moreover, in the Dijkstra algorithm, each link is only associated with one criterion—length—and there is no equivalent attribute like 'flow' in the Physarum model reacting to the change of the link cost. As a result, many classical algorithms for the traffic equilibrium problem must have two separate processes: path finding and flow shift. In contrast, our Physarum algorithm is very straightforward; once the link cost L is updated, with the help of Eq. (10), the flow is redistributed and reallocated dynamically in the next iteration. Physarum algorithm is suitable for solving the network optimization problems in dynamic environment because it can utilize the computational (or intermediate) results in the previous iterations and respond to the changes by adjusting the tube thickness (Q).

5 Traffic Assignment Problem

5.1 Problem Formulation

Consider a strongly connected network $G(V, E)$, where V and E are sets of nodes and links, respectively. Let $\mathbb{OD}$ be all the OD pairs in network G, for which travel demand $d^{s,t}$ is generated, where $(s, t) \in \mathbb{OD}$. Let $q_p^{s,t}$ represent the path flow originated at node s and destined to node t, then we have

$$\sum_{p \in P^{s,t}} q_p^{s,t} = d^{s,t}, \ \forall (s, t) \in \mathbb{OD}. \tag{11}$$

where $P^{s,t}$ is a set of cycle-free paths connecting s with t. All path flows must be non-negative to guarantee a meaningful solution

$$q_p^{s,t} \geq 0, \ \forall p \in P^{s,t}, \ \forall (s, t) \in \mathbb{OD}. \tag{12}$$

Let f_a denote the flow along the link a. Then the total flow on the link a is the sum of all paths that include the link

$$f_a = \sum_{(s,t)\in\mathbb{OD}} \sum_{p\in P^{s,t}} q_p^{s,t} \delta_{ap}^{s,t}, \ \forall a \in E. \tag{13}$$

where $\delta_{ap}^{s,t} = 1$ if the link a is a segment of the path p connecting s with t. Otherwise, $\delta_{ap}^{s,t} = 0$.

In a transportation network, each user non-cooperatively seeks to minimize their own cost by taking the path with least perceived cost from their origin to destination. The network is said to be in equilibrium if no user can reduce their cost by shifting to other alternative routes. The traffic equilibrium assignment problem can be mathematically formulated as:

$$Min \ \ z(x) = \sum_{a\in E} \int_0^{f_a} t_a(x)dx,$$

$$s.t.$$
$$f_a = \sum_{(s,t)\in\mathbb{OD}} \sum_{p\in P^{s,t}} q_p^{s,t} \delta_{ap}^{s,t}, \ \forall a \in E,$$
$$\sum_{p\in P^{s,t}} q_p^{s,t} = d^{s,t}, \ \forall (s,t) \in \mathbb{OD}, \tag{14}$$
$$q_p^{s,t} \geq 0, \ \forall p \in P^{s,t}, \forall (s,t) \in \mathbb{OD}.$$

where z is the objective function, f_a denotes the total traffic flow on link a, and $t_a(x)$ is a cost function of link a that is convex and monotonically increasing, $q_{s,t}$ represents the total traffic demand from s to t, and $f_p^{s,t}$ denotes the flow on path p between origin node s and destination node t.

5.2 Proposed Method

In[1] the traffic network, link cost is a function of the traffic flow. Every time, when the flow volume on a specific link changes, the link cost needs to be updated accordingly, which in turn makes the traffic flow be redistributed in the updated network. The process continues until the system converges to the user equilibrium.

The Physarum algorithm enjoys several nice properties when applied to the traffic assignment problem. First of all, the network Poisson equation shown in Eq. (10) makes the Physarum solver capable of redistributing the flow in accordance with the updated link cost L_{ij}. Every time the cost is updated, the Physarum solver adapts itself to redistribute the flow in a way such that the selected paths are optimal after the link cost is updated.

---

[1]This subsection is taken from the paper published in IEEE Transactions on Cybernetics. For more details, please see [47].

Since the cost function of shipping Q_{ij} units along link (i, j) is denoted as $t_{ij}(Q_{ij})$, we propose the following equation to update the cost of the link (i, j):

$$L_{ij}^{h+1} = \frac{L_{ij}^h + t_{ij}\left(Q_{ij}^{h+1}\right)}{2} \tag{15}$$

where Q_{ij}^{h+1} represents the flow along link (i, j) at the $h + 1_{th}$ iteration, L_{ij}^{h+1} and L_{ij}^h denote the cost on the link at the $h + 1_{th}$ and h_{th} iteration, respectively. The composite cost function L_{ij}^{h+1} combines the current cost L_{ij}^h with the cost in the next iteration denoted by $t_{ij}\left(Q_{ij}^{h+1}\right)$. The composite cost function guides the algorithm to allocate the flow in a way that balances the cost and flow.

The cost function defined in Eq. (15) has several benefits. Specifically, $t_{ij}\left(Q_{ij}^{h+1}\right)$ denotes the cost on link (i, j) at the $h + 1$ iteration provided that Q_{ij}^{h+1} is allocated to link (i, j). Thus, we update the link cost L_{ij}^{h+1} in the way that combines the current link cost L_{ij}^h and the future link cost $t_{ij}\left(Q_{ij}^{h+1}\right)$, which, in turn, affects the flow distribution further in the next iteration according to Eq. (10). When the two costs are the same ($L_{ij}^h = t_{ij}(Q_{ij}^{h+1})$), the algorithm converges.

Algorithm 3: Physarum Solver in Traffic Network Equilibrium Assignment Problem $(L, n, \mathbb{OD})$

1: // n is the size of the network;
2: // $\mathbb{OD}$ is the set of all OD pairs in the network;
3: // L_{ij} is the length of the link connecting node i with node j;
    $D_{ij} \leftarrow (0, 1]$ $(\forall i, j = 1, 2, \ldots, n)$;
    $Q_{ij} \leftarrow 0$ $(\forall i, j = 1, 2, \ldots, n)$;
    $p_i \leftarrow 0$ $(\forall i = 1, 2, \ldots, n)$;
4: **repeat**
5:    Calculate the pressure associated with each node according to Eq. (10)

$$\sum_i \left(\frac{D_{ij}}{L_{ij}} + \frac{D_{ji}}{L_{ji}}\right)(p_i - p_j) = \begin{cases} -d^{s,t}, & \forall (s, t) \in \mathbb{OD}, \\ +d^{s,t}, & \forall (s, t) \in \mathbb{OD}, \\ 0, & \text{otherwise.} \end{cases}$$

6:    $Q_{ij} \leftarrow D_{ij} \times (p_i - p_j)/L_{ij}$ // Using Eq. (1);
7:    $D_{ij} \leftarrow Q_{ij} + D_{ij}$ // Using Eq. (6)
8:    **Update the cost on each link;**
9:    **for** $i = 1 : n$ **do**
10:      **for** $j = 1 : n$ **do**
11:        $L_{ij} = \frac{L_{ij} + t_{ij}(Q_{ij})}{2}$
12:      **end for**
13:    **end for**
14: **until** the required RGAP is met

The general flow of the proposed algorithm is shown in Algorithm 1. There are several possible termination criteria in Algorithm 1, such as the maximum number of iterations is reached, flux through each tube remains unchanged, etc. The algorithm described in the present paper stops when its solution to the traffic assignment problem satisfies the required precision.

5.3   Convergence of the Proposed Method

In this section, we prove that the developed Physarum algorithm converges to the optimal solution that minimizes the objective function (14).

Lemma 1 *When the developed Physarum algorithm converges, the traveling time* L_{st} *among any OD pair* (s, t) *is a constant.*

Proof As indicated in Ref. [42], $D_{ij}^{h+1} = D_{ij}^h$ when the Physarum algorithm converges.

From Eq. (6), it holds that $D_{ij}^h = \left| Q_{ij}^h \right|$. Thus, we have:

$$D_{ij}^h = \begin{cases} Q_{ij}^h, & \text{if } Q_{ij}^h \geq 0, \\ -Q_{ij}^h, & \text{if } Q_{ij}^h < 0. \end{cases} \tag{16}$$

- If $Q_{ij}^h \geq 0$, since D_{ij} and L_{ij} are nonnegative variables, we derive $p_i \geq p_j$ from Eq. (1). Besides, since $D_{ij}^h = Q_{ij}^h$, Eq. (1) further implies $L_{ij} = p_i - p_j$.
- If $Q_{ij}^h < 0$, we have $p_i < p_j$. By substituting $D_{ij}^h = -Q_{ij}^h$ into Eq. (1), we have $L_{ij} = p_j - p_i$.

In summary, it holds that $L_{ij} = |p_i - p_j|$, where p_i and p_j are derived from Eq. (10). The linear system defined in Eq. (10) determines the node potential uniquely. In other words, the potential at each node is a constant, so we have $L_{ij}^{h+1} = L_{ij}^h$ when $h \to \infty$. Likewise, for any OD pair (s, t), L_{st} is a constant.

Lemma 2 *When the developed Physarum algorithm converges, the travel time along any link is equal to the cost derived from the cost function.*

Proof From Lemma 1, we have $L_{ij}^{h+1} = L_{ij}^h$. By substituting it into Eq. (15), we have $L_{ij}^{h+1} = t_{ij} \left(Q_{ij}^{h+1} \right)$ when $h \to \infty$.

Lemma 3 *The optimal solution of the mathematical minimization program formulated in Eq. (14) is equivalent to the user equilibrium conditions.*

Proof Sheffi [48] proved the equivalence between the equilibrium conditions and the optimal solution to the minimization program of the traffic assignment problem in pp. 63–66. Here, we do not repeat the proof for the sake of brevity. Thus, the solution satisfying the equilibrium conditions is the one that minimizes the objective function defined in Eq. (14).

Lemma 4 *The objective function formulated in Eq. (14) has a unique minimum.*

Proof The Hessian matrix of the objective function (14) can be calculated by using a representative term of the matrix. The derivative is firstly computed with respect to the flow on the m-th and n-th link, respectively.

$$\frac{\partial z(x)}{\partial x_m} = t_m(x_m) \tag{17}$$

and the second derivative is:

$$\frac{\partial^2 z(x)}{\partial x_m \partial x_n} = \frac{\partial t_m(x_m)}{\partial x_n} = \begin{cases} \frac{dt_n(x_n)}{dx_n}, & \text{for } m = n, \\ 0, & \text{otherwise.} \end{cases} \tag{18}$$

Obviously, all the off-diagonal elements of the Hessian matrix, $\nabla^2 z(x)$, are zero, and all the diagonal elements are given by $\frac{dt_a(x_a)}{dx_a}$. Thus, we have:

$$\nabla^2 z(x) = \begin{bmatrix} \frac{dt_1(x_1)}{dx_1} & 0 & 0 & \cdots \\ 0 & \frac{dt_2(x_2)}{dx_2} & 0 & \cdots \\ 0 & 0 & \ddots & \cdots \\ \vdots & \vdots & \vdots & \frac{dt_a(x_a)}{dx_a} \end{bmatrix} \tag{19}$$

Because all entries in the diagonal are strictly positive, this matrix is positive definite. Thus, the objective function (14) is strictly convex. In other words, the traffic assignment problem has a unique minimum.

From Lemmas 1 and 2, it entails that all the used paths have the same cost. Thus, the first condition of the user equilibrium is satisfied.

As indicated in Ref. [42], the Physarum model always converge to the shortest path, and the conductivity along the edges outside the shortest path converges to zero. Hence, there is no flow along the non-optimal paths because their traveling time is inferior to that of the shortest paths. Thus, the second condition of the user equilibrium is met. From Lemmas 3 and 4, we know that the solution satisfying the two user equilibrium conditions is equivalent to the one that minimizes the program (14). Hence, the proposed Physarum model converges to user equilibrium.

5.4 Measure of Solution Quality

In this chapter, we use the Relative Gap (RGAP) to measure the convergence:

$$RGAP = 1 - \frac{\sum_{s\in\mathbb{O}}\sum_{t\in\mathbb{D}} d^{st} \cdot p^{st}_{min}}{\sum_{i=1}^{n}\sum_{j=1}^{n} f_{ij} \cdot t_{ij}} \tag{20}$$

where d^{st} represents the travel demand between node s and t, p^{st}_{min} denotes the shortest path of OD pair (s, t), f_{ij} and t_{ij} denote the flow and travel time, respectively, along the link (i, j) when the algorithm terminates.

5.5 Numerical Examples

In this section, we use two examples to demonstrate the efficiency of the proposed algorithm, and compare its performance against the Frank-Wolfe (FW) algorithm, Conjugate Frank-Wolfe (CFW) algorithm, Bi-Conjugate Frank-Wolfe (BFW) algorithm, and Gradient Projection (GP) algorithm. All the five algorithms are implemented using C++. In FW, BFW, and CFW, we implement linear search to balance the flow. All tests are performed in a Windows 7 computer with Intel Core i5-2510 CPU, 4 Core, 2.5 GHz; 4 GB RAM. The RGAP is set as 10^{-5}. We adopt the following cost function developed by the US Bureau of Public Roads (BPR) for all the links [49]:

$$t_{ij} = \alpha_{ij}\left(1 + 0.15\left(\frac{v_{ij}}{C_{ij}}\right)^4\right) \tag{21}$$

where t_{ij}, α_{ij}, v_{ij}, and C_{ij} denote the travel time (cost), free-flow travel time, flow, and capacity along link (i, j), respectively.

Example 1 The network shown in Fig. 6 is adapted from Ref. [50], which has 13 nodes, 19 links, and 10 OD pairs. The origin-destination demands, in vehicles per hour, are shown as blow:

$$q^{1,2} = 660, \ q^{1,3} = 800, \ q^{1,10} = 800, \ q^{1,11} = 600,$$
$$q^{4,2} = 412.5, \ q^{4,3} = 495, \ q^{4,8} = 700, \ q^{4,9} = 300,$$
$$q^{4,10} = 300, \ q^{4,13} = 600.$$

and the link characteristics are shown in Table 1.

The optimal solution found by the proposed Physarum algorithm is shown in Fig. 7, and the result is consistent with that of the other four algorithms. From the flow along each link indicated in Fig. 7, we can compute the travel time along each link according to the BPR function defined in Eq. (21). With the calculated travel

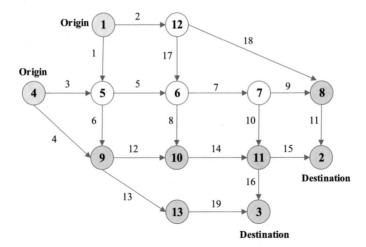

Fig. 6 Test network 1: Nguyen-Dupuis's 13-node network [50], the grey colored circles represent the origins, and the blue colored circles denote the destinations

Table 1 Link characteristics for Nguyen-Dupuis's 13-node network shown in Fig. 6 [50]

Link	Free-flow travel time (min/trip)	Capacity (veh/h)	Link	Free-flow travel time (min/trip)	Capacity (veh/h)
1	7	300	11	9	500
2	9	200	12	10	550
3	9	200	13	9	200
4	12	200	14	6	400
5	3	350	15	9	300
6	9	400	16	8	300
7	5	500	17	7	200
8	13	250	18	14	300
9	5	250	19	11	200
10	9	300			

cost, we compute the travel time along the alternate paths that connect the same OD pair. Specifically, the cost along the path $1 \rightarrow 12 \rightarrow 8 \rightarrow 2$ is $1421.1 + 134.1390 + 20.0464 = 1575.3$, and the traveling time is the same as that of the other two paths, namely, $1 \rightarrow 5 \rightarrow 6 \rightarrow 7 \rightarrow 8 \rightarrow 2$, and $1 \rightarrow 12 \rightarrow 6 \rightarrow 7 \rightarrow 8 \rightarrow 2$. Likewise, this also applies to the paths that connect other OD pairs. Obviously, all the used paths connecting the same OD pair have the same travel time.

Figure 8 illustrates the computational comparisons between the proposed method and the other four methods. Among all the algorithms, GP consumes the least amount of time to converge to the predefined precision, and the proposed algorithm is the second fastest algorithm among them. Since this is a tiny example, there is no explicit

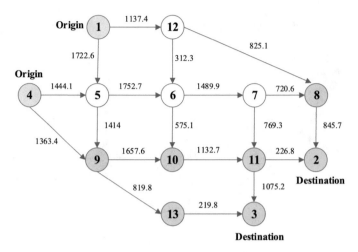

Fig. 7 The optimal solution for the traffic assignment in test network 1

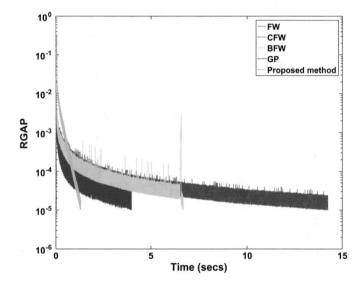

Fig. 8 Computational effort versus accuracy in test network 1

difference in the running time of GP and the proposed algorithm. In the subsequent examples, we will demonstrate the superiority of the proposed algorithm against GP in searching for the equilibrium solution. Figure 8 also reveals the FW algorithm's quick start/never-end feature. In fewer than 2 s, FW reduced the relative gap from 0.89 to 10^{-3}, a difference of 0.899. Yet in 14 s, it could reduce RGAP further by only 9×10^{-4}. FW spent nearly 7 times as long to achieve less than one-hundredth of the reduction. Besides, we also observe the zigzag pattern in the FW-based algorithms.

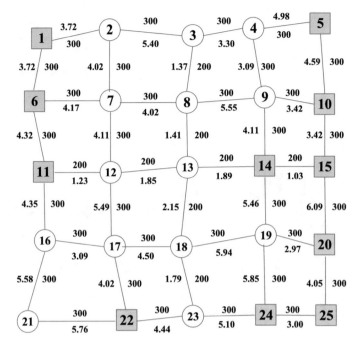

Fig. 9 Test network 2: A transportation network with 25 nodes, the green colored squares represent the origins, and the yellow colored squares denote the destinations

This happens because the search direction is perpendicular to the gradient of the objective function when the algorithm approaches the equilibrium solution [51]. But GP and the proposed algorithm does not have this issue, and such characteristics make them more appropriate to address the traffic assignment problems in practical applications.

Example 2 The network shown in Fig. 9 has 40 two-way links and ten OD pairs. The link cost is asymmetric in this network. The nodes in squares represent the source and sink nodes, and all the remaining 21 nodes are transshipment nodes. The numbers along each link represent the link capacity and free flow travel time, respectively. For example, the free flow travel time on link $(1, 2)$ is 3.72 and the link capacity is 300. Suppose the travel demand matrix is given as:

$$q^{1,22} = 700, \ q^{1,25} = 500, \ q^{2,22} = 800, \ q^{5,24} = 600,$$
$$q^{5,25} = 800, \ q^{6,10} = 800, \ q^{6,24} = 2800, \ q^{11,22} = 800,$$
$$q^{14,22} = 800, \ q^{15,20} = 800.$$

Figure 10 shows the comparison results for a target RGAP of 10^{-5}. Among FW-based algorithms, BFW demonstrates the best performance, followed by CFW. All the FW-based algorithms suffer from the problem that the search direction is

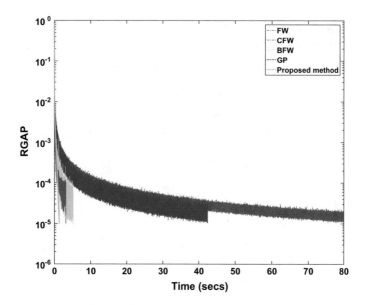

Fig. 10 Computational effort versus accuracy in test network 2

perpendicular to the gradient of the objective function when the level of relative gap reaches 10^{-4}. The proposed method runs much faster than the FW, CFW, BFW, and GP algorithms, and it converges to the equilibrium solution in 2 s. As can be observed, the proposed algorithm outperforms GP in searching for the equilibrium solution. While GP takes 3.2 s to converge, the developed algorithm only costs 1.3 s to obtain the equilibrium solution. Besides, GP also shows a zigzag pattern when the value of relative gap is less than 10^{-4}. In contrast, the convergence curve of the developed algorithm is very smooth and it does not suffer from such problems.

6 Supply Chain Network Design

In this section, we employ Physarum model to address a multi-criteria sustainable supply chain network design problem.

6.1 Multi-criteria Sustainable Supply Chain Network Design Model [52]

Consider the supply chain network shown in Fig. 11: a firm corresponding to node 1 aims at delivering the goods or products to the nodes at the very bottom which

Fig. 11 The supply chain
network topology

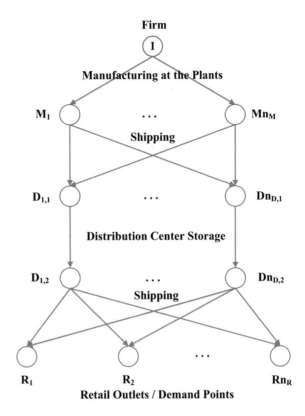

correspond to the retail outlets. The links connecting the source node with the destination nodes represent the activities of production, storage and transportation of good or services. Different network topologies corresponds to different supply chain network activities.

As shown in Fig. 11, the firm takes into consideration n_M manufacturers, n_D distribution centers when n_R retailers with demands $d_{R_1}, d_{R_2}, \ldots, d_{R_{n_R}}$ must be served. The node 1 in the first layer is linked with the possible n_M manufacturers, which are represented as $M_1, M_2, \ldots, M_{n_M}$. These edges in the manufacturing level are associated with the possible distribution center nodes, which are expressed by $D_{1,1}, D_{2,1}, \ldots, D_{n_D,1}$. These links mean the possible shipment between the manufacturers and the distribution centers. The links connecting $D_{1,1}, D_{2,1}, \ldots, D_{n_D,1}$ with $D_{1,2}, D_{2,2}, \ldots, D_{n_D,2}$ reflect the possible storage links. The links between $D_{1,2}, D_{2,2}, \ldots, D_{n_D,2}$ and $R_1, R_1, \ldots, R_{n_R}$ denote the possible shipment links connecting the storage centers with the retail outlets.

Let a supply chain network be represented by a graph $G(N, L)$, where N is a set of nodes and L is a set of links. Each link in the network is associated with a cost function and the cost reflects the total cost of all the specific activities in the supply chain network, such as the transport of the product, the delivery of the product, etc.

The cost related with link a is expressed by $\widehat{c}_a$. A path p connecting node 1 with a retail node shown in Fig. 11 denotes the whole activities related with manufacturing the products, storing them and transporting them, etc. Assume w_k denotes the set of source and destination nodes $(1, R_k)$ and P_{w_k} represents the set of alternative associated possible supply chain network processes joining $(1, R_k)$. Then P means the set of all paths joining $(1, R_k)$ while x_p denotes the flow of the product on path p, then the following Eq. (22) must be satisfied:

$$\sum_{p \in P_{w_k}} x_p = d_{w_k}, \quad k = 1, \ldots, n_R. \tag{22}$$

Let f_a represent the flow on link a, then the following conservation flow must be met:

$$f_a = \sum_{p \in P} x_p \delta_{ap}, \quad \forall a \in L. \tag{23}$$

Eq. (23) means that the inflow must be equal to the outflow on link a.

These flows can be grouped into the vector f. The flow on each link must be a nonnegative number, i.e. the following Eq. (24) must be satisfied:

$$x_p \geq 0, \quad \forall a \in L. \tag{24}$$

Suppose the maximum capacity on link a is expressed by u_a, $\forall a \in L$. It is required that the actual flow on link a cannot exceed the maximum capacity on this link:

$$\begin{aligned} f_a &\leq u_a, \quad \forall a \in L, \\ 0 &\leq u_a, \quad \forall a \in L. \end{aligned} \tag{25}$$

The total cost on each link, for simplicity, is represented as a function of the flow of the product on all the links [53–56]:

$$\widehat{c}_a = \widehat{c}_a(f), \quad \forall a \in L. \tag{26}$$

The total investment cost of adding capacity u_a on link a can be expressed as follows:

$$\widehat{\pi}_a = \widehat{\pi}_a(u_a), \quad \forall a \in L. \tag{27}$$

Summarily, the supply chain network design optimization problem is to satisfy the demand of each retail outlet and minimize the total cost, including the total cost of operating the various links and the capacity investments:

$$Minimize \sum_{a \in L} \widehat{c}_a(f) + \sum_{a \in L} \widehat{\pi}_a(u_a) \tag{28}$$

subject to constraints (22)–(25).

Hereupon, we also take into account the cost associated with the total amount of emissions generated both in the capital phase and operation phase. The generated emissions can occur in each phase, including the manufacturing stage, storing stage, and shipping stage. Suppose $e_a(f_a)$ represents the emission-generation function on link a in the operation phase and it is a function in relation with the product flow on this link. In addition, let $\widehat{e}_a(u_a)$ denote the emission-generation function on link a in the capital investment period. Similarly, it is a function of the product flow on that link. As a result, this objective can be expressed in the following form:

$$Minimize \quad \sum_{a \in L} e_a(f_a) + \widehat{e}_a(u_a) \tag{29}$$

Combing these two objectives shown in Eqs. (30) and (29), we can construct the following objective function:

$$Minimize \quad \sum_{a \in L} c_a(f_a) + \widehat{\pi}_a(u_a) + \omega \left(\sum_{a \in L} e_a(f_a) + \widehat{e}_a(u_a) \right)$$
$$s.t. \tag{30}$$
$$f_a \leq u_a$$
$$0_a \leq u_a$$

where ω is a nonnegative constant assigned to the emission-generation attribute. It is a factor to reflect how much the firm is willing to pay for per unit of emissions and it might also be explained as the tax imposed by the government [57].

6.2 Physarum-Inspired Solution to Multi-criteria Sustainable Supply Chain Network Design Problem

In[2] the design of sustainable supply chain, it is required that the flow is less than its actual capacity. In our view, in the optimal solution, the capacity u_a must be equal to the actual flow f_a along the same link a. Suppose the capacity u_a on link a is more than the flow f_a, this will incur extra cost. On the contrary, if the capacity u_a on link a is equal to the flow f_a, it not only satisfies the constraint indicated in Eq. (25), but also decreases the total cost. From this standpoint, in the optimal solution, the capacity on each link should be equal to its actual flow. In other words, in the optimal solution, $f_a = u_a$. As a matter of fact, in the Physarum model, the flow on link a is equal to the flow Q_{ij}.

As a matter of fact, the sustainable supply chain network design problem is a system optimum (SO) problem from the perspective of flow theory in the transportation systems, and its objective is to minimize the total cost in the supply chain network.

---

[2]This subsection is taken from the paper published in Annals of Operations Research. For more details, please see Ref. [58].

Fig. 12 The baseline supply
chain network topology for
all the examples. Adopted
from Ref. [52]

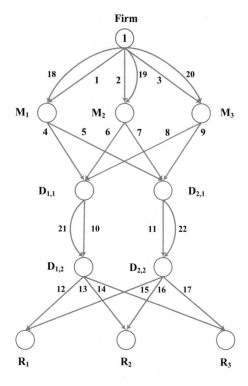

To date, Physarum has been successfully employed to handle the user equilibrium
(UE) problem in the transportation network. To solve the UE problem, we need to
use a revised cost function to replace the length of each link. The procedures to solve
the user equilibrium problem are already described in Algorithm 3 in the last section.

However, different from UE solution, in the sustainable supply chain network,
we aim at minimizing the total cost. For the purpose of using Physarum to solve this
issue, we transform the SO state into the corresponding UE state using the following
Eq. (31) [59, 60].

$$\widetilde{t}_a (x_a) = t_a (x_a) + x_a \frac{dt_a (x_a)}{dx_a}, \quad \forall a \in L \tag{31}$$

where x_a represents the flow on link a, $t_a(x_a)$ denotes the cost function per unit of
flow on link a while $\widetilde{t}_a (x_a)$ denotes the transformed cost function per unit of flow.

In the sustainable supply chain network, L_{ij} means the cost when the flow is Q_{ij}.
Hence, the following Eq. (32) is built to express the cost per unit of flow:

$$LF_{ij} = \frac{\hat{c}_a (Q_{ij}) + \hat{\pi}_a (Q_{ij}) + \omega \left(e_a (Q_{ij}) + \hat{f}_a (Q_{ij}) \right)}{Q_{ij}} \tag{32}$$

Algorithm 4: Physarum-Inspired Model for Constructing the Optimal Sustainable Supply Chain Network Design $(L, 1, N, R)$

// N is the size of the network;
// L_{ij} is the link connecting node i with node j;
// 1 is the starting node while R is the set of retail outlets;
$D_{ij} \leftarrow (0, 1]$ $(\forall i, j = 1, 2, \ldots, N)$;
$Q_{ij} \leftarrow 0$ $(\forall i, j = 1, 2, \ldots, N)$;
$p_i \leftarrow 0$ $(\forall i = 1, 2, \ldots, N)$;
$L_{ij} \leftarrow 0.001$ $(\forall i, j = 1, 2, \ldots, N)$;
$count \leftarrow 1$;
repeat
    Calculate the pressure associated with each node according to Eq. (10)

$$\sum_i \frac{D_{ij}}{L_{ij}} \left(p_i - p_j\right) = \begin{cases} +\sum_{i=1}^{n_R} d_{R_i} & for\ j = 1, \\ -d_{R_j} & for\ j = R_1, R_2, \cdots R_{n_R}, \\ 0 & otherwise \end{cases}$$

$Q_{ij} \leftarrow D_{ij} \times \left(p_i - p_j\right)/L_{ij}$ // Using Eq. (1);
$D_{ij} \leftarrow Q_{ij} + D_{ij}$ // Using Eq. (6)
Update the cost on each link;
for $i = 1 : N$ **do**
    for $j = 1 : N$ **do**
        if $Q_{ij} \neq 0$ **then**
            $L_{ij} = L_{ij} + LF_{ij} + Q_{ij} * \left.\frac{dLF_{ij}}{dQ_{ij}}\right|_{Q_{ij}=Q_{ij}}$;
        end if
    end for
end for
$L = L/2$;
$count \leftarrow count + 1$
until a termination criterion is met

According to the above method, we can construct the procedures for constructing the optimal sustainable supply chain network, which is shown in Algorithm 4.

6.3 Numerical Examples

In this section, two numerical examples are used to demonstrate the procedures of the proposed method for solving the multi-criteria sustainable supply chain network design problem.

The baseline for all the examples are shown in Fig. 12. In this figure, the numbers along these links represent the label of each link. It can be noticed that there are three alternative manufacturing plants and each of them has two possible technologies. Each manufacturer is in association with two possible distribution centers. Similarly, each distribution center is associated with two possible storage centers. The firm

Table 2 Total cost and emission functions for the numerical examples. Adopted from Ref. [52]

Link a	$\widehat{c}_a(f)$	$\widehat{\pi}_a(u_a)$	$e_a(f_a)$	$\widehat{e}_a(f_a)$
1	$f_1^2 + 2f_1$	$0.5u_1^2 + u_1$	$0.05f_1^2 + f_1$	$1.5u_1^2 + 2u_1$
2	$0.5f_2^2 + f_2$	$2.5u_2^2 + u_2$	$0.1f_2^2 + f_2$	$2u_2^2 + 2u_2$
3	$0.5f_3^2 + f_2$	$u_3^2 + 2u_3$	$0.15f_3^2 + 2f_3$	$2.5u_3^2 + u_3$
4	$1.5f_4^2 + 2f_4$	$u_4^2 + u_4$	$0.05f_4^2 + 0.1f_4$	$0.1u_4^2 + 0.2u_4$
5	$f_5^2 + 3f_5$	$2.5u_5^2 + 2u_5$	$0.05f_5^2 + 0.1f_5$	$0.05u_5^2 + 0.1u_5$
6	$f_6^2 + 2f_5$	$0.5u_6^2 + u_6$	$0.1f_6^2 + 0.1f_6$	$0.05u_6^2 + 0.1u_6$
7	$0.5f_7^2 + 2f_7$	$0.5u_7^2 + u_7$	$0.05f_7^2 + 0.2f_7$	$0.1u_7^2 + 0.2u_7$
8	$0.5f_8^2 + 2f_8$	$1.5u_8^2 + u_8$	$0.05f_8^2 + 0.1f_8$	$0.1u_8^2 + 0.3u_8$
9	$f_9^2 + 5f_9$	$2u_9^2 + 3u_9$	$0.05f_9^2 + 0.1f_9$	$0.1u_9^2 + 0.2u_9$
10	$0.5f_{10}^2 + 2f_{10}$	$u_{10}^2 + 5u_{10}$	$0.2f_{10}^2 + f_{10}$	$1.5u_{10}^2 + 3u_{10}$
11	$f_{11}^2 + f_{11}$	$0.5u_{11}^2 + 3u_{11}$	$0.25f_{11}^2 + 3f_{11}$	$2u_{11}^2 + 3u_{11}$
12	$0.5f_{12}^2 + 2f_{12}$	$0.5u_{12}^2 + u_{12}$	$0.05f_{12}^2 + 0.1f_{12}$	$0.1u_{12}^2 + 0.2u_{12}$
13	$0.5f_{13}^2 + 5f_{13}$	$0.5u_{13}^2 + u_{13}$	$0.1f_{13}^2 + 0.1f_{13}$	$0.05u_{13}^2 + 0.1u_{13}$
14	$f_{14}^2 + 7f_{14}$	$2u_{14}^2 + 5u_{14}$	$0.15f_{14}^2 + 0.2f_{14}$	$0.1u_{14}^2 + 0.1u_{14}$
15	$f_{15}^2 + 2f_{15}$	$0.5u_{15}^2 + u_{15}$	$0.05f_{15}^2 + 0.3f_{15}$	$0.1u_{15}^2 + 0.2u_{15}$
16	$0.5f_{16}^2 + 3f_{16}$	$u_{16}^2 + u_{16}$	$0.05f_{16}^2 + 0.1f_{16}$	$0.1u_{16}^2 + 0.1u_{16}$
17	$0.5f_{17}^2 + 2f_{17}$	$0.5u_{17}^2 + u_{17}$	$0.15f_{17}^2 + 0.3f_{17}$	$0.05u_{17}^2 + 0.1u_{17}$
18	$0.5f_{18}^2 + 1f_{18}$	$u_{18}^2 + 2u_{18}$	$0.2f_{18}^2 + 2f_{18}$	$2u_{18}^2 + 3u_{18}$
19	$0.5f_{19}^2 + 2f_{19}$	$u_{19}^2 + u_{19}$	$0.25f_{19}^2 + 3f_{19}$	$3u_{19}^2 + 4u_{19}$
20	$1.5f_{20}^2 + 1f_{20}$	$u_{20}^2 + u_{20}$	$0.3f_{20}^2 + 3f_{20}$	$2.5u_{20}^2 + 5u_{20}$
21	$0.5f_{21}^2 + 2f_{21}$	$u_{21}^2 + 3u_{21}$	$0.1f_{21}^2 + 3f_{21}$	$1.5u_{21}^2 + 4u_{21}$
22	$f_{22}^2 + 3f_{22}$	$0.5u_{22}^2 + 2u_{22}$	$0.2f_{22}^2 + 4f_{22}$	$2.5u_{22}^2 + 4u_{22}$

has to satisfy the demand from three possible retail outlets. The basic data for the following examples is shown in Table 2.

Example 3 In this example, the demands for each retail outlet is

$$d_{R_1} = 45, d_{R_3} = 35, d_{R_3} = 5$$

The cost functions and emission functions are shown in Table 2. In this example, we assume that the firm does not care about the emission generated in its supply chain. Therefore, $\omega = 0$. Figure 13 shows us the flux changing trend during the iterative process. It can be seen that the Physarum converges to the optimal solution after 25 iterations. Table 3 provides us the specific flow on each link. As expected, the flow associated with each link is equal to its capacity. According to Eq. (30), we can obtain that the total cost is 10716.33 and the result is inconsistent with that in Ref. [52]. From Table 3, it can be noted that link 14 has zero capacity and zero flow. Thus, in the final optimal sustainable supply chain network, link 14 will be removed.

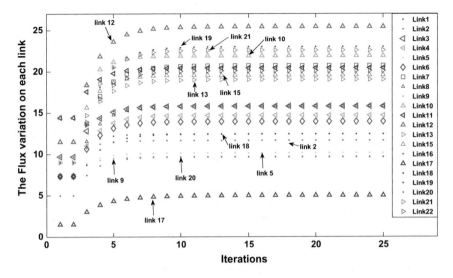

Fig. 13 The flux variation during the iterative process in the Example 3

Table 3 The optimal solution to Example 3

Link a	f_a^*	u_a^*	Link a	f_a^*	u_a^*
1	12.43	12.43	12	25.44	25.44
2	11.67	11.67	13	19.03	19.03
3	15.81	15.81	14	0.00	0.00
4	14.69	14.69	15	19.56	19.56
5	10.16	10.16	16	15.97	15.97
6	13.94	13.94	17	5.00	5.00
7	20.70	20.70	18	12.43	12.43
8	15.83	15.83	19	22.98	22.98
9	9.66	9.66	20	9.69	9.69
10	21.90	21.90	21	22.57	22.57
11	20.43	20.43	22	20.10	20.10

Example 4 In this example, it has the same data as Example 3 except that the parameter $\omega = 5$, which express the degree of the firm is concerned about the environment. The optimal solution to this example is given in Table 4. The total cost as shown in Eq. (28) for this example is 11288.27. The total emission cost is 7735.71. The result is different from that in Ref. [52]. We find that the results in Ref. [52] is not reasonable. To be specific, for the node M_1, its inflow is equal to the sum of the flow on link 18 and link 1, which is equal to 33.22 (19.32+13.90) in Nagurney's solution. As for the outflows associated with node M_1, it is composed of two separate flows on link 4 and link 5, which is equal to 33.23 (19.43+13.80). Obviously, the inflows are not

Table 4 The optimal solution to Example 4

Link a	f_a^*	u_a^*	Link a	f_a^*	u_a^*
1	19.33	19.33	12	26.65	26.65
2	15.68	15.68	13	20.65	20.65
3	13.45	13.45	14	1.69	1.69
4	19.45	19.45	15	18.35	18.35
5	13.78	13.78	16	14.35	14.35
6	13.78	13.78	17	3.31	3.31
7	13.24	13.24	18	13.90	13.90
8	15.76	15.76	19	11.34	11.34
9	8.99	8.99	20	11.30	11.30
10	24.20	24.20	21	24.79	24.79
11	19.66	19.66	22	16.35	16.35

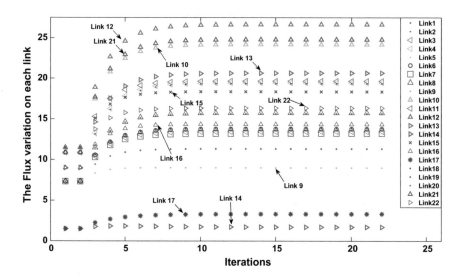

Fig. 14 The flux variation during the iterative process in the Example 4

equal to the outflows. This contradicts the law of flow conservation. Similarly, it is also observed that such kind of phenomenon can be found in the node $D_{2,1}$.

In the Physarum model, it takes 22 iterations to converge to the optimal solution. In the optimal solution, it can be noted that all the links have positive capacity and flows. In addition, the flows are equal to the capacity on all the links. In Example 3, links 1 and 18 have the same flow. However, in Example 4, the flow on link 1 has increased 50% while the flow on link 18 only increases about 10%. This is due to the emission cost on link 1 is less than link 18. Such kind of behavior, also can be found on links 2 and 19 (Fig. 14)

7   Conclusion

In this chapter, we developed a generalized Physarum model to address several network optimization problems. In summary, we make the following contributions:

- We proposed an innovative strategy to enable Physarum model to address the network optimization problems with multiple sources and sinks in both directed and undirected networks.
- We compared Physarum model with Dijkstra algorithm in the manners of finding the shortest path. By performing such a comparison, we highlighted the unique features present in Physarum model as well as the benefits brought by these features.
- We approached the optimal solution to equilibrium traffic assignment problem by leveraging the adaptivity and continuity features in Physarum model. We developed a composite function to relate the link cost and link flow, and prove the converge of the system to the user equilibrium as well.
- By leveraging the equivalent transformation between user equilibrium and system optimal, we utilized the Physarum model to identify the optimized network for the firm to transport products to the retail outlets for the purpose of minimizing the total cost.

References

1. F. Wang, X. Lai, N. Shi, A multi-objective optimization for green supply chain network design. Decis. Support Syst. **51**(2), 262–269 (2011)
2. S. Nannapaneni, S. Mahadevan, S. Rachuri, Performance evaluation of a manufacturing process under uncertainty using Bayesian networks. J. Clean. Prod. **113**, 947–959 (2016)
3. H. Baroud, J.E. Ramirez-Marquez, K. Barker, C.M. Rocco, Stochastic measures of network resilience: applications to waterway commodity flows. Risk Anal. **34**(7), 1317–1335 (2014)
4. X. Zhang, S. Mahadevan, A game theoretic approach to network reliability assessment. IEEE Trans. Reliab. **66**(3), 875–892 (2017)
5. X. Zhang, S. Mahadevan, S. Sankararaman, K. Goebel, Resilience-based network design under uncertainty. Reliab. Eng. Syst. Saf. **169**, 364–379 (2017)
6. W.-B. Du, X.-L. Zhou, O. Lordan, Z. Wang, C. Zhao, Y.-B. Zhu, Analysis of the Chinese airline network as multi-layer networks. Transp. Res. Part E: Logist. Transp. Rev. **89**, 108–116 (2016)
7. P. Angeloudis, D. Fisk, Large subway systems as complex networks. Phys. A: Stat. Mech. Appl. **367**, 553–558 (2006)
8. M. Gen, A. Kumar, J.R. Kim, Recent network design techniques using evolutionary algorithms. Int. J. Prod. Econ. **98**(2), 251–261 (2005)
9. E. Cipriani, S. Gori, M. Petrelli, Transit network design: a procedure and an application to a large urban area. Transp. Res. Part C: Emerg. Technol. **20**(1), 3–14 (2012)
10. T. Santoso, S. Ahmed, M. Goetschalckx, A. Shapiro, A stochastic programming approach for supply chain network design under uncertainty. Eur. J. Oper. Res. **167**(1), 96–115 (2005)
11. A. Chen, K. Subprasom, Z. Ji, A simulation-based multi-objective genetic algorithm (SMOGA) procedure for bot network design problem. Optim. Eng. **7**(3), 225–247 (2006)
12. S. Nannapaneni, S. Mahadevan, A. Dubey, D. Lechevalier, A. Narayanan, S. Rachuri, Automated uncertainty quantification through information fusion in manufacturing processes

13. K.L. Hoffman, M. Padberg, G. Rinaldi, Traveling salesman problem, in *Encyclopedia of Operations Research and Management Science* (Springer, 2013), pp. 1573–1578
14. M.S. Tree, *Minimum Spanning Tree* (2007)
15. R.K. Ahuja, K. Mehlhorn, J. Orlin, R.E. Tarjan, Faster algorithms for the shortest path problem. J. ACM **37**(2), 213–223 (1990)
16. M.R. Garey, D.S. Johnson, The rectilinear steiner tree problem is np-complete. SIAM J. Appl. Math. **32**(4), 826–834 (1977)
17. Z. Gao, J. Wu, H. Sun, Solution algorithm for the bi-level discrete network design problem. Transp. Res. Part B: Methodol. **39**(6), 479–495 (2005)
18. J.N. Hagstrom, R.A. Abrams, Characterizing Braess's paradox for traffic networks, in *Intelligent Transportation Systems, 2001. Proceedings. 2001 IEEE* (IEEE, 2001), pp. 836–841
19. K. Holmberg, D. Yuan, A Lagrangian heuristic based branch-and-bound approach for the capacitated network design problem. Oper. Res. **48**(3), 461–481 (2000)
20. K. Büdenbender, T. Grünert, H.-J. Sebastian, A hybrid Tabu search/branch-and-bound algorithm for the direct flight network design problem. Transp. Sci. **34**(4), 364–380 (2000)
21. T. Mak, P.Y. Cheung, K.-P. Lam, W. Luk, Adaptive routing in network-on-chips using a dynamic-programming network. IEEE Trans. Ind. Electron. **58**(8), 3701–3716 (2011)
22. G. Carello, F. Della Croce, M. Ghirardi, R. Tadei, Solving the hub location problem in telecommunication network design: a local search approach. Networks **44**(2), 94–105 (2004)
23. J.-F. Cordeau, G. Laporte, F. Pasin, An iterated local search heuristic for the logistics network design problem with single assignment. Int. J. Prod. Econ. **113**(2), 626–640 (2008)
24. J. Kennedy, Particle swarm optimization, in *Encyclopedia of Machine Learning* (Springer, 2011), pp. 760–766
25. K. Deb, A. Pratap, S. Agarwal, T. Meyarivan, A fast and elitist multiobjective genetic algorithm: NSGA-II. IEEE Trans. Evol. Comput. **6**(2), 182–197 (2002)
26. M. Dorigo, M. Birattari, T. Stutzle, Ant colony optimization. IEEE Comput. Intell. Mag. **1**(4), 28–39 (2006)
27. A. Tero, R. Kobayashi, T. Nakagaki, A mathematical model for adaptive transport network in path finding by true slime mold. J. Theoret. Biol. **244**(4), 553–564 (2007)
28. C. Gao, C. Yan, A. Adamatzky, Y. Deng, A bio-inspired algorithm for route selection in wireless sensor networks. IEEE Commun. Lett. **18**(11), 2019–2022 (2014)
29. C. Gao, C. Yan, Z. Zhang, Y. Hu, S. Mahadevan, Y. Deng, An amoeboid algorithm for solving linear transportation problem. Phys. A: Stat. Mech. Appl. **398**, 179–186 (2014)
30. C. Yan, C. Gao, J. Yu, Y. Deng, K. Nan, The optimal path tour problem. Int. J. Unconv. Comput. **10** (2014)
31. A. Adamatzky, *Physarum Machines: Computers from Slime Mould*, vol. 74 (World Scientific, 2010)
32. A. Tero, S. Takagi, T. Saigusa, K. Ito, D.P. Bebber, M.D. Fricker, K. Yumiki, R. Kobayashi, T. Nakagaki, Rules for biologically inspired adaptive network design. Science **327**(5964), 439–442 (2010)
33. X. Zhang, Z. Zhang, Y. Zhang, D. Wei, Y. Deng, Route selection for emergency logistics management: a bio-inspired algorithm. Saf. Sci. **54**, 87–91 (2013)
34. X. Zhang, A. Adamatzky, X.-S. Yang, H. Yang, S. Mahadevan, Y. Deng, A Physarum-inspired approach to supply chain network design. Sci. China Inf. Sci. **59**(5), 052203 (2016)
35. X. Zhang, S. Huang, Y. Hu, Y. Zhang, S. Mahadevan, Y. Deng, Solving 0–1 knapsack problems based on amoeboid organism algorithm. Appl. Math. Comput. **219**(19), 9959–9970 (2013)
36. X. Zhang, A. Adamatzky, H. Yang, S. Mahadaven, X.-S. Yang, Q. Wang, Y. Deng, A bio-inspired algorithm for identification of critical components in the transportation networks. Appl. Math. Comput. **248**, 18–27 (2014)
37. X. Zhang, F.T. Chan, H. Yang, Y. Deng, An adaptive amoeba algorithm for shortest path tree computation in dynamic graphs. Inf. Sci. **405**, 123–140 (2017)
38. Y. Liu, C. Gao, Z. Zhang, Y. Lu, S. Chen, M. Liang, L. Tao, Solving np-hard problems with physarum-based ant colony system. IEEE/ACM Trans. Comput. Biol. Bioinform. **14**(1), 108–120 (2017)

39. X. Zhang, F.T. Chan, A. Adamatzky, S. Mahadevan, H. Yang, Z. Zhang, Y. Deng, An intelligent physarum solver for supply chain network design under profit maximization and oligopolistic competition. Int. J. Prod. Res. **55**(1), 244–263 (2017)
40. C. Gao, S. Chen, X. Li, J. Huang, Z. Zhang, A Physarum-inspired optimization algorithm for load-shedding problem. Appl. Soft Comput. **61**, 239–255 (2017)
41. T. Nakagaki, H. Yamada, Á. Tóth, Intelligence: Maze-solving by an amoeboid organism. Nature **407**(6803), 470–470 (2000)
42. V. Bonifaci, K. Mehlhorn, G. Varma, Physarum can compute shortest paths. J. Theoret. Biol. **309**, 121–133 (2012)
43. V. Bonifaci, Physarum can compute shortest paths: a short proof. Inf. Process. Lett. **113**(1–2), 4–7 (2013)
44. F. Harary, The determinant of the adjacency matrix of a graph. SAIM Rev. **4**(3), 202–210 (1962)
45. D. Straszak, N.K. Vishnoi, Natural algorithms for flow problems, in *Proceedings of the Twenty-Seventh Annual ACM-SIAM Symposium on Discrete Algorithms* (Society for Industrial and Applied Mathematics, 2016), pp. 1868–1883
46. D.B. Johnson, A note on Dijkstra's shortest path algorithm. J. ACM (JACM) **20**(3), 385–388 (1973)
47. X. Zhang, S. Mahadevan, A bio-inspired approach to traffic network equilibrium assignment problem. IEEE Trans. Cybern. **48**(4), 1304–1315 (2018)
48. Y. Sheffi, *Urban Transportation Network: Equilibrium Analysis with Mathematical Programming Methods* (Prentice Hall, 1985)
49. US Bureau of Public Roads, *Traffic Assignment Manual* (US Department of Commerce, Washington, DC, 1964)
50. P. Delle Site, F. Filippi, C. Castaldi, Reference-dependent stochastic user equilibrium with endogenous reference points. EJTIR **13**(2), 147–168 (2013)
51. M. Mitradjieva, P.O. Lindberg, The stiff is moving-conjugate direction frank-wolfe methods with applications to traffic assignment. Transp. Sci. **47**(2), 280–293 (2013)
52. A. Nagurney, L.S. Nagurney, Sustainable supply chain network design: a multicriteria perspective. Int. J. Sustain. Eng. **3**(3), 189–197 (2010)
53. A. Nagurney, *Supply Chain Network Economics: Dynamics of Prices, Flows and Profits* (Edward Elgar Publishing, 2006)
54. A. Nagurney, A system-optimization perspective for supply chain network integration: the horizontal merger case. Transp. Res. Part E: Logist. Transp. Rev. **45**(1), 1–15 (2009)
55. A. Nagurney, J. Dong, D. Zhang, A supply chain network equilibrium model. Transp. Res. Part E: Logist. Transp. Rev. **38**(5), 281–303 (2002)
56. A. Nagurney, T. Woolley, Environmental and cost synergy in supply chain network integration in mergers and acquisitions, in *Multiple Criteria Decision Making for Sustainable Energy and Transportation Systems* (Springer, 2010), pp. 57–78
57. K. Wu, A. Nagurney, Z. Liu, J.K. Stranlund, Modeling generator power plant portfolios and pollution taxes in electric power supply chain networks: a transportation network equilibrium transformation. Transp. Res. Part D: Transp. Environ. **11**(3), 171–190 (2006)
58. X. Zhang, A. Adamatzky, F.T. Chan, S. Mahadevan, Y. Deng, Physarum solver: a bio-inspired method for sustainable supply chain network design problem. Ann. Oper. Res. **1–2**, 533–552 (2017)
59. M.G. Bell, Y. Iida, *Transportation Network Analysis* (1997)
60. S. Bingfeng, G. Ziyou, *Modeling Network Flow and System Optimization for Traffic and Transportation System.* (China Communications Press, 2013) (in Chinese)

Maze-Solving Cells

Daniel Irimia

Abstract Moving cells have surprising abilities to navigate efficiently through complex microscale human-engineered mazes. Most often, they follow chemical gradients established by diffusion from a source to a sink. Cells steer towards the steepest gradients and move along the shortest paths towards the source. Occasionally, the gradients are self-generated by the moving cells themselves. While the cells also respond to the same gradients they create, a self-sustained loop is established, which guides the cells towards the shortest path towards exit from confined spaces. Precision measurements of human cell maze-navigation performance could ultimately enhance our capabilities to diagnose, monitor, and treat various diseases that range from inflammation to cancer.

All cells in our body move and often must navigate complex territories. Leukocytes responding to microbes enter tissues packed with other cells, fibers, blood vessels and nerves. Epithelial cells moving to close wounds constantly re-arrange their position relative to mechanical obstacles and their cellular neighbors. Streams of cells navigate between precisely defined locations during embryo development and give rise to tissues and organs. Recent advances of in vivo imaging technologies significantly accelerate our understanding for how and why cells move. However, several challenges limit the utility of in vivo systems. These include poor control of microscale environment conditions in vivo, limited resolution in space and time, requirements for fluorescently tags on the target cells, and the insidious interference from untagged cells. In response to these limitations, systems for monitoring human cell movement ex vivo, in conditions that replicate key features of the in vivo microenvironment, are being developed and essential for advancing this understanding of cell movement.

Emerging microfluidic tools designed to probe cell migration are today capable of achieving unprecedented levels of precision in the control of the microenvironment around moving cells and the quantification of migration parameters. Microfluidic

D. Irimia (✉)
Center for Engineering in Medicine, Department of Surgery, Massachusetts General Hospital, Harvard Medical School, Shriners Burns Hospital, Boston, MA, USA
e-mail: daniel_irimia@hms.harvard.edu

tools also have the advantage of significantly higher experimental throughput compared to in vivo experiments. They are increasingly capable to replicate in vivo conditions, thus enabling unique insights into how cells navigate complex environments.

One key advance towards the goal of understanding cell movement was our observation that cells move at uniform speed when confined in microscale channels with cross-section smaller than that of the moving cells [1]. Subsequently, devices with intricate networks of channels helped us study how cells can find their way through increasingly complex mazes. Precise measurements of various parameters relevant to the navigation of cells, including persistence and directionality, could for the first time be performed directly on individual cells, circumventing the need for complex statistics. This chapter will review and discuss the most important bioengineering insights into the ability of cells to navigate through mazes and the implications of these findings to our understanding of health and disease.

1   Limitations of Traditional Cell Migration Assays Probing Cell Navigation

Traditional migration assays explore the migration of cells on flat surfaces, e.g. guided by the release and diffusion of a chemoattractant from the tip of a micropipette. The directional migration of cells towards the highest chemoattractant concentration at the tip of the micropipette is often described as a "biased random walk". The migration is characterized by tortuous trajectories and large variations in speed [2]. The cells advancing towards the pipette tip stop, go, and change the migration speed constantly. The levels of randomness decrease the precision of the measurements and make comparisons between conditions difficult. Standard metrics for describing motility include the speed and directionality. The speed of migration, defined as displacement over time, is calculated as an average. The directionality is a measure of efficiency in advancing towards a target. Because cells constantly change their direction of migration along meandering trajectories, sophisticated statistical methods are often required to estimate the directionality parameter, limiting the precision of these metrics. Other traditional assays offer even less information. The most popular cell migration assay, the transwell assay (Boyden chamber) is an end-point assay that only provides a measure of the fraction of cells to moved. For cells that proliferate, the doubling of cell number over time interferes with these measurements. The vulnerability of the micropipette chemoattractant gradients to interference from tiny perturbations in the environment, and the continuously evolving nature of gradients across the transmembrane assay further complicate these measurements.

These measurements of cell migration in traditional assay are often inconsistent with in vivo observations, where various cells navigate very efficiently through complex tissue microenvironments towards their targets. Moreover, the cells moving through live tissues are experiencing the natural variability and complexity of the microenvironment and have close interaction with other moving cells on their paths. Thus, a direct comparisons between in vivo and ex vivo measurements using traditional assays is challenging. White blood cells like neutrophils are quite efficient when navigating complex and heterogeneous environments in vivo [3]. However, they are significantly less efficient when moving across flat surfaces, when they take tortuous paths, with frequent divagations from straight lines and frequent stops. Similarly, epithelial cancer cells invade preferentially migration along lymph vessels, collagen fibers, or white matter tracts and in vivo observations documented their persistent migration along these guiding structures [4]. However, epithelial cancer cell migration on flat surfaces, in the presence of chemical gradients e.g. growth factors, appears rather random, with a bias towards the higher concentrations.

The status-quo for cell migration assays is being challenged since the early 2000s by the emergence of a new set of tools for building and controlling the cellular microenvironment at microscale. Borrowed from the electronics industry and enabled by novel biocompatible materials, microfluidic tools enabled the manufacturing of physical structures that were smaller than the size of a typical mammalian cell [5]. These structures could control the cells directly through mechanical interactions and indirectly by controlling the chemical microenvironment around cells. In one early example, the two sides of one 10 μm sized cell were exposed to fluids of different colors, "painting" stripes across the cell [6]. The first microfluidic assays for cell migration attempted to address these issues by providing exquisite control of the shape and stability of chemical gradients [7]. However, the variations in speed and directionality during leukocyte chemotaxis on flat surfaces in microfluidic devices were still comparable to those in traditional assay. The key advance towards precision measurements of moving cells came from an unexpected insight, when we confined cells in channels that had a cross-section that was smaller than the cross-section of the resting cells.

2   Straight Channels Rectify Cell Migration

More than a decade ago, we were surprised to uncover that when moving neutrophils were confined within small channels, their migration towards sources of chemoattractant was enhanced [1]. The confinement reduced the variability of migration speed over time and improved the precision of neutrophil migration measurements. Shortly after, we uncovered that several epithelial cancer cell types including MDA-MB-231 cells (breast adenocarcinoma cell line), human derived mammary epithelial cells, PC-9ZD cells (lung carcinoma cell line), and human-derived pancreatic carcinoma cells, also display persistent, unidirectional motility when confined inside channels smaller than their size [8]. Inside channels, both leukocytes and human epithelial cancer cells can accomplish incredible feats, like the ability to migrate persistently in one direction for hours. Remarkably, cancer cells display this uniform motility phenotype spontaneously, in the absence of an externally-imposed guiding gradient.

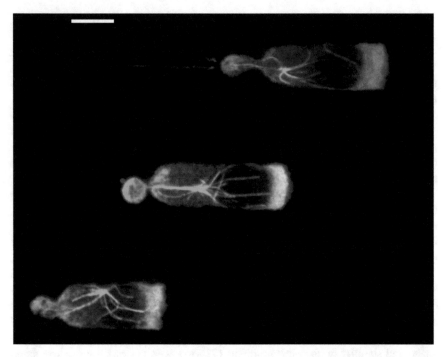

Fig. 1 Mechanical confinement inside microscale channels stimulates the migration of leukemia cell line (HL60) at uniform speed. Fixed cells, actin stained in red, microtubules in green [1]. Scale bar is 5 μm. Reproduced with permission from the Royal Society of Chemistry

The small cross-section of the channels is a key factor in regulating the motility of the cells inside [1]. These channels are assembled in between biocompatible silicone and glass. Human cells from a leukemia cell line (HL-60) migrate at uniform speed through 10×3 μm cross-section channels [1]. The morphology of the cells remains relatively constant over time. The protrusion of the membrane at the front is coordinated with the contraction at the back. This migration phenotype is different than the classical four-step motility model (protrusion, attachment contraction, and detachment) for haptokinetic cell migration. Inside the motile cells, actin is predominantly localized at the leading edge, filling the cross-section of the channel (Fig. 1). Microtubules are centered behind the nucleus, with the longer ones reaching past the nucleus and stopping at the leading edge. Counterintuitively, the migration of leukemia cells, as well as of human neutrophils, is significantly slower in larger than in smaller channels. During migration inside larger channels e.g. 10×10 μm, the speed of migration varies over time. The morphology of cells changes constantly, and the overall phenotype of the cells is comparable to that of cells chemotaxing on two-dimensional surfaces.

We observed a similar dependence between the persistence of cell migration and the cross-section of the channels for epithelial cells [8]. The optimal size of the channels for epithelial cell migration was 10×10 μm and the migration in larger channels was slower and less persistent. These observations validate the benefits of studying cell motility inside small channels. In patients and animal models, cancer cells often migrate along lymphatic vessels [9], towards lymph nodes. They are often found along the periphery of blood vessels [10]. Glioblastoma cells are well known to migrate preferentially along white matter tracts [11]. Like in vivo, cancer cells inside the channels benefit from the mechanical support of the rigid walls. Moving cells contact the extracellular matrix proteins on the walls throughout their entire circumference, enabling 3D-like adhesions. Unlike in vivo, the migration occurs without the need to degrade this matrix.

In addition to reflecting the mechanical features of in vivo environment, several practical benefits emerge from the quantification of cell motility in channels. First, the position of cells moving at uniform speed is predictable during migration in channels and facilitates tracking. Second, thousands of cells could be tracked at the same time in parallel channels. While the cells in parallel channels cannot overlap, there is no need to resolve situations of cell overlap. Finally, in the absence of temporal variations of speed, the precision of migration speed is higher than in any other assay. These features ultimately enable more precise comparisons between cells, and helps quantify the effect of various compounds aiming at slowing or accelerating cell migration.

Despite simplicity, the persistent migration of cells through channels teaches us two important things about the ability of cells to interpret, respond, and change their microenvironment to fulfill their physiologic roles. In the case of human leukocytes, we identified signature motility patterns that are uniquely associated with chemoattractants that guide leukocyte motility in various conditions [12]. We found that the signature motility patterns are independent of the chemoattractant dose or receptor expression. Formylated peptide chemoattractant fMet-Leu-Phe and leukotriene B4 induce a migration phenotype in human neutrophils that is directional and persistent. This phenotype is effective at attracting the neutrophils towards the source of these molecules. By contrast, complement component 5a and interleukin 8 induce both chemoattraction and repulsion in equal proportions. This migration phenotype results in the dispersal of neutrophils throughput the area where the chemokines are present. Overall, the combination of three characteristics: speed, persistence, and directionality, in addition to the percentage of cells migrating in response to a chemoattractant define unique signatures for each chemoattractant and cell type. These signature patterns could have not been defined in vitro using traditional techniques.

Importantly, the moving cells guide themselves from one end to the other of a channel, without turning or slowing down. The mechanisms for this persistent, self-guided navigation emerged slowly and required several additional steps in technical innovation and systematic experiments. Essential for this progress was the design of microfluidic mazes and the observation that epithelial cells navigate these mazes along the shortest path towards exit, starting with no other external cues than the maze itself.

3 Cell Make Binary Decisions in Bifurcating Channels

The first step towards dissecting the mechanisms of cell migration in confined spaces was enabled by the addition of bifurcations to the design of the channels. This slight increase in the complexity of the microfluidic channels enabled us to measure the directional decisions of the moving cells [13]. Whereas the persistent migration in straight channels may be explained by either the mechanical guidance along the walls of the channels or the pursuit of a chemical gradient, the addition of bifurcations decouples the two contributions in unambiguous ways. The quantification also becomes more precise, with the moving cells making essentially binary choices at each bifurcation.

When neutrophils are moving through channels and presented with bifurcations, they consistently steer towards the shorter branch (Fig. 2). We observed that more than 90% of the neutrophils arriving at asymmetric bifurcations maze chose the shorter of two branches leading towards the same source of chemoattractant. The shorter branch presents the cells with a steeper chemoattractant gradient. In control experiments, neutrophils migrating through symmetrical bifurcations chose randomly between the two branches. Moreover, when presented with three symmetrical bifurcations in series, individual human neutrophils do not have preferences for either branches and their decisions appear entirely random [13]. Remarkably, at the point where branches recombine, all neutrophils migrated directly towards the

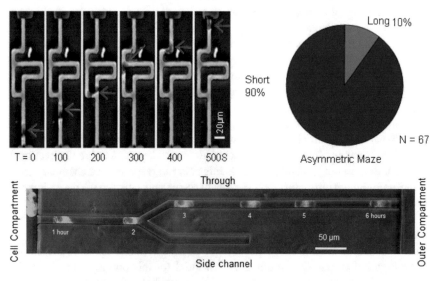

Fig. 2 Human neutrophils (top) and cancer epithelial cells (bottom) make binary directional decisions when navigating through channels with bifurcations. Neutrophils migrate preferentially along the shortest path towards the source of chemoattractant [13]. MDA-MB-231 epithelial cells avoid the dead end and migrate preferentially along the through path towards the outer compartment [14]. Reproduced with permission from the Royal Society of Chemistry

source of chemokine (increasing chemoattractant concentration). No cells travelled back toward the entrance of the channel (decreasing chemoattractant concentration).

Neutrophils move with the same average velocity before and after encountering the bifurcations. This suggests that the mechanisms for cell steering and those for cell migration may be independent. In addition, the rapid decisions at bifurcations before the rest of the cell escapes confinement suggest that the chemical gradient sensing mechanisms are localized only at the leading edge of the cells. Our observations also suggest that the sensing and motility mechanisms are independent of each other. However, this mechanism is difficult to reconcile with current biochemical models for cell polarization. For example, one widely accepted chemotaxis model postulates the existence of a global inhibitor that diffuses between the front and the back of the cells [15]. If such an inhibitor was present the timescale for decisions would be longer and larger in longer cells. Our observations contradict this model. Instead, our observations favor a model of mechanical integrator of discrete, cooperative process at the leading edge. Such an adaptive sensing mechanism provides the cell with cues for maintaining and adjusting its polarization towards the fastest concentration increases [16].

Asymmetrically-bifurcating channels also allowed us to decouple the speed and directionality in moving cancer epithelial cells [14]. Migrating cells make one directional decision at the bifurcation without changing their migration speed. The cells steered either toward the "through" path leading to the outer compartment or toward the "side" path leading to a dead-end (Fig. 2). We found that, unlike the neutrophils, which require a chemoattractant to guide them through the bifurcation, the epithelial cancer cells can navigate the bifurcations effectively despite the absence of an externally imposed gradient. Impressively, cancer cells steered toward the "through" path three times more often than toward the "side".

We designed branching channels with various configurations with the goal of differentiating between potential mechanisms of guided epithelial cell migration. Inside U-shaped channels with a tiny, cell impassible extension to the outer compartment, more than 75% of the cells stopped at the site of the extension rather than continuing migration through the large branch. Abruptly reducing the cross-section of channels to half resulted in the reversal of the direction of migration in less than 10% of the cells. Other designs probed the role of hydraulic resistance and mechanical guidance during epithelial cell migration [14]. These results together indicate that a chemical gradient along the channels rather than mechanical cues is the dominant factor in guiding cell migration through channels. Additional support for the role of chemical cues during cell orientation emerged from observations of the long-distance interactions between multiple cells inside the same channel. Inside complex networks of through and dead-end channels [14], the frequency of a cell entering dead-end channels was significantly lower whenever another cell was already present in these channels. The ability of cells in channels to influence the direction of migration of other cells at a distance suggests that the differences are due to biochemical modifications of the microenvironment by the epithelial cells.

Two additional experiments provide key clues for the nature of these modifications. First, the ability of epithelial cells to navigate correctly through channels and avoid dead-end channels is lost in media lacking epidermal growth factor (EGF). Second, navigation abilities are also lost when EGF receptor signaling is perturbed in the presence of specific inhibitory compounds. Navigation abilities are preserved in the presence of inhibitors for other receptors and signaling pathways.

Following the experimental results from straight and bifurcating channels, it became intriguing to test if cells could navigate through mazes with complex design. This situation clearly brings the ex vivo assays closer to the complexity of in vivo situations. The mazes also uncovered three main navigation strategies for human neutrophils, lymphocytes, and epithelial cancer cells.

4   Cells Navigate Complex Mazes Along the Shortest Path

Inside mazes, moving cells are most often guided by chemoattractant gradients established by diffusion from end-reservoirs (source) to loading chambers (sink). Neutrophils follow externally imposed chemoattractant gradients accurately [17]. Lymphocytes display a patrolling mode that allows them to switch between efficient navigation to a survey of the entire mazes [18]. By contrast, epithelial cancer cells can find the shortest path from entrance to exit in the absence of externally imposed chemoattractant gradients, guided by self-generated gradients [14].

We compared human neutrophils with two cell types that are common models of cell polarization and cell motility, the leukemia cell lines (HL60) and soil amoeba *Dictyostelium discoideum* (Dicty). We found that human neutrophils, their leukemia counterpart, and Dicty cells can navigate millimeter long mazes of interconnected channels in just slightly over 1 h [17]. The narrow, $5 \times 10\ \mu m$ cross-section of the channels, mimics some of the biomechanical features encountered by neutrophils in tissues and the confined spaces encountered by Dicty in the soil.

We quantified the navigation performance of cells using heat maps that helped compare the magnitude of the chemical gradient (Fig. 3a) and the most commonly traversed paths of the maze (Fig. 3b). A "cellular flux" parameter for each edge was defined as the number of cells that crossed the edge in the specified direction, divided by the total number of cells analyzed. The heat map of cellular flux for Dicty cells (Fig. 3b) appeared similar to that of the chemical gradients (Fig. 3a). The heat map of cellular flux for HL60 cells was much more uniform, suggesting more random choices along the migration path. The cellular flux across edges in the direction of the gradient increased with the slope of chemical gradients along the edges.

At every junction in the maze, cells were forced to choose among up to four paths. These decisions were analyzed as a set of binary decisions in which cells choose one edge over the others. Dicty cells had a significantly higher probability of making the optimal choice than HL60 cells. The differences are more pronounced for small differences in gradient slopes. Interestingly, a strong negative correlation between cell speed and correctness was found for Dicty cells, while the HL60 cells performed equally poor regardless of their migration speed.

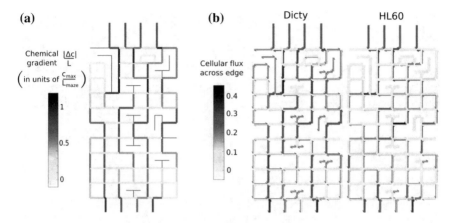

Fig. 3 Dictyostelium and HL60 cells navigate through mazes guided by chemoattractants. **a** The map of gradients for a chemoattractant diffusing from a reservoir (top) to a sink (bottom). Darker colors indicate steeper gradients. **b** The map for the flux of cells in different sections of the maze indicates preferential paths for traversing the maze. For Dicty, these paths are a better match to the gradient than for HL60 cells. Darker colors indicate higher frequency of cells passing through each section. Edges not traversed by any cells are shown as blue [17]. Reproduced with permission from the Public Library of Science

Other leukocytes e.g. human T lymphocytes are less efficient at navigating through mazes. Their migration patterns change depending on their activation levels [18]. In their resting state, T lymphocytes navigate rather efficiently through orthogonal mazes of small channels and follow relevant chemoattractant gradients. However, after activation, the migration patterns of T lymphocytes change and cells turn more frequently. They are less accurate at following the direction of the gradients compared to the resting cells. They also explore larger areas of the mazes, a process that may facilitate the contact between lymphocytes and infected cells. Interaction with antigen presenting cells within the foreign tissue may further activate the T lymphocytes. The activation will stimulate them again to migrate throughout the inflamed and chemokine containing tissue. In vivo live imaging and static histological observations in infected tissues support this hypothesis [19]. Overall, the maze experimental system offers a well-controlled alternative, which could be useful for testing compounds to modulate these patterns.

One striking advance that emerged from the study of epithelial cancer cell motility inside the mazes was the discovery that epithelial cells could employ navigation strategies that do not require any pre-existent chemoattractant gradient [14]. We expected that in the absence of pre-existent chemical gradients, cells will make random decisions at the bifurcations (Fig. 4). If cells only move forward, the calculated chance of a cell to move along the shortest path to exit through five bifurcations is less than ~10% if the decisions at the bifurcations are random. Instead, we observed that epithelial cells reach the exit in significantly larger numbers and along the shortest path from microscopic mazes filled with uniform concentrations of media. More than

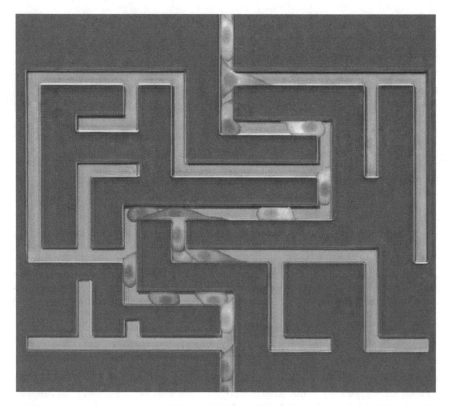

Fig. 4 One epithelial cancer cell navigates through a microscale maze of microfluidic channels. Successive positions at 30 min interval are shown overlaid on the image of the microfluidic maze [14]. No gradient is pre-imposed on the maze. Reproduced with permission from the Royal Society of Chemistry

90% of MDA-MB-231 cells, 75% of the human derived mammary epithelial cells, 70% of PC-9ZD cells, and 40% of the human-derived pancreatic carcinoma cells reach an exit along the short path. These numbers indicate that moving epithelial cells navigate the mazes and repeatedly make correct decisions towards the exit. The cells in mazes avoid dead-ends and avoid turning back towards the entrance. Such experiments are not possible using traditional migration assays. In such assays the supply of EGF diffusing from the culture media exceeds the rate of cellular EGF uptake. The concentrations of EGF around cells are homogeneous in space, with minimal gradient forming.

The strategy employed for navigation through mazes by the epithelial cells in the absence of pre-existent gradients is novel and relies on three processes that are tightly connected. First, epithelial cells take up significant amounts of EGF. This leads to partial depletion of the cell microenvironment. Second, the consumed EGF is replenished only by diffusive transport from the environment. The channels restrict the total flux of EGF towards the cells. Finally, epithelial cells respond to the local gradients that result from EGF uptake and flux. Epithelial cells move in the direction of the steepest gradients and highest EGF concentrations. In combination, the three processes enable the epithelial cells to migrate and exit confinement along the shortest path. Most importantly, they overcome the absence of pre-existent gradients and generate their own cues towards the closest exit.

These results are important for our understanding of cancer cell invasion. While the self-guidance strategy does not require pre-existing chemical gradients, persistent migration is not bound by the spatial limits of any gradient. This characteristic of self-guidance strategy is clearly distinct from the classical chemotaxis. During chemotaxis, cells move directionally only in the limited area where a spatial gradient is present. As soon as the cells reach the area of highest concentration, the directional movement ceases. By contrast, the self-guidance strategy predicts that a gradient can be initiated by the cells and will continuously move with the cells. The overlap of two functions in epithelial cells (making the gradient and responding to it) clearly distinguishes the self-guidance strategy from all other mechanisms of cell migration. The self-guidance strategy is distinct from autocrine or paracrine signaling, when gradients of attractant molecules are being produced and direct the migration of cells towards each other. By contrast, the self-guidance strategy drives the cells to move away from their peers and disperse. The self-guidance strategy could contribute to the uniform distribution of normal epithelial cells in monolayers. It is also important for the redistribution of epithelial cells after epithelial injuries.

In the context of cancer, the overlap between EGF uptake and cellular confinement drives the formation of chemical gradients that guide the migration of malignant epithelial cells. The various networks of channels may be regarded as replicating conditions close to those inside tissues and around tumors e.g. lymphatic vessels, perivascular spaces, and perineural spaces. These spaces are well known to facilitate the dissemination of malignant cells. Moreover, we have previously shown that the migration of cells through channels correlates with their ability to form lung metastases in a mouse model [20]. Together, these results suggest that our experimental setup of channel networks could be a relevant in vitro model for cancer cell invasion.

One additional development of the strategies to measure cell navigation through mazes is towards assays that can probe the interactions between epithelial cells in monolayers [21]. When layers of epithelial cells advance through micropillar arrays, they do this as a collectively advancing front. Whenever the integrity of the advancing front is maintained, this is an indication that the epithelial phenotype is preserved. Whenever individual cells break contacts with neighbors, disperse fast, along straight trajectories, this is an indication that a mesenchymal phenotype is present (Fig. 5). The transition from epithelial to mesenchymal phenotype can be captured by the scattering of individual cells. Additional validation is possible using tagged antibodies for epithelial (e.g. E-cadherin) and mesenchymal (e.g. vimentin) molecular markers. However, the validation is a terminal assay, that requires the fixation and permeabilization of the cells. The transition between phenotypes, commonly named epithelial-to-mesenchymal transition (EMT), is of great interest for understanding disease processes during inflammation (e.g. tissue fibrosis), cancer (e.g. malignant transformation of cells), or embryogenesis.

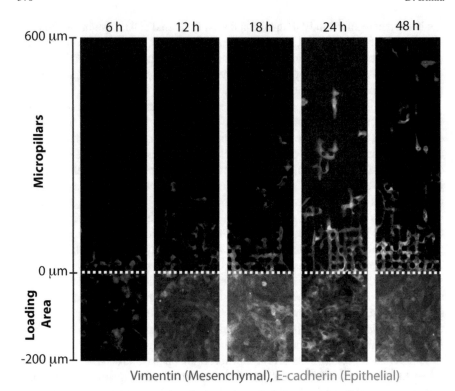

Vimentin (Mesenchymal), E-cadherin (Epithelial)

Fig. 5 Epithelial to mesenchymal transition (EMT) is quantified in real time and at single cell resolution by probing the advancement of MDA-MB-231 cells through arrays of posts inside a microfluidic device. Individually migrating mesenchymal cells detach and scatter from a collectively migrating epithelial front. The epithelial and mesenchymal phenotypes are then validated by staining using tagged antibodies for E-cadherin (green) and vimentin (red), respectively [21]. Reproduced with permission from Springer Nature

Using microfluidic mazes, the epithelial-to-mesenchymal and mesenchymal-to-epithelial dynamics can be measured in real time, using automated tracking, at single-cell resolution. Initial observations suggested that cells could switch frequently between the epithelial and mesenchymal states. Moreover, a 'sorting' mechanism could take place, resulting in the physical segregation of mesenchymal and epithelial cells. These behaviors can also be 'tuned' by altering pillar spacing, further indicative of phenotypic plasticity of migration. These behaviors can be perturbed by small molecule inhibitors, revealing that individually migrating cells exhibited diminished chemosensitivity when compared against their collectively migrating counterparts.

5 Future Directions

Microfluidic devices for precision measurements of migration phenotype can be useful in the clinic. For example, for testing the functional status of leukocytes microfluidic devices enabled our group to define a normal range of human neutrophil velocity in healthy individuals [22, 23]. They also helped optimize a treatment that restores defective neutrophil directionality following burn injuries [24]. Restoration of neutrophil abilities to navigate through simple channels with bifurcations progressed in parallel with increasing the overall capacity of the body to respond appropriately to infections.

Further exploration and better understanding of the directionality strategies in epithelial cells could lead to more effective therapeutic approaches to accelerate wound healing or delay cancer metastasis. New therapeutic strategies for suppressing individual invasion and dissemination may be tested using this assay, based on suppressing migration associated pathways as well as enhancing a reverse mesenchymal to epithelial transition. Ultimately, the abilities of cells to navigate through mazes, directly form relevant biological samples, inside devices that are user-friendly, could bring change to the way we diagnose and monitor health and disease conditions and enhance our understanding of how all the cells in the human body work together in harmony.

References

1. D. Irimia, G. Charras, N. Agrawal, T. Mitchison, M. Toner, Polar stimulation and constrained cell migration in microfluidic channels. Lab Chip **7**, 1783–1790 (2007)
2. S.H. Zigmond, Ability of polymorphonuclear leukocytes to orient in gradients of chemotactic factors. J. Cell Biol. **75**, 606–616 (1977)
3. S. Nourshargh, P.L. Hordijk, M. Sixt, Breaching multiple barriers: leukocyte motility through venular walls and the interstitium. Nat. Rev. Mol. Cell Biol. **11**, 366–378 (2010)
4. P. Friedl, E. Sahai, S. Weiss, K.M. Yamada, New dimensions in cell migration. Nat. Rev. Mol. Cell Biol. **13**, 743–747 (2012)
5. S. Takayama et al., Subcellular positioning of small molecules. Nature **411**, 1016 (2001)
6. S. Takayama et al., Selective chemical treatment of cellular microdomains using multiple laminar streams. Chem. Biol. **10**, 123–130 (2003)
7. N. Li Jeon et al., Neutrophil chemotaxis in linear and complex gradients of interleukin-8 formed in a microfabricated device. Nat. Biotechnol. **20**, 826–830 (2002)
8. D. Irimia, M. Toner, Spontaneous migration of cancer cells under conditions of mechanical confinement. Integr. Biol. Quant. Biosci. Nano Macro **1**, 506–512 (2009)
9. E. Sahai et al., Simultaneous imaging of GFP, CFP and collagen in tumors in vivo using multiphoton microscopy. BMC Biotechnol. **5**, 14 (2005)
10. H. Gerhardt, H. Semb, Pericytes: gatekeepers in tumour cell metastasis? J. Mol. Med. (Berl.) **86**, 135–144 (2008)
11. A. Giese, M. Westphal, Glioma invasion in the central nervous system. Neurosurgery **39**, 235–250; discussion 250–232 (1996)
12. L. Boneschansker, J. Yan, E. Wong, D.M. Briscoe, D. Irimia, Microfluidic platform for the quantitative analysis of leukocyte migration signatures. Nat. Commun. **5**, 4787 (2014)

13. V. Ambravaneswaran, I.Y. Wong, A.J. Aranyosi, M. Toner, D. Irimia, Directional decisions during neutrophil chemotaxis inside bifurcating channels. Integr. Biol. (Camb.) **2**, 639–647 (2010)
14. C. Scherber et al., Epithelial cell guidance by self-generated EGF gradients. Integr. Biol. Quant. Biosci. Nano Macro **4**, 259–269 (2012)
15. A. Levchenko, P.A. Iglesias, Models of eukaryotic gradient sensing: application to chemotaxis of amoebae and neutrophils. Biophys. J. **82**, 50–63 (2002)
16. D. Irimia, G. Balazsi, N. Agrawal, M. Toner, Adaptive-control model for neutrophil orientation in the direction of chemical gradients. Biophys. J. **96**, 3897–3916 (2009)
17. M. Skoge et al., A worldwide competition to compare the speed and chemotactic accuracy of neutrophil-like cells. PLoS ONE **11**, e0154491 (2016)
18. N.G. Jain et al., Microfluidic mazes to characterize T-cell exploration patterns following activation in vitro. Integr. Biol. (Camb.) **7**, 1423–1431 (2015)
19. R. Dorries, The role of T-cell-mediated mechanisms in virus infections of the nervous system. Curr. Top. Microbiol. Immunol. **253**, 219–245 (2001)
20. A. Wolfer et al., MYC regulation of a "poor-prognosis" metastatic cancer cell state. Proc. Natl. Acad. Sci. U.S.A. **107**, 3698–3703 (2010)
21. I.Y. Wong et al., Collective and individual migration following the epithelial-mesenchymal transition. Nat. Mater. **13**, 1063–1071 (2014)
22. K.L. Butler et al., Burn injury reduces neutrophil directional migration speed in microfluidic devices. PLoS ONE **5**, e11921 (2010)
23. A.N. Hoang et al., Measuring neutrophil speed and directionality during chemotaxis, directly from a droplet of whole blood. Technology **1**, 49 (2013)
24. T. Kurihara et al., Resolvin D2 restores neutrophil directionality and improves survival after burns. FASEB J. Off. Publ. Fed. Am. Soc. Exp. Biol. **27**, 2270–2281 (2013)

When the Path Is Never Shortest: A Reality Check on Shortest Path Biocomputation

Richard Mayne

Abstract Shortest path problems are a touchstone for evaluating the computing performance and functional range of novel computing substrates. Much has been published in recent years regarding the use of biocomputers to solve minimal path problems such as route optimisation and labyrinth navigation, but their outputs are typically difficult to reproduce and somewhat abstract in nature, suggesting that both experimental design and analysis in the field require standardising. This chapter details laboratory experimental data which probe the path finding process in two single-celled protistic model organisms, *Physarum polycephalum* and *Paramecium caudatum*, comprising a shortest path problem and labyrinth navigation, respectively. The results presented illustrate several of the key difficulties that are encountered in categorising biological behaviours in the language of computing, including biological variability, non-halting operations and adverse reactions to experimental stimuli. It is concluded that neither organism examined are able to efficiently or reproducibly solve shortest path problems in the specific experimental conditions that were tested. Data presented are contextualised with biological theory and design principles for maximising the usefulness of experimental biocomputer prototypes.

1 Introduction

This chapter addresses the use of biocomputer prototypes for addressing various minimal path problems (MPPs), which include physical solving of graph theoretical tasks such as shortest path problems (SPPs; synonymous with calculation of the Steiner minimum spanning tree of a set of vertices), the Travelling Salesperson Problem (TSP) and labyrinth navigation, by living systems. Measurement of an organism's ability to solve such puzzles, especially mazes, is not new: rodent navigation through geometrically-confined spaces was a staple of psychological research during the previous century and various forms of maze puzzle remain a diverting brain-teaser for children. It has not been until the comparatively recent advent of digital computing,

R. Mayne (✉)
Unconventional Computing Laboratory, University of the West of England, Bristol, UK
e-mail: Richard.Mayne@uwe.ac.uk

© Springer International Publishing AG, part of Springer Nature 2018
A. Adamatzky (ed.), *Shortest Path Solvers. From Software to Wetware*,
Emergence, Complexity and Computation 32,
https://doi.org/10.1007/978-3-319-77510-4_14

however, that we have begun to question the practical applications for experimental biocomputing prototypes that are able to address graph theory problems.

Of particular note is the explosion in research on the navigational abilities of the macroscopic amoeba-like organism *P. polycephalum* during the past decade which have experimentally demonstrated that this single-celled organism is capable of solving the TSP [36], navigating through various labyrinths and geometric puzzles on the first pass [20, 23] and calculating minimum spanning tree of a series of vertices [1], all via adaptation of its somatic morphology as a result of foraging behaviours (which will be expanded upon in the following section). Other notable examples of graphical biocomputation from recent years include, but are not limited to:

1. TSP solving through induced genetic transformation within live bacterial cells, wherein edges are conceptualised as segments of DNA linking gene nodes. Despite the variety of means by which computation can be achieved in transgenic bacteria, output is usually interpreted optically, e.g. through a change in colony colour (expression of coloured/fluorescent proteins), or expression of antibiotic resistance genes [7, 9].
2. Ant swarm migration along optimised single pathways—deduced by pathfinder ants according to an edge weight of attractant and repellent gradients—towards new nesting sites [21]. These dynamics may be put to more tangible computing applications such as addressing the Towers of Hanoi problem [24].
3. Maze navigation via the shortest path by cultured epithelial tumour cells, apparently guided by self-generated chemical gradients in a manner suggested to underlie cancer cell invasion [27].
4. Navigation through complex virtual reality labyrinths by rats undergoing simultaneous neural measurement [10].

Research in this area of biocomputation is justified for the following reasons. Firstly, as we approach the limitations of conventional digital hardware (i.e. the finite nature of the miniaturisation barrier presented by silicon-based computing substrates and associated problem of waste energy thermalisation), we are led to question the value of novel substrates, techniques and applications for computing technologies. Secondly, as alluded to in point 3 of the above enumerated list, interpretation of natural behaviours as expressions of computing aid our understanding of the biosciences and by extension, our ability to experimentally manipulate them. Finally, development of bio-inspired algorithms for use on conventional computing architectures is a richly diverse and varied area of research with virtually limitless applications; examples of successful algorithms relevant to this chapter include ant colony systems for optimised calculation of a range of problems including the TSP, labyrinth navigation and foraging route optimisation [8, 22, 33, 34] and multi-agent *P. polycephalum*-inspired models for solving SPPs and hence planning transport networks [14].

This apparently glowing appraisal of MPP biocomputation (also called 'bioevaluation') somewhat misrepresents the abilities of biological organisms for approaching problems that we are accustomed to tackling via the use of conventional computers, i.e. machines operating according to principles of the Turing model. As was eloquently argued by Stepney [28], biological substrates can only be said to compute in a distinctly non-Turing fashion, i.e. they are non-designed entities that are, to all intents and purposes, non-halting, nonsymbolic, stochastic systems. We are, furthermore, currently far from having elucidated the biological processes (intracellular, intercellular and extracellular signalling events) that constitute biocomputer input/output operations and interactions therein that we are choosing to call a form of computation. As such, the majority of experimental biocomputer prototypes will suffer from poor reproducibility, be slow,[1] costly to operate and require large amounts of operator time investment to set up, program and monitor for output. All of these factors are distinctly far-removed from our usual conception of computation. How, then, can live substrates be said to 'compute' the solution to MPPs given their aforementioned detriments?

The purpose of this chapter is to examine two case studies documenting research which exemplifies the theoretical and experimental limitations of utilising biocomputing substrates for calculating MPPs and enforce a 'reality check' on MPP biocomputation in so doing. In plainer terms, the aim of this chapter is to delineate the differences in the way in which biological substrates can be said to 'compute' the solution to MPPs, in comparison to the electronic substrate (algorithmic) equivalent. The conclusions drawn highlight the comparative strengths and weaknesses of biocomputing substrates and suggest experimental considerations for designing MPP-oriented biocomputers.

2   Case Study 1: The *Physarum* problem

2.1   Background

The *P. polycephalum* plasmodium (vegetative life cycle form) (Fig. 1) is a remarkable and fascinating protistic creature that is, at the time of writing, one of the most intensively researched-upon biological computing substrates. Comprising a macroscopic amoeba-like cell possessing millions of nuclei encapsulated within a single cell membrane, the plasmodial (or 'acellular') slime moulds are archetypal model organisms for excitable, motile cells as well as a go-to organism for educators wishing to demonstrate simple culture techniques with non-pathogenic organisms.

It was discovered in 2000 that *P. polycephalum* could navigate through a maze puzzle on the first pass [20]: this precipitated a biocomputing revolution[2] that saw the development of slime mould sensors, computer interfaces and circuitry, to name

---

[1]Biological time is many orders of magnitude slower than electrical time, implying that biocomputers capitalise on their comparative parallelism in order to beat the efficiency of conventional substrates.

[2]A Google Scholar (http://scholar.google.com) query with the search terms 'Physarum' and 'computing' return approximately 2,750 hits between the years 2000–2017 (search date September 2017).

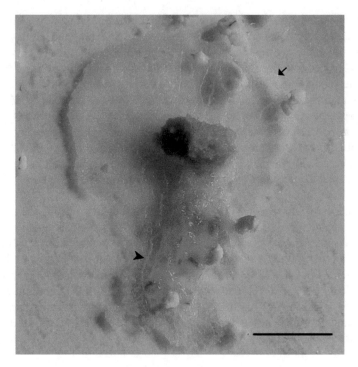

Fig. 1 Photograph of a *P. polycephalum* plasmodium cultivated on 2% non-nutrient agar gel, engulfing a few oat flakes. The organism is composed of caudal tubular regions (arrowhead) and a fan-shaped advancing anterior margin (arrow). Scale bar 10 mm

but a few examples of what are informally known as 'Physarum machines'. Whilst it is inappropriate to expand further on the range of biocomputing applications that have been found for *P. polycephalum*, we refer the reader to Refs. [3, 5, 17], and Jones and Safonow's chapter in this volume, for a comprehensive overview of slime mould computing.

As maze navigation is a MPP, this novel experiment quickly led researchers to question what other problems of graph theory that could be applied to slime mould. In 2007, Adamatzky [1] demonstrated that slime mould may solve SPPs, guided by nutrient gradients, thus demonstrating a clear advantage over previous reaction-diffusion computing substrates which cannot address this class of problem without engineered collisions. Slime mould mechanisms for adapting its inbuilt foraging behaviour to solving permutations of SPPs have been exploited in various biocomputer prototypes, including:

- Solving U-shaped trap problems [23].
- Colour sensing [4].
- Various logic gates [18].
- Constructing proximity graphs and beta skeletons [2].

- Designing transport networks [29].
- Constructing the convex and concave hulls about spatially-distributed nutrient sources [15].

The abilities of slime mould are represented with different degrees of 'enthusiasm' by the researchers who work on them: some maintain cautiously that slime mould biocomputation of MPPs are 'approximations' and hence that we are assigning conventional computing terminology on the organism for ease of comparison [3], whereas at the other end of the spectrum, some authors claim that the organism is capable of 'intelligent' behaviour [20].

In this section, I report that although slime mould SPP bioevaluation is both fascinating and worthy of further research, the manner in which the organism is able to represent the shape of a dataset is incompatible with our familiar, algorithmic understanding of the concept of 'solving a SPP'.

The experiments outlined in this section rely on observing the migration of slime mould through a down-scaled two-dimensional representation of a human living space, guided by chemoattractant gradients and zones of photorepulsion. The specific application of these experiments was to 'bio-evaluate' the layout of this domicile with regard to how 'efficiently' the space is subdivided, but their design is essentially the same as all of the previously mentioned Physarum machines created to address MPPs, i.e. analysis of the organism's migration between spatially-distributed nutrient sources. Success of the organism with regards to SPP navigation was interpreted in terms of the following criteria:

a. Ability of the organism to navigate between a finite number of attractant sources (vertices) in an order that represents a shortest path solution. Results were compared output from the same problem when addressed by a conventional computer (using Dijkstra's algorithm), i.e. if the shortest path between a set of four vertices in a virtual two-dimensional space, [A, B, C, D], is solved (via calculation of edge weights) by a computer as [D] → [B] → [A] → [C], slime mould in corresponding experiments whose conditions mimic those in the simulation will be judged to have correctly calculated the SPP if it navigates between vertices in the order [D, B, A, C]. This is opposed to simply comparing edge lengths between laboratory experiments (physically measuring the slime mould's length) and computer simulations.
b. Reproducibility, i.e. the ability of laboratory experimental slime mould to consistently navigate the same route between distributed vertices.

This methodology was chosen to best represent the differences between the way in which SPPs are 'bioevaluated', rather than algorithmically 'computed'. Total edge length travelled by slime mould in each experiment was also measured (for comparison but not as a primary determinant of path navigation 'success'), as were general observations on culture morphology and behaviour.

The rationale of this experiment was that the organism was expected to plot a route through the living space: we expect, informed by previous work on the topic, that the slime mould will attempt to link the discrete nutrient sources by forming tubular strands of protoplasm between them in a shape approximating the shortest path between them, whilst avoiding illuminated areas (hence why the shortest path solution outcome is not measured here as purely the total edge length).

The application of this Physarum machine was therefore to grant insight into a natural route around the space, e.g. if the organism were to visit the kitchen first, we could reason that this is not the most efficient use of space as the kitchen may not be the most frequently-visited room or indeed the room one is most likely to visit on entering the building. Although this application is somewhat removed from computer science, it represents an active area of unconventional computing, i.e. 'bioinspiration', which looks to analyse natural behaviour and apply it to creative problems; its inclusion here is purely to exemplify how the organism calculates its path.

2.2   Methods

A sample of a *P. polycephalum* plasmodium colonising an oat flake (a preferred slime mould nutrient source) was placed onto a section of 2% non-nutrient agarose gel. In this experiment, the agar section (hereafter 'wet layer') represents a geometric environment (or, graph) wherein vertices in the forthcoming SPP were represented by oat flakes. The wet layer was situated within a 900 mm square plastic Petri dish made of clear polystyrene.

A two-dimensional spatial representation of a living space was constructed as follows. An architectural draft of a two bedroom flat was etched onto two pieces of acrylic, one clear and one opaque. Both pieces of acrylic were then laser cut around each 'room' (two bedrooms, living space, kitchen, bathroom, entrance hall-way and two connecting hallways), leaving two 'backing layers' with holes in and two sets of eight cut-out pieces. The opaque cut outs were then slotted into the clear background, with the exception of the hallways, resulting in the draft comprising opaque sections (rooms) with clear sections separating them (walls and hallways) respectively (hereafter, the acrylic portion is known as the 'dry layer') (Fig. 2).

The dry layer was affixed to the bottom of the Petri dish housing the wet layer and the whole environment was placed overlying an electroluminescent plate producing 196 Lux, in order to represent the two-dimensional living space in a format the slime mould could interpret: opaque 'room' sections of the dry layer cast shadows onto the wet layer, whereas the clear spaces were illuminated, thus creating the organism's preferred dark zones and repellent illuminated zones. The agar gel in the wet layer was cut to the outline of the draft to constrain the organism's movements to within the 'building' and uncolonised oat flakes (discrete chemoattractant sources) were arranged to sit in the centre of each room, with the initial colonisation point being placed in the space representing the entrance hallway. The experiment was shielded from external light sources and the organism was left to propagate around its environment, with photographs being taken every 4 h for 48 h. The experiment was repeated in triplicate.

Fig. 2 Photograph of laser cut acrylic representation of a two bedroom, single floor living space, referred to in text as the 'dry layer'. All named rooms are cut from opaque acrylic and the spaces between them (walls and hallways) are clear acrylic

2.3 Results

A completed representative experiment, photographed after 48 h had elapsed, is shown in Fig. 3a. Path lengths radiating from the inoculation point [A] are overlaid; paths not shown are (in mm): [BD] 600, [BE] 675, [BF] 445, [CE] 685, [CF] 515, [DF] 375. The route taken by the plasmodium is shown in Fig. 3b and follows [A]→[B]→[A]→[F]→[E]→[D]→[C]→[B] to complete a circuit linking all of the oat flakes. The total length of this specific route is 2060 mm, although the physical length of the organism greatly exceeds this (see below). For comparison, the minimal length solution to this problem, as calculated by Dijkstra's algorithm, is shown in Fig. 3c and follows [A]→[F]→[E]→[D]→[C]→[B], with a total path length of 1795 mm.

Morphologically, the plasmodium shown in Fig. 3a contains a number of redundant links: this is best exemplified in the organism's [CD] link, which vaguely resembles the figure '8' due to two bifurcations which concatenate at the nodes. As was mentioned in the previous paragraph, the organism's travel distance between nodes exceeds the algorithmically-generated value, due to the organism not propagating in

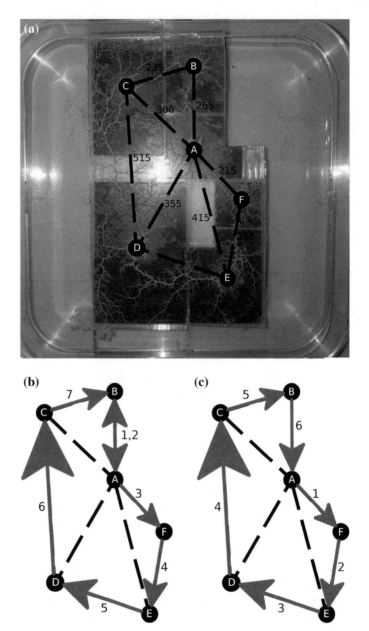

Fig. 3 Slime mould addresses a SPP in a simple graph guided by attractants and repellents. **a** Photograph showing experiment (details in text) after 48 h, overlaid with paths radiating from inoculation point, [A]. Path lengths in millimetres, rounded to nearest 5. **b** Path taken by slime mould, as indicated by arrows and numbering. **c** Shortest path, as calculated by Dijkstra's algorithm

straight lines and the existence of multiple accessory branches in its protoplasmic network. The organism's deviation from straight lines tended to increase proportionally with the inter-node distance and larger 'rooms' tended to have more accessory branches within them. The organism also tended to avoid crossing illuminated zones but did cross 5 wall spaces and one hallway.

In all experiments, the slime mould assumed a ring around the points in the manner shown in Fig. 3a (i.e. a concave hull around the vertices), but did not always take the same route: the other two routes taken were [AFEFABCDEF] and [ABCDEFA] (data not shown).

It is clear from the examplar data that the organisms did not calculate the shortest path, as defined by the criteria delineated in Sect. 2.1: the navigation sequence between vertices was not identical to the sequence calculated to be the shortest by Dijkstra's algorithm (although one did take an optimal route, i.e. the inverse of the algorithmically calculated route) and different routes were taken by each organism in each repeat.

2.4 Discussion

2.4.1 Straight Lines, Redundancy and Accessory Branches

The experiment shown in Fig. 3a highlights why the total length of the organism's protoplasmic tube network (i.e. edge length) was not used as a primary determinant of the experimental outcome: slime moulds do not travel in perfectly straight lines. In some cases, such as in the edge [AF], the fit is good but far from perfect. Conversely, [DC] edge measures about 534 mm (measuring the thickest tube only and no redundant paths), approximately 4% greater than the true shortest route between these paths. In comparison to conventional algorithmic approaches to solving SPPs, this amount of divergence is sufficient to negate the hypothesis that the organism could be said to be computing the 'absolute' shortest path, especially as the error accumulates with each edge. Whilst it is beyond the remit of this investigation to debate at length on why slime moulds no not travel in perfectly straight lines, we may assume, parenthetically, that there is no distinct evolutionary advantage to doing so and that factors such as organisation of intracellular motile machinery and reception of diffused chemical signals involve stochastic elements (a common feature of biological processes that makes live substrates particularly challenging to model [26]).

Another important factor influencing the organisms' total network length was the existence of the edge weighting factors other than node spacing. Although every attempt was made to control the independent variables in the experiments presented, a multitude of 'background' factors that are extremely difficult to control are likely to have contributed to the variation in the organisms' routes observed across all experiments. Exemplar influencing factors include the organisms' health and nutritional status, fluctuations in temperature, presence of microbes and distribution of moisture throughout the wet layer. It is likely that such factors also played an important role in determining the overall distribution of the organisms' protoplasm in particular areas of the wet layer, i.e. redundant links and accessory branches, which are thought to be constructed in areas of high nutrient availability and along gradients of attraction (chemical or otherwise), respectively [16]. As such, descriptions of slime mould addressing MPPs cannot be favourably interpreted in purely algorithmic terms, so our biocomputing vocabulary must be altered accordingly.

2.4.2   Ordering of Vertex Visits

The path the organism took in the experiment shown in Fig. 3 involved doubling back, increasing its total path length by 2 edges (equating to over 10% in terms of physical path length). This highlights how live biocomputers are always in a state of flux and require constant observation in order to assess the state of the computation; observing the output of the plasmodium at the 48 h mark gives no indication as to how the organism's network was constructed. The nature of biological substrates is, insofar as we have ascribed an exogenous purpose to its foraging behaviour, non-halting; this could be considered both a benefit (it allows for assessment of the state of computation and overall system dynamics at any point) and a detriment (user input is required to determine when the operation has finished).

   Whilst the organism did navigate via an optimal route in one of the repeats, albeit not in the order calculated by Dijkstra's algorithm, the reproducibility of results from this small sample was poor by both biological and computing standards.

   To address the question of why the slime mould in the above example took the longer path with the addition of the [ABA] diversion, one may be tempted to assume that the organism lacks the necessary 'intelligence' to adequately distinguish between the benefits and detriments of the potential edges [AB] and [AF] and so opted for one at random, then found that the conditions at [B] were less favourable than at [A], so doubled back to explore other paths. Whilst this is certainly possible, it is apparent that the data are insufficient to properly measure the effect of the aforementioned non-visible weighting factors regarding environment favourability, organism status and the (likely nonlinear) relationship between these and the physical distances separating vertices. The complex interplay between attraction and repulsion is best illustrated here by the example of the [DC] path in Fig. 3a: the organism traverses the repellent 'hallway' zone, presumably as the benefits of migrating across the gap outweigh the energy costs of circumventing it.

   It is essentially impossible to control all of the variables in experiments such as those described here, hence variation in biocomputer must be anticipated and accounted for in experimental designs. As variation is the basis by which all organisms were able to evolve, an intuitively-designed biocomputer will capitalise on variation, despite this being anathema to the traditional concept of computing.

2.4.3 Shortest Path Approximations are only Constructed in Nutrient-Limited Environments

The experiment described above represents a nutrient-sparse environment for slime mould: can the organism construct similar graphs in nutrient-rich environments? In Fig. 4a a *P. polycephalum* plasmodium inoculated onto a lattice delineated by oat flakes is shown. The figure shows that *P. polycephalum* forms a more interconnected graph in nutrient-rich environments, which is perhaps more akin to a Gabriel graph than a Steiner minimum spanning tree (although 2 points near the centre have not been linked, for an unknown reason). It was demonstrated in 2009 [2] that *P. polycephalum* may approximate any of the proximity graphs in the Toussaint hierarchy, dependent on relative edge weighting. For comparison, Fig. 4b shows slime mould growth on nutrient enriched (i.e. a uniform attractant field) agar; the entire plate is morphologically more similar to the amorphous advancing anterior margins usually observed in nutrient-limited substrates. This highlights that slime mould biocomputation of SPPs only occurs within a specific set of conditions relating to nutrient availability, meaning that reproducibility of SPPs is dependent on a fairly narrow window of initial conditions relating to the organism's nutritional status and the spacing of nutrient sources.

(a)   **(b)**

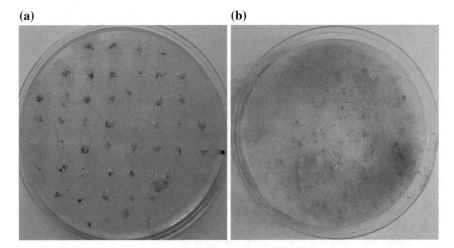

Fig. 4 Photographs of *P. polycephalum* propagating in nutrient rich environments, 48 h post-inoculation. **a** An excessive amount of discrete food sources (oat flakes) are provided. **b** Growing on enriched (oatmeal) agarose substrate. Adapted from [17]

2.5   Summary

During the recent advent of slime mould computing research, much hype was generated in the media around the use of slime mould for addressing shortest path problems, particularly with respect to route planning. The organism's malleability and easily-interpretable output led to rich and varied works, including a particularly whimsical paper in which it was suggested that slime mould should play a role in planning interplanetary missions [6]. In spite of this, I have demonstrated here that to label slime mould foraging behaviours in nutrient-limited environments as calculation of a SPP is somewhat inaccurate without applying a certain amount of abstraction to the manner in which the term 'shortest path' is interpreted. This is not to devalue slime mould research; clearly the organism is undertaking some immensely complex massively-parallel operations, research upon which is most assuredly important. What I am suggesting, however, is that directly comparing this to conventional path finding algorithms is at best unhelpful.

3   Case Study 2: Banging Your *Paramecium* against a brick wall

3.1   Background

P. caudatum is a single celled protistic freshwater microorganism covered in thousands of minute hair-like appendages called 'cilia' (Fig. 5). Cilia beat rhythmically in order to generate fluid currents in adjacent media, thus generating motive force and enhancing feeding on dispersed particulates. Cilia-based motility in *P. caudatum* therefore represents a novel mechanism for addressing MPPs in aquatic environments.

Whilst historical literature has indicated the use of basic puzzles to assess chemotaxis and thermotaxis in *P. caudatum* (usually, a T-shaped puzzle where the organism is given a binary choice to navigate directly ahead or around a 90° bend) [32], very little attention has been paid to the organism's ability to address problems of graph theory, despite their behaviour in confined environments (microfluidic circuitry, capillary tubes) being reasonably well characterised [12].[3]

In this section I will demonstrate how *P. caudatum* is particularly ill-adapted for addressing MPPs in geometrically-constrained labyrinth puzzles and by extension illustrate some of the practical limitations of designing MPP-solving biocomputers. Single *P. caudatum* cells were placed in small labyrinth puzzles in the presence of a chemoattractant gradient at the exit, according to the principle that the organism

---

[3]It is essential in this context to mention ingenious work of Reidel-Kruse et al. [25] who developed multiple 'games' in tiny enclosed environments wherein paramecium behaviour was influenced by user input, including 'PAC-mecium' and 'ciliaball'.

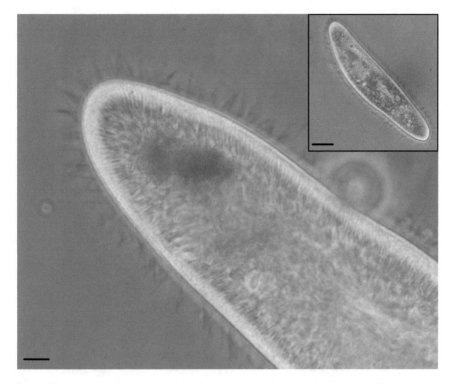

Fig. 5 Photomicrographs of *P. caudatum*, phase contrast optics. (Main) Anterior tip of the organism, where hair-like cilia may be easily seen coating the cell's membrane. Scale bar 10 μm. (Inset) Lower magnification image showing the whole cell. Scale bar 25 μm

congregates in regions of highest nutrient density [31]. Successful navigation towards the puzzle's exit within a specific timeframe (10 min) was judged to be evidence in support of the hypothesis that *P. caudatum* are able to solve this variety of MPP.

3.2 Methods

P. caudatum were cultivated in an in-house modification of Chalkley's medium which was enriched with 10 g of desiccated alfalfa and 20 grains of wheat per litre, at room temperature. Cultures were exposed to a day/night cycle but were kept out of exposed sunlight. Organisms were harvested in logarithmic growth phase by gentle centrifugation at $400 \times G$ before being transferred to fresh culture media. Cells used in experiments were transferred via a micropipette to the testing environment.

The testing environments, these being labyrinth puzzles designed to accommodate *P. caudatum*, were fabricated as follows. Labyrinths were generated in open-SCAD using an open-source Python script [30] which rendered graphic files in STL

Fig. 6 Projection of labyrinth puzzle mould used in *P. caudatum* navigation experiments. The route between the entrance/exit reservoirs is shown in red

format (Fig. 6). The size of the completed labyrinth was approximately $10 \times 7\,mm$; the dimensions of the labyrinth's walls were chosen to accommodate approximately 8–10 cell widths ($750\,\mu m^2$), which was reasoned to be ample room to allow a single *P. caudatum* cell to manoeuvre and reduce the likelihood of collisions with the environment's walls. Maze designs were modified to have two distinct reservoirs at the entrance and exits to the labyrinths before being inverted using Solidwoks 2017 (Dassault Systèmes, France). Moulds were then printed in PLA using an Objet 260 FDM 3D printer (Stratasys, USA) at a resolution of $50\,\mu m$. The moulds were then cleaned in isopropyl alcohol, rinsed three times in deionised water and air dried. Labyrinths were cast in clear polydimethylsiloxane (PDMS) (Sylgard 184, Dow Corning, USA) by pouring elastomer solution onto the mould, removing any air in a vacuum chamber and finally polymerising in an oven at $40\,°C$ for $48\,h$.

Completed testing environments were stuck to large glass microscope coverslips (depth $0.11\,mm$) and a small piece of a solid chemoattractant (desiccated alfalfa) was placed in one of the labyrinth's reservoirs. The maze was then filled with tap water that had been resting for $48\,h$ from the end containing the solid chemoattractant source, taking care not to dislodge it from its reservoir. This method was chosen in order to generate a gradient of chemoattractants along the maze. The environment was then allowed to rest for 15 min in a sealed Petri dish, in order to allow bubbles to disappear but prevent fluid evaporation. Individual *P. caudatum* cells were then transferred to the unoccupied reservoir using a micropipette. Observations were made using a stereomicroscope and video footage was collected using a Brunel Eyecam (Brunel Microscopy, UK). Each experiment was run for 10 min, after which the water began to evaporate to a noticeable degree. Labyrinths were not sealed in order to not expose the organisms to alterations in fluid pressure or dissolved oxygen content.

The timescale for the experiment was judged to be sufficiently long for the organism to navigate the puzzle (based on typical *P. caudatum* movement speed reaching in excess 1 mm per second), whilst disallowing the eventuality of the organism arriving at the exit via a random walk. The experiment was repeated 10 times.

3.3   Results

The experiment detailed in Fig. 7 is representative of all experiments conducted. Immediately after inoculation into the maze puzzle, the *P. caudatum* cell spent several seconds rotating on the spot. Following, the organism would migrate in approximately the correct direction, i.e. towards the aperture leading to the maze, before colliding with a wall. Collisions would cause the the organism to migrate in a reverse direction at apparently random angles. This would begin an erratic oscillation in anteroposterior migration which prevented the organism from progressing far into the maze; the furthest a *P. caudatum* cell was observed to have migrated within the 10 min experiment was approximately 4 mm end-to-end, amounting to two right angle corners successfully traversed.

3.4   Discussion

3.4.1   Spontaneous Alternation Behaviour Is Incompatible with Constrained-Geometry Puzzles

Spontaneous alternation behaviour (SAB) in *Paramecium* spp. is a well documented phenomenon that occurs when the organism collides with a solid object or otherwise meets an unfavourable stimulus. As Witcherman noted in his classic treatise on *Paramecium* [35], this behaviour is not quite taxis nor quite kinesis as it interferes with true directional movement with a non-specific reaction to move away in any other direction. This response was anticipated, although attempts to engineer the labyrinth's passages as widely as possible had been made as earlier experiments with narrower channels produced much the same effect (data not shown).

The evolutionary advantages of SAB demonstrate the momentary requirements of motile freshwater microorganisms that sit in the middle of their food chain (i.e. they predate smaller organisms and are prey to larger organisms), as it would appear

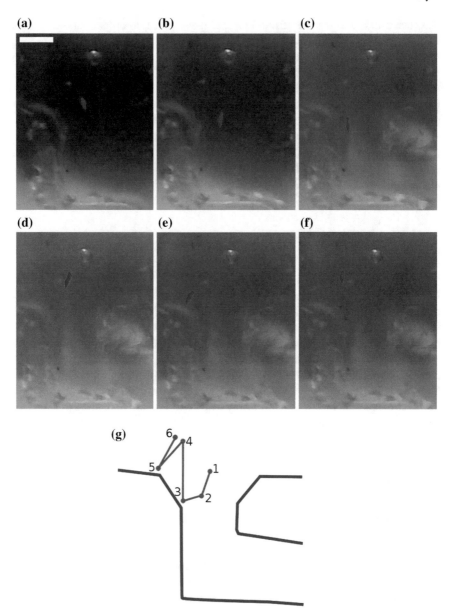

Fig. 7 Figure to demonstrate motion of *P. caudatum* in a labyrinth. **a–f** Experimental photographs to show movement of organism (false coloured red) about geometric constraints in a PDMS labyrinth (see text). Images taken at approx. 1 s intervals. Scale bar (in panel *a*) 500 μm. **g** Schematic to demonstrate organism movement (red, numbered sequentially) about labyrinth boundaries (blue)

to favour keeping the organism away from harm at the cost of reducing the efficiency of its search for food. This does, however, result in the organism being virtually incapable of efficiently traversing environments such as the miniature labyrinths used in these experiments, although this of course does not imply that they are unable to address MPPs in unconstrained geometries.

Previous literature has indicated that *P. caudatum* responds to extremely confined environments (microfluidic circuitry or otherwise sealed systems) by exhibiting a somersaulting behaviour which allows it to assume sinusoidal paths by bending over on its self, rather than reverting SAB [12]. It must be noted, however, that this behaviour is quite different to any previously described mode of *Paramecium* movement and is unlikely to occur in anything but the most confined environments.

3.4.2 Challenges Associated with Biocomputing in Aqueous Environments

Generating a chemoattractant gradient in an aqueous medium is not a straightforward task. Chemical attraction was chosen as the input over, for example, light or electrical gradients due to the technical limitation of microscopic illumination and the necessity to use electrical fields of a potentially harmful magnitude to be detectable at opposite ends of the labyrinth, respectively. It was reasoned that the organisms were able to sense a chemoattractant given their propensity to always initially migrate in the direction of the stimulus. Nutrient density at the labyrinth start point was low and the organisms only exhibited patterns of movement associated with migration, as opposed to the slower-swimming 'feeding behaviours' that can be observed in areas of sufficiently high nutrient substrate concentration [19]. Nevertheless, it is apparent that in this instance, any chemoattractive effects that were induced were insufficient to guide the organism through the maze without collisions. Representing multiple vertices as attractant fields in unconstrained geometric spaces would represent a significant technical challenge.

Another issue encountered, which is not adequately represented by standard photography, was the impact that being in a three dimensional environment had on the organisms' pathfinding abilities: *P. caudatum* cells were observed to collide with both the upper and lower levels of the experimental environment (the 'roof' and 'floor', as demarcated by the area filled with fluid) in the majority of experiments, to much the same effect as wall collisions. Even though these labyrinth puzzles are pseudo-two dimensional puzzles, it must be remembered that the search for an efficient path encompasses the need to find the most efficient route in all three dimensions. Even had the organisms successfully navigated to the labyrinth exit, care would have to be taken in stating that the organism had navigated the 'shortest path' through, due to the phenomenon of helical swimming in all *Paramecium* species and z-axis deviations [13]. This emphasises the need to take into account the third dimension when conducting such experiments.

3.5 Summary

Despite being an intensely researched-upon model organism that exhibits many behaviours that can be interpreted as expressions of natural computing (orchestrated manipulation of micro-scale objects [19], basic learning/memory [11]), their mode of motion and hazard avoidance essentially precludes their use in experiments which use geometries constrained to the degrees described. This implies that the only way *P. caudatum* may be coaxed into addressing MPPs is either in architectureless space (which is complex to monitor microscopically) or otherwise in extremely confined environments, which would in turn still be problematic to achieve in practice.

Although it could perhaps have been predicted that *P. caudatum* cells are a good medium for implementing graph theory biocomputation, it is nevertheless a fact that *Paramecium* species have long-since been used as model aquatic organisms for investigating various taxes, meaning that results gained with these organisms may be partially representative of a large class of organisms. Although there are doubtless other species of ciliate better suited to such applications, this section serves to emphasise that being placed in confined environments is not representative of *P. caudatum's* natural habitat (i.e. large bodies of static or running freshwater) and by extension that live substrates are not always as tolerant to abuse (in designing MPPs or any other form of biocomputer) as archetypal substrates such as slime mould may suggest.

4 Conclusions

It is a singular temptation to view emergent biological phenomena in our world and dream of how they may be harnessed, mimicked or emulated for computing applications. Whilst I have endeavoured to prevent the tone of this chapter from being overtly negative or unduly sceptical (biocomputing is a wonderful science whose advancement is of great necessity!), I have attempted to highlight the experimental considerations that make laboratory experiments involving morphological and topological operations with live substrates difficult to implement and even harder to interpret, reproduce and refine. The data presented here suggest that shortest path biocomputation is not something that can easily be achieved and that a certain amount of optimism and open-mindedness is required to interpret such experiments as anything other than rough approximations of MPP solutions in the rare experiments that are ostensibly successful.

In conclusion, the experiments outlined here involve taking organisms out of their natural environment and encouraging them to do distinctly unnatural things. Interpreting the output of a biocomputer almost always requires a certain degree of abstraction; this does not devalue the experiments or make the phenomena under investigation less interesting, but it must be remembered that approximation is a distinctly *unconventional* paradigm in computer science. The best biocomputing experiments will put minimal stress on the organism and observe processes that occur naturally, only intervening subtly in order to tweak a parameter or make a measurement.

Finally, I will note on behalf of all experimentalists in the field that the divergence between the observed results of biocomputation and our pre-conceived notions of their expected outcomes in comparison with the algorithmic equivalents highlights why it is essential to make physical biocomputing prototypes in unconventional computing research, rather than computer models alone.

Acknowledgements The author thanks both reviewers for their invaluable insights and suggestions and Matthew Hynam for providing the laser cut architectural drafts used in slime mould experiments.

References

1. A. Adamatzky, Physarum machines: encapsulating reaction-diffusion to compute spanning tree. Naturwissenschaften **94**(12), 975–980 (2007). https://doi.org/10.1007/s00114-007-0276-5, http://www.ncbi.nlm.nih.gov/pubmed/17603779
2. A. Adamatzky, Developing Proximity graphs by physarum polycephalum: does the plasmodium follow the toussaint hierarchy? Parallel Process. Lett. **19**(1), 105–127 (2009). https://doi.org/10.1017/CBO9781107415324.004
3. A. Adamatzky, *Physarum Machines: Computers from Slime Mould* (World Scientific, 2010)
4. A. Adamatzky, Towards slime mould colour sensor: recognition of colours by Physarum polycephalum. Org. Electron. **14**(12), 3355–3361 (2013). https://doi.org/10.1016/j.orgel.2013.10.004
5. A. Adamatzky (ed.), *Advances in Physarum Machines* (Springer International Publishing, 2016). https://doi.org/10.1007/978-3-319-26662-6
6. A. Adamatzky, R. Armstrong, B. De Lacy Costello, Y. Deng, J. Jones, R. Mayne, T. Schubert, G. Sirakoulis, X. Zhang, Slime mould analogue models of space exploration and planet colonisation. J. Br. Interplanet. Soc. **67**, 290–304 (2014)
7. J. Baumgardner, K. Acker, O. Adefuye, S.T. Crowley, W. DeLoache, J.O. Dickson, L. Heard, A.T. Martens, N. Morton, M. Ritter, A. Shoecraft, J. Treece, M. Unzicker, A. Valencia, M. Waters, A. Campbell, L.J. Heyer, J.L. Poet, T.T. Eckdahl, Solving a hamiltonian path problem with a bacterial computer. J. Biol.l Eng. **3**(1), 11 (2009). https://doi.org/10.1186/1754-1611-3-11
8. M. Dorigo, L.M. Gambardella, Ant colony system: a cooperative learning approach to the traveling salesman problem. IEEE Trans. Evol. Comput. **1**(1), 53–66 (1997). https://doi.org/10.1109/4235.585892
9. M. Esau, M. Rozema, T.H. Zhang, D. Zeng, S. Chiu, R. Kwan, C. Moorhouse, C. Murray, N.T. Tseng, D. Ridgway, D. Sauvageau, M. Ellison, Solving a four-destination traveling salesman problem using escherichia coli cells as biocomputers. ACS Synth. Biol. **3**, 972–975 (2014). https://doi.org/10.1021/sb5000466
10. C.D. Harvey, F. Collman, D.A. Dombeck, D.W. Tank, Intracellular dynamics of hippocampal place cells during virtual navigation. Nature **461**(7266), 941–946 (2009). https://doi.org/10.1038/nature08499
11. T. Hennessey, W. Rucker, C. Mcdiarmid, Classical conditioning in paramecia. Anim. Learn. Behav. **7**(4), 417–423 (1979)
12. S. Jana, A. Eddins, C. Spoon, S. Jung, Somersault of Paramecium in extremely confined environments. Sci. Rep. **5**, 13, 148 (2015). https://doi.org/10.1038/srep13148
13. S. Jana, S.H. Um, S. Jung, Paramecium swimming in capillary tube. Phys. Fluids **24**(4) (2012). https://doi.org/10.1063/1.4704792

14. J. Jones, Mechanisms inducing parallel computation in a model of physarum polycephalum transport networks. Parallel Process. Lett. **25**, 1540, 004 (2015). https://doi.org/10.1142/S0129626415400046

15. J. Jones, R. Mayne, A. Adamatzky, Representation of shape mediated by environmental stimuli in Physarum polycephalum and a multi-agent model. Int. J. Parallel Emerg. Distrib. Syst. **2**, 166–184 (2015). https://doi.org/10.1080/17445760.2015.1044005

16. R. Mayne, Biology of the Physarum polycephalum plasmodium: preliminaries for unconventional computing, in *Advances in Physarum Machines*, ed. by A. Adamatzky, vol. 21, chap. 1 (Springer, 2016), pp. 3–22, http://link.springer.com/10.1007/978-3-319-26662-6

17. R. Mayne, *Orchestrated Biocomputation: Unravelling the Mystery of Slime Mould "Intelligence"* (Luniver Press, Bristol, UK, 2016)

18. R. Mayne, A. Adamatzky, Slime mould foraging behaviour as optically coupled logical operations. Int. J. Gen. Syst. **44**(3), 305–313 (2015). https://doi.org/10.1080/03081079.2014.997528

19. R. Mayne, J.G. Whiting, G. Wheway, C. Melhuish, A. Adamatzky, Particle sorting by paramecium cilia arrays. Biosystems **156–157**, 46–52 (2017). https://doi.org/10.1016/j.biosystems.2017.04.001

20. T. Nakagaki, H. Yamada, A. Toth, Intelligence: maze-solving by an amoeboid organism. Nature **407**, 470 (2000). https://doi.org/10.1038/35035159

21. S.C. Pratt, E.B. Mallon, D.J.T. Sumpter, N.R. Franks, Quorum sensing, recruitment, and collective decision-making during colony emigration by the ant Leptothorax albipennis. Behav. Ecol. Sociobiol. **52**(2), 117–127 (2002). https://doi.org/10.1007/s00265-002-0487-x

22. K. Ramsch, C. Reid, M. Beekman, M. Middendorf, A mathematical model of foraging in a dynamic environment by trail-laying argentine ants. J. Theor. Biol. **306**, 32–45 (2012)

23. C.R. Reid, T. Latty, A. Dussutour, M. Beekman, Slime mold uses an externalized spatial "memory" to navigate in complex environments, in *Proceedings of the National Academy of Sciences of the United States of America* vol. 109(43), 17, 2012, pp. 490–494. https://doi.org/10.1073/pnas.1215037109

24. C.R. Reid, D.J.T. Sumpter, M. Beekman, Optimisation in a natural system: argentine ants solve the towers of hanoi. J. Exp. Biol. **214**(1), 50–58 (2011). https://doi.org/10.1242/jeb.048173

25. I.H. Riedel-Kruse, A.M. Chung, B. Dura, A.L. Hamilton, B.C. Lee, Design, engineering and utility of biotic games. Lab Chip **11**(1), 14–22 (2011). https://doi.org/10.1039/C0LC00399A

26. L. Saiz, J.M.G. Vilar, Stochastic dynamics of macromolecular-assembly networks. Mol. Syst. Biol. 1–11 (2006). https://doi.org/10.1038/msb4100061

27. C. Scherber, A.J. Aranyosi, B. Kulemann, S.P. Thayer, M. Toner, O. Iliopoulos, D. Irimia, Epithelial cell guidance by self-generated EGF gradients. Integr. Biol. **4**(3), 259 (2012). https://doi.org/10.1039/c2ib00106c

28. S. Stepney, The neglected pillar of material computation. Phys. D Nonlinear Phenom. **237**, 1157–1164 (2008). https://doi.org/10.1016/j.physd.2008.01.028

29. A. Tero, S. Takagi, T. Saigusa, K. Ito, D.P. Bebber, M.D. Fricker, K. Yumiki, R. Kobayashi, T. Nakagaki, Rules for biologically inspired adaptive network design. Science **327**(5964), 439–442 (2010). https://doi.org/10.1126/science.1177894, http://www.ncbi.nlm.nih.gov/pubmed/20093467

30. Thingiverse: openscad maze generator. https://www.thingiverse.com/thing:24604. Accessed 01 May 2017. Produced by user 'dnewman'

31. J. Van Houten, Two mechanisms of chemotaxis inparamecium. J. Comp. Physiol. **127**(2), 167–174 (1978). https://doi.org/10.1007/BF01352301

32. J. Van Houten, H. Hansma, C. Kung, Two quantitative assays for chemotaxis inparamecium. J. Comp. Physiol. **104**(2), 211–223 (1975). https://doi.org/10.1007/BF01379461

33. M. Vela-Pérez, M. Fontelos, J. Velásquez, Ant foraging and geodesic paths in labyrinths: analytical and computational results. J. Theor. Biol. **320**, 100–112 (2013)

34. K. Vittori, G. Talbot, J. Gautrais, V. Fourcassie, A. Araujo, G. Theraulaz, Path efficiency of ant foraging trails in an artificial network. J. Theoreical Biol. **239**, 507–515 (2006)
35. R. Witcherman, *The biology of Paramecium*, 2nd edn., chap. 5: Movement, behaviour and motor response, Plenum, 1982, pp. 211–238
36. L. Zhu, M. Aono, S.J. Kim, M. Hara, Amoeba-based computing for traveling salesman problem: Long-term correlations between spatially separated individual cells of Physarum polycephalum. BioSystems **112**(1), 1–10 (2013). https://doi.org/10.1016/j.biosystems.2013.01.008

Shortest Path Finding in Mazes by Active and Passive Particles

Jitka Čejková, Rita Tóth, Artur Braun, Michal Branicki,
Daishin Ueyama and István Lagzi

Abstract Maze solving and finding the shortest path or all possible exit paths in mazes can be interpreted as mathematical problems which can be solved algorithmically. These algorithms can be used by both living entities (such as humans, animals, cells) and non-living systems (computer programs, simulators, robots, particles). In this chapter we summarize several chemistry-based concepts for maze solving in two-dimensional standard mazes which rely on surface tension driven phenomena at the air-liquid interface. We show that maze solving can be implemented by using: (i) active (self-propelled) droplets and/or (ii) passive particles (chemical entities).

J. Čejková (✉)
University of Chemistry and Technology Prague, Prague, Czechia
e-mail: jitka.cejkova@vscht.cz

R. Tóth · A. Braun
Laboratory for High Performance Ceramics, Empa, Dübendorf, Switzerland
e-mail: ritatoth@hotmail.com

A. Braun
e-mail: artur.braun@alumni.ethz.ch

M. Branicki
School of Mathematics, University of Edinburgh, Edinburgh, UK
e-mail: m.branicki@ed.ac.uk

D. Ueyama
Faculty of Engineering, Musashino University, Tokyo, Japan
e-mail: d_ueyama@musashino-u.ac.jp

I. Lagzi
Department of Physics, Budapest University of Technology and Economics
and MTA-BME Condensed Matter Research Group, Budapest, Hungary
e-mail: istvanlagzi@gmail.com

© Springer International Publishing AG, part of Springer Nature 2018
A. Adamatzky (ed.), *Shortest Path Solvers. From Software to Wetware*,
Emergence, Complexity and Computation 32,
https://doi.org/10.1007/978-3-319-77510-4_15

1 Introduction

Solving mazes and finding of the shortest path or all possible exit paths in mazes are scientifically challenging problems, although sometimes it seems to be "just to find the trivial way from the start to the exit of a maze". Here, we focus on mazes in Euclidean geometry in two or three dimensions, possibly with multiple solutions but without any loops or inaccessible areas. Algorithms for solving such mazes can be classified according to several criteria. For example, some maze-solving strategies require no prior knowledge about the maze structure and its complexity. On the other hand, other algorithms require information about the global maze structure. Another classification is based on on the type of the desired outcome, e.g., whether or not the solver identifies the optimal or all possible solutions to exiting the maze. Of course, the type of algorithm chosen depends on specific needs and situations; some algorithms may be plausibly envisaged as implementable by living entities, other algorithms can be only used by artificial physical or chemical solvers [1–4, 8, 9, 12–17].

Consider first humans and animals as the 'solvers'. In this case one can assume that the solvers are rather sophisticated in the sense that they possess the necessary sensing skills for spatially resolved data acquisition and the cognitive abilities for signal processing, comparison, and decision making, leading to efficient (though not necessarily optimal) path finding strategies. For a trained human maze solving may be often an easy easy task if the global maze structure is available. However, if only a local information about a maze is available at any time, or if the solver has rudimentary cognitive skills, the same problem could be difficult and challenging. Here, we outline three simple maze-solving strategies which do not require no prior knowledge on the maze structure and its complexity, namely the *random mouse, the wall follower*, and the *Trémaux algorithms*.

The simple *random mouse* solver abandons targeted goal finding and relies instead on a random trial-and-error strategy which results in a random walk through the maze which is terminated once the exit is reached. The disadvantage of this algorithm is that it is extremely slow and the time for this algorithms to execute (i.e., find the exit) is unbounded [19].

The more efficient *wall follower* algorithm is based on always following the same wall in the maze (e.g., by keeping one hand in contact with one wall). For non-degenerate mazes (i.e., with an 'entry' and 'exit', and assuming that one does not start in inside an enclosed area) this algorithm is guaranteed to complete the task of finding the exit in finite time. In general, the execution time of this strategy depends on the orientation (i.e., the choice of hand/wall).

The *Trémaux's* algorithm represents another random strategy which results in a version of the self-avoiding random walk. In this method the solver makes random decisions about where to go from the starting point while recording the past path (think of drawing a line on the floor to mark the path). If the solver reaches the end point of any corridor, it returns to the closest crossroad where he chooses any corridor that it has not visited yet. Paths are either unmarked (i.e., unvisited), marked once or marked twice (i.e., leading to a dead end).

Clearly, the above exit-finding strategies can be utilised by humans or implemented algorithmically, for example, in order to aid robot navigation. Robots are able to randomly make decisions which corridor to choose (and thus to use the random mouse algorithm), follow the wall (and use the wall follower algorithm) or label the floor and then check the numbers of lines (by using Trémaux's algorithm). However, an important question concerns the issue of using simple non-programmable objects for maze solving.

Recently, we have shown that oil droplets can self-propel (by picking up momentum due to the thermodynamic setting of the environment of the drops) and thus follow chemical gradients mimicking the chemotactic behaviour of living cells [5, 10]. While it is implausible to assume that simple droplets could be programmed to execute any of the aforementioned algorithms, their chemotactic behaviour can be exploited to the similar end. It turns out that if a chemical signal is placed at the exit, the chemotactic droplets are capable of solving the maze by following the shortest path predefined by the concentration gradient. In fact, diffusive and Marangoni flows can be used to construct 'physical' maze solvers [1]. We will discuss this problem more in detail in Sect. 2 where we focus on *maze solving by active particles*, and Sect. 3 which outlines *maze solving by passive particles*.

2 Maze Solving by Active Particles

Here, we introduce methods for maze solving using active particles that can sense their environment and respond to environmental stimuli. Active particle in this context is an autonomous self-propelled object that can convert energy from the environment into directed or persistent motion. Moreover, such an object has the ability to behave chemotactically. *Chemotaxis* is an oriented movement of cells or animals in concentration gradients and it is universal process by exploited by animate matter in nature. It has been found that non-living objects also exhibit similar properties and such motion is usually referred to as *artificial chemotaxis*. It has been shown that there are several chemical systems that exhibit such behaviour. We focus on organic droplets at the liquid-air interface that perform oriented chemotactic movement towards the source of "chemoattractant", i.e., they implement positive artificial chemotaxis [7].

We have shown that microliter sized decanol droplets suspended at the liquid-air interface in decanoate water solution perform a self-propelled motion and follow salt or hydroxide concentration gradients [5]. The decanol droplet in a homogeneous solution of decanoate exhibits a weak random motion. However, when a source of salt is added, after an induction period, it starts to move directionally and follows the salt concentration gradient. The salt concentration gradient in decanoate solution establishes a significant surface tension gradient (with a surface tension difference of ~40 mN/m between no salt and saturated salt cases) at the liquid-air interface

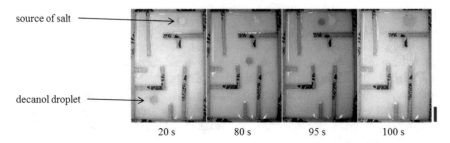

source of salt

decanol droplet

20 s   80 s   95 s   100 s

Fig. 1 Maze solving by a decanol droplet (5 μL, purple colour) in decanoate water solution (10 mM). The exit represented by a stationary nitrobenzene droplet containing solid sodium chloride salt (yellow colour). Nitrobenzene droplet is immiscible with water and shows no self-propelled motion, thus allows a gradual leaching (diffusion) of salt ions to the surrounding solution creating a concentration gradient around the sodium chloride crystal. The maze was designed by using a microscope slide and adhesive double sided tape (3M) for walls of the maze. Scale bar at the bottom right represents 1 cm. (Reprinted with permission from [5]. Copyright 2014 American Chemical Society.) See supplementary movie http://youtu.be/P5uKRqJIeSs

and it generates surface tension difference between the leading and the trailing edge of the oil droplet. In turn, this surface tension difference induces a fluid flow inside the droplet as well as in the bulk water phase contributing to the directional and self-propelled motion of the droplet.

There exist other chemical systems exhibiting a similar chemotactic behaviour. In particular, organic droplets (mineral oil or dichloromethane) containing fatty acid, 2-hexyldecanoic acid (HDA), are able to move towards the acidic region (pH $\sim$ 1) [10]. The mechanism of the self-propelled motion is similar to the one described above for the decanol droplet but, in this case, the surface tension difference is due to the pH gradient that affects the protonation rate of fatty acid molecules.

Placing the chemoattractant at the exit of the maze and the self-propelled droplet at the entrance, results in a maze-solving ability of the chemotactic droplets (see Figs. 1 and 2). The droplets follow the shortest path predefined by concentration gradient of the chemoattractant (salt or pH); the diffusion and Marangoni flows represent the physical maze solvers in this setting. The chemoattractant, i.e., the chemical substance placed at the exit, diffuses and induces a surface tension gradient which, in turn, drives the Marangoni flow. If the droplet is placed anywhere in the maze, it follows the steepest gradient and reaches the source of the chemical signal. The unique property of these chemotactic droplets is not the ability to find the target, but to follow the track leading to the target, because not all droplets have this ability to chemotactically follow the chemical gradient [6]. The efficiency (solution time) of maze solving scales with the intensity of generated Marangoni flow, which is somehow proportional to the length of the shortest path.

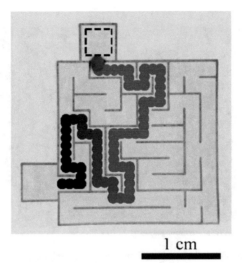

1 cm

Fig. 2 Maze solving by an organic droplet (1 μL) containing 20% of 2-Hexyldecanoic acid in 0.05 M alkaline solution of potassium hydroxide. Maze was fabricated from polydimethylsiloxane (PDMS) using photolithography (with thickness and depth of 1.4 and 1 mm, respectively). The images were created by overlaying experimental images at different times (spheres indicate the position of the droplet at different times), and the time window for this process is 1 min. (Reprinted with permission from [10]. Copyright 2010 American Chemical Society)

3 Maze Solving by Passive Particles

Another approach to exploring mazes replies on utilizing the global fluid flow in the channel network. If one can induce and maintain a fluid flow in the maze with the largest velocities between the entrance and the exit of maze, the shortest path can be found by following (visualizing) this flow. There exist several approaches to establishing this kind of fluid flow. For example, it has been reported that applying a pressure difference between the entrance and the exit of a maze can generate a fluid flow through the microfluidic channel network [9].

On the other hand, Marangoni flows can also be generated at the liquid-air interface of a 2D channel network. Marangoni flow results in a mass transfer of the liquid medium at the liquid-air interface due to a surface tension difference (surface tension gradient). The fluid flows from the lower surface tension region to the greater surface tension region at the interface; consequently, due to the conservation of mass, the near surface flow induces a reverse flow at the bottom of the channel.

In a typical experiment, we filled the maze with an alkaline solution (pH = 11) containing fatty acid (HDA) with a concentration of 0.2%. In this pH the head group of the fatty acid molecules are deprotonated, and they are oriented at the liquid-air interface and the deprotonated form of HDA acts as surfactant (reducing the surface tension at the liquid-air interface). To generate and maintain the Marangoni flow in the maze we placed either a small acidic hydrogel block (pH ∼ 1) or a cold small

(a)   **(b)**

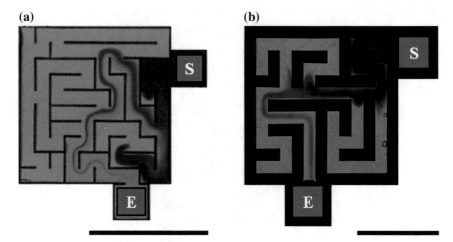

Fig. 3 Maze solving by induced Marangoni flow at the liquid-air interface. The letter E indicates the exit of the maze, **a** gel soaked with acid and **b** cold small stainless steel sphere. The letter S (as start) shows the entrance of the maze, where phenol red dye particles are added. In a typical experiment, the shortest path can be found and visualized in the order of minutes. Scale bars correspond to 1 cm. (Reprinted **a** from [18] under Creative Commons Attribution 4.0 International Public License and **b** from [11] with permission of The Royal Society of Chemistry)

stainless steel sphere to the exit, the temperature difference between the exit and the starting point was 60 K, which translates into 20.0 mN/m surface tension difference. [11, 18]. After the addition of this object, a small amount (∼0.3 mg) of dry phenol red dye powder was placed at the liquid-air interface at the starting point (entrance). These small dye particles act as passive tracers and can be seen travelling towards the exit at the liquid-air interface (low pH and low temperature regions). The dye particles transported by the induced Marangoni flow gradually dissolve in the liquid phase and the colour showed their paths through the maze (Fig. 3).

4   Conclusions

We have outlined several mechanisms through which simple droplets of organic solvents can efficiently navigate their way through a maze based on chemical signals. Such droplets actively take part in the maze-solving process in the sense that they self-propel autonomously in the direction of a concentration gradient of a salt or pH. This movement is referred to as artificial chemotaxis. Our approach presented here is similar to the artificial potential fields algorithms, in which the motion planning utilizes a potential field providing attraction to the goal and repulsion from obstacles. We have demonstrated that passive particles are also able to solve a maze utilizing the Marangoni flow. Marangoni flows can be induced either by a pH or a temperature gradient between the entrance and the exit of the maze. The corresponding flow velocities are the most intense along the shortest path connecting the entrance to the exit; therefore, the most passive particles are dragged along this way. If the particles dissolve during their travel, they can also aid path visualisation; such particles also help determine other paths solving the maze, and these longer paths are associated with less intense colour since less particles move along those paths.

Acknowledgements J. Č. was financially supported by the Czech Science Foundation (Grant No. 17-21696Y). Other authors acknowledge the financial support of the Hungarian Research Fund (OTKA K104666). Financial support for R. T. by the Marie Heim-Vogtlin Program under project no PMPDP2-139698 is gratefully acknowledged. D. U. and I. L. gratefully acknowledge the financial support of the National Research, Development and Innovation Office of Hungary (TÉT12JP-1-2014-0005).

References

1. A. Adamatzky, Hot ice computer. Phys. Lett. A **374**, 264–271 (2009)
2. A. Adamatzky, Slime mold solves maze in one pass, assisted by gradient of chemo-attractants. IEEE Trans. Nanobiosci. **11**, 131–134 (2012)
3. A. Adamatzky, Physical maze solvers. All twelve prototypes implement 1961 Lee algorithm, in *Emergent Computation: A Festschrift for Selim G. Akl*, ed. by A. Adamatzky (Cham, Springer International Publishing, 2017), pp. 489–504
4. A. Braun, R. Tóth, I. Lagzi, Künstliche Intelligenz aus dem Chemiereaktor. Nachr. Chem. **63**, 445–446 (2015)
5. J. Čejková, M. Novák, F. Štěpánek, M.M. Hanczyc, Dynamics of chemotactic droplets in salt concentration gradients. Langmuir **30**, 11937–11944 (2014)
6. J. Čejková, T. Banno, F. Štěpánek, M.M. Hanczyc, Droplets as liquid robots. Artif. Life **23**, 528–549 (2017)
7. J. Čejková, S. Holler, N.T. Quyen, C. Kerrigan, F. Štěpánek, M.M. Hanczyc, Chemotaxis and chemokinesis of living and non-living objects, in *Advances in Unconventional Computing*, ed. by A. Adamatzky (Springer, 2017), pp. 245–260
8. A.E. Dubinov, A.N. Maksimov, M.S. Mironenko, N.A. Pylayev, V.D. Selemir, Glow discharge based device for solving mazes. Phys. Plasmas **21**, 093503 (2014)
9. M.J. Fuerstman, P. Deschatelets, R. Kane, A. Schwartz, P.J.A. Kenis, J.M. Deutch, G.M. Whitesides, Solving mazes using microfluidic networks. Langmuir **19**, 4714–4722 (2003)
10. I. Lagzi, S. Soh, P.J. Wesson, K.P. Browne, B.A. Grzybowski, Maze solving by chemotactic droplets. J. Am. Chem. Soc. **132**, 1198–1199 (2010)
11. P. Lovass, M. Branicki, R. Tóth, A. Braun, K. Suzuno, D. Ueyama, I. Lagzi, Maze solving using temperature-induced Marangoni flow. RSC Adv. **5**, 48563–48568 (2015)
12. T. Nakagaki, H. Yamada, A. Tóth, Maze-solving by an amoeboid organism. Nature **407**, 470–470 (2000)
13. T. Nakagaki, H. Yamada, A. Tóth, Path finding by tube morphogenesis in an amoeboid organism. Biophys. Chem. **92**, 47–52 (2001)
14. Y.V. Pershin, M. Di Ventra, Solving mazes with memristors: a massively parallel approach. Phys. Rev. E **84**, 046703 (2011)
15. D.R. Reyes, M.M. Ghanem, G.M. Whitesides, A. Manz, Glow discharge in microfluidic chips for visible analog computing. Lab Chip **2**, 113–116 (2002)
16. O. Steinbock, A. Tóth, K. Showalter, Navigating complex labyrinths: optimal paths from chemical waves. Science **267**, 868–871 (1995)
17. O. Steinbock, P. Kettunen, K. Showalter, Chemical wave logic gates. J. Phys. Chem. **100**, 18970–18975 (1996)

18. K. Suzuno, D. Ueyama, M. Branicki, R. Tóth, A. Braun, I. Lagzi, Maze solving using fatty acid chemistry. Langmuir **30**, 9251–9255 (2014)
19. Y. Yu, G. Pan, Y. Gong, K. Xu, N. Zheng, W. Hua, X. Zheng, Z. Wu, Intelligence-augmented rat cyborgs in maze solving. PLoS ONE **11**, e014775 (2016)

The Electron in the Maze

Simon Ayrinhac

Abstract A physical method to solve a maze using an electric circuit is presented. The temperature increase due to Joule heating is observed with a thermal camera and the correct path is instantaneously enlightened. Various mazes are simulated with Kirchhoff's circuit laws. Finally, the physical mechanisms explaining how the electric current chooses the correct path are discussed.

1 Resolving Mazes with Electricity

1.1 Preliminary Considerations About Electrical Circuits and Thermography

The maze-solving problem and the shortest path problem are inspiring problems in algorithmics and they involve many fields of science, such as robotics or optimization. In addition to numerical methods, many experimental methods have been proposed to solve these problems, including fluids [1], memristors [2], living organisms (ants [3], honey bees [4], amoeba or "blobs" [5], nematodes [6], plants [7]) or plasma [8]. In this chapter, a solution by a simple physical method using an electric current is proposed.

First, mazes and labyrinths should be distinguished. Labyrinths have only one way, which is very complicated and which generally leads to the center, as can be seen in drawings on the floor of several cathedrals (see Fig. 1). In contrast, mazes

S. Ayrinhac (✉)
Sorbonne Université, Muséum d'Histoire Naturelle,
UMR CNRS 7590, IRD, Institut de Minéralogie,
de Physique des Matériaux et de Cosmochimie,
IMPMC, 75005 Paris, France
e-mail: simon.ayrinhac@sorbonne-universite.fr

© Springer International Publishing AG, part of Springer Nature 2018
A. Adamatzky (ed.), *Shortest Path Solvers. From Software to Wetware*,
Emergence, Complexity and Computation 32,
https://doi.org/10.1007/978-3-319-77510-4_16

409

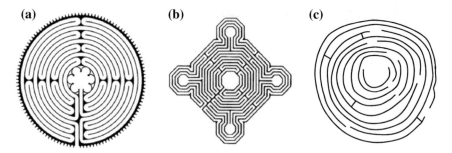

Fig. 1 Examples of labyrinths: **a** The labyrinth on the floor of the cathedral at Chartres (France); **b** the logo of *Monuments historiques* (national heritage sites) in France; **c** a handwritten labyrinth that was designed according to the intriguing instructions "You have two minutes to design a maze that takes one minute to solve". Reproduced from the *Inception* movie (real. C. Nolan, 2010)

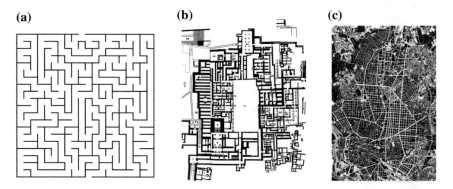

Fig. 2 Examples of mazes: **a** a computer generated maze; **b** a plan of the Palace of Knossos (now a ruin near the town of Heraklion in Crete), the historical location of the myth of the Minotaur; **c** map of a city with ways (streets and avenues)

possess a complex branching (see Fig. 2). Although labyrinths are fascinating from a symbolic point of view, mazes are more interesting.[1]

This chapter presents a simple physical method to solve a maze, using an electric current. The maze can be done by an electric circuit that is constituted by, for example, copper tracks printed on an epoxy card [9].

Basically, two points of the maze are connected with a battery: if the entrance and the output of the maze are connected, then the electric current flows and the maze is solved. If they are unconnected, then the circuit is open and no current flows. A simple ohmmeter (usually a multimeter in a particular mode) gives the answer: if the resistance between two points is very low, then the path is continuous; in contrast, if the resistance is very high, then the path is broken. However, in this method, the exact path followed by the current is unknown.

---

[1]The title of this chapter is a tribute to the American sci-fi writer Robert Silverberg and his novel "The Man in the Maze".

Thermography is a contactless and nondestructive method that can reveal the good path. The power P dissipated by a resistor, with electrical resistance R, is given by Joule's law

$$P = RI^2. \tag{1}$$

For a resistor obeying Ohm's law $U = RI$, where U is the voltage, the electrical energy provided by the battery is integrally converted into heat. When the electric charges flow, the temperature increase in the tracks is due to Joule heating.

The temperature increase ΔT is limited by thermal losses in the circuit. A first origin of thermal losses is conduction, which depends on the surrounding materials and the contact areas (controlled by the size of the circuit). A second origin is radiation produced by a hot body. A third origin is the convective heat transfer between an object and the surrounding fluid—in this case, the atmosphere. Given that the radiation heat transfer is negligible at low temperature, the following simple scaling law is relevant for a standard circuit on printed circuit board (PCB) [10]

$$\Delta T \propto I^2. \tag{2}$$

The increase in temperature is visualized by a thermal camera that detects infrared radiation (IR).

Thermal cameras are often used for educational purposes [11–15] to provide a clear visualization of invisible phenomena or to illustrate complex phenomena. There a wide range of topics in physics [16] or in chemistry [17] where a thermal camera may come in handy. With this kind of apparatus, qualitative as well as quantitative applications are possible. Although prices have decreased significantly in recent years, thermal cameras are still rather expensive. However, there are other devices suitable for thermal imaging applications: such as a simple webcam with an IR filter [18] or a smartphone-based device such as FLIR ONE or Seek Thermal.

The main purpose of an infrared camera is to convert IR radiation intensity in a temperature measurement and to show the spatial variations in a false-color visual image. Intensity is integrated from a spectral band, generally in the long-wave infrared (7.5–13 μm). The temperature is given by a formula which takes into account three phenomena [16]: the true thermal emission from the object, the thermal radiation emitted by its surroundings and reflected by the object, and the atmospheric absorption. A suitable temperature measurement needs the knowledge of some parameters, for example the emissivity, humidity, distance and ambient temperature.

Thermal imaging can find wide application in electronics. For example, electrical components in the microelectronic boards of computers produce heat that can damage the circuits. To avoid failures, processors or power transistors need to be cooled by fans or Peltier modules, for example. IR imaging is a non-contact and non-destructive technique that can be used to test and survey electronic boards, allowing a diagnostics of possible malfunctions. Given that these boards are often made of different materials, the differences in components emissivity make quantitative temperature measurements difficult (an explanation of emissivity will be given later).

Temperature measurement depends strongly on the emissivity ε of materials. Unfortunately, for metals, the emissivity is very low, and they are hard to see directly in thermography. Emissivity is defined as the ratio of the amount of the radiation emitted from the surface to that emitted by a blackbody at the same temperature [16]. A blackbody is a perfect absorber for all incident radiations. It appears black when cooled at 0 K and when heated up it emits light at all wavelengths and the resulting spectrum (given by Planck's law [16]) depends only on the temperature of the blackbody. Kirchhoff's law of thermal radiation states that $\varepsilon = \alpha$ where α is the absorption coefficient [19]. This formula means that, for an opaque body, the more a body absorbs, the more it emits light. A metal is a good reflector, so it has a bad absorption and, therefore, a poor emissivity.

1.2 Experiments on Circuits

A regular maze (see Fig. 3) is printed on a epoxy card with tracks made of copper. A transparent plastic sheet, which has a higher emissivity contrary to metal, is placed over the circuit to ensure that the temperature increase is seen by thermography. The transparent cover sheet allows the maze to be seen in both in infrared light and visible light. With a thermal camera, the correct track appears to be immediately illuminated despite the complexity of the circuit (see Fig. 4).

Our maze is designed to highlight the following special features (see Fig. 5):

- In case of branching with a path twice as long as the parallel branch, the shorter path appears to be more brightly illuminated compared to the longer path. This happens because the trace resistance R is proportional to the length of the resistor ℓ, such as $R = \ell/\sigma A$, where σ is the electrical conductivity and A the

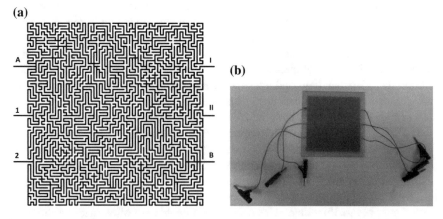

Fig. 3 a The maze used in the experiment; **b** the maze on a printed circuit board with dimensions 15×15 cm^2. The conductive copper tracks have a thickness of 35 μm and a width of 800 μm. The circuit is covered by a plastic transparent sheet

(a)

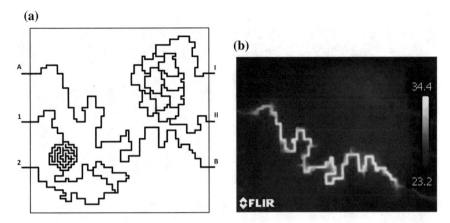

Fig. 4 **a** The studied maze without dead ends; **b** thermal image (320 × 240 pixels) of the circuit connected to the battery at points labelled (A) and (B). The infrared light captured by the camera immediately shows the correct path! The colour bar on the right of the image is the temperature scale in degrees Celsius. This scale is calculated with an emissivity parameter of 0.95. The spectral range of the camera is in the long-wave (LW) region, i.e. 7.5–13 μm. Note that the minimum temperature on the scale is not the room temperature

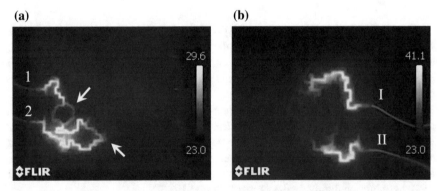

Fig. 5 **a** The battery is connected to the circuit by the points labelled (1) and (2). This picture demonstrates two effects: the difference between two paths with a path with double length compared to another (lower arrow) and the Wheatstone Bridge (upper arrow). **b** The battery is connected to the circuit by the points labelled (I) and (II). Because there are many good branches with equal lengths, it is difficult to identify the shortest path

cross-sectional area. The voltage U is equal in the two branches and gives $I_\ell = 2I_{2\ell}$, so with Eq. (2), the temperature increase in the shortest path is four times higher than in the longer path $\Delta T_\ell = 4\Delta T_{2\ell}$.

- If the branching is configured as a Wheatstone bridge, then the parallel branch does not appear. See Fig. 11 and the associated text for explanations.
- In case of multiple paths, the branching is complicated and the shortest path is hard to be seen.

(a)   **(b)**

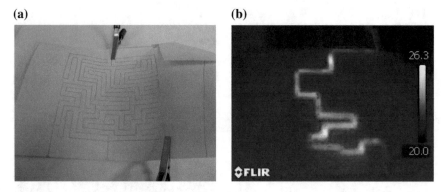

Fig. 6 **a** A maze is drawn on a paper sheet with a pen delivering conductive ink (around 1 Ω/cm). It takes about 10 min to draw the whole maze. This maze was previously presented in detail in Ref. [9]. Despite the care taken in the drawing, the tracks are not perfectly regular and the paper is much more fragile compared to a PCB, and can be torn easily; **b** the correct path appears illuminated with an infrared camera. The temperature increase is clearly seen in this case because the paper has a better emissivity compared to metallic conductive tracks. Due to the sideways spreading of the heat the correct path in the image looks "blurred". This method is cheaper and faster compared to printing the same maze on a PCB

This kind of demonstration is possible provided that several conditions are met: the tracks should have the same section and they should be built with the same material, the branching should not be too complex (i.e., one-solution mazes) and the correct path should exist among many dead ends. So, the ideal circuit is an intermediate between a labyrinth and a maze.

The circuit can be drawn by an ink pen on a paper sheet (see Fig. 6). However, despite the care taken in the drawing, the tracks are not perfectly regular and the paper is much more fragile than a PCB and can be torn easily. Nevertheless, this method is cheaper and faster than printing the same maze on a PCB. To further reduce costs, the IR camera can be replaced by a temperature sensitive liquid crystal film (around 15$), to obtain qualitatively the same result (see Fig. 7).

1.3 Simulated Circuits

To investigate more complex topologies, various circuits were simulated in the permanent regime using Kirchhoff's laws [20]: the algebraic sum of currents at a node is zero (Kirchhoff node rule), and the directed sum of the voltages around a loop is zero (Kirchhoff loop rule). The operation involves a solution of a linear system with n equations, involving resistances, currents and applied voltages, where n is the number of branches in the electrical network.

(a)   **(b)**

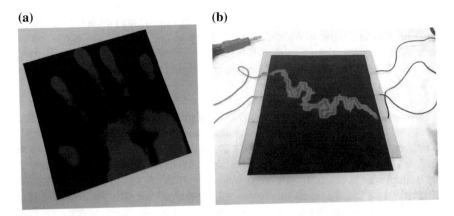

Fig. 7 **a** A thermochromic liquid crystal film (the size of the sheet is $15 \times 15 \, cm^2$), sensitive to temperature, with a hand print. The transition from black to color occurs between 20–25 °C. The colour change is reversible and quick, with a response time about 10 ms [21]. **b** The correct path of the maze appears illuminated with the liquid crystal film placed on the PCB

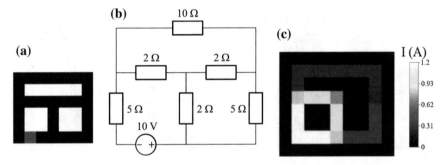

Fig. 8 **a** A circuit drawn in an image file (7×6 pixels), imitating copper tracks on a PCB. **b** The equivalent electrical circuit. **c** Distribution of the intensities obtained by the application of Kirchhoff's laws (see text). The color bar at the right indicates the intensity values in each branch

The studied circuits are directly generated by drawing the tracks in an image file (see the example in Fig. 8). The battery voltage is 10 V and the track electrical resistance is 1 Ω by unit of length (equal to the track width). This resistance value is arbitrary. The nodes are not be taken into account in the calculation of resistance. In the nodes, the current is calculated by the averaging of the surroundings currents. The resulting picture represents the current I in amperes at each point of the circuit. Note that the resulting picture is not a thermography image rendering.

In a grid-like circuit (Fig. 9), the current is spread over the whole circuit. In this case, all the paths are equivalent and the current appears equal in all of the paths. In another example with disordered tracks (Fig. 10), the shortest path appears clearly. In some cases, the current can fall to zero in a part of the circuit. This is due to the creation of an "electrical bridge", also called Wheatstone bridge, as illustrated in the

Fig. 11. The process to follow the shortest path between two points connected by the battery is to choose at each node the branch where the intensity is maximum. Generally speaking, the shortest path is the path where the intensity is maximized. This idea is sustained by the basic electric conception that more current follows the path of less resistance.

The resistive grid was early used to explore some physical problems, such as solution of partial differential equations [22], or mobile robot path planning [23]. In robot path planning, a collision-free environment can be modelled with a resistive grid of uniform resistance, and obstacles are represented by regions of infinite resistance. The path planning can be evaluated in real space if the robot moves through a maze, for example, or in the configurational space where the dimensions are the degrees of freedom of the robot, considering a robot manipulator arm, for example. The path from start to goal is found using voltage measurements from successive nodes. In the limit of the continuous case, the electromagnetism equations imply that for steady currents in regions with no sources the voltage obeys Laplace's equation, if we

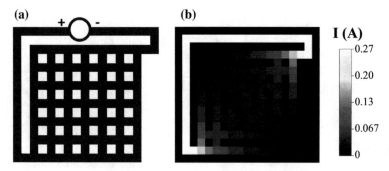

Fig. 9 **a** A grid-like circuit. The battery is located in the upper branch. **b** Distribution of the intensities inside the studied circuit. The color bar at the right indicates the intensity values in each branch

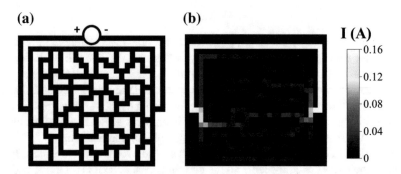

Fig. 10 **a** A circuit maze with disordered tracks. The battery is located in the upper branch. **b** Distribution of the intensities inside the studied circuit. The color bar at the right indicates the intensity values

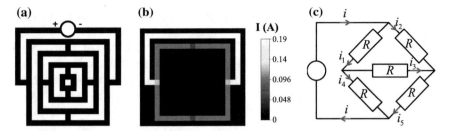

Fig. 11 **a** A complex circuit. The battery is located in the upper branch. **b** In the simulated circuit, the central part does not appear. This is due to the creation of an electrical "bridge". **c** Diagram of an electrical circuit containing an electrical "bridge". Resistors and currents are labelled R and i_n, respectively. If the circuit is well balanced, all R are equal, and it can be demonstrated that $i_3 = 0$. This configuration is called a Wheatstone bridge and it is often used to measure resistance

assume that conductivity is uniform and constant. The two-dimensional Laplace's equation

$$\frac{\partial^2 V(x, y)}{\partial x^2} + \frac{\partial^2 V(x, y)}{\partial y^2} = 0 \tag{3}$$

may be solved to calculate the value of the voltage V at every point (x, y). The direction of the movement is given locally by the direction of the voltage gradient ∇V. Globally, this approach produces an optimal path solution, depending on the limit conditions to avoid spurious local minima.

2 Discussion of the Physical Mechanisms

After the demonstration, a physical question remains: How does the electric current choose the correct path amongst many others?

A common explanation is that the battery produces a potential difference ΔV at the two extremities of the circuit, and the resulting electric field **E** possesses a magnitude constant and a direction along the wire. This is especially puzzling in a maze circuit, where the electric field must follow the multiple bends of the circuit. In contrast, the electric field **E** and then the potential difference ΔV is produced by distributions of point charges. So where are the charges producing the electric field inside the wires? This question is challenging because *electrokinetics* and *electrostatics* are two topics that are usually treated separately in physics textbooks; charges distributions on a one side, and electrical circuits on the other side. Consequently, for most students the two topics are unconnected, leading to many misconceptions [24]; that is, commonly held beliefs that have no basis in actual scientific knowledge.

This point was extensively examined in the literature [25–29]. The electric charges responsible for the electric field inside the conductor are located on the surfaces of the wire. This fact is known from the pioneering works of Weber and Kirchhoff [30], but it has been completely forgotten during the last 150 years. The quantity of surface charge is very small: the order of magnitude of the charge necessary to turn an electric current of 1 A around a corner is equal to about the charge of one electron [25]. A quantitative estimate in a typical circuit [29] for the magnitude of the surface charge density is 10^{-12}–10^{-10} C m^{-2}. Comparatively, the quantity of charge moving inside the wires is much higher, about 10^{-6} C m^{-2}s^{-1}, which corresponds to 1 A.

Although the quantity of surface charges is small, their role is essential [27]: they ensure that the equality of the potential in the conductors is at equilibrium, they permit the circulation of the charges and they produce an electric field outside the wires. Two types of surface charges can be distinguished [29]: at the boundary of two conductors with different resistivities and at the surface of the conductors.

The free electrons in the metal are pushed by the electric force arising from the electric field. If there is a curve, they pile-up on the surface, and their electric field changes the pathway of the incoming moving charges. There is a *feedback mechanism* between the surface charges and the charges moving inside. The book of Chabay and Sherwood [20] provides an excellent and very accessible overview of this problem for undergraduate students.

This feedback mechanism explains how the charges avoid dead-ends of the circuit maze. First, they pile-up on the extremity of the dead-end; the build-up of negative charge pushes the arriving electrons, and then the flowing current reaches zero. Finally, the only path left for the charges to follow is the solution path of the maze. This phenomenon is analogous to liquid propagating in a microfluidic network [1].

Electric circuits are often compared to hydraulic circuits from a pedagogical point of view (the most complete comparison can be found in Table 1 in Ref. [31]). However there are fundamental differences between electrons and water: electrons do not interact with one another, and energy is not carried by the free electrons. Energy is carried outside the circuit by the electromagnetic fields forming the Poynting vector $\mathbf{S} = \frac{1}{\mu_0}\mathbf{E} \times \mathbf{B}$. This formula combines the magnetic field $\mathbf{B}$ due to electric current inside the wires (moving charges) and the electric field $\mathbf{E}$ due to surface charges.

Solving the maze with an electric current reveals the existence of a transient period between the beginning of the experiment and the moment that the current is stabilized in the solved maze.[2] Simulations performed by Preyer in a simple RC circuit [28] can give us a better understanding of phenomena observed in the transient state. Just after the connection of the battery at two points of the maze, the electric field spreads through the circuit at the speed of light. During this step, the surface charges build on the tracks. The surface charges locally change the electric field and the current, as they influence each other. This feedback mechanism occurs in the transient state and is at work when the uniform current flow is established, which means that the maze is solved.

All of the changes occur at the speed of light c inside the material around the circuit (usually air), which is also the speed that the information propagates between different parts of the circuit. The drift velocity v of charges moving inside the wires

---

[2]In the following discussion, we do not consider the thermal equilibrium of the system, which requires more time than electric equilibration.

is much slower than c, typically a few microns a second. The simple propagation of the electric field through the whole circuit needs a time $\tau \approx \ell/c$, where ℓ is the characteristic length of the maze, but the time τ' needed for the feedback mechanism to operate is much longer [28] with $\tau' > 2\ell/c$, which can be considered as the minimum time needed to solve the maze.

With this physical method, the maze resolution is fast. This explains why this is considered to be the fastest and the cheapest method among many other physical methods [32], especially if the circuit is drawn by ink pen on a paper sheet.

Acknowledgements The thermal camera was provided by the "unité de Formation et de Recherche de Physique" at Sorbonne Université, Faculté des Sciences et Ingénierie. The author is indebted to M. Fioc for his valuable comments on the manuscript.

References

1. M.J. Fuerstman, P. Deschatelets, R. Kane, A. Schwartz, P.J. Kenis, J.M. Deutch, G.M. Whitesides, Solving mazes using microfluidic networks. Langmuir **19**(11), 4714–4722 (2003)
2. Y.V. Pershin, M. Di Ventra, Solving mazes with memristors: A massively parallel approach. Phys. Rev. E **84**, 046703 (2011)
3. L.O. Stratton, W.P. Coleman, Maze learning and orientation in the fire ant (solenopsis saevissima). J. Compa. Physiol. Psychol. **83**(1), 7 (1973)
4. S. Zhang, A. Mizutani, M.V. Srinivasan, Maze navigation by honeybees: Learning path regularity. Learn. Mem. **7**(6), 363–374 (2000)
5. T. Nakagaki, H. Yamada, Á. Tóth, Intelligence: Maze-solving by an amoeboid organism. Nature **407**(6803), 470–470 (2000)
6. J. Qin, A.R. Wheeler, Maze exploration and learning in c. elegans. Lab on a Chip **7**(2), 186–192 (2007)
7. A. Adamatzky, Towards plant wires. Biosystems **122**, 1–6 (2014)
8. D.R. Reyes, M.M. Ghanem, G.M. Whitesides, A. Manz, Glow discharge in microfluidic chips for visible analog computing. Lab on a Chip **2**(2), 113–116 (2002)
9. S. Ayrinhac, Electric current solves mazes. Phys. Educ. **49**(4), 443 (2014)
10. J. Adam, New correlations between electrical current and temperature rise in PCB traces, in *Twentieth Annual IEEE Semiconductor Thermal Measurement and Management Symposium, 2004* (IEEE, 2004), pp. 292–299
11. M. Vollmer, K.-P. Möllmann, F. Pinno, D. Karstädt, There is more to see than eyes can detect. The Phys. Teach. **39**(6), 371–376 (2001)
12. K.-P. Möllmann, M. Vollmer, Infrared thermal imaging as a tool in university physics education. Eur. J. Phys. **28**(3), S37 (2007)
13. C. Xie, E. Hazzard, Infrared imaging for inquiry-based learning. The Phys. Teach. **49**, 368 (2011)
14. J. Haglund, F. Jeppsson, E. Melander, A.-M. Pendrill, C. Xie, K.J. Schönborn, Infrared cameras in science education. Infrared Phys. Technol. **75**, 150–152 (2016)
15. E. Netzell, F. Jeppsson, H. Jesper, K. Schönborn, Visualising energy transformations in electric circuits with infrared cameras. Sch. Sci. Rev. **98**(364), 19–22 (2017)
16. M. Vollmer, K.-P. Möllmann, *Infrared Thermal Imaging: Fundamentals Research and Applications* (Wiley, 2011)
17. C. Xie, Visualizing chemistry with infrared imaging. J. Chem. Educ. **88**(7), 881–885 (2011)
18. N.A. Gross, M. Hersek, A. Bansil, Visualizing infrared phenomena with a webcam. Am. J. Phys. **73**, 986–990 (2005)

19. U. Besson, Paradoxes of thermal radiation. Eur. J. Phys. **30**(5), 995 (2009)
20. R.W. Chabay, B.A. Sherwood, *Matter and Interactions*, vol. 2 (Wiley, New York, 2011)
21. J. Stasiek, M. Jewartowski, T.A. Kowalewski, The use of liquid crystal thermography in selected technical and medical applications—Recent development. J. Cryst. Process Technol. **4**(1), 46–59 (2014)
22. G. Liebmann, Solution of partial differential equations with a resistance network analogue. Br. J. Appl. Phys. **1**(4), 92 (1950)
23. L. Tarassenko, A. Blake, Analogue computation of collision-free paths. In *Proceedings. 1991 IEEE International Conference on Robotics and Automation, 1991* (IEEE, 1991), pp. 540–545
24. S. Rainson, G. Tranströmer, L. Viennot, Students' understanding of superposition of electric fields. Am. J. Phys. **62**(11), 1026–1032 (1994)
25. W.G.V. Rosser, Magnitudes of surface charge distributions associated with electric current flow. Am. J. Phys. **38**(2), 265–266 (1970)
26. A.M. Heald. Electric fields and charges in elementary circuits. Am. J. Phys. **52**(6), 522–526 (1984)
27. J.D. Jackson, Surface charges on circuit wires and resistors play three roles. Am. J. Phys. **64**(7), 855–870 (1996)
28. N.W. Preyer, Transient behavior of simple RC circuits. Am. J. Phys. **70**, 1187–1193 (2002)
29. R. Müller, A semiquantitative treatment of surface charges in DC circuits. Am. J. Phys. **80**(9), 782–788 (2012)
30. A.K.T. Assis, J.A. Hernandes, *The Electric Force of a Current: Weber and the Surface Charges of Resistive Conductors Carrying Steady Currents* (Apeiron, 2007). This book is freely available at the following address: http://www.ifi.unicamp.br/~assis/
31. K.W. Oh, K. Lee, B. Ahn, E.P. Furlani, Design of pressure-driven microfluidic networks using electric circuit analogy. Lab on a Chip **12**(3), 515–545 (2012)
32. A. Adamatzky, Physical maze solvers. All twelve prototypes implement 1961 Lee algorithm. arXiv:1601.04672 (2016)

Maze Solvers Demystified and Some Other Thoughts

Andrew Adamatzky

Abstract There is a growing interest towards implementation of maze solving in spatially-extended physical, chemical and living systems. Several reports of proto-types attracted great publicity, e.g. maze solving with slime mould and epithelial cells, maze navigating droplets. We show that most prototypes utilise one of two phenomena: a shortest path in a maze is a path of the least resistance for fluid and current flow, and a shortest path is a path of the steepest gradient of chemoattractants. We discuss that substrates with so-called maze-solving capabilities simply trace flow currents or chemical diffusion gradients. We illustrate our thoughts with a model of flow and experiments with slime mould. The chapter ends with a discussion of exper-iments on maze solving with plant roots and leeches which show limitations of the chemical diffusion maze-solving approach.

1 Introduction

To solve a maze[1] is to find a route from the source site to the destination site. In [6] we reviewed experimental laboratory prototypes of maze solvers. We speculated that the experimental laboratory prototypes of maze solvers, despite looking different, use the same principles in their actions: mapping and tracing. A maze is mapped in parallel by developing chemical, electrical, or thermal gradients.[2] A path from a given source site to the destination site is traced in the mapped maze using living cells, fluid flows or electrical current. The traced paths are visualised with morphological structures of living cells, dyes, droplets, thermal sensing or glow-charge. The experimental laboratory maze solvers vary in their speeds substantially. The solvers based on glow-

---

[1] A labyrinth is a maze with a single path to an exit/destination.
[2] This is a material implementation of 1961 Lee algorithm, where each site of a maze gets a label showing a number of steps someone must make to reach the site from the destination site [21, 30].

A. Adamatzky (✉)
Unconventional Computing Lab, UWE, Bristol BS16 1QY, UK
e-mail: andrew.adamatzky@uwe.ac.uk

© Springer International Publishing AG, part of Springer Nature 2018
A. Adamatzky (ed.), *Shortest Path Solvers. From Software to Wetware*,
Emergence, Complexity and Computation 32,
https://doi.org/10.1007/978-3-319-77510-4_17

421

discharge [14, 28] or thermal visualisation of a path [10], and the solver utilising
crystallisation [2] produce the traced path in a matter of milliseconds or seconds.
Prototypes employing assembly of conductive particles [27], dyes [15], droplets [11,
19] and waves [8, 9] give us results in minutes. Living creatures—slime mould [4]
and epithelial cells [31]—require hours or days to trace the path.

Chemical, physical and living maze solvers are conventional examples of unconventional computers. In the present chapter we do not provide all technical details of
the experimental laboratory prototypes, these can be found in [6], but rather share
our thoughts on maze solvers in a context of unconventional computing and discuss
some experiments with inconclusive results.

Whilst mentioning 'unconventional computing' we might provide a definition
of the field. The field is vaguely defined as the computing with physical, chemical and living substrates (as if conventional computers compute with 'non-physical'
substrates!). In our recent opinion paper [7] unconventional computists provided
several definitions, e.g. challenging impossibilities (Cristian Calude), going beyond
discriminative knowledge (Kenichi Morita), intrinsic parallelism and nonuniversal-
ity (Selim Akl), and continuous computation (Bruce MacLennan). José Félix Costa
defines the unconventional computing as 'physics of measurement' which echoes
with our own opinion of the unconventional computing as an art of interpretation [3].
Take, for example, the famous, and still very much relevant, book by Stéphane Leduc
"Théorie physico-chimique de la vie et générations spontanées" published in Paris
in 1910. Not only did this book laid a foundation of the Artificial Life but somewhat contributed to the field of unconventional computing. Namely, have a look at
the Fig. 1a. This is a structure that emerged when Leduc placed drops on potassium
ferrocyanide on the gelatine gel. Neighbouring diffusing drops applied pressure to
each other and diffusion stopped at the bisectors between the drops. Leduc presented
this as a chemical model of multi-cellular formation. Unaware of the Leduc's experiments Adamatzky and Tolmachiev rediscovered a similar formation in 1996 and
reinterpreted it as a chemical processor which computes Voronoi diagram of a planar set of points [34]: the data points are represented by drops of potassium iodide
diffusing in a thin-layer agar with palladium chloride (Fig. 1b, c). These our historical
reminiscences smoothly flow into the next section of the chapter on fluid mappers.

2 Fluid Mappers. Shortest Path is a Path of the Least Hydrodynamic Resistance

In 1900 Hele-Shaw and Hay developed an analogy between stream-lines of a fluid
flow in a thin layer and the lines of magnetic induction in a uniform magnetic
field [16]: pressure gradient of a fluid flow is equivalent to magnetic intensity and rate
of the flow is analogous to magnetic induction. As Hele-Shaw and Hay wrote [16]:

> The method described is the only one hitherto known which enables us to determined the
> lines of induction in the substance of a solid magnetic body.

(a)

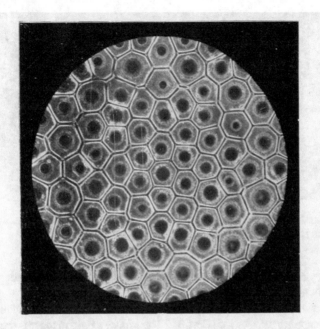

FIG. 7.— Tissu de cellules artificielles formé par la diffusion dans une solution
de gélatine à 10 °/₀ de gouttes d'une solution de ferrocyanure de potassium à 10°/₀.

(b)   **(c)**

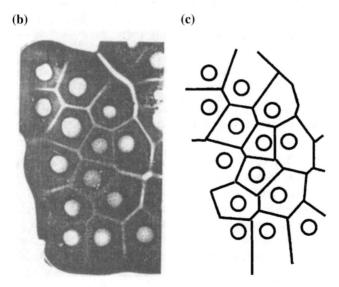

Fig. 1 Unconventional computing is an art of interpretation. **a** Cellular structure produced by
Stéphane Leduc in 1910 with drops of potassium ferrocyanide diffusing in gelatine [20]. **b, c** Chemical processor made by Adamatzky and Tolmachiev in 1996 [34]: photo of a completed reaction (**b**)
and corresponding Voronoi diagram (**c**)

Fig. 2 Stream-lines of a fluid flow around a domain with low permeability, designs were proposed in 1904. From [17]

In 1904 they applied their approach to solve a "problem of the magnetic flux distortion brought about by armature teeth" [17] (Fig. 2).

Hele-Shaw and Hay's idea was picked up by Arthur Dearth Moore who developed fluid flow mapping devices [25] (Fig. 3).[3] The Moore's fluid mapper is made of a cast slab, covered by a glass plate, with input (source) and output (sink) ports, fluid flow lines are visualised by traces from dissolving crystals of potassium permanganate or methylene blue. He shown that his fluid mappers can simulate electrostatic and magnetic fields, electric current, heat transfer and chemical diffusion [25]. This is a description of the mapper in Moore's own words [26]:

> When a given potential field situation is to be portrayed, the lower member of the fluid mapper is built to scale, with suitable boundaries, open or closed; islands, if any; one or more sources or sinks; and so on. Each source or sink is connected by a rubber tube to a tank, so that raising or lowering a tank will induce flow in the flow space. When the operation is conducted so that the flow is not affected by inertia, the flow pattern set up can quite accurately duplicate either the equipotential lines, or else the flux lines, of the potential field under consideration.

Moore mentioned 'islands', which could play a role of obstacles or even maze walls, when a collision-free shortest path is calculated or a maze solved, however, there is no published evidence that Moore applied his inventions to solve mazes. Maybe he did. The fluid mappers became popular, for a decade, and have been used to solve engineering problems of underground gas recovery and canal seepage.[4]

In 1952 Moore's method was applied to study current flow for various positions of electrocardiographic leads: an outline of a human body was made of a plaster

---

[3]Moore has also invented hydrocal, a hydraulic computing device for solving unsteady problem in heat transfer [24] at the same time when Luk'yanov's invented his famous hydraulic differential equations solver [22].

[4]http://quod.lib.umich.edu/b/bhlead/umich-bhl-851959?rgn=main;view=text.

Fig. 3 A short news story about Moore's works was published in "The Michigan Daily" newspaper on June 29, 1950 [1]. ©The Michigan Daily

and covered with a glass plate to allow only a thin layer of fluid inside, locations of a source and sinks of fluid flow corresponded to positions of electrocardiographic electrodes, variations of resistance of organs were modelled by varying the depth of the plaster slab [23] (Fig. 4a). In 1954 a fluid mapper was evaluated in designs of fume exhaust hoods [12]: it was possible to plot hood characteristics, stream, pressure and velocity lines with the help of the experimental fluid mapper (Fig. 4b).

First published evidence of experimental laboratory fluid maze solver is dated back to 2003. In a fluidic maze solver developed in [15] a maze is the network of micro-channels. The network is sealed. Only the source site (inlet) and the destination site (outlet) are open. The maze is filled with a high-viscosity fluid. A low-viscosity coloured fluid is pumped under pressure into the maze, via the inlet.

(b)

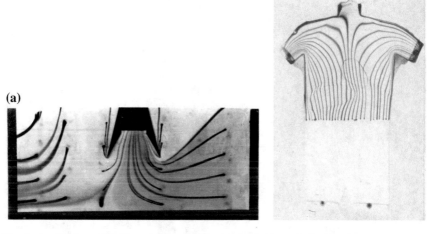

(a)

Fig. 4 Applications of fluid mappers. **a** Fluid mapper used in optimisation of a canopy exhaust hood in 1954 [12]. **b** Imitation of a current flow in a human body with a thin-layer fluid flow, with domains of low permeability corresponding to lungs and liver, the fluid enters the model from the left leg and leaves the model through the arms. The experiments are conducted in 1952. From [23]

Due to a pressure drop between the inlet and the outlet liquids start leaving the maze via the outlet. A velocity of fluid in a channel is inversely proportional to the length of the channel. High-viscosity fluid in the channels leading to dead ends prevents the coloured low-viscosity fluid from entering the channels. There is no pressure drop between the inlet and any of the dead ends. Portions of the 'filler' liquid leave the maze. They are gradually displaced by the colour liquid. The colour liquid travels along maximum gradient of the pressure drop, which is along a shortest path from the inlet to the outlet. When the coloured liquid fills the path the viscosity along the path decreases. This leads to an increase of the liquid velocity along the path. The shortest path—least hydrodynamic resistance path—from the inlet to the outlet is represented by channels filled with coloured fluid. Visualisation of the fluid flow indicating a shortest path in a maze is shown in Fig. 5. Fluids solve mazes at any scale, not just micro-fluidics, as has been demonstrated by Masakazu Matsumoto, where water explores the maze and milk traces the shortest path (Fig. 6) and in our own experiments with milk and coffee in a labyrinth (Fig. 7).

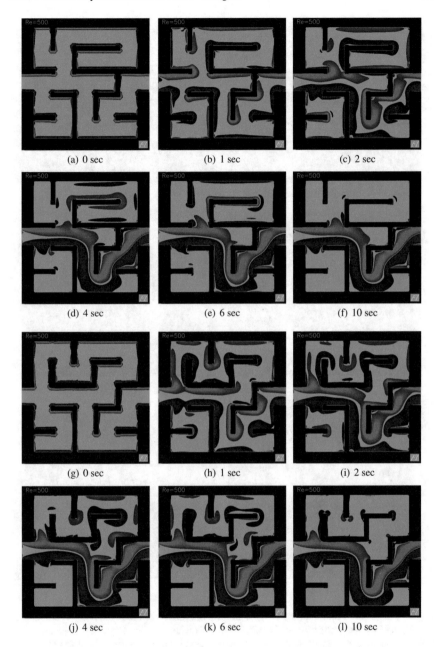

(a) 0 sec (b) 1 sec (c) 2 sec

(d) 4 sec (e) 6 sec (f) 10 sec

(g) 0 sec (h) 1 sec (i) 2 sec

(j) 4 sec (k) 6 sec (l) 10 sec

Fig. 5 Fluid flow through the maze. Entrance is on the left, exit is on the right. Labyrinth is generated in Maze Generator http://www.mazegenerator.net/ and modified to maze. Flow simulation is done in Flow Illustrator http://www.flowillustrator.com/ for visual flow control $dt = 0.01$ and Reynolds number 500. Maze is black, red coloured areas are part of fluid making clockwise rotation and green coloured areas—counter-clockwise. See videos of these computational experiments https://youtu.be/FUBYr3cOoC8 (labyrinth) and https://youtu.be/0jFPXBhQBS0 (maze). Time shown as per video generated by the Flow Illustrator

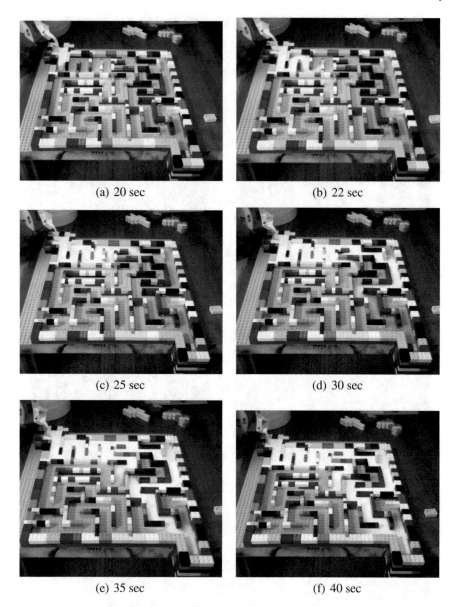

(a) 20 sec   (b) 22 sec

(c) 25 sec   (d) 30 sec

(e) 35 sec   (f) 40 sec

Fig. 6 Snapshots of milk and water solving maze. Snapshots from the video of experiments by Masakazu Matsumoto https://youtu.be/nDyGEq_ugGo who used one Lego 6177, one Lego 628, half-a-litre of milk and two litres of water. Printed with kind permission from Masakazu Matsumoto

(a)   **(b)**

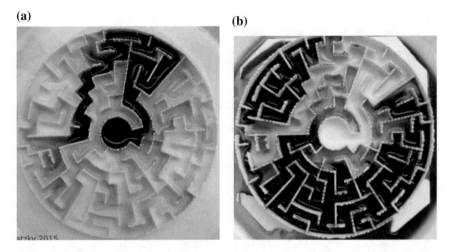

atzky 2015

Fig. 7 Labyrinth solving with coffee and milk: **a** the path is traced by coffee, **b** the path is traced by milk

3 Electrical Mappers. Shortest Path Is a Path of the Least Electrical Resistance

Approximation of a collision-free path with a network of resistors was first proposed in [32, 33]. A space is represented as a resistor network, obstacles are insulators. An electrical power source is connected to the destination and the source sites. The destination site is the electrical current source. Current flows in the network but does not enter obstacles. A path can be traced by a gradient descent in electrical potential. That is for each node a next move is selected by measuring the voltage difference between the current node and each of its neighbours, and moving to the neighbours which shows maximum voltage. As shown by Simon Ayrinhac (originally in [10], a shortest path can be visualised without discretisation of the space. A maze is filled with a continuous conductive material. Corridors are conductors, walls are insulators. An electrical potential difference is applied between the source and the destination sites. The electrical current 'explores' all possible pathways in the maze. An electrical current is stronger along the shortest path. Local temperature in a locus of a conducting material is proportional to a current strength through this locus. A temperature profile can be visualised with thermal camera [10] or glow-discharge [28] or temperature sensitive liquid crystal sheets (Fig. 8).

(a)   (b)

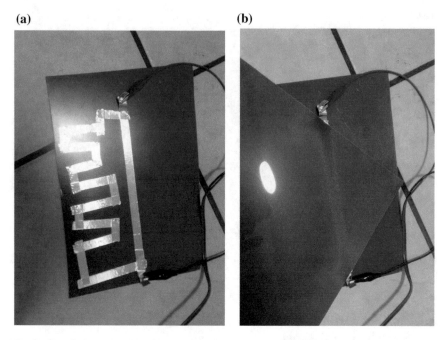

Fig. 8 Calculating a shortest path with electrical current. **a** Circuit. **b** Visualisation of a shortest path with the temperature sensitive liquid crystal sheet (Edmund Optics Inc., USA). Current applied through is 3.2 A.

4 Diffusion Mappers. Shortest Path Is a Path of the Steepest Gradient of Chemoattractants

A source of a diffusing substance is placed at the destination site. After the substance propagates all over the maze a concentration of the substance develops. The concentration gradient is steepest towards the source of the diffusion. Thus starting at any site of the maze and following the steepest gradient one can reach the source of the diffusion. The diffusing substance represents one-destination-many-sources shortest paths. To trace a shortest path from any site, we place a chemotactic agent at the site and record its movement towards the destination site. There are three experimental laboratory prototypes of visualising a shortest path in a diffusion field: by using travelling droplets, crawling epithelial cells and growing slime mould.

A path along the steepest gradient of potassium hydroxide has been visualised by István Lagzi and colleagues with a droplet of a mineral oil or dichloromethane mixed with 2-hexyldecanoic acid [19]. Daniel Irimia and colleagues used epithelial

(a) **(b)**

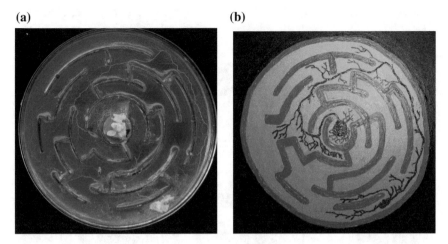

Fig. 9 A path between central chamber of exit of a maze is represented by thickest protoplasmic tube. **a** Photo of experimental setup with the maze solved. **b** Painting of the setup where the path to the maze's central chamber is more visible

cells to visualise the steepest gradient of the epidermal growth factor [31]. Let us discuss our own experiments on visualising a path in a maze with slime mould.

The slime mould maze solver based on chemo-attraction is proposed in [4]. An oat flake is placed in the destination site. The slime mould *Physarum polycephalum* is inoculated in the source site. The oat flakes, or rather bacterias colonising the flake, release a chemoattractant. The chemo-attractant diffuses along the channels (Fig. 10). The slime mould explores its vicinity by branching protoplasmic tubes into openings of nearby channels. When a wave-front of diffusing attractants reaches the slime mould, the cell halts its lateral exploration. The slime mould develops an active growing zone propagating along the gradient of the attractant's diffusion. The problem is solved when the slime mould reaches the source site. The thickest tube represents the shortest path between the destination site and the source site (Fig. 9). Mechanisms of tracing the gradient by the slime mould are confirmed via numerical simulation a two-variable Oregonator partial-differential equations in a two-dimensional space (Fig. 11). Not only nutrients can be placed at the destination site

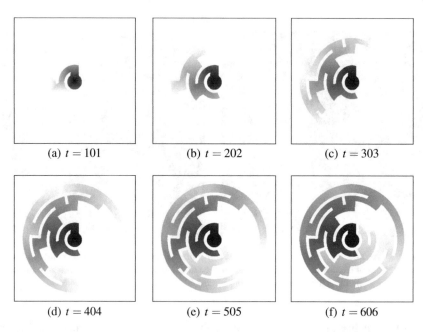

(a) $t = 101$ (b) $t = 202$ (c) $t = 303$

(d) $t = 404$ (e) $t = 505$ (f) $t = 606$

Fig. 10 Numerical simulation of a diffusing chemo-attractant. The grey-level is proportional to a concentration of the chemo-attractant. Time steps indicated are iteration of numerical integration. See details in [4]

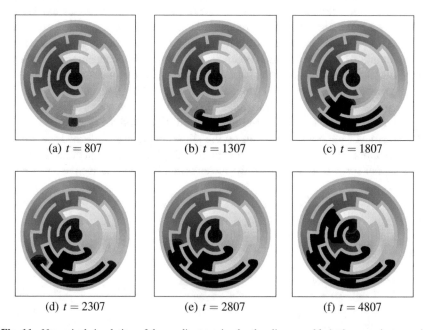

(a) $t = 807$ (b) $t = 1307$ (c) $t = 1807$

(d) $t = 2307$ (e) $t = 2807$ (f) $t = 4807$

Fig. 11 Numerical simulation of the gradient tracing by the slime mould. Active growing zone is shown by red colour. The slime mould's body is shown by blue colour. See details of the model in [4]

but any volatile substances that attract the slime mould, e.g. roots of the medicinal plant *Valeriana officinalis* [29]. Note that in our experiments reported in [4] slime mould did not calculate the shortest path inside the maze but just one of the paths, while Oregonator based model always produces the shortest path.

5 Thoughts on Inconclusive Experiments

In 2009 we attempted to understand what is going on in the slime mould's 'mind' when it traces gradients of chemoattractants. We positioned 16 electrodes at the bottom of a plastic maze (Fig. 12a) with reference electrode in the central chamber, poured some agar above, inoculated the slime mould in the central chamber and placed an oat flake at the outer channel of the maze. Configurations of the slime mould growing in the maze are shown Fig. 12b, c. Electrical potential differences between each of 16 electrodes and the reference electrode recorded during several days are shown in Fig. 12d, e. We found that active growing zones of the slime mould show a higher level of the electrical potential difference and there are signs of an apparent communication between the zones performing parallel search at different parts of the maze. However, it still remains unclear how exactly 'suppression' of growing zones propagating along the longest routes is implemented.

A spectrum of leeches' behaviour traits is extensively classified [13]. A leech positions itself at the water surface in resting state. The leech swims towards the source of a mechanical or optical stimulation. The leech stops swimming when it comes into contact with any geometrical surface. Then, the leech explores the surface by crawling. When a leech finds a warm region the leech bites. We attempted to solve a maze, a template printed of nylon and filled with water, with young leeches *Hirudo verbana* (see details of experimental setup in [5]). We tried fresh blood, temperature and vibration as sources of physical stimuli which would attract leeches to the target site. The leeches did not show attraction to the blood when placed over 5 cm away from the source. Being placed in the proximity of the source the leeches crawled or swam to the target site (Fig. 13a). We have also conducted scoping experiments with leeches in presence of thermal gradients. To form the gradient we immersed, by 5 mm, a tip of a soldering iron, heated to 40 °C, in the water inside a central chamber of the maze. In half of the experiments, leeches escaped from the template, in a quarter of the experiments leeches moved to the domains proximal to the source of a higher temperature and in a quarter of experiments leeches moved towards the source of thermal stimulation. Trajectories of the leeches movement in the presence of a source of vibration did not show any statistically significant preference towards

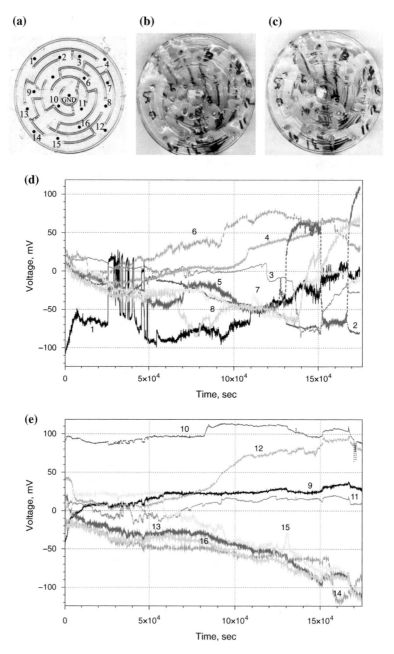

Fig. 12 Illustrations of experiments on electrical activity of slime mould *Physarum polycephalum* during maze solving: **a** positions of electrodes, reference electrode is labeled GND, **b** two days after inoculation, **c** three days after inoculation, **d** electrical activity recorded on channels 1–8, sampling rate is one per second, **e** electrical activity recorded on channels 9–16, plots of electrical potential on electrodes are shown by different colours and also numbered

(a)   **(b)**

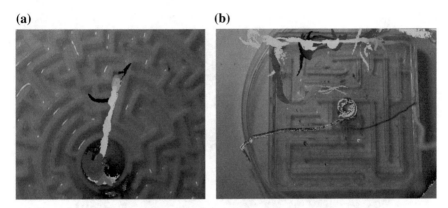

Fig. 13 A leech partially solves maze. **a** Trajectory of the leech approaching the target, blue pixels show starting position of a leech, red pixels—final position. **b** Trajectory of a leech in a template with vibrating motor

movement into areas with highest level of vibration, in some cases a leech was moving towards the vibrating motor from the start of an experiment but then swam or crawled away (Fig. 13b). Inconclusive results with vibration-assisted maze solving could be due to a reflection of waves at maze walls.

We have undertaken a few scoping experiments on plants navigating mazes guided only by gravity force and physical structure of a maze, see illustrations in Fig. 14a. When seeds are placed in or near a central chamber of a maze their roots somewhat grow towards the exit of the labyrinth. However, they often become stuck midway and rarely reach the exit. Yokawa and colleagues [35] demonstrated that by using volatiles it is possible to navigate the roots in simple binary mazes, more complicated mazes have not been tested. Few more experiments on a collision-free path approximation by plant roots have been done on a 3D templates of Bristol (UK) city and USA. The seeds of lettuce, in experiments with a template of Bristol, were placed in large open spaces, corresponding to squares. The templates were kept in a horizontal position. We found that root apexes prefer wider streets, they rarely enter side streets (Fig. 14b). Potential prototypes of shortest path solvers with roots could be the case of future studies, at this moment we only know that roots navigate around obstacles (Fig. 14b) and elevations (Fig. 14c).

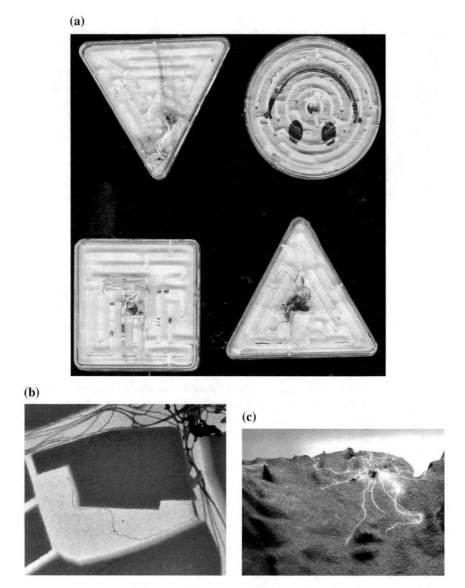

Fig. 14 **a** Lettuce seedlings grow inside plastic mazes. **b**, **c** Roots propagate in a 3D model of Bristol. **d** Maize roots navigate around elevations on 3D template of USA

6 Conclusion

To solve a maze we need a mapper and a tracer. The tracer's role is straightforward, we would say easy, just follow a map made by the mapper. This is the mapper who does all 'computation'. Does it? And here we come to a disturbing thought that a computation exists only in our mind. Nature does not compute. It is us who invented the concept of computation. As Stanley Kubrick told in his interview to "Playboy" magazine in 1968 [18]:

> The most terrifying fact about the universe is not that it is hostile but that it is indifferent; but if we can come to terms with this indifference and accept the challenges of life within the boundaries of death — however mutable man may be able to make them — our existence as a species can have genuine meaning and fulfilment.

Designing and re-designing experimental laboratory prototypes of unconventional computing devices might be our way to cope with the Nature's indifference.

References

1. A. Adamatzky, Pellets, soap end problems. Michigan Daily **6**(2), 2 (1950)
2. A. Adamatzky, Hot ice computer. Phys. Lett. A **374**(2), 264–271 (2009)
3. A. Adamatzky, *Physarum Machines: Computers from Slime Mould*, vol. 74 (World Scientific, 2010)
4. A. Adamatzky, Slime mold solves maze in one pass, assisted by gradient of chemo-attractants. IEEE Trans. Nanobiosci. **11**(2), 131–134 (2012)
5. A. Adamatzky, On exploration of geometrically constrained space by medicinal leeches Hirudo verbana. Biosystems **130**, 28–36 (2015)
6. A. Adamatzky, Physical maze solvers. All twelve prototypes implement 1961 lee algorithm, in *Emergent Computation* (Springer, 2017), pp. 489–504
7. A. Adamatzky, S. Akl, M. Burgin, C.S. Calude, J.F. Costa, M.M. Dehshibi, Y.-P. Gunji, Z. Konkoli, B. MacLennan, B. Marchal et al., East-west paths to unconventional computing. Prog. Biophys. Mol. Biol. (2017)
8. A. Adamatzky, Collision-free path planning in the Belousov-Zhabotinsky medium assisted by a cellular automaton. Naturwissenschaften **89**(10), 474–478 (2002)
9. K. Agladze, N. Magome, R. Aliev, T. Yamaguchi, K. Yoshikawa, Finding the optimal path with the aid of chemical wave. Phys. D: Nonlinear Phenom. **106**(3–4), 247–254 (1997)
10. S. Ayrinhac, Electric current solves mazes. Phys. Educ. **49**(4), 443 (2014)
11. J. Cejkova, M. Novak, F. Stepanek, M.M. Hanczyc, Dynamics of chemotactic droplets in salt concentration gradients. Langmuir **30**(40), 11937–11944 (2014)
12. J.D. Clem, The use of the fluid mapper in an investigation of flow into symmetrical openings obstructed by plane surfaces. Ph.D. thesis (Georgia Institute of Technology, 1954)
13. M.H. Dickinson, C.M. Lent, Feeding behavior of the medicinal leech, Hirudo medicinalis l. J. Compar. Physiol. A: Neuroethol. Sensory Neural Behav. Physiol. **154**(4), 449–455 (1984)
14. A.E. Dubinov, A.N. Maksimov, M.S. Mironenko, N.A. Pylayev, V.D. Selemir, Glow discharge based device for solving mazes. Phys. Plasmas **21**(9), 093503 (2014)
15. M.J. Fuerstman, P. Deschatelets, R. Kane, A. Schwartz, P.J.A. Kenis, J.M. Deutch, G.M. Whitesides, Solving mazes using microfluidic networks. Langmuir **19**(11), 4714–4722 (2003)
16. H.S. Hele-Shaw, Lines of induction in a magnetic field. Proc. R. Soc. Lond. **67**(435–441), 234–236 (1900)

17. H.S. Hele-Shaw, A. Hay, P.H. Powell, Hydrodynamical and electromagnetic investigations regarding the magnetic-flux distribution in toothedcore armatures. J. Inst. Electr. Eng. **34**(170), 21–37 (1905)
18. S. Kubrick, *Stanley Kubrick: Interviews* (University Press of Mississippi, 2001)
19. I. Lagzi, S. Soh, P.J. Wesson, K.P. Browne, B.A. Grzybowski, Maze solving by chemotactic droplets. J. Am. Chem. Soc. **132**(4), 1198–1199 (2010)
20. S. Leduc, *Théorie physico-chimique de la vie et générations spontanées* (Poinat, 1910)
21. C.Y. Lee, An algorithm for path connections and its applications. IRE Trans. Electron. Comput. **3**, 346–365 (1961)
22. V.S. Luk'yanov, Hydraulic instruments for technical calculations. Izveslia Akademia Nauk SSSR **2**
23. R. McFee, R.M. Stow, F.D. Johnston, Graphic representation of electrocardiographic leads by means of fluid mappers. Circulation **6**(1), 21–29 (1952)
24. A.D. Moore, The hydrocal. Ind. Eng. Chem. **28**(6), 704–708 (1936)
25. A.D. Moore, Fields from fluid flow mappers. J. Appl. Phys. **20**(8), 790–804 (1949)
26. A.D. Moore, Fluid mappers as visual analogs for potential fields. Ann. New York Acad. Sci. **60**(1), 948–962 (1955)
27. A. Nair, K. Raghunandan, V. Yaswant, S.S. Pillai, S. Sambandan, Maze solving automatons for self-healing of open interconnects: modular add-on for circuit boards. Appl. Phys. Lett. **106**(12), 123103 (2015)
28. D.R. Reyes, M.M. Ghanem, G.M. Whitesides, A. Manz, Glow discharge in microfluidic chips for visible analog computing. Lab Chip **2**(2), 113–116 (2002)
29. V. Ricigliano, J. Chitaman, J. Tong, A. Adamatzky, D.G. Howarth, Plant hairy root cultures as plasmodium modulators of the slime mold emergent computing substrate Physarum polycephalum. Front. Microbiol. **6** (2015)
30. F. Rubin, The Lee path connection algorithm. IEEE Trans. Comput. **100**(9), 907–914 (1974)
31. C. Scherber, A.J. Aranyosi, B. Kulemann, S.P. Thayer, M. Toner, O. Iliopoulos, D. Irimia, Epithelial cell guidance by self-generated EGF gradients. Integr. Biol. **4**(3), 259–269 (2012)
32. L. Tarassenko, A. Blake, Analogue computation of collision-free paths, in *1991 IEEE International Conference on Robotics and Automation, 1991. Proceedings* (IEEE, 1991), pp. 540–545
33. L. Tarassenko, G. Marshall, F. Gomez-Castaneda, A. Murray, Parallel analogue computation for real-time path planning, in *VLSI for Artificial Intelligence and Neural Networks* (Springer, 1991), pp. 93–99
34. D. Tolmachiev, A. Adamatzky, Chemical processor for computation of Voronoi diagram. Adv. Funct. Mater. **6**(4), 191–196 (1996)
35. K. Yokawa, F. Baluska, Binary decisions in maize root behavior: Y-maze system as tool for unconventional computation in plants. IJUC **10**(5–6), 381–390 (2014)

Index

A
Agent, 240
    algorithm, 249
    cellular automata, 241
    moving, 242
Ant Colony Optimization (ACO), 265, 273
Ant Colony System, 273, 285
AntNet, 274
Argentine ants, 269
Asynchronous system, 102
Automorphism, 37
Automorphism group, 47

B
Bicriteria path problem, 75
Bifurcation, 269, 370
Braesss paradox, 330

C
Cancer epithelial cell, 371, 373, 375
CA-w model, 242
CDL, 208, 210
Cellular automata, 181, 200
Cellular description language, 210
Checkboard pattern, 240
Chemoattractant, 383, 403, 431
Chemotaxis, 403
Collective behaviour, 268
Collision-free path, 429
Complexity measures, 103
Composite collective decision-making, 286
Confinement, 367
Cyclic subgroup, 48

D
Decaoate, 403
Decnol, 403
Deneubourg choice function, 270
Deneubourg model, 269
Detection, 278
Dichloromethane, 404
Diffuse update algorithm, 108
Dijkstra algorithm, 339
Directionality, 366
Discrimination thresholds, 278
Distributed Bellmann-Ford method, 106
Distributed learning automata, 212
Dynamic network, 103

E
Elitist Ant Systems, 273, 285
Emissivity, 412
Epithelial cell, 431
Error rates, 278
Eusocial, 265, 267

F
Fatty acid, 405
Fechner, 278
Fluid flow, 422
Fluidic maze solver, 425
Fluid mapper, 422

G
Gathering task, 28
Genetic algorithm, 253
Gradients, 365, 366

H

2-hexyldecanoic acid, 404, 405
Hirudo verbana, 433
Hölldobler, B., 268
Hydrocal, 424
Hydrodynamic resistance, 426
Hydrogel, 405

I

Infinite grid, 59
Influenza virus algorithm, 149
Information capacity, 278
Isometry, 37

L

L. niger, 271
Laplace's equation, 417
Lapse rate, 278, 285
Lasius fuliginosus, 270
Learning, 286
Learning ACO algorithms, 286
Learning automata, 212
Lee algorithm, 208
Leech, 433
Linepithema humile, 269

M

Marangoni flows, 404
Max-Min ant system, 273
Maze, 365, 372, 421
    fluidic
        solver, 425
    Lee algorithm, 421
    milk and coffee solver, 426
    random mouse solver, 402
    slime mould sover, 431
    Trémaux solver, 402
    wall follower solver, 402
Maze solving
    multi-agent model, 309
Memory, 286
Microfluidic, 367
Motility patterns, 369
Multi-agent model
    maze adaptation, 309
    maze tortuosity, 309
    mechanisms of computation, 300
    path planning, 308
    pattern formation, 300

scale parameter, 301
Multi-cellular formation, 422
Multiobjective optimisation, 147

N

Network design problem, 330
Neutrophils, 371
Nitrobenzene, 404
Nondeterministic polynomial, 266
NP-problems, 266, 273

O

Optimal gathering, 34
    on infinite grids, 59
    on trees, 52
Optimisation problem, 32

P

Paramecium caudatum, 390
Path planning
    collision-free paths, 311
Pattern
    checkerboard, 243
Pattern formation, 266
Percolation, 3
Persistence, 369
Pheromones, 267, 269
Pheromone trail, 269
Physarum algorithm, 341
Physarum polycephalum, 227, 297, 333, 380,
        381, 431
    computation by, 297
Polydimethylsiloxane, 405
Potassium ferrocyanide, 422
Psychometric function, 278, 279
Psychophysical theory, 278

R

Rank-Based ant system, 273
Resistive grid, 416
Resistor, 429
Robot, 28
    computational cycle, 28

S

Self-generated gradients, 365, 372, 374, 375
Self-guided, 375
Self-organization, 265, 266
Shortest path, 240, 245, 249, 250

Short path selection, 269
Slime mould, 431
Slime mould maze solver, 431
Speed, 366
Speedup, 255
Stigmergy, 267
Symmetry breaking, 269, 270

T
Template, 243
Termination, 249
Thermography, 411
Travelling Salesman Problem (TSP), 265, 273, 379
Tree topologies, 52

U
Unconventional computing, 422
U-Turns, 271

V
Vehicle routing, 146
Voronoi diagram, 422

W
Weber point, 34, 39, 43, 50, 52
Weber's Law, 276
Wilson, E. O., 268
Wireless network, 3

Printed in the United States
By Bookmasters